Global Environmental Issues
Selections from *CQ Researcher*

CQ PRESS

A Division of SAGE
Washington, D.C.

CQ Press
2300 N Street, NW, Suite 800
Washington, DC 20037
Phone: 202-729-1900; toll-free, 1-866-4CQ-PRESS (1-866-427-7737)
Web: www.cqpress.com

Contents

Annotated Contents

The 12 *CQ Researcher* reports reprinted in this book have been reproduced essentially as they appeared when first published. In the few cases in which important developments have since occurred, updates are provided in the overviews highlighting the principle issues examined.

ENVIRONMENTAL PROTECTION

Climate Change

Delegates from around the globe arrived in Copenhagen, Denmark, for the U.N. Climate Change Conference in December 2009 hoping to forge a significant agreement to reduce greenhouse gas emissions and temper climate change. But despite years of diplomatic preparation, two weeks of intense negotiations and the clamor for action from thousands of protesters outside the meeting, the conferees adopted no official treaty. Instead, a three-page accord—cobbled together on the final night by President Barack Obama and the leaders of China, India, Brazil and South Africa—established only broad, nonbinding goals and postponed tough decisions. Yet defenders of the accord praised it for requiring greater accountability from emerging economies such as China, protecting forests and committing billions in aid to help poorer nations. But the key question remains: Will the accord help U.N. efforts to forge a legally binding climate change treaty for the world's nations?

Protecting Wetlands

The nation's millions of acres of wetlands are valuable natural resources. Ponds, lakes, swamps, bogs, bays and marine estuaries not only shelter countless fish, birds and animals but also filter pollutants from water and soak up floodwaters. Since the United States was settled, more than half of its wetlands have been lost, and crucial areas like Louisiana's coast and the Florida Everglades are eroding daily. Although the United States is now gaining more wetlands every year than it is losing, scientists say too many acres of crucially needed wetlands are still being lost. For several decades national policy has called for protecting wetlands, but the powerful construction, energy and agriculture industries say current environmental regulations make projects too expensive. Conservationists, sports enthusiasts and many state officials argue that stronger regulations are still urgently needed. Meanwhile, recent Supreme Court decisions have intensified debate over how broadly the federal government can oversee activities affecting wetlands.

Oceans in Crisis

The world's oceans are in a dire state. Large predatory species are being decimated—including sharks, whales, tuna, grouper, cod, halibut, swordfish and marlin—and replaced by species with less commercial and nutritive value. In fact, a growing body of evidence suggests that the world's marine ecosystems have been altered so dramatically they are undergoing evolution in reverse, returning to a time when algae and jellyfish dominated the seas. The crisis is having an increasingly profound effect on humans. Fishing cultures from Newfoundland to West Africa are vanishing, and toxic algal blooms have closed beaches and recreational areas from Florida to the Black Sea. The damage is being caused by overfishing, climate change and destruction of habitat due to coastal development and pollution. Scientists and policy makers widely agree that a broad-based approach known as ecosystem-based management would help restore the oceans' productivity, but significant research and strong international cooperation are needed to bring about such a shift.

Disappearing Species

The polar bear may vanish as global warming melts Arctic ice. India's tigers and Africa's gorillas are also at severe risk, and thousands of other valuable animals, plants and insects are disappearing as tropical rain forests around the world are logged. Twelve percent of birds, one-fifth of mammals and nearly a third of amphibians that have been assessed so far have been found to be imperiled. Many biologists conclude that humans are setting off a mass extinction that's exterminating 30,000 species a year—possibly as much as 10,000 times faster than natural evolution. Now honeybees have mysteriously gone missing from American farms, while coral reefs are dwindling and deep-sea fisheries are depleted. Many scientists fear that Earth faces an irreversible biological catastrophe even more severe than climate change, and that conservation efforts could be too late to preserve much of the planet's irreplaceable biodiversity.

THE POLITICAL ECONOMY OF THE ENVIRONMENT
Carbon Trading
Carbon emissions trading—the buying and selling of permits to emit greenhouse gases caused by burning fossil fuels—is becoming a top strategy for reducing pollution that causes global climate change. Some $60 billion in permits were traded worldwide in 2007, a number expected to grow much larger if the next U.S. administration follows through on pledges to reduce America's carbon emissions. Advocates say carbon trading is the best way to generate big investments in low-carbon energy alternatives and control the cost of cutting emissions. But carbon trading schemes in Europe and developing countries have a mixed record. Some industries are resisting carbon regulations, and programs intended to help developing countries onto a clean energy path have bypassed many poor nations, which are the most vulnerable to the impacts of climate change. Some experts argue that there are simpler, more direct ways to put a price on carbon emissions, such as taxes. Others say curbing climate change will require both taxes and trading, plus massive government investments in low-carbon energy technologies.

Looming Water Crisis
Since the beginning of the twenty-first century drought has marched across much of the globe, hitting China, the Mediterranean, southeast Australia and the U.S. Sun Belt. The amount of water used by humans has tripled since 1950, and irrigated cropland has doubled. About one-fifth of the world's population lacks sufficient water, a figure that could reach 40 percent by 2025 by some estimates, in part because of growing world economies. In the poorest societies more than a billion people lack access to clean water, and dirty water kills 5,000 children—enough to fill 12 jumbo jets—every day. By century's end drought is expected to spread across half the Earth's land surface due to climate change, causing hunger and higher food prices. The United Nations says it would cost an extra $10 billion or more annually to provide clean water and sanitation for all. Some recommend privatizing water supplies, while others suggest that charging more for water to encourage conservation would help to avert future crises.

Rapid Urbanization
About 3.3 billion people—half of Earth's inhabitants—live in cities, and the number is expected to hit 5 billion by 2030. Most urban growth today is occurring in developing countries, where about a billion people live in city slums. Delivering services to crowded cities has become increasingly difficult, especially in the world's 19 "megacities"—those with more than 10 million residents. Moreover, most of the largest cities are in coastal areas, where they are vulnerable to flooding caused by climate change. Many governments are striving to improve city life by expanding services, reducing environmental damage and providing more jobs for the poor, but some still use heavy-handed clean-up policies like slum clearance. Researchers say urbanization helps reduce global poverty because new urbanites earn more than they could in their villages. The global recession that began in 2008 could reverse that trend, however, as many unemployed city dwellers return to rural areas. But most experts expect rapid urbanization to resume once the economic storm has passed.

Energy Nationalism
A world thirsting for imported oil and gas is seeking new supplies in Central Asia and Africa, where many nations have nationalized their energy resources. In a dramatic reversal from the 1970s, government-owned or controlled petroleum companies today control 77 percent of the world's 1.1 trillion barrels of oil reserves. While the emergence of these rising petrostates has helped diversify the world's energy sources, many are considered oil "hot spots"—vulnerable to disruption from international terrorists or domestic dissidents. In addition, many of the petrostates are blending politics and energy into

a new energy nationalism, rewriting the rules of the world's energy markets and restricting international oil corporations' operations. Russia's confrontational energy policies alarm its neighbors, and critics say a booming China is combing the world for access to oil and gas resources without concern for suppliers' corruption or human-rights violations. Many also worry that growing competition for dwindling oil supplies will lead to greater risks of international conflict.

Race for the Arctic
With oil prices soaring, revelations that the Arctic could contain up to 22 percent of the world's undiscovered oil and gas have given extra impetus to an international race to claim the region's $1 trillion in oil and other riches. Russia kick-started the race last summer when it stunned the world by planting its flag on the North Pole seabed—two miles below the Arctic Ocean. Global warming has dramatically shrunk the ice covering the ocean, raising the prospect of new, shorter transcontinental shipping routes and spurring the United States, Canada, Russia, Denmark and Norway to begin gathering data to prove they own large swaths of offshore Arctic territory. But environmentalists warn that tougher international rules are needed—possibly an Arctic treaty—to prevent energy exploration from exacerbating global warming and damaging the fragile region. The Inuit and other indigenous groups also fear their concerns will be ignored in the dash to extract riches from the region.

CONSUMPTION
Buying Green
Americans will spend an estimated $500 billion this year on products and services that claim to be good for the environment because they contain non-toxic ingredients or produce little pollution and waste. While some shoppers buy green to help save the planet, others are concerned about personal health and safety. Whatever their motives, eco-consumers are reshaping U.S. markets. To attract socially conscious buyers, manufacturers are designing new green products and packaging, altering production processes and using sustainable materials. But some of these products may be wastes of money. Federal regulators are reviewing green labeling claims to see whether they mislead consumers, while some critics say that government mandates promoting environmentally preferable products distort markets and raise prices. Even if green marketing delivers on its pledges, many environmentalists say that sustainability is not a matter of buying green but of buying less.

Reducing Your Carbon Footprint
As climate change rises closer to the top of the government's policy agenda, more and more consumers are trying to change their behavior so they pollute and consume less. To reduce their individual "carbon footprints," many are cutting gasoline and home-heating consumption, choosing locally grown foods and recycling. While such actions are important in curbing global warming, the extent to which consumers can reduce or reverse broad-scale environmental damage is open to debate. Moreover, well-intentioned personal actions can have unintended consequences that cancel out positive effects. To have the greatest impact, corporate and government policy must lead the way, many environmental advocates say.

Oil Jitters
Vastly increased demand for oil in rapidly modernizing China and India, warfare and instability in the Middle East and the weakening U.S. dollar have revived fears of a new energy crisis. Gasoline shortages—and the accompanying lines at gas stations—were thought to have ended with the Jimmy Carter administration. As 2008 began, however, American drivers were paying more than $3 a gallon, and crude oil hit a milestone—$100 a barrel. Some oil experts warn of even bigger price shocks to come as oil-producing nations use more and more of their own oil and energy demand jumps 50 percent by 2030. Some experts predict an oil "production crunch" within four to five years that will have severe geopolitical and economic impacts, and one expert says the energy supply-demand gap could create "social chaos and war" by 2020. In any event, the days of cheap, plentiful oil appear to be over, and motorists may have to learn how to conserve energy.

Preface

Will the growing competition for energy trigger new international conflicts? Will the Copenhagen Accord slow global warming? Can individual action significantly reduce global climate change? These questions—and many more—are at the heart of the debate about global environmental politics and policy. Students must first understand the facts and contexts of these and other policy issues if they are to analyze and articulate well-reasoned positions.

The first edition of *Global Environmental Issues* includes twelve reports by *CQ Researcher*, an award-winning weekly policy brief that explains difficult concepts and provides balanced coverage of competing perspectives. Each article analyzes past, present and possible political maneuvering and is designed to promote in-depth discussion and further research to help readers formulate their own positions on crucial international issues.

This collection is organized into three subject areas: environmental protection, the political economy of the environment and consumption—to cover a range of topics found in most global environmental politics and policy courses. Citizens, journalists and business and government leaders also can turn to the collected articles to become better informed on key issues, actors and policy positions.

CQ RESEARCHER

CQ Researcher was founded in 1923 as Editorial Research Reports and was sold primarily to newspapers as a research tool. The magazine was renamed and redesigned in 1991 as *CQ Researcher*. Today, students are its primary audience. While still used by hundreds of journalists and newspapers, many of which reprint portions of the reports, the *Researcher*'s main subscribers are high school, college and public libraries. In 2002, the *Researcher* won the American Bar Association's coveted Silver Gavel award for magazine excellence for a series of nine reports on civil liberties and other legal issues.

Researcher staff writers—all highly experienced journalists—sometimes compare the experience of writing a *Researcher* report to drafting a college term paper. Indeed, there are many similarities. Each report is as long as many term papers—about 11,000 words—and is written by one person without any significant outside help. One of the key differences is that writers interview leading experts, scholars and government officials for each issue.

Like students, staff writers begin the creative process by choosing a topic. Working with the *Researcher*'s editors, the writer identifies a controversial subject that has important public policy implications. After a topic is selected, the writer embarks on one to two weeks of intense research. Newspaper and magazine articles are clipped or downloaded, books are ordered and information is gathered from a wide variety of sources, including interest groups, universities and the government. Once the writers are well informed, they develop a detailed outline and begin the interview process. Each report requires a minimum of ten to fifteen interviews with academics, officials, lobbyists and people working in the field. Only after all interviews are completed does the writing begin.

CHAPTER FORMAT

Each issue of *CQ Researcher*, and therefore each selection in this book, is structured in the same way. Each begins with an overview, which briefly summarizes the areas that will be explored in greater detail in the rest of the chapter. The next section chronicles important and current debates on the topic under discussion and is structured around a number of key questions, such as "Do carbon offsets slow climate change?" and "Should water be privatized?" These questions are usually the subject of much debate among practitioners and scholars in the field. Hence, the answers presented are never conclusive but detail the range of opinion on the topic.

Next, the "Background" section provides a history of the issue being examined. This retrospective covers important legislative measures, executive actions and court decisions that illustrate how current policy has evolved. Then the "Current Situation" section examines contemporary policy issues, legislation under consideration and legal action being taken. Each selection concludes with an "Outlook" section, which addresses possible regulation and court rulings, as well as domestic and international government initiatives.

Each report contains features that augment the main text: two to three sidebars that examine issues related to the topic at hand, a pro versus con debate between two experts, a chronology of key dates and events and an annotated bibliography detailing major sources used by the writer.

CUSTOM OPTIONS

Interested in building you ideal CQ Press Issues book, customized to your personal teaching needs and interests? Browse by course or search for specific topics or issues from our online catalog of *CQ Researcher* issues at http://custom.cqpress.com.

ACKNOWLEDGMENTS

We wish to thank many people for helping to make this collection a reality. Tom Colin, managing editor of *CQ Researcher*, gave us his enthusiastic support and cooperation as we developed this ninth edition. He and his talented staff of editors and writers have amassed a first-class library of *Researcher* reports, and we are fortunate to have access to that rich cache.

Some readers may be learning about *CQ Researcher* for the first time. We expect that many readers will want regular access to this excellent weekly research tool. For subscription information or a no-obligation free trial of Researcher, please contact CQ Press at www.cqpress.com or toll-free at 1-866-4CQ-PRESS (1-866-427-7737).

We hope that you will be pleased by the first edition of *Global Environmental Issues*. We welcome your feedback and suggestions for future editions. Please direct comments to Elise Frasier, Acquisitions Editor for International Relations and Comparative Politics, College Publishing Group, CQ Press, 2300 N St. NW, Suite 800, Washington, D.C. 20037, or efrasier@cqpress.com.
—The Editors of CQ Press

Contributors

Thomas J. Colin, managing editor of *CQ Researcher*, has been a magazine and newspaper journalist for more than 30 years. Before joining Congressional Quarterly in 1991, he was a reporter and editor at the Miami Herald and National Geographic and editor in chief of Historic Preservation. He holds a bachelor's degree in English from the College of William and Mary and in journalism from the University of Missouri.

Kathy Koch, assistant managing editor of *CQ Researcher*, specializes in education and social policy issues. She has freelanced in Asia and Africa for various U.S. newspapers, including The Christian Science Monitor and USA Today. She also covered environmental legislation for the CQ Weekly and reported for newspapers in South Florida. She graduated in journalism from the University of North Carolina at Chapel Hill.

Tom Arrandale freelances from Livingston, Mont., on environmental issues. He is a columnist for Governing magazine and has written for Planning magazine, High Country News and Yellowstone Journal. He authored The Battle for Natural Resources (CQ Press, 1983). He visits Yellowstone National Park regularly to hike, snowshoe and photograph wildlife. He graduated from Dartmouth College with a history degree and from the University of Missouri's School of Journalism with a master's degree.

Brian Beary—a freelance journalist based in Washington, D.C.—specializes in EU-U.S. affairs and is the U.S. correspondent for Europolitics, the EU affairs daily newspaper. Originally from Dublin, Ireland, he worked in the European Parliament for Irish MEP Pat "The Cope" Gallagher in 2000 and at the EU Commission's Eurobarometer unit on public opinion analysis. A fluent French speaker, he appears regularly as a guest international relations expert on various television and radio programs. Apart from his work for Congressional Quarterly, Beary also writes for the European Parliament Magazine and the Irish Examiner daily newspaper. His last report for CQ Global Researcher was "Future of Turkey."

Peter Behr is a Washington freelance writer who worked for more than 25 years at The Washington Post, where he reported on energy issues and served as business editor. A former Nieman Fellow at Harvard University, Behr was a public policy scholar at the Woodrow Wilson International Center for Scholars and is now writing a book about the U.S. electric power grid. His report on "Energy Nationalism" appeared in the July 2007 issue.

Thomas J. Billitteri is a freelance journalist in Fairfield, Pa., who has more than 30 years' experience covering business, nonprofit institutions and related topics for newspapers and other publications. He has written previously for *CQ Researcher* on teacher education, parental rights and mental health policy. He holds a BA in English and an MA in journalism from Indiana University.

Alan Greenblatt is a freelance writer in the Washington area and former staff writer at Governing magazine. He previously covered elections, agriculture and military spending for CQ Weekly, where he won the National Press Club's Sandy Hume Award for political journalism. He graduated from San Francisco State University in 1986 and received a master's degree in English literature from the University of Virginia in 1988. His recent *CQ Researcher* reports include "Future of the GOP" and "State Budget Crisis."

Reed Karaim, a freelance writer living in Tucson, Arizona, has written for The Washington Post, U.S. News & World Report, Smithsonian, American Scholar, USA Weekend and other publications. He is the author of the novel, If Men Were Angels, which was selected for the Barnes & Noble Discover Great New Writers series. He is also the winner of the Robin Goldstein Award for Outstanding Regional Reporting and other journalism awards. Karaim is a graduate of North Dakota State University in Fargo.

Peter Katel is a veteran journalist who previously served as Latin America bureau chief for Time magazine, in Mexico City, and as a Miami-based correspondent for Newsweek and The Miami Herald's El Nuevo Herald. He also worked as a reporter in New Mexico for 11 years and wrote for several non-governmental organizations, including International Social Service and The World Bank. He has won several awards, including the Inter American Press Association's Bartolome Mitre Award. He is a graduate of the University of New Mexico in University Studies.

Jennifer Weeks is a *CQ Researcher* contributing writer in Watertown, Mass., who specializes in energy and environmental issues. She has written for The Washington Post, The Boston Globe Magazine and other publications, and has 15 years' experience as a public policy analyst, lobbyist and congressional staffer. She has an A.B. degree from Williams College and master's degrees from the University of North Carolina and Harvard University.

CLIMATE CHANGE

BY REED KARAIM

Excerpted from the CQ Global Researcher. Reed Karaim. (February 2010). "Climate Change." *CQ Global Researcher*, 25-50.

Climate Change

BY REED KARAIM

THE ISSUES

It was the global gathering many hoped would save the world. For two weeks in December, delegates from 194 nations came together in Copenhagen, Denmark, to hammer out an international agreement to limit global warming. Failure to do so, most scientists have concluded, threatens hundreds of millions of people and uncounted species of plants and animals.

Diplomatic preparations had been under way for years but intensified in the months leading up to the conference. Shortly before the sessions began, Yvo de Boer, executive secretary of the United Nations Framework Convention on Climate Change — the governing body for negotiations — promised they would "launch action, action and more action," and proclaimed, "I am more confident than ever before that [Copenhagen] will be the turning point in the fight to prevent climate disaster." [1]

But delegates found themselves bitterly divided. Developing nations demanded more financial aid for coping with climate change. Emerging economic powers like China balked at being asked to do more to limit their emissions of the greenhouse gases (GHGs) — created by burning carbon-based fuels — blamed for warming up the planet. The United States submitted proposed emissions cuts that many countries felt fell far short of its responsibility as the world's dominant economy. As negotiations stalled, frustration boiled over inside the hall and on the streets outside, where tens of

Erosion is washing away beachfront land in the Maldives. The island nation in the Indian Ocean faces possible submersion as early as 2100, according to some climate change predictions. President Mohamed Nasheed said the voluntary emission cuts goal reached in Copenhagen last December was a good step, but at that rate "my country would not survive."

Reuters/Reinhard Krause

thousands of activists had gathered to call world leaders to action. A historic opportunity — a chance to reach a global commitment to battle climate change — seemed to be slipping away.

Then, on Dec. 18 — the final night of the conference — leaders from China, India, Brazil, South Africa and the United States emerged from a private negotiating session with a three-page, nonbinding accord that rescued the meeting from being judged an abject failure.

But the accord left as much confusion as clarity in its wake. It was a deal, yes, but one that fell far short of the hopes of those attending the conference,

and one largely lacking in specifics. The accord vowed to limit global warming to 2 degrees Celsius (3.6 Fahrenheit) above pre-Industrial Revolution levels, provide $30 billion in short-term aid to help developing countries cope with the effects of climate change — with more promised longerterm — and included significant reporting and transparency standards for participants, including emerging economic powers such as China and India.

The accord did not, however:

• Include earlier language calling for halving global greenhouse gas emissions by 2050;

• Set a peak year by which greenhouse gases should begin to decline;

• Include country-specific targets for emission reductions (signatories began filling in the numbers by the end of January) (*See Current Situation, p. 41.*);

• Include a timetable for reaching a legally binding international treaty, or

• Specify where future financial help for the developing world to cope with climate change will come from. [2]

Called back into session in the early morning hours, delegates from much of the developing world reacted with dismay to a deal they felt left their countries vulnerable to catastrophic global warming.

"[This] is asking Africa to sign a suicide pact — an incineration pact — in order to maintain the economic dependence [on a high-carbon economy] of a few countries," said Lumumba Di-Aping, the Sudanese chair of the G77 group of 130 poor countries. [3]

British Prime Minister Gordon Brown, however, hailed the deal as a "vital first step" toward "a green and

Major Flooding, Drought Predicted at Century's End

Significant increases in runoff — from rain or melting snow and ice — are projected with a high degree of confidence for vast areas of the Earth, mainly in northern regions. Up to 20 percent of the world's population lives in areas where river flood potential is likely to increase by the 2080s. Rainfall and runoff are expected to be very low in Europe, the Middle East, northern and southern Africa and the western United States.

Projected Changes in Annual Runoff (Water Availability), 2090-2099
(by percentage, relative to 1980-1999)

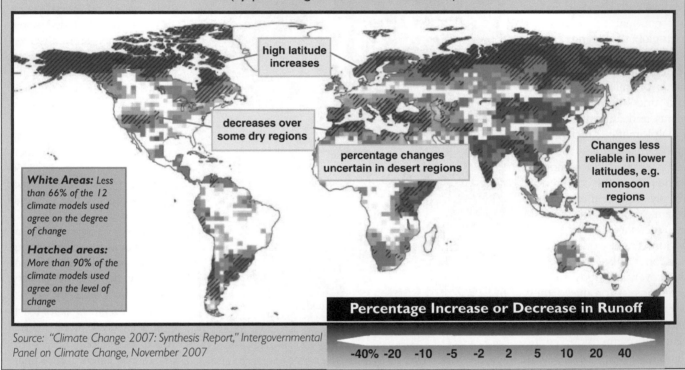

high latitude increases

decreases over some dry regions

percentage changes uncertain in desert regions

Changes less reliable in lower latitudes, e.g. monsoon regions

White Areas: Less than 66% of the 12 climate models used agree on the degree of change

Hatched areas: More than 90% of the climate models used agree on the level of change

Percentage Increase or Decrease in Runoff

-40% -20   -10   -5   -2   2   5   10   20   40

Source: "Climate Change 2007: Synthesis Report," Intergovernmental Panel on Climate Change, November 2007

---

low-carbon future for the world." [4] A total of 55 countries, including the major developed nations, eventually signed onto the deal.

But at the Copenhagen conference, delegates agreed only to "take note" of the accord, without formally adopting it.

Since then, debate has raged over whether the accord represents a step backward or a realistic new beginning. "You had the U.S., China and India closing ranks and saying it's too hard right now to have a binding agreement," says Malini Mehra, an Indian political scientist with 20 years of involvement in the climate change debate. "It's really worse than where we started off."

Others are more upbeat. Michael Eckhart, president of the American Council on Renewable Energy, points out that the convention had revealed how unworkable the larger effort — with 194 participants — had become. "The accord actually sets things in motion in a direction that is realistic," he says. "To have these major nations signed up is fantastic."

Copenhagen clearly demonstrated how extremely difficult and complex global climate negotiations can be. Getting most of the world's nations to agree on anything is no easy task, but climate change straddles the biggest geopolitical fault lines of our age: the vast economic disparity between the developed and developing worlds,

questions of national sovereignty versus global responsibility and differences in political process between democratic and nondemocratic societies.

Climate change also involves a classic example of displaced hardship — some of the worst effects of global warming are likely to be felt thousands of miles from those nations that are most responsible for the higher temperatures and rising seas, making it easier for responsible parties to delay action. Finally, tackling the problem is likely to take hundreds of billions of dollars.

None of this is comforting to those already suffering from climate change, such as Moses Mopel Kisosion, a Maasai herdsman who journeyed from Kenya to tell anyone who would listen how increasingly severe droughts are destroying his country's traditional way of life. (*See story on climate refugees, p. 34.*) But it does explain why reactions to the Copenhagen Accord — which even President Barack Obama acknowledged is simply a "beginning" — have varied so widely. [5]

For some U.S. environmental groups, the significance of the accord was in the commitment Obama secured from emerging economies to provide greater transparency and accountability, addressing one of the U.S. Senate's objections to earlier climate change proposals. The Senate never ratified the previous international climate agreement, known as the Kyoto Protocol.

Carl Pope, executive director of the Sierra Club, called the accord "historic — if incomplete," but said, "Now that the rest of the world — including countries like China and India — has made it clear that it is willing to take action, the Senate must pass domestic legislation as soon as possible." [6]

But to nongovernmental organizations focused on global poverty and economic justice, the accord represented an abdication of responsibility by the United States and other developed countries. Tim Jones, chief climate officer for the United Kingdom-based anti-poverty group World Development Movement, called the accord "a shameful and monumental failure that has condemned millions of people around the world to untold suffering." [7]

Easily lost in the heated rhetoric, however, is another part of the Copenhagen story: The conference illustrated how a consensus now unites most of the globe about the threat climate change poses. And although skeptics continue to speak out (*see p. 36*), the scientific community has overwhelmingly concluded that average global temperatures are rising and that manmade emissions — particularly carbon dioxide from burning coal, oil and

Carbon Emissions Rising; Most Come from China

Global emissions of carbon dioxide (CO_2) — the most common greenhouse gas (GHG) blamed for raising the planet's temperature — have grown steadily for more than 150 years. Since 1950, however, the increases have accelerated and are projected to rise 44 percent between 2010 and 2030 (top graph). While China emits more CO_2 than any other country, Australians produce the most carbon emissions per person (bottom left). Most manmade GHG comes from energy production and transportation (pie chart).

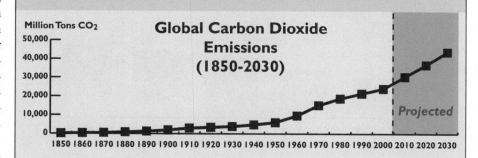

Global Carbon Dioxide Emissions (1850-2030)

Million Tons CO_2

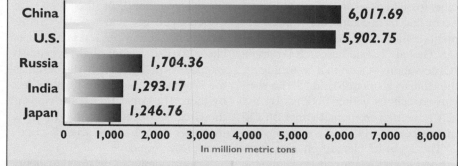

Countries Emitting the Most CO_2 of Major Economies (2006)

China	6,017.69
U.S.	5,902.75
Russia	1,704.36
India	1,293.17
Japan	1,246.76

In million metric tons

Largest Per Capita CO_2 Emitters (2006)

Tons of CO_2 per person

Australia 20.58 U.S. 19.78 Canada 18.81 Saudi Arabia 15.70 Russia 12.00 South Korea 10.53

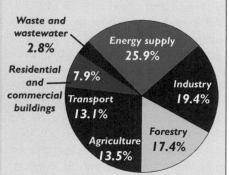

Sources of Manmade GHG Emissions

Waste and wastewater 2.8%
Residential and commercial buildings 7.9%
Energy supply 25.9%
Industry 19.4%
Transport 13.1%
Forestry 17.4%
Agriculture 13.5%

* *Projected*

Sources: *"Climate Change 101: International Action,"* Pew Center on Global Climate Change, undated; *"Climate Change 2007: Synthesis Report,"* Intergovernmental Panel on Climate Change, November 2007; Union of Concerned Scientists

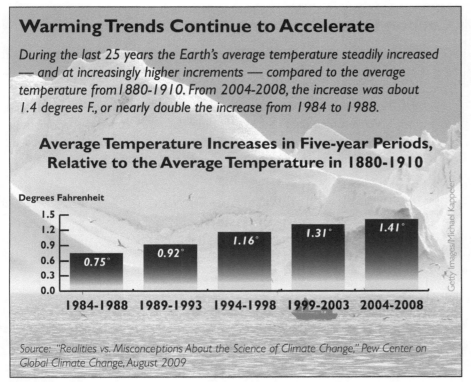

Warming Trends Continue to Accelerate

During the last 25 years the Earth's average temperature steadily increased — and at increasingly higher increments — compared to the average temperature from 1880-1910. From 2004-2008, the increase was about 1.4 degrees F., or nearly double the increase from 1984 to 1988.

Average Temperature Increases in Five-year Periods, Relative to the Average Temperature in 1880-1910

Degrees Fahrenheit

1984-1988	0.75°
1989-1993	0.92°
1994-1998	1.16°
1999-2003	1.31°
2004-2008	1.41°

Source: "Realities vs. Misconceptions About the Science of Climate Change," Pew Center on Global Climate Change, August 2009

other fossil fuels — are largely to blame. According to a comprehensive assessment released in June 2009 by the U.S. Global Change Research Program, "Observations show that warming of the climate is unequivocal." [8] The conclusion echoes earlier findings by the U.N.'s Intergovernmental Panel on Climate Change (IPCC). [9]

The costs of climate change, both economic and in human lives, already appear significant. Disasters tied to climate change kill around 300,000 people a year and cause roughly $125 billion in economic losses, according to the Global Humanitarian Forum, a Geneva-based think tank led by former U.N. Secretary General Kofi Annan. [10] Evidence widely cited during the conference strengthens the conclusion the world is heating up. The World Meteorological Organization (WMO) reported that the last decade appeared to be the warmest on record, continuing a trend. The years 2000 through 2009 were "warmer than the 1990s, which were warmer than the 1980s, and so on," said Michel Jarraud, the

secretary general of the WMO, as Copenhagen got under way. [11] Other reports noted that sea levels appeared likely to rise higher than previously estimated by 2100, with one estimating seas could rise more than six feet by then. The Antarctic ice shelves and the Greenland ice sheet are also melting faster than the U.N. scientific body previously found. [12]

Copenhagen also provided evidence of a growing international political consensus about climate change. About 120 heads of state attended the final days of the conference, hoping to sign their names to an agreement, an indication of the seriousness with which the global community now views the issue.

"It was remarkable the degree to which Copenhagen galvanized the public," says David Waskow, Oxfam America's climate change policy adviser, who attended the conference. "That's true with the literally millions who came out to show their support for strong action on climate change around the world. It's true with the number of heads of state who showed up, and

even in terms of the number of developing countries making substantial offers to tackle their emissions."

As observers try to determine where the world is headed on climate change and how the Copenhagen Accord helps or hinders that effort, here are some of the questions they are considering:

Is the Copenhagen Accord a meaningful step forward in the fight against global warming?

No one claims that a three-page accord that leaves out hard emission-reduction targets or a firm timetable is the final answer to global climate change. But does it bring the world closer to adequately addressing the problem?

Accord supporters range from the dutiful to the enthusiastic. But the unifying thread is a feeling that the accord is better than no deal at all, which is where the conference seemed to be headed until the 11th-hour negotiations.

"If the standard is — were we going to get a blueprint to save the world? The fact is, we were never going to meet it. None of the documents circulating were a feasible basis for agreement among the major players," says Michael A. Levi, director of the Program on Energy Security and Climate Change for the U.S. Council on Foreign Relations. "What we ended up with is something that can be useful if we use it the right way. It has pieces that empower all sorts of other efforts, like increased transparency, some measure of monitoring and reporting. It sets a political benchmark for financing. It can be a meaningful step forward."

Levi also notes that countries signing the accord agreed to fill in their targets for emissions cuts (as the major signatories and other nations did at the end of January), addressing one of the main criticisms of the deal.

But the Indian political scientist Mehra says even if countries abide by their commitments to cut emissions, the accord will not meet its target of holding global warming to 2 degrees Celsius

(3.6-degrees Fahrenheit), which U.N. scientists consider the maximum increase that could avoid the worst effects of climate change, including a catastrophic rise in sea levels and severe damage to world food production.

She cites an IPCC conclusion that says in order to meet the 2-degree goal industrialized countries must reduce their emissions to 25-40 percent of 1990 levels by 2020 and by 50 percent by 2050. "What we actually got in the various announcements from the developed nations are far below that, coming in at around 18 percent," Mehra says.

Indeed, research by Climate Interactive — a joint effort by academic, nonprofit and business entities to assess climate policy options — found that the countries' commitments would allow temperatures to rise about 3.9 degrees Celsius (7 degrees Fahrenheit) by 2100 — nearly twice the stated goal. [13] "If you're looking at an average of 3 to 4 degrees, you're going to have much higher rises in significant parts of the world. That's why so many of the African negotiators were so alarmed by this," says Mehra. "It's worse than where we started because it effectively sets in stone the lowest possible expectations."

But other analysts point out that President Obama and other leaders who backed the accord have acknowledged more must be done. [14] They add that focusing on the initial emissions goals ignores the areas where the deal breaks important ground. "A much bigger part of the story, I think, is the actual money the developed world is putting on the table, funds for mitigation and adaptation," says Mike Hulme, a professor at the University of East Anglia in Great Britain who has been studying the intersection between climate and culture. "This is as much part of the game as nominal reduction targets."

The accord calls for $10 billion a year to help poorer, more vulnerable coun-

tries cope with climate change over the next three years, rising to $100 billion a year by 2020. The money will come from "a wide variety of sources, public and private, bilateral and multilateral, including alternative sources of finance," according to the agreement. [15]

Equally important, say analysts, is the fact that the agreement sets new standards of participation and accountability for developing economies in the global warming fight. "The developing countries, particularly China, made a step forward and agreed not only to undertake some actions to re-

leaders lower expectations for three months coming into this, and then actually having them undershoot those expectations was unbelievable," says Jason Blackstock, a research scholar at the International Institute for Applied Systems Analysis in Austria, who studies the intersection of science and international affairs. He places some of the blame at the feet of President Obama: "This is clearly not one of his top issues, and that's disappointing."

But Thomas Homer-Dixon, who holds an international governance chair at the Balsillie School of International

Protesters outside the U.N. Climate Change Conference in Copenhagen on Dec. 10, 2009, call for rich countries to take responsibility for their disproportionate share in global warming. Greenhouse gas emissions by industrial countries are causing climate changes in poor countries thousands of miles away. The nonbinding Copenhagen Accord calls for $10 billion a year for the next three years to help them deal with climate change.

duce emissions, but to monitor and report those. I think that's significant," says Stephen Eule, a U.S. Chamber of Commerce climate expert and former George W. Bush administration climate official.

However, to many of the accord's critics, the accord mostly represents a failure of political leadership. "It was hugely disappointing. Watching world

Affairs in Waterloo, Canada, and studies climate policy, believes critics are underestimating the importance of leaders from around the globe sitting down face-to-face to tackle the problem. "Symbolically, that photograph of the leaders of those countries sitting around the table with their sleeves rolled up was enormous," he says. "All of a sudden we're having a direct conversation among the actors that matter, both in the developed and developing world."

He also credits the conference for tackling difficult questions such as how much money developed countries need to transfer to the developing world to fight climate change and how much countries have to open themselves up to international inspection. "There's been sort of an agreement not to talk about the hard stuff," he says, "and now, at Copenhagen, it was finally front and center."

But to those who believe that the time for talk is running out, the dialogue meant nothing without concrete results. "This [deal], as they themselves say, will not avert catastrophic climate change," said Kumi Naidoo, Greenpeace International's executive director. "That's the only thing on which we agree with them. Everything else is a fudge; everything else is a fraud, and it must be called as such." [16]

Is the U.N.'s climate change negotiating framework outdated?

Although delegations from most of the world's nations came to Copenhagen, the final deal was hammered out by the leaders of only five countries. Those nations — the United States, China, India, Brazil and South Africa — provide a snapshot of the changing nature of geopolitical power.

Although they had been involved in larger group discussions of about 30 nations, the traditional European powers and Japan were not involved in the final deal. The five key players represented the world's largest economy (the United States), the largest emitter of greenhouse gases and second-biggest economy (China) and significant emerging economies in South America (Brazil), Africa (South Africa) and India, with the world's second-largest population.

The five-nation gathering could be seen as an effort to fashion a thin cross-section of the global community. But the U.N.-sponsored Copenhagen

conference was supposed to embody the entire world community. To some observers, the fact that the accord was fashioned outside the official sessions appeared to be an attempt to undermine the U.N. effort.

Anne Petermann, co-director of the Global Justice Ecology Project, an international grassroots organization, notes the Bush administration also worked outside the U.N., setting up a smaller meeting of major economies to discuss climate change. "It wasn't particularly surprising the U.S. negotiated an accord that was completely outside the process," she says. "This wasn't the first time that the U.S. had come in with a strategy of undermining the U.N. Framework Convention."

To other analysts, however, the ability of the small group of leaders to come together where the larger conference had failed shows that the U.N. effort no longer fits the crisis. "The Framework Convention is actually now an obstacle to doing sensible things on climate change," says East Anglia's Hulme. "Climate change is such a multi-faceted problem that we need to find subgroups, multiple frameworks and initiatives to address it."

To others, the U.N. effort remains both the best chance for the world to reach a binding climate change agreement and essential to proceeding. "Because you've really got to have a global solution to this problem, it's essential that all the interested parties, including the most vulnerable countries, be around the table," says Oxfam's Waskow. "There's no question the U.N. Framework Convention, which has been working on this for many years, is the right place for that."

But Homer-Dixon, of the Balsillie School of International Affairs, believes the U.N. Framework process "has too many parties." He expects that on the negotiating side "we're going to migrate to something like the G-20 [economic forum], which includes all the major emitters. It would make sense to have the G-20 responsible."

However, Kassie Siegel, the climate law expert for the Center for Biological Diversity, a U.S. environmental group, thinks critics underestimate the U.N. effort. "Both the U.N. Framework Convention and the Intergovernmental Panel on Climate Change have been building capacity since 1992," she says. "There's not any other institution that came close to their experience on this issue. The U.N. Framework process is the best and fastest way forward."

Supporters also note that the United States and other signatories to the Copenhagen Accord have called for efforts to continue toward reaching a binding agreement at the next U.N. climate gathering in Mexico City at the end of this year. "I don't think the U.N. negotiations are irrelevant because the U.S. is still engaged in the Framework Convention," says Nicola Bullard, a climate change analyst and activist with Focus on the Global South, a nongovernmental group in Bangkok, Thailand.

But Eckhart believes the results in Copenhagen mean that key countries will now focus most of their efforts outside the U.N. framework. "I doubt Mexico City is still relevant," he says. "What can they get done in Mexico City that they couldn't get done in Copenhagen?"

The relationship between the Copenhagen Accord and the U.N. Framework Convention is somewhat ambiguous. Jacob Werksman, a lawyer specializing in international environmental and economics law at the World Resources Institute, concludes the conference's decision to only "take note" of the accord means that some provisions, including the call for setting up a Copenhagen Green Climate Fund to manage billions of dollars in aid through the U.N. mechanism, cannot occur without a conference decision to accept the accord.

U.N. Secretary General Ban Ki-moon has called on all U.N. countries to back the accord. [17] (See "At Issue," p. 43.) But some analysts believe the U.N. Framework Convention can't legally adopt it until the Mexico City conference, which would push the Climate Fund and possibly other accord provisions down the road another year — a delay climate change activists say the world can't afford.

Would a carbon tax reduce emissions more effectively?

Obscured by the immediate furor over Copenhagen is a longer-term debate over whether the developed world is taking the right tack in its approach to reducing emissions.

The most popular approach so far has been the so-called cap-and-trade programs. [18] Progressively lower caps on overall emissions allow power companies and other entities to trade their emission quotas, creating a market-based approach to cutting greenhouse gases. Several European nations have embraced "cap-and-trade," and the climate change legislation that passed the U.S. House last June takes such an approach. But the system has been criticized for its complexity and susceptibility to manipulation and abuse.

Some analysts believe a carbon tax — a levy on carbon-emitting fuels, coupled with a system to rebate most of the tax back to consumers, is a more straightforward and effective way to control emissions. Robert Shapiro, former undersecretary of commerce during the Clinton administration and chair of the U.S. Climate Task Force, advocates such a program and works to educate the public on the need for action on climate change.

Shapiro's plan would use 90 percent of the carbon tax revenue to cut payroll taxes paid by workers and businesses, with the remaining 10 percent going to fund research and development of clean energy technology. The tax would provide a price incentive for discouraging the use of carbon emitting fuels and encouraging the use of green energy, while the tax cut would keep the approach from unduly burdening lower-income Ameri-

cans. "A carbon tax would both directly reduce greenhouse gas emissions and provide powerful incentives for technological progress in this area," Shapiro wrote. "It offers the best way forward in both the national and global debate over climate change." [19]

Residents grin and bear flooding in Jakarta, Indonesia, in December 2007. Similar scenes would be played out in coastal cities and communities around the world if climate change causes glaciers and polar ice caps to melt, which many researchers predict. Analysts say the worst effects of climate change are expected to be felt in Asia.

However, carbon tax opponents argue it would be no more effective than cap-and-trade and would lead to a huge expansion of government. Analysts at the Heritage Foundation, a conservative U.S. think tank, wrote that a carbon tax "would cause significant economic damage and would do very little to reduce global temperatures." Even coupling it with a payroll tax cut, they continue, "would do little to offset the high energy prices that fall particularly hard on low-income households." The real agenda of a carbon tax, they charge, is "about raising massive amounts of revenue to fund a huge expansion in government." [20]

Several Scandinavian countries have adopted carbon taxes, with mixed results. Norway has seen its per capita CO_2 emissions rise significantly. But Denmark's

2005 emissions were 15 percent below what they were in 1990, and the economy still remained strong. [21]

But to Bullard, at Focus on the Global South, a carbon tax is the approach most likely to spur changes in personal behavior. "Reducing consumption is really important, reducing our own dependence on fossil fuels," she says. "I think it's very important to have a redistributive element so that working people and elderly people don't end up with a huge heating bill. But it's really a simpler and more effective route than a complicated solution like cap-and-trade."

However, Bill McKibben — an American environmentalist and the founder of 350.org, an international campaign dedicated to scaling back GHG emissions — says a carbon tax faces an almost insurmountable political hurdle in the United States. "Even I can't convince myself that America is going to sit very long with something called a carbon tax," he says.

McKibben thinks "cap and rebate" legislation recently introduced by Sens. Maria Cantwell, D-Wash., and Susan Collins,

R-Maine, would be more palatable to voters. It would cap total emissions — a limit that would be tightened over time — with the government auctioning off available carbon credits. The money raised would be rebated to consumers to offset any higher energy bills. [22]

Congressional efforts, however, have focused on cap-and-trade. But as wariness grows in the U.S. Senate toward the ramifications of cap-and-trade, Shapiro believes a carbon tax could prove a more appealing option. "A real public discussion and debate about a carbon tax tied to offsetting cuts in payroll or other taxes," he said, "could be the best news for the climate in a very long time." [23] ■

BACKGROUND

Road to Copenhagen

The road to Copenhagen was a long one. In one sense, it began with the Industrial Revolution in the 18th and 19th century, which brought with it the increased burning of coal and the beginning of large-scale carbon dioxide emissions in Europe and America. It also started with scientific speculation in the 1930s that manmade emissions could be changing the planet's climate.

Those first studies were widely discounted, a reflection of the difficulty humanity has had coming to grips with the idea it could be changing the global climate. But by the mid-1980s, thanks in large part to the work of David Keeling at the Mauna Loa Observatory in Hawaii, the world had a nearly three-decade record of rising carbon dioxide levels in the atmosphere. [24] Scientists were also reporting an overall warming trend in the atmosphere over the last 100 years, which they considered evidence of a "greenhouse effect" tied to CO_2 and other manmade emissions.

Humankind began a slow, often painful struggle to understand and deal

Climate Change Could Force Millions to Relocate

"Climate Refugees" from Africa to the Arctic could be affected.

Maasi herdsman Moses Mopel Kisosion had never been outside Kenya before. He'd never ridden on a plane. But he flew across parts of two continents to deliver a message to anyone who would listen at the Copenhagen climate conference in December.

Climate change, he believes, is destroying the ability of his people, the Kajiado Maasi, to make a living. "I am a pastoralist, looking after cattles, walking from one place to another looking for grass and pastures," Kisosion said. "And now, for four years, we have a lack of rain, so our animals have died because there's no water and no grass. . . . We are wondering how our life will be because we depend on them." [1]

The Maasi are hardly alone in worrying if they will be able to continue living where they are. From small South Pacific island nations to the Arctic, hundreds of millions of people might have to relocate to survive as a result of climate change. If global warming predictions prove accurate, some researchers believe the world could soon find itself dealing with a tidal wave of "climate refugees."

A study by the U.N. Office for the Coordination of Humanitarian Affairs and the Internal Displacement Monitoring Centre found that "climate-related disasters — that is, those resulting from hazards that are already being or are likely to be modified by the effects of climate change — were responsible for displacing approximately 20 million people in 2008." [2]

Norman Myers, a British environmentalist, sees the situation worsening as the effects of climate change grow. In a 2005 study, he concluded that up to 200 million people could become climate refugees. [3] But he recently revised his estimate significantly. "We looked at the best prognosis for the spread of desertification and sea level rise, including the associated tsunamis and hurricanes, and we meshed those figures with the number of people impoverished or inhabiting coastal zones," says Myers. "We believe we could see half a billion climate refugees in the second half of the century."

The human displacement is likely to take place over several decades, experts say, and determining who is a climate refugee and who is simply a political or economic refugee could be difficult. International organizations have just begun the discussion about their status and what kind of assistance they might require.

The European Commission is funding a two-year research project, "Environmental Change and Forced Migration Scenarios," based on case studies in 24 vulnerable countries. [4] An African Union Summit in Kampala, Uganda, also met last October to consider how it would address the growing number of displaced Africans. [5]

Wahu Kaara, a Kenyan political activist, says the need for action is pressing. Kenya has recorded four major droughts in the last decade, significantly higher than the average over the previous century. "Very many people are dislocated and have to move to where they can salvage their lives," she says. "We have seen people die as they walk from one place to another. It's not a hardship; it's a catastrophe. They not only have lost their animals, they have lost their lives, and the framework of their lives for those who survive."

While Africa already may be suffering population movement due to climate change, the worst consequences are likely to be felt in Asia, analysts say. Rising sea levels threaten low-lying coastal areas, which constitute only 2 percent of the land surface of the Earth but shelter 10 percent of its population. About 75 percent of the people living in those areas are in Asia. [6]

The Maldives, a nation of low-lying islands in the Indian Ocean that could be submerged if predictions prove accurate, has taken the lead in trying to organize smaller island nations in the global warming debate. President Mohamed Nasheed initially supported the Copenhagen Accord and its 2-degree Celsius target for limiting global warming as a beginning. But before the deal was struck, he declared, "At 2 degrees, my country would not survive." [7]

Rising sea levels threaten every continent, including the Americas. Until recently, Kivalina Island, an eight-mile long barrier island in northern Alaska, had survived the punishing storms that blew in from the ocean because of ice that formed and piled up on the island. [8]

Inupiat hunters from the island's small village began noticing changes in the ice years ago, says the island's tribal administrator, Colleen Swan, but the change has accelerated in recent years. "In early September and October, the ice used to start forming, but now it doesn't form anymore until January and it's not building up," she says. "When that happened, we lost our barrier from fall sea storms, and our island just started falling apart. We started losing a lot of land beginning in 2004."

The U.S. Army Corps of Engineers is building a seawall to protect what's left of Kivalina, but Swan says it is expected to buy only 10 or 15 years. "People in the United States are still debating whether climate change is happening. The U.N. is focusing on the long-term problem of emissions," Swan says, "but we're in the 11th hour here. The bottom line is we need someplace to go."

— *Reed Karaim*

A house tumbles into the Chukchi Sea in Shishmaref, Alaska. Like other victims of climate change, residents may have to abandon the tiny community due to unprecedented erosion caused by intense storms.

[1] Moses Mopel Kisosion spoke in a video blog from KilmaforumC9, the "people's forum" on climate change held in Copenhagen during the official conference. It is available online at http://en.cop15.dk/blogs/view+blog?blogid=2929.

[2] "Monitoring disaster displacement in the context of climate change," the U.N. Office for the Coordination of Humanitarian Affairs and The Internal Displacement Monitoring Centre, September 2009, p. 12.

[3] Norman Myers, "Environmental Refugees, an Emergent Security Issue," presented at the 13th Economic Forum, Prague, May 2005.

[4] "GLOBAL: Nowhere to run from nature," IRIN, Nov. 9, 2009, www.irinnews.org/report.aspx?ReportId=78387.

[5] "AFRICA: Climate change could worsen displacement — UN," IRIN, Nov. 9, 2009, www.irinnews.org/report.aspx?ReportId=86716.

[6] Anthony Oliver-Smith, "Sea Level Rise and the Vulnerability of Coastal Peoples," U.N. University Institute for Environment and Human Security, 2009, p. 5, www.ehs.unu.edu/file.php?id=652.

[7] "Address by His Excellency Mohamed Nasheed, President of the Republic of Maldives, at the Climate Vulnerable Forum," Nov. 9, 2009, www.actforclimatejustice.org/2009/11/address-by-his-excellency-mohamed-nasheed-president-of-the-republic-of-maldives-at-the-climate-vulnerable-forum/.

[8] See John Schwartz, "Courts As Battlefields in Climate Fights," *The New York Times*, Jan. 26, 2010.

with a global challenge. From the beginning, there were doubters, some well-intentioned, some with a vested interest in making sure that the world continued to burn fossil fuels. Even as the scientific consensus on climate change has grown stronger, and many nations have committed themselves to tackling global warming, the issue continues to provoke and perplex.

Climate and Culture

In her book *Field Notes from a Catastrophe, Man, Nature and Climate Change*, American writer Elizabeth Kolbert visits, among other spots, Greenland's ice fields, a native village in Alaska and the countryside in northern England, surveying how global warming is changing the Earth. In the opening section, she admits her choices about where to go to find the impact of climate change were multitudinous.

"Such is the impact of global warming that I could have gone to hundreds if not thousands of other places," Kolbert writes, "From Siberia to the Austrian Alps to the Great Barrier Reef to the South African *fynbos* (shrub lands)." [25]

Despite mounting evidence, however, climate change remains more a concept than a reality for huge parts of the globe, where the visible impacts are still slight or nonexistent. Research scholar Blackstock, whose work focuses on the intersection between science and international affairs, points out that for many people this makes the issue as much a matter of belief as of fact.

"It really strikes to fundamental questions on how we see the human-nature interface," he says. "It has cultural undertones, religious undertones, political undertones." Blackstock thinks many climate scientists have missed this multifaceted dimension to the public dialogue. "Pretending this is just a scientific debate won't work," he says. "That's important, but we can't have that alone."

The heart of the matter, he suggests, is how willing we are to take responsibility for changes in the climate and how we balance that with other values. This helps to explain the varying reactions in the United States, which has been reluctant to embrace limits on carbon emissions, and Europe, which has been more willing to impose measures. "You're seeing the cultural difference between Europe and America," Blackstock says, "the American values of individualism and personal success versus the communal and collective good, which Europe has more of a sense of being important."

Other analysts see attitudes about climate deeply woven into human culture. The University of East Anglia's Hulme, author of *Why We Disagree About Climate Change*, notes that climate and weather have been critical to humanity for most of its history. The seasons, rains and hot or cold temperatures have been so essential to life — to the ability to obtain food and build stable communities — that they have been attributed to deities and formed the basis for religious ceremonies. Even in the modern age, Hulme says, "People have an instinctive sense that weather and climate are natural phenomena, that they work at such scales and complexity that humans could not possibly influence them."

He points out that weather was once the realm of prophets, "and part of our population is still resistant to the idea that science is able to predict what the weather will be. This deep cultural history makes climate change a categorically different phenomenon than other scientifically observed data."

Climate is also often confused with weather. England, for example, has a temperate, damp climate, but can have dry, hot years. The human inclination is to believe what's before our eyes, so every cold winter becomes a reason to discount global warming.

Sander van der Leeuw, director of the School of Human Evolution and Social Change at Arizona State University in Tempe, Ariz., notes that facing climate change also means contemplating the costs of consumerism. "Those of us in the developed world have the most invested in this particular lifestyle," he says. "If that lifestyle has to change, we'll be facing the most wrenching dislocations."

Van der Leeuw, who worked for the European Union on climate change issues in the 1990s, is actually optimistic about the progress the world has made on climate change in the face of these challenges. "It's a very long process," he says, "but I'm encouraged by my students. It's wonderful to see how engaged they are, how open to thinking differently on these issues. I know we have very little time, but history is full of moments where we've reacted in the nick of time."

However, there are still those who doubt the basic science of climate change.

The Doubters

To enter the world of the climate change skeptics is to enter a mirror reflection of the scientific consensus on the issue. Everything is backwards: The Earth isn't warming; it may be cooling. If it is warming, it's part of the planet's natural, long-term climate cycles. Manmade carbon dioxide isn't the heart of the problem; it's a relatively insignificant greenhouse gas. But even if carbon dioxide is increasing, it's beneficial for the planet.

And that scientific consensus? It doesn't exist. "What I see are a relatively small number, perhaps a few hundred at most, of extremely well-funded, well-connected evangelistic scientists doing most of the lobbying on this issue," says Bob Carter, a geologist who is one of Australia's more outspoken climate change skeptics.

Many scientists who take funds from grant agencies to investigate global warming, he says, "don't speak out with their true views because if they did so, they would lose their funding and be intimidated."

It's impossible to know if people are keeping views to themselves, of course. But professional science has a method of inquiry — the scientific method — and a system of peer review intended to lead to knowledge that, as much as possible, is untainted by prejudice, false comparison or cherry-picked data. The process isn't always perfect, but it provides our best look at the physical world around us.

In December 2004, Naomi Oreske, a science historian at the University of California, San Diego, published an analysis in *Science* in which she reviewed 928 peer-reviewed climate studies published between 1993 and 2003. She did not find one that disagreed with the general consensus on climate change. [26]

The U.S. National Academy of Sciences, the Royal Society of London, the Royal Society of Canada, the American Meteorological Society, the American Association for the Advancement of Science and 2,500 scientists participating in the IPCC also have concluded the evidence that humans are changing the climate is compelling. "Politicians, economists, journalists and others may have the impression of confusion, disagreement or discord among climate scientists, but that impression is incorrect," Oreske wrote, after reviewing the literature. [27]

The debate over climate change science heated up last fall, when, shortly before the Copenhagen conference, hackers broke into the University of East Anglia's computer network and made public hundreds of e-mails from scientists at the school's climate research center — some prominent in IPCC research circles. Climate change skeptics were quick to point to the "Climategate" e-mails as evidence researchers had been squelching contrary opinions and massaging data to bolster their claims.

Continued on p. 38

Chronology

1900-1950s
Early research indicates the Earth is warming.

1938
British engineer Guy Stewart Callendar concludes that higher global temperatures and rising carbon dioxide levels are probably related.

1938
Soviet researchers confirm that the planet is warming.

1957
U.S. oceanographer Roger Revelle and Austrian physicist Hans Suess find that the oceans cannot absorb carbon dioxide as easily as thought, indicating that manmade emissions could create a "greenhouse effect," trapping heat in the atmosphere.

1958
U.S. scientist David Keeling begins monitoring atmospheric carbon dioxide levels, creating a groundbreaking record of their increase.

1960s
Climate science raises the possibility of global disaster.

1966
U.S. geologist Cesare Emiliani says ice ages were created by tiny shifts in Earth's orbit, backing earlier theories that climate reacts to small changes.

1967
Leading nations launch 15-year program to study the world's weather.

1968
Studies show Antarctica's huge ice sheets could melt, raising sea levels.

1970s-1980s
Research into climate change intensifies, and calls for action mount.

1975
A National Aeronautics and Space Administration (NASA) researcher warns that fluorocarbons in aerosol sprays could help create a greenhouse effect.

1979
The National Academy of Sciences finds that burning fossil fuels could raise global temperatures 6 degrees Fahrenheit in 50 years.

1981
U.S. scientists report a warming trend since 1880, evidence of a greenhouse effect.

1985
Scientists from 29 nations urge governments to plan for warmer globe.

1988
NASA scientist James Hansen says global warming has begun; he's 99 percent sure it's manmade.

1988
Thirty-five nations form a global panel to evaluate climate change and develop a response.

1990s
As the world responds to global warming, industry groups fight back.

1990
The carbon industry-supported Global Climate Coalition forms to argue that climate change science is too uncertain to take action.

1995
The year is the hottest since the mid-19th century, when records began being kept.

1997
More than 150 nations agree on the Kyoto Protocol, a landmark accord to reduce greenhouse gases. The U.S. signs but never ratifies it.

2000s
The political battle over climate change action escalates worldwide.

2000
Organization of Petroleum Exporting Countries (OPEC) demands compensation if global warming remedies reduce oil consumption.

2006
National Academy of Sciences reports the Earth's temperature is the highest in 12,000 years, since the last Ice Age.

2007
A U.N. report concludes that global warming is "unequivocal" and human actions are primarily responsible.

2009
The 194 nations attending the Copenhagen Climate Change Conference cannot agree on a broad treaty to battle global warming. After two weeks of contentious discussion, five nations create a nonbinding climate change accord, which 55 nations eventually sign, but which falls far short of delegates' hopes.

2010
The U.N effort to get a global, legally binding climate change treaty is scheduled to continue in November-December in Mexico City.

Continued from p. 36

Reviews by *Time, The New York Times* and the Pew Center on Climate Change, however, found the e-mails did not provide evidence to alter the scientific consensus on climate change. "Although a small percentage of the e-mails are impolite and some express animosity toward opponents, when placed into proper context they do not appear to reveal fraud or other scientific misconduct," the Pew Center concluded. [28]

Some skeptics are scientists, but none are climate researchers. Perhaps the most respected scientific skeptic is Freeman Dyson, a legendary 86-year-old physicist and mathematician. Dyson does not dispute that atmospheric carbon-dioxide levels are rapidly rising and humans are to blame. He disagrees with those who project severe consequences. He believes rising CO_2 levels could have some benefits, and

if not, humanity could bioengineer trees that consume larger amounts of carbon dioxide or find some other technological solution. He is sanguine about the ability of the Earth to adapt to change and is suspicious of the validity of computer models.

"The climate-studies people who work with models always tend to overestimate their models," Dyson has said. "They come to believe models are real and forget they are only models." [29]

Unlike Dyson, many climate change skeptics are connected to groups backed by the oil, gas and coal industries, which have worked since at least 1990 to discredit global warming theories. A 2007 study by the Union of Concerned Scientists found that between 1998 and 2005 Exxon-Mobil had funneled about $16 million to 43 groups that sought to manufacture uncertainty about global warming with the public. [30]

The tactics appear to be patterned after those used by the tobacco industry to discredit evidence of the hazards of smoking. According to the study, ExxonMobil and others have used ostensibly independent front groups for "information laundering," as they sought to sow doubts about the conclusions of mainstream climate science.

Several prominent climate change skeptics — including physicist S. Fred Singer and astrophysicists Willie Soon and Sallie Baliunas — have had their work published by these organizations, some of which seem to have no other purpose than to proliferate the information. "By publishing and re-publishing the non-peer-reviewed works of a small group of scientific spokespeople, ExxonMobil-funded organizations have propped up and amplified work that has been discredited by reputable climate scientists," the study concludes. [31]

Climate Scientists Thinking Outside the Box

"Geoengineering" proposes futuristic solutions that sound like science fiction.

Imagine: A massive squadron of aircraft spewing sulfur particles into the sky. An armada of oceangoing ships spraying sea mist into the air. A swarm of robotic mirrors a million miles out in space reflecting some of the sun's harmful rays away from the Earth. Thousands of giant, air-filtering towers girdling the globe.

The prospect of devastating global warming has led some scientists and policy analysts to consider the kind of planet-altering responses to climate change that were once the province of science fiction. The underlying concept, known as "geoengineering," holds that manmade changes in the climate can be offset by futuristic technological modifications.

That idea raises its own concerns, both about the possibility of unintended consequences and of technological dependence. But from an engineering perspective, analysts say the sulfur particle and sea vapor options — which would reflect sunlight away from the Earth, potentially cooling the planet — appear feasible and not even that expensive.

"Basically, any really rich guy on the planet could buy an ice age," says David Keith, a geoengineering expert at the University of Calgary, estimating that sulfur injection could cost as little as $1 billion or so a year. "Certainly, it's well within the capability of most nations."

"Technologically, it would be relatively easy to produce small particles in the atmosphere at the required rates," says Ken Caldiera, a climate scientist at the Carnegie Institution for Science's Department of Global Ecology in Stanford, Calif. "Every climate-model simulation performed so far indicates geoengineering would be able to diminish most climate change for most people most of the time."

To spread sulfur, planes, balloons or even missiles could be used. [1] For sea vapor, which would be effective at a lower altitude, special ships could vaporize seawater and shoot it skyward through a rotor system. [2]

A global program of launching reflective aerosols higher into the atmosphere would cost around $5 billion annually — still small change compared to the economic costs of significant global warming, says Caldiera. Other geoengineering options are considerably more expensive. The cost of launching the massive (60,000 miles by 4,500 miles) cloud of mirrors into space to block sunlight would cost about $5 trillion. [3] Building air-scrubbing towers would also be expensive and would require improved technology. [4]

But cost is not what worries those studying geoengineering. "Everyone who's thinking about this has two concerns," says Thomas Homer-Dixon, a political scientist at Canada's Balsillie School of International Affairs in Waterloo, Ontario. "One is unintended consequences — because we don't understand climate systems perfectly — something bad could happen like damage to the ozone layer. The second is the moral-hazard problem: If we start to do this, are a lot of people going to think it means we can continue the carbon party?"

Keith thinks the consequences could be managed. "One of the advantages of using aerosols in the atmosphere is that you can modulate them," he says. "If you find it's not working, you can stop and turn the effect off.' But he shares a concern with Caldiera and Homer-Dixon that geoengineering could be used as an excuse to avoid reducing carbon-dioxide emissions.

Geoengineering also raises geopolitical concerns, in part because it could be undertaken unilaterally. Unlike lowering greenhouse gas emissions, it doesn't require a global agreement, yet its effects would be felt around the planet — and not evenly.

That could aggravate international tensions: Any sustained bad weather in one nation could easily raise suspicion that it was the victim of climate modifications launched by another country. "If China, say, were to experience a deep drought after the deployment of a climate-intervention system," says Caldiera, "and people were starving as a result, this could cause them to lash out politically or even militarily at the country or countries that were engaged in the deployment."

Such scenarios, along with the fear of undercutting global negotiations to reduce emissions, make serious international consideration of geoengineering unlikely in the near term, says Homer-Dixon. But if the direst predictions about global warming prove accurate that could change. "You could see a political clamor worldwide to do something," he says.

Some scientists believe stepped-up geoengineering studies need to start soon. "We need a serious research program, and it needs to be international and transparent," says Keith. "It needs to start small. I don't think it needs to be a crash program, but I think there's an enormous value in doing the work. We've had enough hot air speculation. We need to do the work. If we find out it works pretty well, then we'll have a tool to help manage environmental risk."

— *Reed Karaim*

[1] Robert Kunzig, "A Sunshade for Planet Earth," *Scientific American*, November 2008.

[2] *Ibid.*

[3] *Ibid.*

[4] Seth Borenstein, "Wild ideas to combat global warming being seriously entertained," *The Seattle Times*, March 16, 2007, http://seattletimes.nwsource.com/html/nationworld/2003620631_warmtech16.html.

Is the world cooling? Is global warming a natural phenomenon? Is more CO_2 really good for the planet? Science and media watchdog groups have published detailed rebuttals to the claims of climate change skeptics. [32] To cite one example, assertions that the Earth is actually cooling often use 1998 as the base line — a year during the El Niño weather system, which typically produces warmer weather. The Associated Press gave temperature numbers to four statisticians without telling them what the numbers represented. The scientists found no true declines over the last 10 years. They also found a "distinct, decades-long" warming trend. [33]

James Hoggan, a Canadian public relations executive who founded DeSmogblog to take on the skeptics, feels climate scientists have done a poor job of responding to the skeptics, too often getting bogged down in the minutiae of detail. "We need to start asking these so-called skeptics a number of basic questions," says Hoggan, the author of *Climate Cover-Up: The Crusade to Deny Global Warming*. "The first one is, 'Are you actually a climate scientist?' The second one is, 'Have you published peer-reviewed papers on whatever claims you're making?' And a third one is, 'Are you taking money directly or indirectly from industry?' "

Untangling the Threads

Since nations first began to seriously wrestle with climate change, most of the effort has gone into fashioning a legally binding international treaty to cut greenhouse gas emissions while helping poorer nations cope with the effects of global warming.

The approach has a powerful logic. Climate change is a worldwide problem and requires concerted action around the planet. Assisting those most likely to be affected — populations in Africa and Asia who are among the poorest on the globe — is also a burden that is most equitably shared.

But the all-in-one-basket approach also comes with big problems. The first is the complexity of the negotiations themselves, which involve everything from intellectual-property rights to hundreds of billions of dollars in international finance to forest management. Global nations have been meeting on these issues for nearly two decades without a breakthrough deal.

Some observers believe the best chance for moving forward is untangling the threads of the problem. "We don't have to try to set the world to rights in one multilateral agreement," says East Anglia's Hulme. "It's not something we've ever achieved in human history, and I doubt we can. It seems more likely it's acting as an unrealistic, utopian distraction."

Analysts cite the 1987 Montreal Protocol, which phased out the use of chlorofluorocarbons that were damaging

Hunger and Thirst

A young Turkana girl in drought-plagued northern Kenya digs for water in a dry river bed in November 2009 (top). Momina Mohammed's 8-month-old son Ali suffers from severe malnutrition in an Ethiopian refugee camp in December 2008 (bottom). Food and water shortages caused by climate change are already affecting many countries in Africa. A Sudanese delegate to the Copenhagen Climate Change Conference called the nonbinding accord reached at the convention "an incineration pact" for poor countries.

burning stoves would cut global warming gases, yet hardly requires the wrenching shift of moving from coal-fired electricity. Hydrofluorocarbons (HFCs) are more than a thousand times more potent as greenhouse gases than CO_2, but are used in comparably minuscule amounts and should be easier to limit.

"Why are we putting all the greenhouse gases into one agreement? CO_2 is very different from black soot, or methane or HFCs," Hulme says. "Tropical forests, why do they have to be tied to the climate agenda? They sequester carbon, yes, but they're also valuable resources in other regards."

Those who support negotiating a sweeping climate change accord believe that untangling these threads could weaken the whole cloth, robbing initiative from critical parts of the deal, such as assistance to developing countries. But Hulme believes the poorer parts of the world could benefit.

"We can tend to the adaptation needs of the developing world without having them hitched to the much greater complexity of moving the economy in the developed world away from fossil fuels," he says.

Other analysts, however, are unconvinced that climate change would be easier to deal with if its constituent issues were broken out. "There are entrenched interests on each thread," says Blackstock, at Austria's International Institute in Applied Systems Analysis. "That's the real problem at the end of the day." ∎

CURRENT SITUATION

Next Steps

The whole world may be warming, but as has been said, all politics is local — even climate change politics.

the ozone layer, as an example of a successful smaller-scale deal.

So far, the effort to control global warming has focused on limiting carbon-dioxide emissions from power plants and factories. But CO_2 accounts for only half of manmade greenhouse gas emissions. [34] The rest comes from a

variety of sources, where they are often easier or cheaper to cut.

Black carbon, mainly produced by diesel engines and stoves that burn wood or cow dung, produces from one-eighth to a quarter of global warming. [35] Promoting cleaner engines and helping rural villagers move to cleaner-

"It's still the legislatures of the nation states that will really determine the pace at which climate policies are driven through," notes the University of East Anglia's Hulme. "In the end, that's where these deals have to make sense."

Nations around the globe are determining their next steps in the wake of Copenhagen. Most greenhouse gases, however, come from a relative handful of countries. The United States and China, together, account for slightly more than 40 percent of the world's man-made CO_2 emissions. [36] If India and the European Union are added, the total tops 60 percent. [37] The post-Copenhagen climate change status is different for each of these major players.

China — China presents perhaps the most complex case of any of the countries central to climate change. It was classified as a developing country in the Kyoto Protocol, so it was not required to reduce carbon emissions. [38] But as the country's economy continued to skyrocket, China became the world's largest carbon dioxide emitter in 2006, passing the United States. [39] (See graph, p. 29.)

But with roughly 700 million poorer rural citizens, promoting economic growth remains the Chinese government's essential priority. Nevertheless, shortly before Copenhagen, China announced it would vow to cut CO_2 emissions by 40 to 45 percent *per unit of gross domestic product* below 2005 levels by 2020. The complicated formula meant that emissions would still rise, but at a slower rate. China subsequently committed to this reduction when confirming its Copenhagen pledge at the end of January.

U.N. climate policy chief de Boer hailed the move as a critical step. But the United States — especially skeptical members of the U.S. Congress — had hoped to see more movement from China and wanted verification standards.

Some participants say China's recalcitrance is why Copenhagen fell short. The British seemed particularly incensed. Ed Miliband, Great Britain's

climate secretary, blamed the Chinese leadership for the failure to get agreement on a 50-percent reduction in global emissions by 2050 or on 80-percent reductions by developed countries. "Both were vetoed by China," he wrote, "despite the support of a coalition of developed and the vast majority of developing countries." [40]

But the Global Justice Ecology Project's Petermann places the blame elsewhere. "Why should China get involved in reducing emissions if the U.S. is unwilling to really reduce its emissions?" she asks.

Jiang Lin, director of the China Sustainable Energy Program, a nongovernmental agency with offices in Beijing and San Francisco, thinks China's leaders take the threat of climate change seriously. "There's probably a greater consensus on this issue in China than the United States," says Jiang. "The Chinese leadership are trained engineers. They understand the data."

Jiang points out that China already is seeing the effects predicted by climate change models, including the weakening of the monsoon in the nation's agricultural northwest and the melting of the Himalayan glaciers. "The Yellow River is drying up," he adds. "This is very symbolic for the Chinese. They consider this the mother river, and now almost half the year it is dry."

The Copenhagen Accord is not legally binding, but Jiang believes the Chinese will honors its provisions. "When they announce they're committed to something, that's almost as significant as U.S. law," he says, "because if they don't meet that commitment, losing facing is huge for them."

While attention has focused on international negotiations, China is targeting improved energy efficiency and renewable power. In 2005, China's National People's Congress set a goal of generating 20 gigawatts of power through wind energy by 2020. The goal seemed highly ambitious, but China expected to meet it by the end of 2009

and is now aiming for 150 gigawatts by 2020. The target for solar energy has been increased more than 10-fold over the same period. [41]

Coal still generates 80 percent of China's power, and the country continues to build coal-fired plants, but Chinese leaders clearly have their eyes on the green jobs that President Obama has promoted as key to America's future. [42] "Among the top 10 solar companies in the world, China already has three," says Jiang, "and China is now the largest wind market in the world. They see this as an industry in which China has a chance to be one of the leaders."

The United States — To much of the world, the refusal of the United States so far to embrace carbon emission limits is unconscionable. U.S. emissions are about twice Europe's levels per capita, and more than four times China's.

"The United States is the country that needs to lead on this issue," says Oxfam's Waskow. "It created a lot of problems that the U.S. wasn't able to come to Copenhagen with congressional legislation in hand."

In the Copenhagen Accord, President Obama committed the United States to reduce its carbon dioxide emissions to 17 percent below 2005 levels by 2020. That equates to about 4 percent below 1990 levels, far less stringent than the European and Japanese pledges of 20 percent and 25 percent below 1990 levels, respectively. However, Congress has not passed global warming legislation. Last year, the House of Representatives passed a bill that would establish a cap-and-trade system, which would limit greenhouse gases but let emitters trade emission allowances among themselves. The legislation faces stiff opposition in the Senate, however.

In 1997, after the Kyoto Protocol was adopted, the Senate voted 95-0 against signing any international accord unless it mandated GHG emission reductions by developing countries as well. Securing such commitments in Copenhagen — especially from China,

More Countries Agree to Emissions Cuts

The nonbinding climate agreement reached in Copenhagen, Denmark, on Dec. 18 was originally joined by 28 countries, which were to send the United Nations by the end of January their individual goals for reducing carbon emissions by 2020. But other nations also were invited to sign on by submitting their own plans to cut emissions. On Feb. 1, the U.N. reported that a total of 55 nations had submitted targets for cutting greenhouse gases. Analysts say while these countries produce 78 percent of manmade carbon emissions, more cuts are needed. The U.N. will try to use the accord as a starting point for a binding treaty at the next international climate conference in Mexico City, Nov. 29-Dec. 10.

Key provisions in the Copenhagen Accord:

- Cut global greenhouse gas emissions so global temperatures won't rise more than 2 degrees Celsius above the pre-Industrial Revolution level.

- Cooperate in achieving a peak in emissions as soon as possible.

- Provide adequate, predictable and sustainable funds and technology to developing countries to help them adapt to climate change.

- Prioritize reducing deforestation and forest degradation, which eliminate carbon-consuming trees.

- Provide $30 billion in new and additional resources from 2010 to 2012 to help developing countries mitigate climate change and protect forests; and provide $100 billion a year by 2020.

- Assess implementation of the accord by 2015.

Sources: "Copenhagen Accord," U.N. Framework Convention on Climate Change, Dec. 18, 2009; "UNFCCC Receives list of government climate pledges," press release, United Nations Framework Convention on Climate Change, Feb. 1, 2010, http://unfccc.int/files/press/news_room/press_releases_and_advisories/application/pdf/pr_accord_100201.pdf

along with improved verification — was considered critical to improving the chances a climate change bill would make it through the Senate.

Some analysts also blamed the lack of U.S. legislation for what was considered a relatively weak American proposal at Copenhagen. "Obama wasn't going to offer more than the U.S. Senate was willing to offer," says the International Institute for Applied System's Blackstock. "He could have done more and said, 'I cannot legally commit to this, but I'll go home and fight for it.' He didn't."

But Obama's negotiating effort in Copenhagen impressed some observers. "He could have stood back and worried about looking presidential," says the American Council on Renewable

Energy's Eckhart. "He didn't. He rolled up his sleeves and got in there and tried to do good for the world."

Early reviews of the Copenhagen Accord were favorable among at least two key Republican senators, Lisa Murkowski of Alaska and Richard Lugar of Indiana. "Whenever you have developing countries, and certainly China and India stepping forward and indicating that they have a willingness to be a participant . . . I think that that is progress," said Murkowski. [43]

Still, analysts remain skeptical whether it will make a real difference on Capitol Hill. "I don't see Congress doing anything, even in line with the position in the Copenhagen Accord unless Obama makes it his 2010 pri-

ority," says 350.org's McKibben. "There's no question it's going to be hard because it's going to require real change."

The administration is planning to regulate some greenhouse gases through the Environmental Protection Agency (EPA). The Center for Biological Diversity has petitioned the EPA to make further use of regulation to reduce greenhouse emissions. "The president has the tools he needs. He has the Clean Air Act," says the center's Siegel. "All he has to do is use it."

However, some Senate Republicans are already calling for a resolution to undo the EPA's limited actions, and polls show a rising number of Americans skeptical about global warming, particularly Republicans. [44] Given the highly polarized nature of American politics, any significant move on climate change is likely to prove a bruising battle. President Obama has made promoting green energy jobs a priority, but with health care and joblessness still leading the administration's agenda, further action on climate change seems unlikely in the next year. Chances for major legislative action shrunk even further with the election of Republican Scott Brown, a climate change skeptic, to the Senate from Massachusetts. Brown's win ended the democrats' 60-vote, filibuster-proof majority. [45]

India — Although India's economy has grown almost as rapidly as China's in recent years, it remains a much poorer country. Moreover, its low coastline and dependence on seasonal monsoons for water also make it sensitive to the dangers of global warming. Jairam Ramesh, India's environment minister, said, "The most vulnerable country in the world to climate change is India." [46]

India's leaders announced recently they will pursue cleaner coal technology, higher emissions standards for automobiles and more energy-efficient building codes. Prior to Copenhagen, India also announced it would cut CO_2 emissions per unit of GDP from 2005 levels, but

Continued on p. 44

At Issue:

Is the Copenhagen Accord a meaningful step forward in halting climate change?

BAN KI-MOON
SECRETARY-GENERAL, UNITED NATIONS

FROM OPENING REMARKS AT PRESS CONFERENCE, U.N. CLIMATE CHANGE CONFERENCE, COPENHAGEN, DEC. 19, 2009

*t*he Copenhagen Accord may not be everything that everyone hoped for. But this decision of the Conference of Parties is a new beginning, an essential beginning.

At the summit I convened in September, I laid out four benchmarks for success for this conference. We have achieved results on each.

- All countries have agreed to work toward a common, long-term goal to limit global temperature rise to below 2 degrees Celsius.
- Many governments have made important commitments to reduce or limit emissions.
- Countries have achieved significant progress on preserving forests.
- Countries have agreed to provide comprehensive support to the most vulnerable to cope with climate change.

The deal is backed by money and the means to deliver it. Up to $30 billion has been pledged for adaptation and mitigation. Countries have backed the goal of mobilizing $100 billion a year by 2020 for developing countries. We have convergence on transparency and an equitable global governance structure that addresses the needs of developing countries. The countries that stayed on the periphery of the Kyoto process are now at the heart of global climate action.

We have the foundation for the first truly global agreement that will limit and reduce greenhouse gas emission, support adaptation for the most vulnerable and launch a new era of green growth.

Going forward, we have three tasks. First, we need to turn this agreement into a legally binding treaty. I will work with world leaders over the coming months to make this happen. Second, we must launch the Copenhagen Green Climate Fund. The U.N. system will work to ensure that it can immediately start to deliver immediate results to people in need and jump-start clean energy growth in developing countries. Third, we need to pursue the road of higher ambition. We must turn our back on the path of least resistance.

Current mitigation commitments fail to meet the scientific bottom line.

We still face serious consequences. So, while I am satisfied that we have a deal here in Copenhagen, I am aware that it is just the beginning. It will take more than this to definitively tackle climate change.

But it is a step in the right direction.

NNIMMO BASSEY
CHAIR, FRIENDS OF THE EARTH INTERNATIONAL

WRITTEN FOR *CQ GLOBAL RESEARCHER*, FEBRUARY 2010

*t*he Copenhagen Accord is not a step forward in the battle to halt climate change. Few people expected the Copenhagen climate talks to yield a strong outcome. But the talks ended with a major failure that was worse than predicted: a "Copenhagen Accord" in which individual countries make no new serious commitments whatsoever.

The accord sets a too-weak goal of limiting warming to 2 degrees Celsius, but provides no means of achieving this goal. Likewise, it suggests an insufficient sum for addressing international solutions but contains no path to produce the funding. Individual countries are required to do nothing.

The accord fails the poor and the vulnerable communities most impacted by climate change. This non-agreement (it was merely "noted," not adopted, by the conference) is weak, non-binding and allows false solutions such as carbon offsetting. It will prove completely ineffective. Providing some coins for developing countries to mitigate climate change and adapt to it does not help if the sources of the problem remain unchecked.

The peoples' demands for climate justice should be the starting point when addressing the climate crisis. Instead, in Copenhagen, voices of the people were shut out and peaceful protests met brutal suppression. Inside the Bella Center, where the conference took place, many of the poor countries were shut out of back-room negotiations. The accord is the result of this anti-democratic process.

The basic demands of the climate justice movement remain unmet. The U.N. climate process must resume, and it must accomplish these goals:

- Industrialized countries must commit to at least 40 percent cuts in emissions by 2020 by using clean energy, sustainable transport and farming and cutting energy demand.
- Emission cuts must be real. They cannot be "achieved" by carbon offsetting, such as buying carbon credits from developing countries or by buying up forests in developing countries so they won't be cut down.
- Rich countries must make concrete commitments to provide money for developing countries to grow in a clean way and to cope with the floods, droughts and famines caused by climate change. Funding must be adequate, not the minuscule amounts proposed in the accord.

Wealthy nations are most responsible for climate change. They have an obligation to lead the way in solving the problem. They have not done so with the Copenhagen Accord.

Continued from p. 42

rejected legally binding targets.

After the negotiations on the accord, Ramesh told the *Hindustan Times* that India had "upheld the interest of developing nations." [47] But some analysts said India had largely followed China's lead, a position that could cost India some prestige with other developing nations, whose cause it had championed in the past.

"The worst thing India did was to align itself uncritically to China's yoke," says Indian political scientist Mehra, "because China acted purely in its own self interest."

The European Union — European leaders are calling for other countries to join them in backing the Copenhagen Accord, but they've hardly tried to hide their disappointment it wasn't more substantial. The European Union had staked out one of the stronger positions on emissions reductions beforehand, promising to cut emissions by 20 percent from 1990 levels to 2020, or 30 percent if other countries took similarly bold action. They also wanted rich nations to make 80 to 95 percent cuts in GHG emissions by 2050. [48]

Some national leaders also had expended political capital on global warming before the conference. French President Nicolas Sarkozy had announced a proposal to create a French "carbon tax" on businesses and households for use of oil, gas and coal. The proposal was blocked by the French Constitutional Council, but Sarkozy's party plans to reintroduce it this year. [49]

In the United Kingdom, Prime Minister Brown's government passed legislation committing to an 80 percent cut in U.K. greenhouse gas emissions by 2050. [50] Brown also pressed publicly for $100 billion a year in aid to the developing world to cope with climate change.

The European efforts were designed to lead by example. But analysts say the approach yielded little fruit in Copenhagen. "The European perspective that they could lead by example was the wrong strategy. This was a negotiation. Countries do not check their national interests at the door when they enter the U.N.," says the Chamber of Commerce's Eule, who worked on climate change in the Bush administration.

Although Europe's leaders finally backed the accord and formally pledged 20 percent emission reductions, they had only limited influence on the deal's final shape. "Europe finds itself now outside the driver's seat for how this is going to go forward," says Hulme at the University of East Anglia. "I think in Brussels [home of the E.U. headquarters], there must be a lot of conversations going on about where Europe goes from here." He believes Europe's stricter emissions regulations could now face a backlash.

Framework Conference chief de Boer, who is a citizen of the Netherlands, captured the resignation that seemed to envelope many European diplomats during his post-Copenhagen comments to the press. Before the climate conference kicked off, de Boer had predicted that Copenhagen would "launch action, action and more action" on climate change.

But in his December 19 press conference, when asked what he hoped could be accomplished in the year ahead, he responded, "Basically, the list I put under the Christmas tree two years ago, I can put under the Christmas tree again this year." ∎

OUTLOOK

Too Late?

The world's long-term climate forecast can be summed up in a word: warmer. Even if the nations of the world were to miraculously agree tomorrow to reduce global greenhouse gas emissions, global warming could continue for some time because of the "lag" in how the climate system responds to GHG emission reductions.

In the last decade, researchers have poured a tremendous amount of effort into trying to foresee where climate change could take us. But the projections come with an element of uncertainty. Still, taken together, the most startling forecasts amount to an apocalyptic compendium of disaster. Climate change could:

- Lead to droughts, floods, heat waves and violent storms that displace tens of millions of people, particularly in Asia and sub-Saharan Africa (*see "Climate Refugees," p. 34*);
- Create a high risk of violent conflict in 46 countries, now home to 2.7 billion people, as the effects of climate change exacerbate existing economic, social and political problems; [51]
- Cause the extinction of about a quarter of all land-based plant and animal species — more than a million — by 2050; [52]
- Effectively submerge some island nations by 2100, [53] and create widespread dislocation and damage to coastal areas, threatening more than $28 trillion worth of assets by 2050; and [54]
- Cause acidification of the oceans that renders them largely inhospitable to coral reefs by 2050, destroying a fragile underwater ecosystem important to the world's fisheries. [55]

If temperatures climb by an average of 3.5 to 4 degrees Celsius (6.3 to 7.2 Fahrenheit) by the end of the century, as some projections predict, it would mean "total devastation for man in parts of the world," says the Global Justice Ecology Project's Petermann. "You're talking about massive glaciers melting, the polar ice caps disappearing. It would make life on this planet completely unrecognizable."

But some analysts, while endorsing the potential dangers of climate change, still back away from the view that it's a catastrophe that trumps all others. "The prospective tipping points for the worst consequences are just that, prospective tipping points, and they're resting on the credibility of scientific models," says East Anglia University's Hulme. "We should take them seriously. But they're not the Nazis marching across Belgium. We need to weigh our response within the whole range of needs facing the human race."

The critical question likely to determine the shape of the planet's future for the rest of this century and beyond is when humans will stop pouring greenhouse gases into the atmosphere. If done soon enough, most scientists say, climate change will be serious but manageable on an international level, although billions of dollars will be needed to mitigate the effects in the most vulnerable parts of the globe.

But if emissions continue to rise, climate change could be far more catastrophic. "It is critically important that we bring about a commitment to reduce emissions effectively by 2020," said IPCC Chairman Rajendra Pachauri, shortly before Copenhagen. [56]

To accomplish Copenhagen's goal of holding warming to 2 degrees Celsius, Pachauri said emissions must peak by 2015. The agreement, however, sets no peaking year, and the emission-reduction pledges by individual nations fall short of that goal, according to recent analysis by Climate Interactive, a collaborative research effort sponsored by the Sustainability Institute in Hartland, Vt. [57]

World leaders acknowledge they need to do more, and some observers remain hopeful the upcoming climate conference in Mexico City could provide a breakthrough that will avert the worst, especially if pressure to act continues to grow at the grassroots level. "Right now there is a massive gulf between where the public is and where the po-

Causes of Climate Change

Rapidly industrializing China has surpassed the United States as the world's largest emitter of carbon dioxide — one of the greenhouse gases (GHG) responsible for rising world temperatures. Although most GHGs are invisible, air pollution like this in Wuhan, China, on Dec. 3, 2009 (above) often includes trapped greenhouse gases. The destruction of tropical rainforests decreases the number of trees available to absorb carbon dioxide. Palm oil trees once grew on this 250-acre plot being cleared for farming in Aceh, Indonesia (below).

litical process is," says India's Mehra. "But I think [in 2010] you will see government positions mature. And I think you will see more politicians who have the conviction to act."

Canadian political scientist Homer-Dixon considers bold action unlikely, however, unless the world's major emitting nations, including the United States and China, start suffering clearly visible, serious climate-change consequences.

"In the absence of those really big shocks, I'm afraid we're probably achieving about as much as possible," he says. "Because of the lag in the system, if you wait until the evidence is clear, it's too late." ∎

Notes

[1] Yvo de Boer, the United Nation's Framework Convention on Climate Change video message before the opening of the Cop15 conference, Dec. 1, 2009, www.youtube.com/climateconference#p/u/11/xUTXsdkinq0.

[2] The complete text of the accord is at http://unfccc.int/resource/docs/2009/cop15/eng/l07.pdf.

[3] John Vidal and Jonathan Watts, "Copenhagen closes with weak deal that poor threaten to reject," *The Guardian*, Dec. 19, 2009, www.guardian.co.uk/environment/2009/dec/19/copenhagen-closes-weak-deal.

[4] *Ibid.*

[5] "Remarks by the President," The White House Office of the Press Secretary, Dec. 18, 2009, www.whitehouse.gov/the-press-office/remarks-president-during-press-availability-copenhagen.

[6] http://action.sierraclub.org/site/Message Viewer?em_id=150181.0.

[7] See Jones' complete comments at http://wdm.gn.apc.org/copenhagen-'deal'-'shameful-and-monumental-failure'.

[8] Jerry Melillo, Karl Thomas and Thomas Peterson, editors-in-chief, "Global Climate Change Impacts in the United States," U.S. Global Change Research Program, executive summary, June 16, 2009, www.education-research-services.org/files/USGCRP_Impacts_US_executive-summary.pdf.

[9] Intergovernmental Panel on Climate Change staff, "Climate Change 2007: Synthesis Report," The U.N. Intergovernmental Panel on Climate Change, Nov. 17 2007, www.ipcc.ch/pdf/assessment-report/ar4/syr/ar4_syr_spm.pdf.

[10] "Climate Change responsible for 300,000 deaths a year," Global Humanitarian Forum,

http://ghfgeneva.org/NewsViewer/tabid/383/vw/1/ItemID/6/Default.aspx.

[11] Andrew C. Revkin and James Kanter, "No Slowdown of Global Warming, Agency Says," *The New York Times*, Dec. 8, 2009, www.nytimes.com/2009/12/09/science/earth/09climate.html.

[12] "Key Scientific Developments Since the IPCC Fourth Assessment Report," in Key Scientific Developments Since the IPCC Fourth Assessment Report, Pew Center on Global Climate Change, June 2009.

[13] "Final Copenhagen Accord Press Release," The Sustainability Institute, Dec. 19, 2009, http://climateinteractive.org/scoreboard/copenhagen-cop15-analysis-and-press-releases.

[14] "Remarks by the President," *op. cit.*

[15] "Copenhagen Accord," draft proposal, United Nations Framework Convention on Climate Change, Dec. 18, 2009, p. 3. http://unfccc.int/resource/docs/2009/cop15/eng/l07.pdf.

[16] Kumi Naidoo, speaking at Copenhagen in a video blog posted by Greenpeace Australia, www.facebook.com/video/video.php?v=210068211237.

[17] Ban Ki-moon, remarks to the General U.N. Assembly, Dec. 21, 2009, www.un.org/News/Press/docs/2009/sgsm12684.doc.htm.

[18] Jennifer Weeks, "Carbon Trading, Will it Reduce Global Warming," *CQ Global Researcher*, November 2008.

[19] Robert Shapiro, "Addressing the Risks of Climate Change: The Environmental Effectiveness and Economic Efficiency of Emissions Caps and Tradable Permits, Compared to Carbon Taxes," February 2007, p. 26, http://67.23.32.13/system/files/carbon-tax-cap.pdf.

[20] Nicolas Loris and Ben Lieberman, "Capping Carbon Emissions Is Bad, No Matter How You Slice the Revenue," Heritage Foundation, May 14, 2009, www.heritage.org/Research/En

ergyandEnvironment/wm2443.cfm.

[21] Monica Prasad, "On Carbon, Tax and Don't Spend," *The New York Times*, March 25, 2008, www.nytimes.com/2008/03/25/opinion/25prasad.html.

[22] "Cantwell, Collins Introduce 'Cap and Rebate' Bill," Clean Skies, Energy and Environment Network, Dec. 11, 2009, www.cleanskies.com/articles/cantwell-collins-introduce-cap-and-rebate-bill.

[23] Robert J. Shapiro, "Carbon Tax More Likely," *National Journal* expert blog, Energy and the Environment, Jan. 4, 2010, http://energy.nationaljournal.com/2010/01/whats-next-in-the-senate.php-1403156.

[24] A concise history of Keeling and his work is at "The Keeling Curve Turns 50," Scripps Institution of Oceanography, http://sio.ucsd.edu/special/Keeling_50th_Anniversary/.

[25] Elizabeth Kolbert, *Field Notes from a Catastrophe: Man, Nature, and Climate Change* (2006), p. 2.

[26] Naomi Oreskes, "Beyond the Ivory Tower: The Scientific Consensus on Climate Change," *Science*, Dec. 3, 2004, www.sciencemag.org/cgi/content/full/306/5702/1686.

[27] *Ibid.*

[28] "Analysis of the Emails from the University of East Anglia's Climatic Research Unit," Pew Center on Global Climate Change, December 2009, www.pewclimate.org/science/university-east-anglia-cru-hacked-emails-analysis.

[29] Quoted by Nicholas Dawidoff, "The Civil Heretic," *The New York Times Magazine*, March 23, 2009, p. 2, www.nytimes.com/2009/03/29/magazine/29Dyson-t.html?pagewanted=1&_r=1.

[30] "Smoke, Mirrors & Hot Air: How Exxon-Mobil Uses Big Tobacco's Tactics to Manufacture Uncertainty on Climate Science," Union of Concerned Scientists, January 2007, p. 1, www.ucsusa.org/assets/documents/global_warming/exxon_report.pdf.

[31] *Ibid.*

[32] Many are summarized in a policy brief by the nonprofit Pew Center on Global Climate Change, "Realities vs. Misconceptions about the Science of Climate Change," August 2009, www.pewclimate.org/science-impacts/realities-vs-misconceptions.

[33] Seth Borenstein, "AP IMPACT: Statisticians Reject Global Cooling," The Associated Press, Oct. 26, 2009, http://abcnews.go.com/Technology/wireStory?id=8917909.

[34] "Unpacking the problem," *The Economist*, Dec. 5-11, 2009, p. 21, www.economist.com/

About the Author

Reed Karaim, a freelance writer living in Tucson, Arizona, has written for *The Washington Post*, *U.S. News & World Report*, *Smithsonian*, *American Scholar*, *USA Weekend* and other publications. He is the author of the novel, *If Men Were Angels*, which was selected for the Barnes & Noble Discover Great New Writers series. He is also the winner of the Robin Goldstein Award for Outstanding Regional Reporting and other journalism awards. Karaim is a graduate of North Dakota State University in Fargo.

specialreports/displaystory.cfm?story_id=14994 848.

35 *Ibid.*

36 It is important to note that if CO_2 emissions are calculated on a per capita basis, China still ranks far below most developed nations. The highest emitter on a per capita basis is Australia, according to the U.S. Energy Information Agency, with the United States second. See www.ucsusa.org/global_warming/science_and_impacts/science/each-countrys-share-of-co2.html.

37 A chart of the top 20 CO_2 emitting countries is at www.ucsusa.org/global_warming/science_and_impacts/science/graph-showing-each-countrys.html.

38 "China ratifies global warming treaty" CNN.com, Sept. 4, 2002, http://archives.cnn.com/2002/WORLD/africa/09/03/kyoto.china.glb/index.html.

39 "China overtakes U.S. in greenhouse gas emissions," *The New York Times*, June 20, 2007, www.nytimes.com/2007/06/20/business/world business/20iht-emit.1.6227564.html.

40 Ed Miliband, "The Road from Copenhagen," *The Guardian*, Dec. 20, 2009, www.guardian.co.uk/commentisfree/2009/dec/20/copenhagen-climate-change-accord.

41 "A Long Game," *The Economist*, Dec. 5-11, 2009, p. 18.

42 *Ibid.* Keith Bradsher, "China Leading Global Race to Make Clean Energy" *The New York Times*, Jan. 31, 2010, p. A1.

43 Darren Samuelsohn, "Obama Negotiates 'Copenhagen Accord' With Senate Climate Fight in Mind," *The New York Times*, Dec. 21 2009, www.nytimes.com/cwire/2009/12/21/21climatewire-obama-negotiates-copenhagen-accord-with-senat-6121.html.

44 Juliet Elperin, "Fewer Americans Believe in Global Warming, Poll Shows," *The Washington Post*, Nov. 25, 2009, www.washingtonpost.com/wp-dyn/content/article/2009/11/24/AR2009112402989.html.

45 Suzanne Goldenberg, "Fate of US climate change bill in doubt after Scott Brown's Senate win," *The Guardian*, Jan. 20, 2010, www.guardian.co.uk/environment/2010/jan/20/scott-brown-climate-change-bill.

46 "India promises to slow carbon emissions rise," BBC News, Dec. 3, 2009, http://news.bbc.co.uk/2/hi/8393538.stm.

47 Rie Jerichow, "World Leaders Welcome the Copenhagen Accord," Denmark.dk, Dec. 21, 2009, www.denmark.dk/en/menu/Climate-Energy/COP15-Copenhagen-2009/Selected-COP15-news/World-leaders-welcome-the-

FOR MORE INFORMATION

Cato Institute, 1000 Massachusetts Avenue, N.W., Washington D.C. 20001; (202) 842-0200; www.cato.org/global-warming. A conservative U.S. think tank that maintains an extensive database of articles and papers challenging the scientific and political consensus on climate change.

Climate Justice Now; www.climate-justice-now.org. A network of organizations and movements from around the world committed to involving people in the fight against climate change and for social and economic justice at the grassroots level.

Climate Research Unit, University of East Anglia, Norwich, NR4 7TJ, United Kingdom; +44-1603-592722; www.cru.uea.ac.uk. Recently in the news when its e-mail accounts were hacked; dedicated to the study of natural and manmade climate change.

Greenpeace International, Ottho Heldringstraat 5, 1066 AZ Amsterdam, The Netherlands; +31 (0) 20 7182000; www.greenpeace.org/international. Has made climate change one of its global priorities; has offices around the world.

Intergovernmental Panel on Climate Change, c/o World Meteorological Organization, 7bis Avenue de la Paix, C.P. 2300 CH- 1211, Geneva 2, Switzerland; +41-22-730-8208; www.ipcc.ch. U.N. body made up of 2,500 global scientists; publishes periodic reports on various facets of climate change, including a synthesis report summarizing latest findings around the globe.

Pew Center on Global Climate Change, 2101 Wilson Blvd., Suite 550, Arlington, VA, 22201; (703) 516-4146; www.pewclimate.org. Nonprofit, nonpartisan organization established in 1998 to promote research, provide education and encourage innovative solutions to climate change.

United Nations Framework Convention on Climate Change, Haus Carstanjen, Martin-Luther-King-Strasse 853175 Bonn, Germany; +49-228-815-1000; http://unfccc.int/2860.php. An international treaty that governs climate change negotiations.

Copenhagen-Accord.htm.

48 "Where countries stand on Copenhagen," BBC News, undated, http://news.bbc.co.uk/2/hi/science/nature/8345343.stm.

49 James Kantor, "Council in France Blocks Carbon Tax as Weak on Polluters," *The New York Times*, Dec. 31, 2009, www.nytimes.com/2009/12/31/business/energy-environment/31carbon.html.

50 Andrew Neather, "Climate Change could still be Gordon Brown's great legacy," *The London Evening Standard*, Dec. 15, 2009, www.thisislondon.co.uk/standard/article-23783937-climate-change-could-still-be-gordon-browns-great-legacy.do.

51 Dan Smith and Janini Vivekananda, "A Climate of Conflict, the links between climate change, peace and war," *International Alert*, November 2007, www.international-alert.org/pdf/A_Climate_Of_Conflict.pdf.

52 Alex Kirby, "Climate Risk to a Million Species,"

BBC Online, Jan. 7, 2004, http://news.bbc.co.uk/2/hi/science/nature/3375447.stm.

53 Adam Hadhazy, "The Maldives, threatened by drowning due to climate change, set to go carbon-neutral," *Scientific American*, March 16, 2009, www.scientificamerican.com/blog/post.cfm?id=maldives-drowning-carbon-neutral-by-2009-03-16.

54 Peter Wilkinson, "Sea level rise could cost port cities $28 trillion," CNN, Nov. 23, 2009, www.cnn.com/2009/TECH/science/11/23/climate.report.wwf.allianz/index.html.

55 "Key Scientific Developments Since the IPCC Fourth Assessment Report," *op. cit.*

56 Richard Ingham, "Carbon emissions must peak by 2015: U.N. climate scientist," Agence France-Presse, Oct. 15, 2009, www.google.com/hostednews/afp/article/ALeqM5izYrubhpeFvOKCRrZmWSYWCkPoRg.

57 "Final Copenhagen Accord Press Release," *op. cit.*

www.globalresearcher.com   **February 2010**   47

CQ Press Custom Books - Page 23

Bibliography

Selected Sources

Books

Hoggan, James, *Climate Cover-Up: The Crusade to Deny Global Warming,* Greystone Books, 2009.

A Canadian public relations executive who founded the anti-climate-skeptic Web site DeSmogblog takes on what he considers the oil and gas industry's organized campaign to spread disinformation and confuse the public about the science of climate change.

Hulme, Mike, *Why We Disagree About Climate Change: Understanding Controversy, Inaction and Opportunity,* Cambridge University Press, 2009.

A professor of climate change at East Anglia University in Great Britain looks at the cultural, political and scientific forces that come into play when we consider climate and what that interaction means for dealing with climate change today.

Kolbert, Elizabeth, *Field Notes from a Catastrophe: Man, Nature and Climate Change,* Bloomsbury, 2006.

A *New Yorker* writer summarizes the scientific evidence on behalf of climate change and looks at the consequences for some of the world's most vulnerable locations.

Michaels, Patrick J., and Robert C. Balling, *Climate of Extremes: Global Warming Science They Don't Want You to Know,* The Cato Institute, 2009.

Writing for a libertarian U.S. think tank, the authors argue that while global warming is real, its effects have been overstated and do not represent a crisis.

Articles

"Stopping Climate Change, A 14-Page Special Report," *The Economist,* Dec. 5, 2009.

The authors provide a comprehensive review of the state of global climate change efforts, including environmental, economic and political conditions.

Broder, John and Andrew Revkin, "A Grudging Accord in Climate Talks," *The New York Times,* Dec. 19, 2009.

The Times assesses the Copenhagen Accord and reports on the final hours of the climate change convention.

Kunzig, Robert, "A Sunshade for Planet Earth," *Scientific American,* November 2008.

An award-winning scientific journalist examines the various geoengineering options that might reduce global warming, their costs and possible consequences.

Schwartz, John, "Courts as Battlefields in Climate Fights," *The New York Times,* Jan. 26, 2009.

A reporter looks at environmental groups' and other plaintiffs' efforts to hold corporations that produce greenhouse gases legally liable for the effects of climate change on vulnerable areas, including Kivalina Island off the coast of Alaska.

Walsh, Bryan, "Lessons from the Copenhagen Climate Talks," *Time,* Dec. 21, 2009.

Time's environmental columnist provides predictions about the future of the climate change battle, based on the final Copenhagen Accord.

Walsh, Bryan, "The Stolen Emails: Has 'Climategate' been Overblown," *Time Magazine online,* Dec. 7, 2007.

The stolen East Anglia University e-mails, the author concludes, "while unseemly, do little to change the overwhelming scientific consensus on the reality of man-made climate change."

Reports and Studies

"Climate Change 101: Understanding and Responding to Global Climate Change," Pew Center on Global Climate Change, January 2009.

This series of reports aims to provide an introduction to climate change science and politics for the layman.

"World Development Report 2010: World Development and Climate Change," World Bank, November 2009, http://econ.worldbank.org/WBSITE/EXTERNAL/EXTDEC/EXTRESEARCH/EXTWDRS/EXTWDR2010/0,,content-MDK:21969137~menuPK:5287816~pagePK:64167689~pi PK:64167673~theSitePK:5287741,00.html.

This exhaustive, 300-page study examines the consequences of climate change for the developing world and the need for developed nations to provide financial assistance to avert disaster.

Bernstein, Lenny, *et al.,* "Climate Change 2007: Synthesis Report," The Intergovernmental Panel of Climate Change, 2007, www.ipcc.ch/pdf/assessment-report/ar4/syr/ar4_syr_spm.pdf.

The international body tasked with assessing the risk of climate change caused by human activity gathered scientific research from around the world in this widely quoted report to conclude, "warming of the climate system is unequivocal."

Thomas, Karl, Jerry Melillo and Thomas Peterson, eds., "Global Climate Change Impacts in the United States," United States Global Change Research Program, June 2009, www.globalchange.gov/publications/reports/scientific-assessments/us-impacts.

U.S. government researchers across a wide range of federal agencies study how climate change is already affecting the United States.

The Next Step:

Additional Articles from Current Periodicals

Carbon Tax

"New Australian Opposition Leader Rules Out Carbon Tax Policy," *Asia Pulse* **(Australia), Dec. 2, 2009.**

Australian Opposition Leader Tony Abbott has ruled out a carbon tax being a part of any new coalition policy on climate change, but believes that some kind of price should be put on carbon emissions in the context of an overall global scheme.

"Sarkozy Demands EU-Wide Carbon Tax," *Business Recorder* **(Pakistan), Jan. 7, 2010.**

French President Nicolas Sarkozy has called for a carbon tax across Europe and tariffs on items that are harmful to the atmosphere.

Chu, Henry, "Making a Case for 'Carbon Tax,' " *Los Angeles Times*, **Nov. 21, 2009, p. A26.**

Some countries have turned to a "carbon tax" to discourage carbon-intensive activities such as driving cars and taking long flights.

Njobeni, Siseko, "Cap and Trade System 'Better Than Carbon Tax,' " *Business Day* **(South Africa), Dec. 10, 2009.**

Carbon taxes do not encourage companies to change their behavior. A cap-and-trade system is preferable.

Climate Refugees

"Sinking Isles, Rising Hopes," *Canberra* **(Australia)** *Times*, **June 27, 2009.**

Over the past decade, some populations have been displaced by drought, land degradation and significant climate-related events.

"UK Should Open Borders to Climate Refugees, Says Bangladeshi Minister," *Guardian* **(England), Dec. 4, 2009.**

Bangladeshi Finance Minister Abul Maal Abdul Muhith has called on Great Britain and other wealthy countries to accept his country's refugees who have been displaced by climate change.

Sanders, Edmund, "Climate Change Refugees on Rise," *Orlando Sentinel*, **Nov. 1, 2009, p. A25.**

Africans will be hardest hit by climate change — and thus will have more climate refugees — because so many livelihoods on the continent depend on farming and livestock.

Copenhagen Talks

"Copenhagen Accord Marks New Starting Point for Global Fight Against Climate Change," **Xinhua News Agency (China), Dec. 26, 2009.**

The Copenhagen Accord effectively embodies the broad consensus of the international community on further efforts to cut carbon emissions.

"Minister Says Copenhagen Accord Does Not Affect India's Sovereignty," **PTI News Agency (India), Dec. 23, 2009.**

Indian Environment Minister Jairam Ramesh says India's national sovereignty won't be affected by climate talks.

Broder, John M., "Obama Offers Targets to Cut Greenhouse Gas," *The New York Times*, **Nov. 26, 2009, p. A1.**

The U.S. will cut greenhouse gas emissions 17 percent below 2005 levels by 2020, Obama will say.

Simamora, Adianto P., and Stevie Emilia, "Copenhagen Talks End, Close to Collapse," *Jakarta* **(Indonesia)** *Post*, **Dec. 20, 2009.**

The Copenhagen talks were a disappointment because no legally binding document or treaty was produced.

Geoengineering

"No High-Tech 'Quick Fix' for Warming," *Canberra* **(Australia)** *Times*, **Sept. 3, 2009.**

The Royal Society — Great Britain's peak science body — says geoengineering is not a readily acceptable solution to climate change problems.

Thernstrom, Samuel, "Could We Engineer a Cooler Planet?" *The Washington Post*, **June 13, 2009, p. A15.**

The National Academy of Sciences is exploring geoengineering as a better way to reduce climate change than legislative action.

Von Radowitz, John, "Climate-Saving Devices May Be Just as Ruinous," *Birmingham* **(England)** *Post*, **Sept. 2, 2009.**

Risky and unproven climate-changing technology — including geoengineering — can have catastrophic consequences if used irresponsibly.

CITING *CQ GLOBAL RESEARCHER*

Sample formats for citing these reports in a bibliography include the ones listed below. Preferred styles and formats vary, so please check with your instructor or professor.

MLA STYLE
Flamini, Roland. "Nuclear Proliferation." CQ Global Researcher 1 Apr. 2007: 1-24.

APA STYLE
Flamini, R. (2007, April 1). Nuclear proliferation. *CQ Global Researcher*, 1, 1-24.

CHICAGO STYLE
Flamini, Roland. "Nuclear Proliferation." *CQ Global Researcher*, April 1, 2007, 1-24.

Voices From Abroad:

JOHN ASHE
Chair, Kyoto Protocol Talks

A reason for hope

"Given where we started and the expectations for this conference, anything less than a legally binding and agreed outcome falls far short of the mark. On the other hand . . . perhaps the bar was set too high and the fact that there's now a deal . . . perhaps gives us something to hang our hat on."

BBC, December 2009

JOHN SAUVEN
Executive Director
Greenpeace UK

Copenhagen = Crime Scene

"Copenhagen is a crime scene tonight, with the guilty men and women fleeing to the airport. It seems there are too few politicians in this world capable of looking beyond the horizon of their own narrow self-interest, let alone caring much for the millions of people facing the threat of climate change."

The Guardian (England), December 2009

MOHAMED NASHEED
President, Maldives

A critical number

"Anything above 1.5 degrees [Celsius], the Maldives and many small islands and low-lying islands would vanish. It is for this reason that we tried very hard during the course of the last two days to have 1.5 degrees in the document. I am so sorry that this was blatantly obstructed by big-emitting countries."

BBC, December 2009

JOSÉ MANUEL BARROSO
President
European Commission

All countries have a role

"Developed countries must explicitly recognise that we will all have to play a significant part in helping to finance mitigation action by developing countries. . . . The counterpart is that developing countries, at least the economically advanced amongst them, have to be much clearer on what they are ready to do to mitigate carbon emissions as part of an international agreement."

Business Day (South Africa), September 2009

NELSON MUFFUH
Senior Climate Change
Advocacy Advisor
Christian Aid
England

Climate change kills 300,000 a year

"Already 300,000 people die each year because of the impact of climate change, most in the developing world. The lack of ambition shown by rich countries in Copenhagen means that number will grow."

The Observer (England), December 2009

NICOLAS SARKOZY
President, France

A vital contract

"The text we have is not perfect. . . . If we had no deal, that would mean that two countries as important as India and China would be freed from any type of contract . . . [and] the United States, which is not in Kyoto, would be free of any type of contract. That's why a contract is absolutely vital."

BBC, December 2009

STANISLAS KAMANZI
Environment and Lands
Minister, Rwanda

Progress regardless of Copenhagen

"Our policy is that every industrialized investment in the country should come up with an environment friendly technology. So, with or without Copenhagen, we are safe with policies in place."

New Times (Rwanda), December 2009

KYERETWIE OPOKU
Member, Forest Watch
Ghana

Relationships are key

"I accept the technological challenges and all that, but the real challenges are restructuring relationships. If we don't resolve these, forget about going to Copenhagen and getting a deal."

Public Agenda (Ghana), October 2009

VICTOR FODEKE
Chief Climate Officer,
Nigeria

Kyoto: the only hope

"The Kyoto Protocol is the only hope of the developing countries; it is the only legally binding instrument requiring developed countries to cut their emission, killing it is dashing the hope of developing countries."

Daily Trust (Nigeria), December 2009

Peter Broelman, Australia

PROTECTING WETLANDS

BY JENNIFER WEEKS

Excerpted from the CQ Researcher. Jennifer Weeks. (October 3, 2008). "Protecting Wetlands." *CQ Researcher*, 793-816.

Protecting Wetlands

BY JENNIFER WEEKS

THE ISSUES

On wet spring nights across the Northeastern United States, wood frogs and salamanders go on the march. These amphibians spend most of their lives buried in forest undergrowth, but they need to breed in watery places where no fish will eat their eggs. So they migrate to vernal pools — ponds that form during the wet seasons and range from a few feet to several acres across. If all goes well, their offspring will hatch and grow large enough to breathe air before the pools dry up in summer. Some species, such as fairy shrimp, spend their entire life cycles in the pools, leaving eggs behind that stay dormant through dry months and hatch when the pools reappear a year later.

Vernal pools are wetlands — areas where the soil is always or usually saturated with water and that support plants and animals adapted to moist conditions. Many states protect vernal pools because they provide habitat for rare animals. For example, in Massachusetts it is illegal to dump materials into state-certified vernal pools, install septic systems nearby or cut down more than half of the trees within a 50-foot radius. [1]

Other wetlands play similar roles. Estuaries (mixed salt- and freshwater zones where rivers flow into the sea) are among Earth's most productive ecosystems.

"Shallow marsh channels are important habitat for fish," says Doug Myers, science director of People for Puget Sound, a Seattle conservation group. "Chinook salmon rear their young in estuarine deltas, coves and lagoons in the Northwest. And birds

Condominiums and shopping centers encroach on the Los Cerritos Wetlands near Long Beach, Calif. The 400-acre site, which once covered 2,400 acres, is considered vital for birds migrating on the Pacific Flyway.

Getty Images/David McNew

migrating along the Pacific Coast stop to feed along the mud flats."

Many wetlands that are far from coastlines also are important. For example, lakes carved by glaciers across the upper Midwest, known as prairie potholes, are critical breeding and nesting areas for millions of ducks, geese and other waterbirds. (*See map, p. 800.*)

Until the 1970s Americans widely regarded wetlands as swampy places that were useless unless they could be drained or filled in. Before settlers arrived, the continental United States contained more than 220 million acres of wetlands. Today less than half of that area (107 million acres) remains. [2] Some of America's most famous and valued wetland areas, such as Florida's Everglades and Louisiana's Gulf Coast, are also its most degraded.

For the past 20 years policymakers have tried to prevent more net losses

of wetlands. President George W. Bush raised the bar in 2004, arguing that the United States could achieve net annual increases by creating and restoring more acres than it developed. But environmentalists, outdoor advocates and regulators say that not all wetlands are equal, and that more action is needed to protect and restore high-quality wetlands.

"We see a lot of threats to wetlands around Puget Sound, including urban growth, shoreline development and polluted stormwater runoff from paved areas," says Myers. "It's death by a thousand cuts." Nutrient pollution from farms (excess fertilizer and animal waste) and septic systems washes into lakes and bays nationwide, generating huge algae blooms that deprive aquatic organisms of sunlight and dissolved oxygen. [3]

And many advocates fear that recent U.S. Supreme Court rulings limiting federal jurisdiction over wetlands have made some more vulnerable to development.

Wetland protection affects a range of industries that often excavate or drain land, including commercial and residential construction, agriculture, mining and energy. Under Section 404 of the Clean Water Act, when a project involves dredging or filling in the "waters of the United States" — a category that includes many wetlands — a permit must be obtained from the U.S. Army Corps of Engineers. The Corps then must consult with the Environmental Protection Agency (EPA), which has veto power over permit decisions.

This process can be lengthy and expensive. A 2002 study of 103 permit applications found that the average general permit for lower-impact activities cost $28,915 to prepare and took 313 days to gain approval. Individual

Despite Wetland Gains, Concerns Remain

In the last half-century, the nation has gone from losing nearly half a million acres of wetlands a year to a net annual gain of 32,000 acres a year from 1998-2004. The quality of the new wetland, however, worries many environmentalists. They note, for example, that while there were significant gains in freshwater ponds, crucial intertidal wetlands (mainly deepwater bay bottoms and open ocean) declined by about 4,740 acres a year.

Average Annual Net Wetland Gain/Loss for the Lower 48 States, 1954-2004

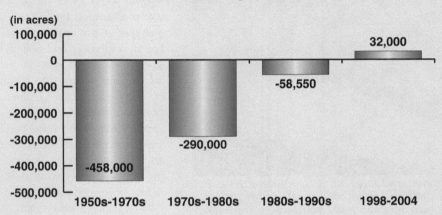

Source: T.E. Dahl, "Status and Trends of Wetlands in the Conterminous United States 1998 to 2004," U.S. Fish and Wildlife Service, December 2005

permits for higher-impact projects cost $271,596 on average and took more than two years. [4] Developers who proceed without permits face civil penalties of up to $32,500 per day and criminal penalties up to $50,000 per day plus three years in prison. [5]

Many trade groups say they support reasonable wetlands protection but that current standards are too broad and the permitting process too cumbersome. "While [the permits'] environmental purposes are laudable, they do add to the cost and delay the completion of the public and private infrastructure that literally forms the foundation of our nation's economy," Associated General Contractors of America CEO Stephen E. Sandherr told the House Transportation and Infrastructure Committee in July 2007. [6]

Contractors, growers and other such groups would like to see the Corps and EPA eliminate or limit federal protection for small, isolated and temporary wetlands.

But environmentalists argue that destroying wetlands could end up costing the country much more, because wetlands provide billions of dollars worth of ecological services that benefit the public. Often referred to as "nature's kidneys," they filter out pollutants from water and trap suspended particles. They also absorb flood waters and release them slowly, like natural sponges. According to one estimate, wetlands cover less than 3 percent of Earth's surface but provide up to 40 percent of annual, renewable ecosystem services such as purifying water and cycling nutrients. [7]

After Hurricane Katrina caused at least $125 billion in damages along Louisiana's Gulf coast in September 2005, several studies indicated the storm surge would have been lower if large swathes of coastal wetlands had not been obliterated by Mississippi River flood-control projects and coastal oil and gas development. [8] In 2007 Louisiana approved a master plan for protecting and restoring its coast that, if fully funded, is expected to cost more than $50 billion and take up to 30 years to complete. [9]

Since the 1980s regulators have used a process known as "mitigation" (preserving, enhancing or creating wetlands to compensate for destroying others) as a tool to balance wetland conservation and development. Initially, owners who wanted to fill in wetlands had to do mitigation projects on the same site or nearby. To make the process more flexible, however, agencies developed mitigation banking, in which developers buy credits from a wetland "bank" (acres restored by a third party) to compensate for acres that they drain or alter.

The National Mitigation Banking Association, a trade group, calls mitigation banking "a unique concept . . . that unites sound economic and environmental practices." [10] But skeptics say the process often helps developers rather than maximizing the quality of U.S. wetlands.

"If a developer fills in wetlands for an urban project and restores something 50 miles away, flooding may be caused in the city where the wetlands used to be. There's no net loss of wetlands, but you have a big loss of [ecological] value" says Jon Kusler, associate director of the Association of State Wetland Managers (ASWM).

As scientists, government officials and business leaders debate how to balance wetland protection with development, here are some issues they are considering:

Does the Clean Water Act protect most wetlands?

Like other keystone environmental laws, the 1972 Clean Water Act (CWA) sought to create clear national standards for environmental quality instead of leaving most responsibility to the states. [11] But environmental groups and industry have argued for years over which wetlands fall under federal control. Conservationists say that most wetlands play important ecological roles and should be protected. Businesses counter that federal jurisdiction expanded in the 1980s and '90s to include unimportant wetlands that Congress never intended to regulate. State and local regulators are often caught in the middle.

Initially the Corps of Engineers interpreted the CWA's limits on discharging dredged or fill material into "navigable waters" narrowly, applying them only to bodies such as rivers and canals that could be used for interstate commerce. However, a federal court ruled in 1975 that the law covered all U.S. waters within the scope of Congress' constitutional power to regulate under the Commerce Clause. [12] In response, the Corps rewrote its regulations to also cover possible construction or degradation affecting tributaries of navigable waters, plus wetlands such as prairie potholes, mud flats and sloughs, "which could affect interstate commerce." [13]

The Supreme Court addressed the issue in its 1985 *Riverside Bayview Homes, Inc. v. United States* ruling, which affirmed that Congress could regulate wetlands adjacent to navigable waters. Since water flowed between these systems, the opinion reasoned, activities that harmed the wetlands could also impair the navigable waters. [14]

But in a seminal 2001 case, *Solid Waste Agency of Northern Cook County [SWANCC] v. U.S. Army Corps of Engineers*, the high court held that federal jurisdiction did not cover certain isolat-

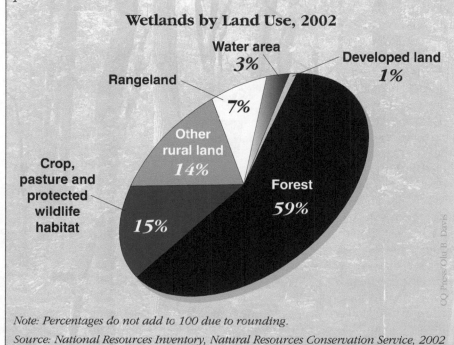

Forests Contain Most of U.S. Wetlands

About 66 million acres — or 59 percent — of the 111 million acres of wetland in the United States are in forests. Another 22 percent are on agricultural and range lands. More than one-third of threatened and endangered species in the U.S. live only in wetlands, and half spend at least part of their lives there. Besides supporting wildlife, wetlands also control pollution and flooding, protect the water supply and provide recreation.

Wetlands by Land Use, 2002

- Water area **3%**
- Developed land **1%**
- Rangeland **7%**
- Other rural land **14%**
- Crop, pasture and protected wildlife habitat **15%**
- Forest **59%**

Note: Percentages do not add to 100 due to rounding.

Source: National Resources Inventory, Natural Resources Conservation Service, 2002

ed wetlands the Corps had sought to protect because they were used or could be used by migratory birds. (The Corps had said the wetlands were important to interstate commerce because people traveled to view and hunt the birds.) [15] Chief Justice William H. Rehnquist's majority opinion found "no persuasive evidence" that Congress intended to regulate "non-navigable, isolated, intrastate waters." This wording suggested to some observers that other wetlands, regardless of whether they were suitable for birds or not, might fall outside federal protection as well.

"The Rehnquist court took the case intending to write a constitutional decision that limited Congress' power under the Commerce Clause," says Vermont

Law School Professor Patrick Parenteau. "They couldn't muster five votes for that position, so they fell back to a vague statutory decision. Ever since, there's been a battle over whether *SWANCC* really announced new, limited principles of constitutional authority."

In the wake of *SWANCC*, many regulators and state courts assumed that small, isolated wetlands lying entirely within one state fell beyond federal protection but that most other types were covered. "People weren't regulating everything they had before, but coverage was still pretty broad," says Kusler of the Association of State Wetland Managers. The Bush administration proposed new guidelines in 2003 that would have narrowed coverage but withdrew them after receiving thousands of critical comments. [16]

Then in 2006 the Supreme Court decided *Rapanos v. United States*, which combined two cases involving tracts in Michigan. [17] One case examined whether the Corps could regulate wetlands next to a man-made ditch that ultimately flowed into navigable waters. The other concerned a wetland that bordered a tributary that ultimately flowed into navigable waters but was separated from the tributary by a four-foot-wide manmade barrier.

In a split verdict (4-1-4), Justice Antonin Scalia and three other justices concluded the Corps had overreached in both cases. Scalia argued that "waters of the United States" should include only bodies of water that were relatively permanent, such as streams and lakes and wetlands with a "continuous surface connection" to those waters — but not wetlands that had only intermittent or distant physical connections to U.S. waters. Four other justices took the opposite position, supporting Corps jurisdiction in both cases.

Justice Anthony Kennedy took a middle stance, rejecting the Corps position but with a perspective different from Scalia's. Kennedy proposed a case-by-case approach for determining federal jurisdiction, based on whether wetlands had a "significant nexus" to traditional navigable waters. A wetland met this test if it significantly affected the chemical, biological or physical quality of navigable waters, either alone or in combination with other wetlands.

Given the split verdict, legal doctrine dictated that Kennedy's opinion was the controlling guideline for lower courts because his concurrence provided the fifth vote for vacating lower court decisions that had supported the Corps.

Property-rights advocates praise *Rapanos* for putting overdue limits on federal wetland controls. "The decision told federal regulators that they just can't regulate everything," says

Priscilla and Jeff Wilson wait out the flood in St. Charles, Mo. after the Mississippi River inundated the Midwest last June. In 2005, after Hurricane Katrina battered Louisiana's coast, several studies indicated the storm surge would have been lower if coastal wetlands had not been obliterated by Mississippi River flood control projects and coastal oil and gas development.

Russ Harding, director of the property rights network at Michigan's Mackinac Center for Public Policy and a former state environmental regulator. "It was a good decision, but I would have hoped for a brighter line and more clarity about what is subject to federal authority."

The decision alarmed conservationists, who worry it will worsen ongoing losses of isolated wetlands. "We're very concerned about how many wetlands are losing federal protection," says Scott Yaich, director of conservation operations for Ducks Unlimited, a sportsmen's group that works to

protect waterfowl habitat. "EPA and the Corps interpreted *SWANCC* to remove 20 million acres of wetlands from jurisdiction under the Clean Water Act. The Corps' post-*Rapanos* guidance removes up to 60 million acres from federal control."

All sides agree that recent Corps guidance, intended to translate abstruse *Rapanos* terms like "significant nexus" and "relatively permanent waters" into policy, has made the permitting process slower and more cumbersome. According to the guidance, wetlands that meet either the Scalia or Kennedy tests fall under federal jurisdiction. Moreover, Corps field offices across the nation are interpreting the new rules in different ways, says Leah Pilconis, senior counsel at the Associated General Contractors of America. "Not knowing exactly what is required and having decisions made inconsistently wastes time and money in the construction process," she says. "It raises interest costs, makes scheduling harder and delays completion dates."

The AGC and other trade associations want the Corps to narrow its criteria for determining federal jurisdiction and reduce paperwork requirements. Scientists have a broader problem with the current approach: They reject the idea that only some wetlands have a "significant nexus" with waters of the United States.

"Wetlands are all connected," says Joy Zedler, a professor of botany and ecology at the University of Wisconsin, who chaired a 2001 National Academy of Sciences study on wetland restoration. "The Supreme Court is free to make non-science-based decisions, so it can say that wetlands

A Wetlands Glossary

Bog: A wetland ecosystem that is highly acidic and has an accumulation of decomposed plants known as peat.

Carbon sink: A system that absorbs and stores carbon dioxide from the atmosphere. Forests, oceans and wetlands all can act as carbon sinks.

Carbon sequestration: Storing carbon in a natural sink or a geologic reservoir underground.

Dredging: Removing sediment from a channel to make it deep enough for navigation.

Estuary: An environment where land, freshwater and seawater (saline) habitats overlap.

Levee: A raised embankment built to keep a river from overflowing its banks.

Marsh: An environment where terrestrial and aquatic habitats overlap; a wetland dominated by grasses.

Mitigation: Actions that are undertaken to reduce the impact of an activity, such as buying credits from a wetland bank to make up for wetlands that are filled in for development.

Mudflat: A muddy, low-lying strip of ground usually submerged, more or less completely, by the rise of the tide.

Peat: Organic material (leaves, bark, nuts) that has decayed partially. It is dark brown with identifiable plant parts and can be found in peatlands and bogs.

Salt marsh: Flat land flooded by tidal saltwater.

Saltwater intrusion: The invasion of freshwater bodies by denser saltwater.

Swamp: A wetland with trees and shrubs.

Taking: A government action that deprives property owners of their rights, either by claiming the property for public use or by passing regulations making it impossible for them to develop the land.

Watershed: All the water that drains into a particular body of water (stream, pond, river, bay, etc.)

Wetland: Land saturated with water and containing plants and animals adapted to living on, near or in water.

---

have to be connected above the ground to U.S. waters. But that position doesn't withstand scientific scrutiny, because there are underground connections. We can't project into the future, because we don't know how rainfall patterns will shift and where floodwater patterns will flow. So it makes sense to avoid damaging systems that may be critically connected."

Are federal agencies doing enough to protect wetlands?

In terms of sheer acreage, federal inventories show that the United States has gained more wetlands than it has lost in recent years — a net increase of about 32,000 acres annually between 1998 and 2004. [18] (*See graph, p. 796.*) The increase grew out of a goal President Bush set on Earth Day in 2004 of protecting, improving and restoring or creating 3 million acres or more of wetlands (at least 1 million acres in each category) by 2009. The White House Council on Environmental Quality says the goal has been met ahead of schedule, although Bush's targets do not reflect wetland losses that occurred at the same time. [19]

Critics say development and pollution are still destroying valuable areas. For example, an investigation by *The St. Petersburg Times* found that between 1990 and 2005 Florida lost 84,000 acres of wetlands to development. Moreover, many replacement wetlands that developers were required to create were expensive failures, and the Corps rarely verified information in permit applications or inspected wetlands after permits were approved. [20]

"Peer-reviewed science is supposed to be the determining factor in our permitting process, but it doesn't translate very easily into regulations," says Jason Lauritsen, assistant director of the Audubon Society's Corkscrew Swamp Sanctuary near Naples, Fla. "The broader the scientific gaps are, the more politics can come into play, and decisions tend to favor property owners."

For years wetland advocates and fiscal conservatives have called for reforming the Corps, which they say is too committed to wasteful, large-scale construction projects that often harm the environment. [21] Many past Corps projects, such as Mississippi River flood control levees, have worsened

flooding by cutting off water and sediment flows from adjoining floodplains, leaving these areas to dry out and sink. This year EPA vetoed a $220 million Corps proposal to drain 67,000 acres of wetlands along the Yazoo River in Mississippi in order to provide flood control in a sparsely populated rural area. [22]

But critics concede that the Corps is gradually becoming a better environmental steward and is carrying out some valuable restoration projects. For example, in the 1960s when the Corps straightened Florida's winding Kissimmee River — the headwaters of the Everglades — it turned it into a 56-mile canal and drained 30,000 acres of wetlands in the process. Today the Corps is restoring the Kissimmee's meandering course and removing structures that controlled its flow. Although the project is still under way, it has already improved water quality and tripled bird counts around the river. (*See sidebar, p. 806.*) [23]

"The Corps is greener than it was 10 or 15 years ago, although it's hard to say exactly how green," says Kusler of the State Wetland Managers Asso-

'Duck Factory' Nurtures Waterfowl

The Prairie Pothole Region — also known as the "Duck Factory" of North America — produces more than half the continent's waterfowl. Created by retreating glaciers 12,000 years ago, the 300,000-square-mile region once contained 25 million wetlands, or about 83 sites per square mile — a density unmatched anywhere in North America.

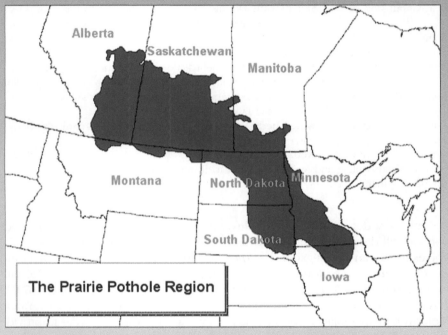

The Prairie Pothole Region

Source: U.S. Fish and Wildlife Service, Kulm Wetland Management District, Kulm, N.D.

ciation. "It came to that role kicking and screaming, but now it ranks restoration goals higher than it used to."

However, he points out, Corps regulations still include many exemptions for activities like draining wetlands, as well as discharges of dredge-and-fill material that occur in the course of ongoing farming, ranching or forestry activities (for example, plowing or maintaining drainage ditches). "The [Clean Water Act's] 404 program still isn't doing an adequate job," Kusler contends.

Expert advisers like the National Research Council say the problem is broader. In the council's view, the United States does not have a clear, well-focused policy for managing water resources. [24] Many federal and state agencies share responsibility for water issues, and their agendas often con-

flict. In these situations, the National Academy of Public Administration observed, agencies make separate decisions about individual projects without considering how these steps affect large ecological systems like river basins. [25] Other agencies also shape wetland policy. The U.S. Fish and Wildlife Service (FWS) manages national wildlife refuges and promotes conservation on private lands. As part of the 404 process, FWS prepares "biological opinions" (BiOps) assessing how proposed developments will affect the habitat of endangered or threatened species.

In Florida, the FWS has objected to many proposed wetland development permits. But in 2005 it fired biologist Andrew Eller, who argued the agency was using misleading data to make it look as though the endangered Florida

panther — whose habitat is threatened by development — was at a lower risk than it actually was. [26] In response to a legal complaint, FWS corrected its information and reinstated Eller. Now, however, in at least some cases the agency relies on developers to provide data for BiOps. [27]

"The standard set of questions on 404 permit applications was developed by FWS. Applicants fill it out, and we analyze it," says Bill Wilen, a senior biologist with the service's National Wetlands Inventory office. "It saves FWS time and effort by highlighting potential impacts."

Another FWS wetland initiative, conserving land in the prairie pothole region — Iowa, Minnesota, the Dakotas and Montana — is falling short of the agency's goals because land prices are rising. Since 1959 FWS has acquired or protected about 3 million acres of wetlands and grasslands in the area, which provides nesting and breeding habitat and stopover space during migration for hundreds of species of birds. In 2007 the Government Accountability Office (GAO) reported that while FWS was managing these lands effectively, at the current pace it would take the service until 2050 to reach its target of protecting 12 million acres.

Most options for speeding up the process require congressional approval, however. For example, Congress could raise the price of the federal Duck Stamp that hunters must purchase, which has been $15 since 1991. It also could appropriate more money from Treasury's Land and Water Conservation Fund, which helps pay for acquiring land, water and wetlands. [28]

According to biologist Wilen, FWS agrees with GAO's conclusions. "We have very sophisticated models that help us decide where we'd like to purchase land, and we prioritize so we can get the best bang for the buck. FWS wants to get it done, but right now we just don't have the cash to go faster," he says.

Rising oil and food costs and expanding biofuels production have been driving up land prices, making it harder to protect wetlands. [29] World oil prices are projected to average $116 per barrel in 2008, up from $72 in 2007, triggering price hikes for agricultural commodities because the fertilizer used to grow them is made from fossil fuels. [30] At the same time, the U.S. and other industrialized nations are scaling up production of biofuels made from plants to reduce their reliance on imported oil. Using grain — mostly corn — to make fuel also drives up prices for commodities.

Rising grain prices drive farmers to plant more crops on marginal acres — including land formerly enrolled in the Agriculture Department's Conservation Reserve Program (CRP), which pays farmers to set environmentally sensitive land aside as wildlife habitat. About 34 million acres are protected under the CRP, or just under 10 percent of U.S. cropland More than 400,000 acres were converted to cropland in North Dakota alone in 2007. [31] "Conservation is in for a long swim against a strong current," said Jim Ringelman, conservation director in the Prairie Pothole Region for Ducks Unlimited. [32]

Does mitigation work?

Since thousands of acres of wetlands are developed each year, the only way to prevent net losses is to restore, improve or create more acres of new wetlands. This process, known as mitigation, has become central to the wetland permitting process. Many

Central Florida's once-meandering Kissimmee River is being "unstraightened" as part of the $10 billion Everglades restoration project. The U.S. Army Corps of Engineers channelized the river in 1971 to control periodic flooding, but the project dried out thousands of acres of critical habitat and deprived the Everglades of seasonal water supply. Remnants of the river's winding path can be seen next to a straightened section.

experts say that mitigation done correctly can improve the environment and that the process is improving with advances in wetland science. But critics reply that it creates too many easy-to-build, low-value wetlands. To work well, they argue, mitigation should replace not only lost acres but the specific functions of natural wetlands.

In the 1970s and '80s the Corps and EPA required developers to avoid damaging wetland functions, minimize damage or replace these functions on-site or nearby. These conditions also applied to farmers who converted wetlands to cropland after 1985 under the Swampbuster program, which restricts wetland drainage for agriculture.

On-site mitigation makes work harder for builders, who have to work around wetlands, and farmers who cultivate all of their acres. "If their holdings are already farmed, mitigating on site just means trading one wet spot for another," says Rick Robinson, environmental affairs director for the Iowa Farm Bureau Federation (IFBF).

Many field studies in the 1980s and '90s found that mitigation projects required under the 404 program were not meeting their targets, and that some were never finished. According to a 2001 National Research Council report, only 70 to 76 percent of the mitigation required in studies it reviewed was implemented, and half did not meet permit requirements. The committee concluded that "there is a substantial net loss in wetland area from the wetlands permitting program," and that wetlands that were built had low value. [33]

Under pressure to make permitting more flexible, federal and state agencies developed "mitigation banking," in which bank operators create or restore "banks" of wetlands and sell mitigation credits to customers seeking wetland development permits. This approach sought to make permitting more efficient, concentrate more resources on large land parcels and allow regulators to target important wetlands for restoration or improvement. *

---

* Mitigation banking is similar to pollution credit programs, which allow industrial polluters to buy unused pollution credits from non-polluters.

Starting in the early 1990s, mitigation banking became an active market. By 2005 the Corps estimated that some 450 mitigation banks were operating across the country, with more planned. [34] Some were run by private companies, others by government agencies with major mitigation responsibilities, such as state transportation departments, or conservation groups. Credit prices ranged from as low as $4,000 per acre in rural areas to $100,000 or more in urban or suburban areas. [35]

"Environmentalists should love mitigation banking," says the Mackinac Institute's Harding. "On-site mitigation isn't very effective because it tends to produce lots of small isolated wetlands that are expensive to create. Banking lets developers move ahead with projects more quickly and creates new viable habitat." In addition, states typically do not allow banks to sell most of their credits until they actually carry out their wetland projects. And bank owners retain long-term responsibility for the health of their wetlands, even after they sell the restoration credits, so they have a financial interest in creating viable wetlands. [36]

To help farmers meet Swampbuster requirements, the Iowa farm bureau created the first mitigation bank in the state in 2002, working with the state Department of Natural Resources (DNR). The bank, located in prairie pothole territory, converted 70 acres of corn and soybean fields into wetlands and has sold most of its available credits for roughly $14,000 apiece. [37]

"It's been going very well," says Robinson. "Agriculture isn't where the big mitigation banking money is, because farmers have smaller environmental impacts than developers and less money to pay, but we've had overwhelming interest from farmers."

But Audubon's Lauritsen sees flaws in the process in Florida. "If you develop a wetland that was wet for three months of the year, mitigation banking doesn't make you replace it with one that is also wet for three months

each year. That has an ecological impact on wildlife," he argues. "We need better techniques for recognizing the functions of different types of wetlands and recovering what we've lost — especially shallow wetlands, which are the easiest to drain and build on."

Better site selection would help, Lauritsen says. "There's a lot of promise in mitigation banking done the right way," he says. "We have a lot of agricultural property, planted with citrus and row crops, which still has wetland soil and is flooded in the spring before the pumps are turned on. You could restore those properties without spending a lot of money, and produce a lot of dividends."

According to the most recent FWS inventory, there was a net loss of many important types of wetlands between 1998 and 2004, including salt marshes and mangrove forests. They were replaced mainly by freshwater ponds, which increased by almost 700,000 acres (12.6 percent), due mainly to the creation of artificial ponds for purposes such as stormwater control or decorative landscaping. "These ponds are not an equivalent replacement for vegetated wetlands," the FWS notes. [38]

"Ponds are popular because they're easy. If you excavate in a wet place, you'll get plants and animals living there," says the University of Wisconsin's Zedler. "Other types like sedge meadows that are wet in some seasons and dry in others are harder to achieve. But ponds are simple systems, and they don't provide all of the ecological services that we're losing."

Although mitigation can be useful, many wetland advocates say, they see little value in a no-net-loss goal unless the focus shifts from quantity to quality. "Saying that it's OK to trade prairie potholes for stormwater retention basins is like making farmers trade their pickup trucks for compact cars and telling them that they haven't had any net loss of transportation," says Yaich of Ducks Unlimited. "They wouldn't be happy." ∎

BACKGROUND

Overlooked Resources

For most of the nation's history, Americans have viewed wetlands as worthless. Settlers saw wetlands, loosely referred to as mires, fens or swamps, as insect breeding grounds that impeded travel and farming.

Starting in the 1820s, the Army Corps of Engineers dammed and dredged many productive marshes and bottomlands (river flood plains) to improve navigation on U.S. rivers. Under the Swamp Land acts of 1849, 1850 and 1860, Congress transferred more than 64 million acres categorized as "wet and unfit for cultivation" to 15 states, mostly in the South and Midwest, to allow the land to be drained for flood control and agriculture. Many of the tracts ended up in private hands or were given to railroad companies. [39] Port development in San Francisco, Seattle and other cities led to the diking and filling of coastal marshes.

Contrary to their image, natural wetlands teemed with life. Marshes around the Chesapeake Bay supported fisheries that produced millions of pounds of crabs and oysters yearly in the late 1800s. Prairie potholes — then called sloughs — across the upper Midwest were rich nesting and feeding grounds for birds. *Little House on the Prairie* author Laura Ingalls Wilder, whose family homesteaded in South Dakota in the 1880s, described bird life on a nearby marsh:

"Millions of rustling grass-blades made one murmuring sound, and thousands of wild ducks and geese and herons and cranes and pelicans were talking sharply and brassily in the wind.

All those birds were feeding among the grasses of the sloughs. They rose on flapping wings and settled again, crying

Continued on p. 804

Chronology

1890-1900s
Conservationists protect some wetlands, but many are lost.

1890
Congress gives Army Corps of Engineers control of dredging and filling "navigable waters."

1905
Florida begins draining Everglades for agriculture and development.

———•———

1930s-1940s
Wetland conservation efforts expand, along with development.

1934
Congress authorizes Duck Stamps to fund wetlands conservation.

1936
Agriculture Department begins helping farmers drain wetlands.

1948
After disastrous floods, Congress permits water to be channeled from the Everglades for flood control.

———•———

1960s-1980s
Wetland protection expands.

1969
National Environmental Policy Act requires impact studies for major federal projects, such as dams.

1972
Clean Water Act gives Corps of Engineers responsibility for regulating development in wetlands. . . . Coastal Zone Management Act seeks to protect coast areas.

1973
Endangered Species Act restricts development of wetlands.

1977
President Jimmy Carter orders federal agencies to minimize wetland loss.

1980
Environmental Protection Agency (EPA) embraces concept of saving some wetlands to compensate for destroying others.

1985
Swampbuster program eliminates subsidies for farmers who convert wetlands for agriculture. . . . Supreme Court rules in *U.S. v. Riverside Bayview Homes* that the Corps can regulate wetlands *adjacent* to navigable waters.

1987
National Estuary Program directs EPA to protect important coastal bays.

1988
President George H. W. Bush endorses "no net loss" of wetlands as national policy.

1989
Louisiana sets aside oil revenues for coastal restoration.

———•———

1990s-Present
Politicians support wetlands, but property-rights advocates and industry argue federal controls are too broad.

1990
Congress calls for a plan to restore Louisiana's coastal wetlands.

1995
U.S. House requires government compensation if endangered species or wetlands regulations reduce landowners' property values by more than 20 percent.

2000
Comprehensive Everglades Restoration Plan is launched.

2001
In *SWANCC v. U.S. Army Corps of Engineers*, Supreme Court restricts the Corps' jurisdiction to navigable waters, their tributaries and adjacent wetlands.

2002
Louisiana and the Corps propose a $14 billion coastal restoration plan. Bush administration downscales it drastically.

2004
President George W. Bush pledges to increase overall wetlands acreage annually.

2005
Hurricanes Katrina and Rita destroy some 240 square miles of coastal wetlands in Louisiana and Mississippi.

2006
In divided verdict, the Supreme Court proposes a "significant nexus" test to determine whether isolated wetlands are connected to navigable waters.

2007
President Bush vetoes a $23 billion water bill, which includes funds for Everglades and Louisiana coastal restoration.

2008
Heavy spring rains cause record flooding along the Mississippi River; Hurricanes Gustav and Ivan cause further flooding and property damage along Gulf Coast. . . . Florida announces plans to buy 187,000 acres of the Everglades from U.S. Sugar Corp. for $1.75 billion to aid restoration efforts.

Why Wetlands Matter

They provide valuable "ecosystem services."

Over the past several decades, U.S. laws and regulations have become more protective of wetlands in recognition of the billions of dollars worth of valuable "ecosystem services" they provide by nurturing wildlife, protecting against storms and serving as "nature's kidneys." Depending on a wetland's type and location, these services may include:

Supporting wildlife — Wetlands provide habitat and feeding grounds for animals, birds, fish and shellfish. About three-quarters of the fish and shellfish harvested commercially in the United States depend on coastal estuaries for shelter at some point in their lives, such as spawning or growth before they migrate to open waters. [2] U.S. commercial fisheries generate roughly $25 billion in annual revenues for fishermen, processors and distributors. [3] In April 2008 the Commerce Department declared that the West Coast salmon industry had collapsed because historically low numbers of salmon were returning from the oceans to spawn; scientists said loss of freshwater habitat was a factor. [4]

Many species of shorebirds feed or nest in wetlands, including tidal mudflats, beaches and freshwater marshes. These areas are crucial stops for millions of birds that migrate north and south along the Atlantic and Pacific flyways, including some that travel to Central America or the Caribbean in winter and to Alaska and Canada in summer. Other species that travel up through the central United States depend on flood plains along the lower Mississippi River or prairie potholes in the upper Great Plains.

Cleaning up the water — Plants slow water down as it flows through wetlands, which causes dirt and other particles to settle out and sink to the bottom. Many types of microorganisms in wetlands, such as bacteria and algae, break down dead organisms and wastes. Wetland plants take up excess nutrients, such as nitrogen and phosphorus from farm runoff, that

otherwise would contribute to algae blooms and create "dead zones" in the water. So-called constructed (artificial) wetlands are often used to filter and clean polluted water from farms, homes and commercial developments.

Controlling floods — Wetlands are natural sponges that can soak up large quantities of rainwater, storm surge or melting snow, store it and then release it slowly over time. This makes them valuable flood-control systems, especially in urban areas where water runs off quickly from paved surfaces. And coastal wetlands can act as speed bumps that slow incoming storms. Some studies have estimated that four miles of well-vegetated wetlands can reduce storm surge by roughly one foot, although much depends on the type of wetlands and the dimensions of the storm. [5]

Regulating global climate — As plants grow they absorb carbon dioxide (CO_2), the main greenhouse gas that traps heat in the atmosphere and contributes to global climate change. When plants die and decompose, their carbon content is released back into the air. But if plant material is buried in soil — especially cool, wet soils like peat, where decomposition rates are slow — the carbon is stored. Wetlands store large amounts of carbon, but there is great uncertainty about how this process will be affected by climate change. (*See "Current Situation," p. 811.*)

---

[1] Unless otherwise cited, this section is based on Joy B. Zedler and Suzanne Kercher, "Wetland Resources: Status, Trends, Ecosystem Services, and Restorability," *Annual Review of Environment and Resources* (2005), pp. 39-74.

[2] "Habitat Connections: Wetlands, Fisheries and Economics," National Oceanographic and Atmospheric Administration, www.nmfs.noaa.gov/habitat/habitatconservation/publications/habitatconections/habitatconnections.htm.

[3] *Business Wire*, Dec. 5, 2005.

[4] " 'Fishery Failure' Declared for West Coast Salmon Fishery," National Oceanographic and Atmospheric Administration, May 1, 2008.

[5] "Wetlands Break Waves, Quell Surge," LaCoast.gov, www.lacoast.gov/WATERMARKS/2006-03/2protectMainland/; Louisiana Sea Grant Program, "Louisiana Hurricane Resources," www.laseagrant.org/hurricane/archive/wetlands.htm.

Continued from p. 802

news to each other and talking among themselves among the grasses, and eating busily of grass roots and tender water plants and little fishes." [40]

After the Civil War, the Corps took on new responsibility for flood control and regulation of dumping and filling in harbors. In 1899 the Rivers and Harbors Act required developers to obtain a permit from the Corps for any activity that would excavate, discharge material into, or obstruct "navigable waters," including coastal waters, lakes,

rivers and streams. The Corps' main focus, however, was construction, not conservation. [41]

States promoted wetland development by creating so-called drainage districts, or local groups with the power to issue bonds, drain land for approved uses and tax all landholders who benefited from this service. Many Midwestern districts installed massive drain-tile systems using underground networks of ceramic pipes that channeled water away from fields to drainage ditches and streams. By 1935, farmers in Illinois had installed

enough drain tiles to circle the world six times. [42] "Michigan wouldn't have been settled if it hadn't been drained," says the Mackinac Institute's Harding.

Early Conservation

By the early 20th century, conservation advocates were campaigning to save important natural resources from over-hunting and development. They found a friend in President Theodore Roosevelt, who used his executive power to designate 51 tracts of land in 17 states and territories as National Bird Reservations between 1903 and 1909. Many tracts included islands, lakes, marshes and other wetlands. Thirty-one of these "Roosevelt Reservations" later became units of the National Wildlife Refuge system. [43]

The invention of steam shovels and mechanized earth-moving equipment facilitated carving up landscapes on a bigger scale. In 1928, a year after flooding along the Mississippi River inundated thousands of miles of land from Illinois to the Gulf of Mexico, Congress put the Corps in charge of flood control and navigation along the river's entire length. The Corps began installing levees, stabilizing riverbanks and realigning channels to control flood waters, a program that stretched through the next several decades. It also built a system of 29 locks and dams on the upper Mississippi extending for 670 miles from Minneapolis to St. Louis, to create a nine-foot-deep channel for easy navigation.

These projects profoundly affected wetlands along the river. Levees prevented river waters from spilling over, drying out previously fertile bottomlands. In other places the navigation system raised water levels, putting floodplains under water. In Louisiana levees prevented the Mississippi and its tributaries from flooding and depositing sediments across the delta plain, accelerating erosion along the state's coastline.

Another catastrophic 1928 storm, in Florida, caused massive Lake Okeechobee to flood thousands of surrounding acres, killing some 2,500 people. In response President Herbert Hoover directed the Corps to dike the lake. This cut off freshwater flow from the lake down through the Everglades during rainy seasons, lowering the water table and drying out soils, which spurred further logging and farming on former wetlands.

"The Everglades were dying," wrote journalist Marjorie Stoneman Douglas in her 1947 bestseller, *The Everglades:*

River of Grass. "The endless acres of saw grass, brown as an enormous shadow where rain and lake water had once flowed, rustled dry." [44]

Just after Douglas' book was published, President Harry S Truman dedicated a 1.3-million-acre tract as Everglades National Park. But at the same time the Corps was designing the Central & Southern Florida Project — a massive flood control system intended to tame areas around the new park for agriculture and development. Launched in 1948, the project created 1,000 miles of levees, 720 miles of canals and 200 water-control structures to regulate freshwater flows and channel rainy-season runoff out to sea, further drying out the greater Everglades ecosystem.

Other federal agencies also promoted development on wetlands. In 1936 the Agriculture Department started sharing farmers' costs for converting wetlands to cropland. The Civilian Conservation Corps and other Depression-era relief initiatives also put unemployed Americans to work draining wetlands.

To slow wetland losses the U.S. Biological Survey (which later became the Fish and Wildlife Service) in 1934 launched the federal Duck Stamp program. Revenues from the program were used to buy wetlands that were important waterfowl habitat and add them to the National Wildlife Refuge system. Over the next 70 years Duck Stamp sales to hunters financed the purchase or lease of more than 5 million acres. [45]

Exploitation to Protection

By the 1960s economic growth was taking a heavy toll on U.S. water resources. Many rivers, lakes and streams were heavily polluted with industrial discharges. Unrestricted private development was encroaching on scenic coastlines, often in ecologically sensitive areas.

For example, farm fertilizer and phosphate detergent discharges into Lake

Erie spawned massive blooms of algae; when the weeds died and decomposed, the process depleted oxygen from the water, leaving the lake so void of fish life that it was widely viewed as biologically dead. In 1963 there was so much industrialization and development around San Francisco Bay that only four of its 276 miles of shoreline were open to the public, and adjoining towns like Berkeley and San Mateo were planning to fill in much of the bay to expand their land areas. [46]

Such developments spurred a new wave of environmentalism and landmark laws including the Clean Water Act of 1972, which created a national permitting system for pollution discharges. Section 404 of the CWA gave the Corps of Engineers authority to regulate dredging and filling in wetlands, subject to oversight by the new Environmental Protection Agency, which had veto power. Congress also passed the Coastal Zone Management Act, which provided federal aid to states that developed comprehensive plans for preserving and restoring their coastlines.

A year later Congress enacted the Endangered Species Act (ESA), which restricted development of areas identified as critical habitat for endangered or threatened species. Section 7 of the act required federal agencies to consult with the Fish and Wildlife Service to ensure that their activities would not jeopardize listed species or damage critical habitat.

While these laws offered new leverage for protecting wetlands, the Agriculture Department and other agencies still supported wetland conversion. President Jimmy Carter, a critic of federal water projects for both their costs and environmental impacts, addressed this conflict in a 1977 executive order that directed all federal agencies to "take action to minimize the destruction, loss or degradation of wetlands, and to preserve and enhance the natural and beneficial values of wetlands in carrying out the agency's responsibilities." [47]

Complex Plans Aim to Restore Everglades

But politics and funding are still obstacles.

The history of Florida's Everglades echoes the broader story of U.S. wetlands. Long viewed as a swamp, drained and developed for the past century, the Everglades now is only about half as big as it was before World War II, and what's left is severely threatened by water pollution and constant development. However, the area still supports 68 threatened or endangered species of plants, animals and birds, including the Florida panther, wood stork, American crocodile and leatherback turtle.

Indeed, the Everglades is not a swamp but an intricate system of interdependent ecosystems including sawgrass marshes, mangrove forests, tropical hardwood hammocks and the marine environment of Florida Bay. The roughly 6,000-square-mile area is vital to South Florida's water supply.

In 2000 Congress approved the Comprehensive Everglades Restoration Plan (CERP), scheduled to take about 40 years and cost more than $10 billion (up from an initial projected cost of $7.8 billion). The massive rescue operation, which includes 60 individual projects, aims to make more water available for the Everglades by capturing flows that are currently diverted and storing water so it can be released as needed. Some elements of the plan are moving forward, but costs are still rising, and many important elements have been delayed. [1]

Two central problems are hurting the Everglades: It does not receive enough water, and what does flow through is polluted. But making the ecosystem healthy again is more complicated than just opening spigots. Scientists are trying to determine how much water the Everglades needs and how fast it should move.

"The Everglades has a very distinctive land form," explains Greg Noe, an ecologist with the U.S. Geological Survey. "It has elevated ridges and sloughs like the channels in corrugated cardboard, which run north-south and then bend west at the lower end." The long interconnected sloughs, or ditches, contain many fish and small aquatic organisms and are important feeding areas for wading birds. Dense sawgrass grows up to 10 feet high on the ridges. There was so much sawgrass that writer Marjory Stoneman Douglas famously dubbed the Everglades a "river of grass" in 1947.

As water levels fall, sloughs dry out and become filled with sawgrass. "One big CERP goal is to add enough water back to restore that landscape, and also to improve water quality so that you don't harm the system by putting more water in," says Noe. "No one knows exactly what the right water depths are — if it's too deep you drown everything out, and if it's too shallow the sloughs fill in with sawgrass."

But Noe believes that the CERP research plan can answer these technical questions. "It's the biggest restoration project in the U.S. and probably the most complicated, because so many organizations are involved, but everyone has a vested interest in getting it right," he says. CERP incorporates an approach called adaptive management, in which scientists carry out selected projects, measure results, then refine their next steps based on these findings. "If people weren't willing to learn and change the plan, I'd be worried," says Noe.

Florida has lagged on cleaning up water that flows through the Everglades. After a 1992 legal settlement with the federal government, the Florida legislature passed the Everglades Forever Act in 1994, which required state regulators to ensure that water entering the Everglades would contain no more than 10 parts of phosphorus per billion by 2006 (phosphorus, a nutrient found in agricultural runoff, is one of the main water pollutants in the Everglades), and take other steps to achieve that standard. But in 2003 the state extended the compliance deadline to 2016 in a bill supported by the sugar industry but derided by environmentalists as the "Everglades Whenever" Act. [2]

The state is still struggling to meet minimum national water-quality standards. A May 2007 memo by Major Gen. Don Riley, director of civil works for the U.S. Army Corps of Engineers, asserted that Florida "is not currently meeting [water quality] requirements for water that would flow into [the Lake Okeechobee watershed] and it is not likely to come into compliance for several decades." [3]

But Florida has outdone the federal government on one aspect of Everglades restoration: funding. Although the cleanup costs are

Carter's order eventually led to the Swampbuster program, enacted in the 1985 farm bill. The program made farmers ineligible for federal aid or subsidies if they converted certain types of wetlands to cropland. Like the 404 program, however, Swampbuster allowed mitigation projects to compensate for activities that destroyed wetlands.

Running for president in 1988, George H.W. Bush pledged that his administration would work to achieve "no net loss" of wetlands in the U.S. During his administration (1989-1993) the Agriculture Department expanded the Conservation Reserve Program for preserving environmentally sensitive wetlands. The 1990 farm bill then created a parallel Wetland Reserve Program, which paid farmers for restoring wetland functions to marginal farmlands.

Behind the scenes, however, Vice President Dan Quayle's staff sought to redefine wetlands so that only areas that remained wet year-round received federal protection. This proposal, which was opposed by the EPA and ultimately failed, could have deregulated up to half of U.S. wetlands. [48]

Legal Resistance

Expanding federal controls slowed the net rate of wetland losses significantly in the 1970s and '80s, but

supposed to be shared, Florida spent nearly $2.4 billion on Everglades restoration projects from 2000 though 2007 while Congress provided only $360 million. However, the 2007 Water Resources Development Act authorized three important CERP projects with total costs of $1.8 billion, including restoration of more than 150,000 acres of wetlands, and President George W. Bush's fiscal 2009 budget proposal includes $215 million for CERP activities. [4]

Great egrets thrive in Florida's Everglades, along with 68 threatened or endangered species of plants, animals and birds, including the Florida panther and American crocodile.

National Park Service Photo/Rodney Cammauf

development boundary" closer to the Everglades, allowing the Lowe's hardware chain to build a store and offices on 20 acres of wetlands.

"It shoots our credibility," said Miami-Dade Mayor Carlos Alvarez, who pledged to veto the plan. [8]

A September assessment by the National Research Council found that in spite of recent progress, CERP was not moving fast enough. "The project is bogged down in budgeting, planning and procedural matters while the ecosystem that it was created to save is in peril," the council warned. [9]

Everglades supporters were elated in June when Gov. Charlie Crist, R-Fla., announced a deal to buy out the holdings of U.S. Sugar, one of two major sugar companies in Florida, including 300 square miles of agricultural land south of Lake Okeechobee. Crist's administration said the struggling company would be allowed to farm the land for six more years, after which the state would use it to restore freshwater flow from the lake to the Everglades.

The $1.75 billion sale may not be finalized until sometime in 2009, but skeptics are pointing out that it may benefit the powerful sugar industry as much as it advances Everglades cleanup. [5] Some farmers argue that agriculture is being scapegoated as the root of the Everglades' problems when development is equally at fault. [6]

There's no shortage of new development proposals around the Everglades. In early 2008 Palm Beach County commissioners approved 10,500 acres of new rock mines north of the Everglades, although the state Department of Environmental Protection was still reviewing whether mining would affect local water quality or Everglades restoration. [7] And Miami-Dade County commissioners voted to move the county's "urban

[1] "South Florida Ecosystem: Some Restoration Progress Has Been Made, but the Effort Faces Significant Delays, Implementation Challenges, and Rising Costs," U.S. Government Accountability Office, Sept. 19, 2007, p. 5; "Cash for Everglades Restoration Dries, Up, The Associated Press, Nov. 21, 2007.

[2] Michael Peltier, "Florida Governor Bush Signs Contentious Everglades Bill," Reuters, May 22, 2003.

[3] Posted online by Public Employees for Environmental Responsibility, www.peer.org/docs/ace/07_14_11_gen_riley_memo.pdf.

[4] Audubon of Florida, "Everglades Report," winter 2008; Daniel Cusick, "Army Corps: Proposed Budget Offers an 'Awkward Kiss on the Cheek' for Everglades," *E&E Report*, Feb. 4, 2008.

[5] Patrik Jonsson, "U.S. Sugar Buyout: Sweet Deal for the Everglades?" *The Christian Science Monitor*, Aug. 20, 2008; Mary Williams Walsh, "Helping the Everglades, or Big Sugar?" *The New York Times*, Sept. 14, 2008, p. BU1.

[6] Jonsson, *op. cit.*

[7] Paul Quinlan and Jennifer Sorentrue, "Miners Get OK to Dig in Western Palm Beach County," *The Palm Beach Post*, April 24, 2008, p. 1B.

[8] "Good News for the Everglades: New Funding and a Mayor's Stand Offer Hope for Florida's Treasure," *Sarasota Herald-Tribune*, May 2, 2008, p. A10.

[9] National Research Council, "Progress Toward Restoring the Everglades," second biennial report, Sept. 29, 2008, http://dels.nas.edu/dels/rpt_briefs/everglades_brief_final.pdf.

some critics said development curbs infringed on landowners' rights by reducing property values. When government committed so-called regulatory takings ("taking" value without actually seizing the property), they argued, it owed owners financial compensation. [49]

Several landowners who initiated lawsuits in the mid-1980s won federal court rulings that wetland regulations were takings, and received monetary damages. [50] In one lawsuit filed by a beachfront developer in South Carolina, the Supreme Court held that government action barring all economic uses of land (assuming that those uses were legal when the property was acquired) automatically constituted a taking. [51] This ruling signaled that if regulators did not make a significant effort to be fair to landowners, they risked having their rules struck down in court.

Some observers predicted that these cases would stimulate more antiregulatory claims, but no such wave developed — possibly because federal agencies got the message. "The 'takings revolution' fizzled, primarily because most wetland permits were granted, so there was little to complain about," says Vermont Law School's Parenteau. And mitigation banking, launched in 1990 and expanded under President Clinton, offered developers a new way to offset wetland losses.

But wetland regulation was further complicated by the Supreme Court's 2001 *SWANCC* ruling that Congress could not regulate isolated wetlands based solely on the presence of migratory birds. [52] The ruling effectively deregulated some isolated wetlands, and the language of the majority opinion raised questions about where federal authority stopped. Most state courts, however, continued to define federal jurisdiction quite broadly, and several states passed bills setting up comprehensive state-level protection for wetlands.

President Bush, who had endorsed his father's goal of "no net loss" in 2002, expanded on it in 2004. "Instead of just limiting our losses, we will expand the wetlands of America," Bush said. Critics endorsed the goal but said reaching it would require many steps, including a narrow reading of the *SWANCC* ruling, stricter enforcement of mitigation requirements and more funding for the National Wildlife Refuge system. [53]

The issue was further muddled by the Supreme Court's fragmented *Rapanos* ruling in 2006. Justice Kennedy's proposal for a "significant nexus" test left lower courts to determine case by case whether wetlands were subject to federal jurisdiction. EPA and the Corps issued guidance in 2007 to help regulators, but these standards required additional paperwork that greatly complicated the process of reviewing permit applications.

"Right now the 404 program is probably in as much trouble as it's been in for years," says Yaich of Ducks Unlimited. "The guidance imposed much more difficult tasks in response to Justice Kennedy's opinion, and the system is bogged down. It's the worst of all worlds — we're dealing with applications very inefficiently, including many permits that are perfectly fine, and we're not protecting enough wetlands." ∎

CURRENT SITUATION

Which Waters?

Legislators who are alarmed about wetlands losing federal protection have introduced bills in both houses of Congress to specify which waters are subject to regulation. "By focusing on the phrase 'navigable waters' in its *SWANCC* and *Rapanos* decisions, the Supreme Court muddied the jurisdictional understanding of the CWA," Rep. James L. Oberstar, D-Minn., chairman of the House Transportation Committee, said in introducing the House version of the Clean Water Restoration Act, in 2007.

Oberstar's legislation and its Senate counterpart seek to clarify the issue by replacing the term "navigable waters of the United States" in the Clean Water Act with "waters of the United States." This step, they argue, makes explicit that Congress intended to provide broad protection for water bodies — even non-navigable streams and isolated wetlands — the way the CWA was commonly interpreted prior to the *SWANCC* ruling in 2001.

Many state regulators, as well as environmental and sportsmen's groups, have endorsed the legislation. "Agency guidance issued in 2007 has left in doubt the protection of 'non-navigable' headwater, intermittent and ephemeral streams," a coalition of hunting and fishing groups wrote to Congress in early 2008. "These streams provide valuable habitat in their own right and are critical to downstream water quality and aquatic habitat. Without Clean Water Act protection, these streams are now vulnerable to sewage and industrial pollution as well as dredging and filling."

The concern about non-navigable waters was supported by a March 2008 memo from the EPA's chief enforcement official, who said the *Rapanos* opinion and EPA's guidance for applying the opinion had led the agency to downgrade or avoid pursuing about 500 potential violations of federal wetland law. The House Transportation Committee and the Oversight and Government Reform Committee both are investigating EPA's handling of wetland enforcement cases.

"This sudden reduction in enforcement activity will undermine the implementation of the Clean Water Act and adversely impact EPA's responsibility to protect the nation's waters," Oberstar and Henry A. Waxman, D-Calif., chairman of the Government Reform Committee, told the agency in July. [54]

But opponents, including trade associations, farm bureaus and conservative think tanks, argue that there was wide disagreement about the scope of the CWA before *SWANCC* and that the Clean Water Restoration Act would broaden federal water law far beyond what Congress originally intended. Under the CWRA, many critics say, any puddle or rain gutter could be subject to federal regulation. (*See "At Issue," p. 809.*)

"Expanding the jurisdictional definition could put the Corps in every ditch and grass waterway in Iowa," says Robinson of the Iowa Farm Bureau Federation. "Farmers understand Swampbuster and USDA's wetland protection programs, but bringing in the Corps will make it much more confusing."

The central problem is finding clear principles to justify which waters should be subject to federal versus state oversight. Science offers little help, since researchers say most wetlands are connected ecologically in various ways even if they are physically isolated, so ultimately the debate comes down to warring interpretations of how far Congress can regulate economic activities.

Continued on p. 810

At Issue:

Should Congress pass the Clean Water Restoration Act?

BRETT HULSEY
SUPERVISOR, DANE COUNTY, WIS.

FROM TESTIMONY BEFORE U.S. HOUSE TRANSPORTATION
AND INFRASTRUCTURE COMMITTEE, APRIL 16, 2008

*m*y county constituents place a high value on the quality of our lakes, streams and drinking water. They want clean, safe water for recreation — for swimming, boating and fishing. They understand that protecting drinking-water sources from pollution makes for better quality water coming out their taps and protects our health and safety at a lower cost. . . .

We have experienced five major floods costing local residents and the county $50 million since 1993. Our citizens want to prevent flood damage in the most environmentally protective and cost-effective ways possible to avoid the costs of repairing homes and infrastructure damaged by flooding. They also want to avoid the costs of cleaning up waters that have been needlessly polluted by others.

We saw the importance of protecting headwater streams and isolated wetlands during the Mississippi River floods of 1993, [which] killed more than 50 and cost at least $16 billion. Our county is at the headwaters of the Yahara River that flows to the Rock River and to the Mississippi, [which] drains 40 percent of the continental United States. After these floods, I worked with [the Federal Emergency Management Agency] and state agencies to purchase more than 10,000 homes and structures and move them out of harm's way. . . .

The [Environmental Protection Agency] estimates that some 20 million acres of wetlands — one-fifth of the remaining wetlands in the lower 48 states — could lose protections based on its interpretation of [the *SWANCC* and *Rapanos* decisions]. This would allow developers to drain wetlands, build new homes that would then be flooded and have to be purchased by local governments and taxpayers. There is a compelling public reason to protect these wetlands and headwater streams from development in the first place. . . .

Some have argued that the Clean Water Restoration Act somehow represents a vast expansion of Clean Water Act protections, but . . . this change in the law would take us back to where the Clean Water Act was before the *SWANCC* and *Rapanos* decisions. It would restore the law's scope, not expand it.

Some preposterous concerns are that this will mean roadside ditch and gutter regulation. The rain gutter on my house was not regulated by the Clean Water Act before these court decisions, and I am confident that it won't be regulated after the Clean Water Restoration Act is enacted. By the way, my gutters flow to a rain barrel and rain gardens, allowing the water to soak into the ground.

LINDA C. RUNBECK
PRESIDENT, AMERICAN PROPERTY COALITION

FROM TESTIMONY BEFORE U.S. HOUSE TRANSPORTATION
AND INFRASTRUCTURE COMMITTEE, APRIL 16, 2008

h.R. 2421 has come in through the back door masquerading as a so-called simple clarification of the Clean Water Act when it is not. Rather, this bill has the potential to transform the Clean Water Act into a full-blown national land use control act. In it, federal agencies are given unlimited jurisdictional boundaries to intrude on every activity where Americans are involved with water and land. . . .

The bill would open the door to federal regulation of even insignificant, small depressions of mostly dry land, isolated wetlands, arroyos in the desert, sand flats, ditches and gutters, areas scarcely recognizable as "waters of the U.S." It doesn't end there: this bill would also, for the first time ever, authorize federal regulation of any "activities affecting water." And to be clear, "activities" might have a direct impact or an indirect impact on waters. So, regulated "activities" could take place on a hilltop or a mountaintop 25 miles from water, and the feds would still have the power to bring that activity to an immediate halt. . . .

The bill also offers a convenient scapegoat for Congress to shift the costs (and the blame) associated with water cleanup onto property owners and local governments. At the same time, Congress fails to provide them any meaningful measuring stick, as the bill contains no national water-quality standards. . . . Absent any . . . cost-benefit analyses or effective assessment mechanisms, this bill has the potential to exhaust the resources of individuals and local governments. Finally, after intruding on the freedoms and pocketbooks of millions of Americans, this bill can provide us no assurance that the nation's water quality will be improved. . . .

In conclusion, given the local nature of ditches, gutters, isolated wetlands, small depressions in fields and prairie potholes, the American people deserve something better than a centralized national land use bill imposed on an unsuspecting American public. Congress should take the time to get it right. The right way would be to: 1) establish goals; 2) authorize completion of a comprehensive assessment of the quality of the nation's waters; 3) establish priorities, costs and a realistic timetable for achieving water quality goals; 4) complete the task of bringing the remaining point sources into compliance and 5) allow local governments and local citizens the opportunity to develop local and regional alternatives that will ensure the broadest public support in order to achieve the desired results.

Continued from p. 808

"Congress has to say clearly and directly how far it intends to push federal jurisdiction under the Commerce Clause," says Vermont Law School's Parenteau. "Otherwise, we'll never have an end to these arguments."

Louisiana Sinking

Louisiana's coastal marshes are the most threatened wetlands in the United States. Since the 1930s, the state has lost more than 1.2 million acres (1,900 square miles) of coastal habitat due to storms, flood control programs and industrial development. Dams and levees prevent the Mississippi River from flooding and depositing sediments on Louisiana's coastal plain. Thousands of miles of barge access canals, excavated by companies drilling offshore for oil and gas, have sliced through coastal marshes and created channels for saltwater intrusion from the Gulf of Mexico.

About 13 square miles of Louisiana's coastline become open water each year, and in 2005 Hurricanes Katrina and Rita accelerated the process, destroying 240 square miles of wetlands. Federal and state agencies have spent almost $800 million since 1990 on coastal restoration projects, such as reintroducing freshwater to declining marshes, protecting shorelines from erosion and restoring barrier islands. [55] But slowing wetland losses will require much larger-scale efforts.

Coastal scientists say that Katrina and Rita have pushed the situation to a crisis point, and that if the state does not start restoring wetlands faster than it loses them — today five square miles are lost for every one created — within 10 years much of the state's coastline will be permanently gone. Without wetlands to soak up flood waters and help buffer coastal areas, future storms are likely to do even worse damage along the Gulf Coast than recent hurricanes.

"People think we still have 20, 30, 40 years left to get this done. They're not

The loss of wetlands habitat where salmon can spawn and rear their young has contributed to the collapse of the West Coast commercial salmon fishery, the Department of Commerce said recently. Historically low numbers of salmon are returning from the ocean to spawn, according to scientists.

Getty Images/Bill Shaefer

even close," said Kerry St. Pe, director of the Barataria-Terrebonne National Estuary Program. "If we aren't building land I can walk on inside of 10 years, we'll be moving communities." [56]

Scientists began warning in the 1970s that levees and canals were damaging Louisiana's coastline. In 1989 the state created a trust fund with revenues from oil and gas development to pay for restoration projects, and in 1990 Congress authorized federal funding and created a task force to manage the ef-

fort. But fishermen, developers and oil and gas companies objected to many projects that hurt their individual economic interests. For example, the state had to fight a two-year legal battle with oyster fishermen who opposed freshwater diversion projects. Ironically, these measures — which siphoned freshwater and sediment from the Mississippi River to rebuild coastal marshes — made many oyster beds more productive. [57]

In 2002 Louisiana and the Corps presented the Bush administration with a 30-year, $14 billion master plan for coastal restoration. The White House cut the plan back to less than $2 billion over a decade. Congress was considering an initial $1.1 billion proposal when hurricanes Rita and Katrina struck in 2005.

The National Research Council, which was then reviewing the scaled-back Louisiana coastal plan, reported in 2006 that while most pieces of the blueprint were sound, a broader plan was needed that showed how the pieces fit together and laid a base for further work. The study also called for better communication with the public about how various projects would affect land use, plus a program to compensate families that had to be moved during restoration. [58]

In late 2007 Congress passed its first water projects authorization bill in eight years, overriding a veto by President Bush, who described the $23 billion legislation as fiscally irresponsible. The bill provided $1.9 billion for Louisiana coastal restoration projects, plus nearly $1 billion for hurricane protection measures. It also created yet another task force to recommend strategies for conserving and restoring the state's coastal ecosystems. [59]

Two more hurricanes, Gustav and Ike, walloped the Gulf Coast in September 2008, causing such massive crop losses from flooding that the state sought $700 million in federal disaster aid specifically for farmers. [60]

"Louisiana sugarcane growers have received agricultural disaster assistance twice over our more than 200 years of production," farmer Wallace Ellender IV told the Senate Agriculture Committee. "The fact that both of those assistance packages were made necessary by intense hurricanes in this decade is a direct result of rampant coastal erosion. Unless we invest in energetic coastal restoration efforts soon, my farm may be a beachfront property in a few short years before slipping quietly beneath the waves." [61]

Wetter or Dryer?

Global climate change is a wild card for wetlands as temperatures rise and U.S. leaders look for ways to reduce atmospheric concentrations of carbon dioxide, the main greenhouse gas (GHG) generated as a result of human activities. GHGs contribute to global warming by trapping heat in the atmosphere and warming Earth's surface. Wetlands are a potential solution, because they absorb large quantities of carbon in plant matter and soils. But climate change could also harm wetlands, either by flooding them or drying them out. [62] Before wetlands can become a solution to climate change, scientists say, we need to know more about how they will be affected by rising temperatures.

Some national and regional policies for reducing GHG emissions allow emitters to "offset," or compensate for, the GHGs they release by paying to grow forests that will pull carbon out of the air. Some wetland advocates would like to see wetlands

receive similar treatment. Under such schemes, an electricity producer might receive credit for restoring a specific number of wetland acres where plants and soils would absorb atmospheric carbon to offset some emissions from the company's coal- or gas-fired power plants. The system would be like a mitigation bank, except that instead of selling credits for wetland restoration the offset provider would sell credits for storing carbon.

"Wetlands are one of the better sequestration sources out there. They're equal to or better than forests," says Myers of People for Puget Sound. His organization is identifying local marshes that were diked and drained for agriculture, where selling carbon sequestration credits could accelerate restoration work and provide additional funding.

But wetland chemistry is complicated. While wetlands store large quantities of carbon, they also emit other greenhouse gases as bacteria break down organic material, and warming may speed these processes up. And as temperatures rise, climate models predict that in many areas rain will fall harder and faster over shorter time periods, so many ecosystems will be subject to extreme wet and dry periods.

More drought could make wetlands less useful as carbon sinks, says U.S. Geological Survey ecologist Greg Noe. "Carbon is very sensitive to drying in wetlands," Noe observes. "For example, peat lands in the Everglades have lost 10 feet of soil over the past 50 years, because as those areas became dryer the soil was exposed to oxygen that sped up the decomposition process."

Overall, however, wetlands are worth studying as carbon sinks, Noe says. "They're not perfect systems because they're very sensitive to changes in hydrology, but they do show a lot of promise," he says. "River systems are one option. There's lots of carbon in them, and it might be fairly easy to restore them — for example, by breaching levees to recreate marshes." ∎

OUTLOOK

Valuing Wetlands

For the past 30 years national policy has declared that wetlands are valuable, but it is still hard to protect individual wetlands when they impede construction projects with dollar values attached.

"We need tools for factoring the value of ecological services into local economies so we can put our money where our mouths are," says the Florida Audubon Society's Lauritsen. "Until people understand the real values that wetlands provide, protecting them will be like putting our fingers in a dike."

Regions that are subject to major floods have the most to gain from restoring wetlands, and some leaders are pressing for action. After Hurricanes Gustav and Ike hit Louisiana in September, Republican Gov. Bobby Jindal said he would draw on state oil and gas revenues for coastal restoration. [63] And after heavy rains caused severe flooding across the Midwest in June, killing 24 people and forcing as many as 40,000 from their homes, some observers argued that more wetlands, not higher levees, were the right strategy for next time.

"No matter how finely tuned our engineering is, Mother Nature did a better job," argued an editorial in Illinois. "The recent floods should prompt a serious push to restore Illinois' lost wetlands — which also soak up pollution — as an alternative to more man-made steel and sand barriers." [64]

Other regions also see wetlands as long-term investments. Federal and state agencies are moving forward with a 50-year program to turn 15,000 acres of former industrial salt ponds around south San Francisco Bay back into tidal marsh. Supporters say the nearly

$1 billion project will produce cleaner water, more habitat for wildlife and better flood protection for low-lying, bayside cities like San Jose. "The bay is precious, and tidal wetlands will help make it more resilient," wrote Sen. Dianne Feinstein, D-Calif. [65]

Polls show that Americans support policies to keep rivers, lakes and streams clean and ensure that drinking water is safe. But conservationists will have to convince Americans that all wetlands are valuable, from the Everglades to vernal pools in their own back yards.

"The public wants to protect wetlands, but Americans think about actually wet land, not drainage ditches that hardly ever hold water," says the Mackinac Institute's Harding. ∎

Notes

[1] "Vernal Pool Information," www.massnature.com.

[2] T. E. Dahl, "Status and Trends of Wetlands in the Coterminous United States 1998 to 2004," U.S. Fish and Wildlife Service, 2006, p. 57.

[3] For background see Mary H. Cooper, "Water Quality," *CQ Researcher*, Nov. 24, 2000, pp. 953-976.

[4] David Sunding and David Zilberman, "The Economics of Environmental Regulation by Licensing: An Assessment of Recent Changes to the Wetlands Permitting Process," *Natural Resources Journal*, vol. 42 (2002), pp. 73-76.

[5] Testimony of Stephen E. Sandherr, Associated General Contractors of America, before the House Transportation and Infrastructure Committee, July 19, 2007, p. 3.

[6] *Ibid.*

[7] Joy B. Zedler and Suzanne Kercher, "Wetland Resources: Status, Trends, Ecosystem Services, and Restorability," *Annual Review of Environment and Resources*, vol. 30 (2005), p. 56.

[8] Erik Stokstad, "Louisiana's Wetlands Struggle for Survival," *Science*, Nov. 25, 2005, p. 1266. For estimated damages from Katrina, see Axel Graumann, *et al.*, "Hurricane Katrina: A Climatological Perspective," U.S. Department of Commerce, National Oceanographic and Atmospheric Administration, updated August 2006.

[9] Louisiana Coastal Protection and Restoration Authority, *Integrated Ecosystem Restoration and Hurricane Protection: Louisiana's Comprehensive Master Plan for a Sustainable Coast* (2007), www.lacpra.org.

[10] www.mitigationbanking.org.

[11] The CWA was originally titled the Federal Water Pollution Control Act when it was passed in 1972, but became known as the Clean Water Act after it was amended in 1977.

[12] *NRDC v. Calloway*, 392 F. Supp. 685 (D.D.C. 1975).

[13] 33 C.F.R., section 328.3(a)(iii).

[14] 474 U.S. 121 (1985).

[15] 531 U.S. 159 (2001).

[16] Felicity Barringer, "In Reversal, E.P.A. Won't Narrow Wetlands Protection," *The New York Times*, Dec. 17, 2003, p. A35.

[17] 126 S. Ct. 2208 (2006).

[18] Dahl, *op. cit.*, p. 46.

[19] "Conserving America's Wetlands 2008: Four Years of Partnering Resulted in Accomplishing the President's Goal," Council on Environmental Quality, April 2008.

[20] Matthew Waite and Craig Pittman, "Satellite Photographs Show Losses," *St. Petersburg Times*, May 22, 2005, p. 11A; Craig Pittman and Matthew Waite, "They Won't Say No," *St. Petersburg Times*, May 22, 2005, p 1A.

[21] For background see David Hosansky, "Reforming the Corps," *CQ Researcher*, May 30, 2003, pp. 497-520.

[22] Michael Grunwald, "A Green Day for Bush," *Time*, Feb. 2, 2008; Chris Talbot, "EPA Vetoes Large Flood-Control Plan," The Associated Press, Sept. 3, 2008.

[23] Florida Department of Environmental Protection, "Kissimmee River Restoration Continues," June 17, 2005; "Working With Nature: Corps Restoration Projects Benefit People, Communities, and Wildlife," Corps Reform Network, undated, www.corpsreform.org/sitepages/downloads/ProjectsInTheField-Reports/CRN-pfFS-Pjcts_Restoration.pdf.

[24] National Research Council, *New Directions in Water Resources Planning for the U.S. Army Corps of Engineers* (1999), p. 7; "Prioritizing America's Water Resources Investments," National Academy of Public Administration, pp. 115-116.

[25] National Academy of Public Administration, *op. cit.*, p. 115. For a study urging the Corps to do more planning in a watershed context, see National Research Council, *Compensating for Wetland Losses Under the Clean Water Act*, 2001.

[26] For details see "Campaigns: Florida Panther," Public Employees for Environmental Responsibility, www.peer.org/campaigns/whistleblower/panther/index.php.

[27] Ted Williams, "Bait and Switch," *Audubon*, March-April 2008.

[28] "Prairie Pothole Region: At the Current Pace of Acquisitions, the U.S. Fish and Wildlife Service Is Unlikely to Achieve Its Habitat Protection Goals for Migratory Birds," U.S. Government Accountability Office, GAO-07-1093, Sept. 2007.

[29] For background on biofuels, see Adriel Bettelheim, "Biofuels Boom," *CQ Researcher*, Sept. 29, 2006, pp. 793-816 and Jennifer Weeks, "Buying Green," *CQ Researcher*, Feb. 29, 2008, pp. 193-216.

[30] "Short-Term Energy Outlook," U.S. Energy Information Administration, Sept. 9, 2008.

[31] Mary Clare Jalonick, "USDA Urges Farmers to Keep Setting Aside Land," The Associated Press, July 29, 2008.

[32] "DU Says CRP Losses Astounding," Ducks Unlimited, Jan. 4, 2008, www.ducks.org/news/1456/DUsaysCRPlossesastou.html.

[33] National Research Council, *op. cit.*, pp. 113-121.

[34] U.S. Environmental Protection Agency, "Mitigation Banking Factsheet," updated April 18, 2008, www.epa.gov/owow/wetlands/facts/fact16.html.

[35] James Salzman and J. B. Ruhl, " 'No Net-Loss' — Instrument Choice in Wetlands Protection,"

About the Author

Jennifer Weeks is a *CQ Researcher* contributing writer in Watertown, Mass., who specializes in energy and environmental issues. She has written for *The Washington Post*, *The Boston Globe Magazine* and other publications, and has 15 years' experience as a public-policy analyst, lobbyist and congressional staffer. She has an A.B. degree from Williams College and master's degrees from the University of North Carolina and Harvard.

Duke Law School, Science, Technology and Innovation Research Paper Series, Sept. 2005, p. 9; Jessica Wilkinson and Jared Thompson, "2005 Status Report on Compensatory Mitigation in the United States," Environmental Law Institute, 2005, p. 28.

[36] National Research Council, *op. cit.*, 2001, pp. 82-92.

[37] "Iowa Wetland Mitigation Bank, Inc.," www.ifbf.org/newsissues/environment/Brochure1106.pdf.

[38] "Status and Trends of Wetlands in the Coterminous United States 1998 to 2004," U.S. Fish and Wildlife Service, 2005, pp. 44, 74-76.

[39] "A Century of Wetland Exploitation," U.S. Geological Survey, Northern Prairie Wildlife Research Center, Aug. 3, 2006.

[40] Laura Ingalls Wilder, *By the Shores of Silver Lake* (1971), p. 77.

[41] Hosansky, *op. cit.*

[42] Alison Carney Brown, "Miles of Tiles," *Chicago Wilderness Magazine*, Spring 2004.

[43] William Reffalt, untitled background article for U.S. Fish & Wildlife Service, www.fws.gov/refuges/centennial/pdf2/pelicanIsland_reffalt.pdf.

[44] Marjorie Stoneman Douglas, *The Everglades: River of Grass, 60th anniversary edition* (2007), p. 349.

[45] www.fws.gov/duckstamps/Info/Stamps/stampinfo.htm.

[46] John Hart, *San Francisco Bay: Portrait of an Estuary* (2003), pp. 33-36.

[47] Executive Order 11990, Protection of Wetlands, May 24, 1977, www.epa.gov/owow/wetlands/regs/eo11990.html.

[48] Joseph Alper, "War Over the Wetlands: Ecologists v. the White House," *Science*, Aug. 21, 1992; John H. Cushman Jr., "Quayle, in Last Push for Landowners, Seeks to Relax Wetland Protections," *The New York Times*, Nov. 12, 1992, p. A16.

[49] For background see Kenneth Jost, "Property Rights," *CQ Researcher*, June 16, 1995, pp. 513-536. For background on the "takings" ruling, Kenneth Jost, "Hawaii Law Tests 'Regulatory Takings' Doctrine," in "Property Rights," *CQ Researcher*, March 4, 2005, pp. 197-220.

[50] *Florida Rock Industries v. United States*, 791 F. 2d 893 (1986), 45 Fed. Cl. 21 (1999); *Loveladies Harbor Inc. v. United States*, 21 Cl.Ct. 153 (1990), 28 F. 3d 1171 (1994).

[51] *Lucas v. South Carolina Coastal Council*, 505 US 193 (1992).

[52] *Solid Waste Agency of Northern Cook County v. United States Army Corps of Engineers, et al.*, 531 U.S. 159 (2001).

[53] Julie M. Sibbing, "Nowhere Near No-Net-Loss," National Wildlife Federation, www.nwf.org/wildlife/pdfs/NowhereNearNoNetLoss.pdf; James Salzman and J.B. Ruhl, "'No Net-Loss' — Instrument of Choice in Wetlands Protection," Duke Law School Science, Technology and Innovation Research Paper Series, Sept. 1, 2005.

[54] "Internal EPA Document Shows 500 Enforcement Cases Adversely Affected," U.S. House of Representatives, Committee on Oversight and Government Reform, July 7, 2008, http://oversight.house.gov/story.asp?ID=2065.

[55] "Coastal Wetlands: Lessons Learned From Past Efforts in Louisiana Could Help Guide Future Restoration and Protection," U.S. Government Accountability Office, GAO-08-130, Dec. 2007.

[56] Bob Marshall, "Last Chance: The Fight to Save a Disappearing Coast," *New Orleans Times-Picayune*, March 4, 2007, p. 1.

[57] Jeffrey Meitrodt and Aaron Kuriloff, "Oyster Farmers Initially Backed Project," *New Orleans Times-Picayune*, May 4, 2003, p. 21; Bob Marshall and Mark Schliefstein, "Los-

ing Ground," *New Orleans Times-Picayune*, March 5, 2007, p. 1.

[58] "Drawing Louisiana's New Map: Addressing Land Loss in Coastal Louisiana," National Research Council, 2006, pp. 3-12.

[59] Bruce Alpert, "Congress Overrides Bush Water Bill Veto," *New Orleans Times-Picayune*, Nov. 8, 2007, p. 2.

[60] Jonathan Tilove, "Officials Tell Senate Farmers Need Aid," *New Orleans Times-Picayune*, Sept. 25, 2008, p. 11.

[61] Testimony before the Senate Agriculture Committee, Sept. 24, 2008, p. 6.

[62] For background see "Wetlands and Global Climate Change," Association of State Wetland Managers, www.aswm.org/science/climate_change/climate_change.htm.

[63] "Jindal Calls for Action on Coastal Restoration," WWLTV.com, Sept. 16, 2008.

[64] "Take Long View in Assessing Flood Control," *The State Journal-Register*, July 23, 2008, p. 4.

[65] Dianne Feinstein, "Bay Restoration at an Exciting Point," *San Jose Mercury News*, Dec. 24, 2007.

FOR MORE INFORMATION

Associated General Contractors of America, 2300 Wilson Blvd., Suite 400, Arlington, VA 22201; (703) 548-3118; www.agc.org. The main trade association for the U.S. commercial construction industry, including highway and municipal projects.

Association of State Wetland Managers, 2 Basin Road, Windham, ME 04062; (207) 892-3399; www.aswm.org. Works to improve wetland regulation and management; open to anyone involved with wetland resources.

Audubon of Florida, 444 Brickell Ave., Suite 850, Miami, FL 33131; (305) 371-6399; www.audubonofflorida.org. State chapter of the National Audubon Society that works to protect and restore ecosystems statewide; operates Corkscrew Swamp Wildlife Sanctuary in Naples, Fla.

Ducks Unlimited, One Waterfowl Way, Memphis, TN 38120; (901) 758-3825; www.ducks.org. A conservation group founded by sportsmen in 1937, during the Dust Bowl, to protect wetlands and other places where waterfowl breed or migrate.

Iowa Farm Bureau Federation, 5400 University Ave., West Des Moines, IA 50266; (515) 225-5400; www.iowafarmbureau.com. Advocates for farmers, farm families and Iowa's rural heritage.

Mackinac Center for Public Policy, 140 West Main St., P.O. Box 568, Midland, MI 48640; (989) 631-0900; www.mackinac.org. A think tank advocating market-oriented solutions to public policy problems in Michigan.

People for Puget Sound, 911 Western Ave., Suite 580, Seattle, WA 58104; (206) 382-7007; www.pugetsound.org. A nonprofit working to protect and restore lands and waters around Puget Sound.

U.S. Army Corps of Engineers, 441 G St., N.W., Washington, DC 20314; (202) 761-0010; www.hq.usace.army.mil. World's largest public engineering, design and construction management agency builds and operates water projects nationwide.

Bibliography

Selected Sources

Books

Ernst, Howard R., *Chesapeake Bay Blues: Science, Politics, and the Struggle to Save the Bay*, Rowman & Littlefield, 2003.
A professor of political science at the U.S. Naval Academy shows how development and pollution devastated the nation's largest estuary and how politics slowed clean-up efforts.

Grunwald, Michael, *The Swamp: The Everglades, Florida, and the Politics of Paradise*, Simon & Schuster, 2006.
A *Time* correspondent, who has written extensively about the U.S. Army Corps of Engineers and water development recounts the environmental history of the Everglades and dissects Floridians' relationship with their environment.

Save San Francisco Bay, *Protecting Local Wetlands: A Toolbox For Your Community*, Save the Bay, 2000.
Although it focuses on California, this handbook provides a thorough overview of federal and state controls, approaches to local wetland regulation and ways tax incentives and other measures can help foster healthy wetlands nationwide.

Articles

"Last Chance: The Fight to Save a Disappearing Coast," *New Orleans Times-Picayune* and Nola.com, March 4-6, 2007, www.nola.com/speced/lastchance/.
A three-part special series warns that many scientists believe the Gulf of Mexico could reach New Orleans' suburbs within a decade unless drastic action is taken to preserve Louisiana's remaining coastal marshes.

Barringer, Felicity, "Death Looms for a Flood-Control Project," *The New York Times*, April 9, 2008, p. A14.
In an unusual exercise of its veto power, the Environmental Protection Agency (EPA) is preparing to veto the Yazoo Pumps, a controversial flood-control project in southern Mississippi that would destroy or damage at least 67,000 acres of wetlands.

Cave, Damien, "Harsh Review of Restoration in Everglades," *The New York Times*, Sept. 30, 2008, p. A18.
A National Research Council report said efforts to rescue the Everglades have failed because of bureaucratic delays, funding shortages and overdevelopment.

Dahl, Thomas, "Beyond No Net Loss: Imagery Aids Wetland Conservation," *Geoworld*, September 2005.
The Fish and Wildlife Service is using high-resolution satellite images and computerized mapping to keep better track of wetland status and trends.

Kay, Jane, "50-Year Plan for Turning South Bay Salt Ponds to Tidal Wetlands," *San Francisco Chronicle*, Dec. 12, 2007, p. A1.
In the largest wetlands restoration on the West Coast, state and federal agencies are launching a 50-year effort to turn salt ponds around San Francisco Bay back into healthy tidal marshes.

Pittman, Craig, and Matthew Waite, "Vanishing Wetlands," *The St. Petersburg Times*, 2005-2006, www.sptimes.com/2006/webspecials06/wetlands/.
An award-winning multi-part investigation finds that faulty permitting and lax oversight are major causes of ongoing wetland losses in Florida.

Tibetts, John H., "Rising Tide: Will Climate Change Drown Coastal Wetlands?" *Coastal Heritage*, winter 2007.
Rising sea levels are already forcing some salt marshes to migrate inland, exposing communities to increased flooding, and climate change is likely to accelerate these effects.

Reports and Studies

"Biodiversity Values of Geographically Isolated Wetlands in the United States," *Natureserve*, Dec. 1, 2005, www.natureserve.org/publications/isolatedwetlands.jsp.
An EPA-funded study by a private conservation organization finds that geographically isolated U.S. wetlands support 86 species of endangered or threatened animals and plants, half of which need isolated wetland habitat to survive.

"Drawing Louisiana's New Map: Addressing Land Loss in Coastal Louisiana," National Research Council, 2006.
A peer-reviewed study of coastal restoration efforts proposed by the Corps of Engineers and the state of Louisiana finds that a broader and more integrated plan is needed.

"Prairie Pothole Region: At the Current Pace of Acquisitions, the U.S. Fish and Wildlife Service Is Unlikely to Achieve Its Habitat Protection Goals for Migratory Birds," U.S. Government Accountability Office, 2007.
Since 1959, the agency has protected about 3 million acres in the vast region in the northern Great Plains, but it needs to move faster to protect all of the wetlands and grasslands that it has identified as important habitat for migratory birds.

Dahl, T.E., *Status and Trends of Wetlands in the Coterminous United States 1998 to 2004*, U.S. Department of the Interior, Fish and Wildlife Service, 2005.
The most recent federal wetlands inventory finds that the U.S. is now achieving small, annual, net gains in wetlands but that most of the increase is in freshwater ponds.

The Next Step:

Additional Articles from Current Periodicals

Clean Water Act

Boyle, Katherine, "Group Using Poll to Make Point on Clean Water Act Protections,' *Environment and Energy Daily*, **Jan. 18, 2008.**

An environmental law firm is lobbying for legislation to ensure that the Clean Water Act applies to all bodies of water in the United States.

Caputo, Anton, "Report Spotlights the Growing Risks to Texas' Wetlands," *San Antonio Express-News*, **Feb. 15, 2008, p. 3B.**

A series of Supreme Court decisions has stripped 3.5 million acres of wetlands in Texas from protection under the Clean Water Act.

Descherer, Chris, "Water System Up the Creek?" *Atlanta Journal-Constitution*, **Dec. 18, 2006, p. 19A.**

The Supreme Court believes that if a stream does not receive water continually throughout the year then it isn't protected under the Clean Water Act.

Everglades

Gibson, William E., "House Overrides Veto of Everglades Funds," *Sun-Sentinel* **(Florida), Nov. 7, 2007, p. 1A.**

The House of Representatives has voted to override President Bush's veto of a water bill that would allow spending on the restoration of the Everglades.

Grosskruger, Paul L., "Saving the Everglades One Small Project at a Time," *The Miami Herald*, **Dec. 8, 2007, p. A27.**

The passage of the Water Resources Development Act gives congressional authorization for Everglades projects that have long lacked public support.

Pittman, Craig, "Restoration of Glades Falls Years Behind," *St. Petersburg Times*, **July 3, 2007, p. 1A.**

Crucial elements of the Everglades restoration project that began in 2000 are already six years behind schedule and the cost is approaching $20 billion.

Mitigation

Keller, Mike, "Banking on Wetlands," *Biloxi Sun Herald* **(Mississippi), Nov. 5, 2006, p. G1.**

Mitigation is opening up a new industry of restoring wetlands in rural areas of Mississippi.

Pittman, Craig, and Matthew Waite, "How Billions Are Made 'Restoring' Florida's Wetlands," *St. Petersburg Times*, **Dec. 17, 2006, p. 1A.**

Congress has steered billions of dollars into mitigation bankers' pockets and has created new rules that could possibly double the size of the industry.

Salinero, Mike, "Wetlands Replacement Plan Muddied By Critics' Unease," *Tampa Tribune*, **July 22, 2007, p. 1.**

A 160-acre wetlands site in Florida is being handled by a mitigation bank that would allow developers who want to destroy wetlands to compensate for the damage by buying credits in the form of wetland acreage.

Wallgren, Christine, "In Wetlands, Builders Can Pay to Play," *Boston Globe*, **Dec. 21, 2006, p. Reg1.**

Critics of wetlands mitigation say the process results in a loss of local control over environmental development, possibly shifting it to state governments.

Wildlife

Marshall, Bob, "Wetlands Wildlife at Risk," *Times-Picayune* **(Louisiana), Aug. 5, 2007, p. 99.**

Wetlands are key components of waterfowl nesting habitat, but the developers' lobby in Washington hasn't assigned any value to such fish and wildlife.

Perkins, Iveory, "$1.5 Million to Help Restore Detroit's Wildlife, Coast," *Detroit News*, **Oct. 17, 2006, p. 2B.**

Coastal wetlands along the Detroit River International Wildlife Refuge are undergoing a $1.5 million restoration project that will clean wetlands and protect the area's habitat.

Presecky, William, "Wetland Proves It's For the Birds," *Chicago Tribune*, **Dec. 14, 2007, p. B9.**

The migration of birds to a man-made wetland in Illinois shows the value of recreating and restoring wetlands, according to local forest officials.

CITING CQ RESEARCHER

Sample formats for citing these reports in a bibliography include the ones listed below. Preferred styles and formats vary, so please check with your instructor or professor.

MLA STYLE

Jost, Kenneth. "Rethinking the Death Penalty." CQ Researcher 16 Nov. 2001: 945-68.

APA STYLE

Jost, K. (2001, November 16). Rethinking the death penalty. *CQ Researcher, 11,* 945-968.

CHICAGO STYLE

Jost, Kenneth. "Rethinking the Death Penalty." CQ Researcher, November 16, 2001, 945-968.

3

OCEANS IN CRISIS

BY COLIN WOODARD

Excerpted from the CQ Global Researcher. Colin Woodard. (October 2007). "Oceans in Crisis." *CQ Global Researcher*, 237-264.

Oceans in Crisis

BY COLIN WOODARD

THE ISSUES

Sall Samba has spent much of his adult life fishing for octopus from his home in Nouadhibou, Mauritania, on Africa's Atlantic coast. Fishing from a wooden canoe, he could bring home 160 pounds on a five-day trip — earning $600 a month in a country where the average wage is only $200. In 2004, he built a home and bought new canoes; times were good.

Not anymore. "You used to be able to catch fish right in the port," the 39-year-old told *The Wall Street Journal* recently. "Now the only thing you can catch is water." [1]

Today Samba and other fishermen must compete with huge industrial trawlers from Russia, China and Spain. But while Samba pulls his catch out of the sea by hand in plastic traps, a single Spanish vessel dragging a massive nylon net catches 260,000 pounds of octopus on a typical 45-day fishing trip.

Some 340 big foreign vessels fish Mauritanian waters because the government recently sold fishing rights to Asian and European nations that have overfished their own territorial waters. Stocks of octopus, which account for half of Mauritania's fish exports, are declining, and Samba has seen his monthly income fall by two-thirds.

Samba's experience is rapidly becoming universal in the world's coastal regions. According to the U.N. Food and Agriculture Organization (FAO), a quarter of the world's commercial fish stocks have been overexploited or depleted, and about half are fully exploited — meaning fishermen are taking as much as can be reliably replenished by the ecosystem. [2] (*See graphic, p. 241.*)

A Russian trawler hauls in a netful of red fish on the Grand Banks in the northwest Atlantic Ocean. The world's oceans have lost more than 90 percent of large predatory fish — such as tuna, swordfish and grouper — over the past half-century, prompting fishermen to hunt smaller species. Scientists and environmentalists blame the loss of ocean biodiversity on overfishing, pollution and climate change.

Greenpeace/©Robert Visser

Moreover, 90 percent of the world's large, predatory fish have been harvested since 1950, prompting fishermen to progressively move to smaller, less valuable species further down the food chain. [3] The shift has triggered the rapid depletion of marine species previously considered unmarketable — such as dogfish, urchins and basking sharks — which in turn has reduced the food available to the surviving stocks of larger species. Small, lower-valued schooling fish like anchovies now dominate world fishery landings.

"We're eating bait and moving on to jellyfish and plankton," says Daniel Pauly, director of the Fisheries Centre at the University of British Columbia, who predicts future generations will associate seafood not with tuna or cod but with simple, gelatinous creatures. "My kids will tell their children: 'Eat your jellyfish.' " [4]

The decimation of global fisheries is blamed largely on powerful, new technologies that allow fishermen to capture fish faster than the ocean can produce them. Radar, fish finders, satellite tracking and navigation systems, onboard processing plants and flash freezers are put aboard ever faster vessels capable of fishing far from shore for long periods.

In addition, most fishing gear is indiscriminate: The vast nets used by trawlers typically kill huge quantities of unmarketable marine life. Each year 7 million metric tons of seabirds, juvenile fish, sea turtles, dolphins, sharks, crabs, starfish, anemones, sponges and other creatures are caught, killed and discarded by mechanized fishing. On average, this "bycatch" accounts for 8 percent of fishermen's catches; but among shrimp fishermen in the tropics, bycatch represents 56 percent of the haul. [5]

Trawl nets and gear dragged along the sea bottom are said to cause lasting damage to the seafloor habitat and, thus, to the ability of marine ecosystems to sustain themselves. The heavy nets plow away the bottom plants, sponges and corals that animals use for cover, while killing large numbers of the invertebrates they feed on.

In the Gulf of Maine, for instance, the average seafloor section is trawled once a year; on the Georges Bank off Massachusetts, it's plowed three to four times a year. The trawls also create muddy clouds thought to reduce the survival of small fish by clogging their gills. [6] Elliott Norse, president of the Marine Conservation Biology Institute in Bellevue, Wash., calls sea bottom trawling "clear cutting the seafloor." Trawling companies contend there's no proof their activities dam-

Caribbean Corals Are Disappearing

Only 10 percent of the coral reefs in the Caribbean are alive, compared with more than 50 percent three decades ago. Experts say corals are dying due to global warming, pollution, sedimentation and over-harvesting of fish and other reef resources — sometimes using dynamite or poison.

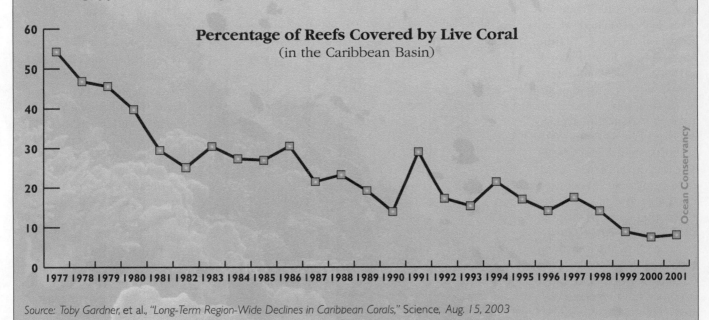

Percentage of Reefs Covered by Live Coral
(in the Caribbean Basin)

Source: Toby Gardner, et al., "Long-Term Region-Wide Declines in Caribbean Corals," Science, Aug. 15, 2003

age the ocean floor and that trawling actually may benefit seafloor species.

It's not just fish that are in crisis, however. Coral reefs, the foundation of most tropical marine life, are declining at an alarming rate. The latest international assessment found that one-fifth of the world's coral reefs "have been effectively destroyed and show no immediate prospects of recovery," while another 24 percent are "under imminent risk of collapse." Live coral cover on Caribbean reefs has declined by 80 percent over the past 30 years. [7] (*See graph above.*)

Without corals, tropical oceans would become biological wastelands, because they don't support the growth of phytoplankton, the microscopic plants that form the base of the marine food chain. Reefs are colonies of coral polyps — anemone-like organisms that build limestone shells around themselves. They filter food particles from the water and capture the sun's energy through photosynthetic micro-organisms inside their

tissue. Corals support the profusion of fish associated with tropical reefs. [8]

Reefs are being damaged in a variety of ways. Clearing coastal mangroves for development dooms reef creatures that feed there and triggers erosion that smothers the coral polyps under plumes of sand and soil. Overfishing results in the harvesting of increasing numbers of ever-smaller fish, lobsters and conch. Fishermen in the Philippines, Micronesia, Jamaica and Indonesia use dynamite and other explosives to stun and kill marine life over a wide area — a one-time bonanza that destroys the reef. Sewage and fertilizer run-off from towns, resorts, fish farms and golf courses trigger the growth of seaweed, kelp and other plants that can smother and eventually kill the reefs. Even far from human activity, reefs are dying from disease and overly warm water temperatures linked to climate change. [9]

Colder waters are affected, too. In the High Arctic, Inuit mothers' breast

milk is dangerous to their babies' health because the polar bears, seals, walruses, fish and whales they eat are contaminated by heavy metals, PCBs and other industrial compounds now found in seawater and stored in the animals' fat. Many Inuit have concentrations of certain pesticides in their bodies that exceed safe levels 20-fold. Beached whales often must be treated as hazardous waste because of the high concentrations toxic substances in their bodies. [10] Both wild and farm-raised salmon have also been shown to have potentially significant contaminant levels. [11]

Fertilizers, sewage and other nutrient pollution have triggered massive algal blooms that can strip the ocean of dissolved oxygen, dooming animals that cannot escape the area. Such oxygen-starved "dead zones" have spread from harbors and river mouths to suffocate entire seas. (*See sidebar, p. 244.*) Each summer, fertilizer runoff from 31 states and parts of Canada flows into the Mississippi River and then to the Gulf of Mexico, creating a New Jersey-size dead zone south of New Orleans where few species can survive. [12]

Non-native, or "invasive," species also can damage marine ecosystems. [13] The species are carried around the globe in the ballast tanks of ocean-going vessels, which pump water in and out of the tanks to maintain seaworthiness. This ballast can contain the eggs, larvae or adult forms of hundreds of species, some of which become established in waters that contains no natural predators.

"Once an exotic species is established, trying to remove it is like trying to put the toothpaste back in the tube," says James T. Carlton, professor of marine sciences at Williams College in Massachusetts. In the early 1990s, a comb jelly snuffed out much of the life in the Black Sea (see p. 244), while a mutant form of a tropical seaweed, *Caulerpa taxifolia*, has smothered vast stretches of the Mediterranean shore since it was accidentally released into the sea by a Monaco aquarium. [14]

Some scientists worry that in many marine ecosystems the more advanced organisms are disappearing while the populations of the most primitive ecosystems are exploding. "Dead zones aren't dead; they are just full of jellyfish and bacteria," notes Jeremy B. C. Jackson, director of the Geosciences Research Division at the Scripps Institution of Oceanography in San Diego, who calls the process "the rise of slime."

In Sweden, summer blooms of *cyanobacteria* turn the surface of the Baltic Sea into a yellow-brown slurry that kills fish, burns people's eyes and makes breathing difficult. Hawaiian condo owners have had to use tractors to remove piles of algae piling up on their beaches, while toxic algal blooms are believed responsible for mass die-offs of sea lions, whales, manatees and dolphins. Red tides — algal blooms that make shellfish poisonous to humans — are 10 times more common than they were 50 years ago,

Most Fish Stocks Are Overexploited

Three-quarters of the world's fisheries were either fully exploited — at or near their maximum sustainable limits — overexploited or depleted in 2005. Fisheries biologists say the stocks cannot recover quickly and are in danger of further decline.

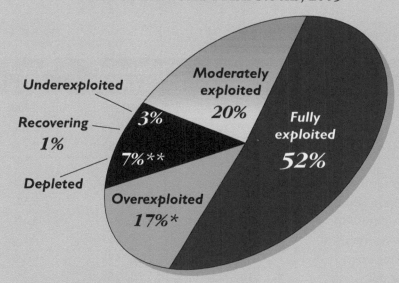

Status of the World's Fish Stocks, 2005

Underexploited 3%
Recovering 1%
Depleted 7%**
Overexploited 17%*
Moderately exploited 20%
Fully exploited 52%

* Exploited beyond the ability of the system to sustain itself over the long term.
** Current catches fall far below historic levels.

China and Peru Catch the Most

China and Peru haul in nearly 27 million tons of fish a year — almost as much as the next eight countries combined.

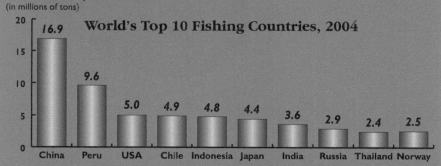

Amount of fish captured*
(in millions of tons)

World's Top 10 Fishing Countries, 2004

China 16.9
Peru 9.6
USA 5.0
Chile 4.9
Indonesia 4.8
Japan 4.4
India 3.6
Russia 2.9
Thailand 2.4
Norway 2.5

* Fish caught in the wild, excluding those grown by aquaculture.

Source: "The State of the World Fisheries and Aquaculture 2006," U.N. Food and Agriculture Organization, www.fao.org/docrep/009/A0699e/A0699E04.htm

Beach Litter Can Be Lethal

Nearly 7.7 million pieces of beach litter were collected in 2006 by some 350,000 Ocean Conservancy coastal cleanup volunteers around the world. About two-thirds of the items were food containers and plastic bags; the rest were smoking related. Experts say 1 million seabirds and 100,000 marine mammals and sea turtles die each year after ingesting or becoming entangled in ocean debris.

Top 10 Ocean Debris Items Worldwide

Debris Items	Number of Items	Percent of Total
Cigarette debris	1,901,519	24.7%
Food wrappers, containers	768,115	10.0%
Caps/lids	704,085	9.1%
Bags	691,048	9.0%
Beverage bottles (Plastic) 2 liters or less	570,299	7.4%
Beverage bottles (Glass)	420,800	5.5%
Cups/plates/forks/knives/spoons	353,217	4.6%
Straw/stirrers	349,653	4.5%
Beverage cans	327,494	4.3%
Cigar tips	186,258	2.4%

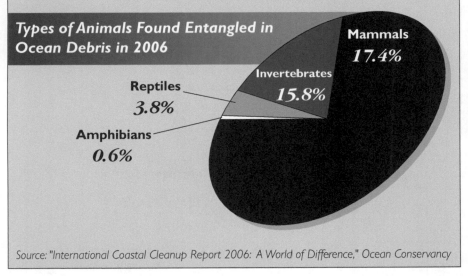

Types of Animals Found Entangled in Ocean Debris in 2006

Mammals 17.4%

Invertebrates 15.8%

Reptiles 3.8%

Amphibians 0.6%

Source: "International Coastal Cleanup Report 2006: A World of Difference," Ocean Conservancy

owing in part to increases in sewage and fertilizer run-off. "We're pushing the oceans back to the dawn of evolution, a half-billion years ago when the oceans were ruled by jellyfish and bacteria," says Pauly at the University of British Columbia. [15]

Experts argue that adopting ecosystem-based approaches to regulating human activity on the seas would help ensure the system as a whole is healthy, rather than just focusing on a particular species. Some fishing interests resist such an approach

— which would involve creation of marine reserves and other protected areas — but its greatest opponent is public and political apathy.

As scientists and governments try to determine how best to protect the world's oceans, here are some of the questions being debated:

Are humans destroying the oceans?

Yes, according to numerous recent scientific studies including a June 2007 assessment of Europe's seas by 100 scientists from 15 countries.

"In every sea, we found serious damage related to the accelerated pace of coastal development, the way we transport our goods and the way we produce our food on land as well as the sea," said Laurence Mee, director of the Marine Institute at the University of Plymouth (in England), who coordinated the project. "Without a concerted effort to integrate protection of the sea into Europe's development plans, its biodiversity and resources will be lost." [16]

A four-year analysis released in November 2006 by an international group of ecologists and economists concluded that if current trends continue, every seafood species currently fished will be commercially extinct by 2050. The study found that every species lost increases the speed at which the larger ecosystem unravels.

"Whether we looked at tide pools or studies over the entire world's oceans, we saw the same picture emerging," said the study's lead author, Boris Worm, assistant professor of biology at Dalhousie University in Halifax, Nova Scotia. "I was shocked and disturbed by how consistent these trends are — beyond anything we expected." [17]

Likewise, two independent, bipartisan U.S. commissions — the Pew Oceans Commission and the U.S. Commission on Ocean Policy (USCOP) — concluded in 2003 and 2004, respectively, that pollution, habitat destruction and overfishing are endangering the world's oceans. [18]

"There is overwhelming scientific evidence that our ocean ecosystems are in serious trouble, serious enough that it really is endangering the future of ocean life itself," says Leon Panetta, former chief of staff in the Clinton White House, who chaired the Pew Commission. "The biggest challenge is to get people to pay attention, because if they do, then we can make our case."

"What is the state of our oceans? Unfortunately we have to report to you that the state is not good, and it is getting worse," Admiral James D. Watkins, chair of USCOP told Congress. Furthermore, the harm humans are inflicting on the oceans, the USCOP report concluded, has "serious consequences for the entire planet." [19]

Marine scientists have been aware of the situation for more than a decade. In 1998 — the U.N. International Year of the Ocean — more than 1,600 marine scientists and conservation biologists from 65 nations issued a joint warning that the seas were in peril and that immediate action was needed to prevent further damage.

"Getting scientists to agree on anything is like herding cats, so having 1,600 experts voice their concerns publicly highlights just how seriously the sea is threatened," said Norse, of the Marine Conservation Biology Institute, who organized the effort. "We must change what we're doing now to prevent further irreversible decline." [20]

However, some researchers and fishing industry groups deny there is a problem, claiming the situation is exaggerated by environmentalists to further fundraising opportunities. "Are we running out of fish? No," said Dan Furlong, executive director of the U.S. Mid-Atlantic Fishery Management Council. Furlong cites U.S. National Marine Fisheries Service assessments showing that of the 230 stocks the agency manages, only 44 are known to be overfished, 136 "are not subject to overfishing," while the status of the remaining 50 are unknown. "In other

Ocean Conservancy

Plastic bottles, food containers and grocery bags make up a large portion of the refuse that ends up in the ocean and washes ashore. More than 100,000 marine mammals alone are killed each year by either ingesting or becoming entangled in debris.

words, the glass is more than half full for those stocks," he says. The public, he says, has been duped by environmentalists who pushed Congress to require that stocks be rebuilt. As a result, he says, "despite significant improvements across a broad range of fisheries, we are cast in the role of doing poorly because we will likely fail to meet . . . the arbitrary, capricious deadline to maximize stocks all at the same time." [21]

Bjorn Lomborg, associate professor of statistics at the University of Aarhus in Denmark, argues in his controversial book *The Skeptical Environmen-*

talist that while there are problems, the oceans are doing fine.

"The oceans are so incredibly big that our impact on them has been astoundingly insignificant," he argues, citing U.N. data suggesting that in the open oceans, far from land, the U.N. has found the seas to be relatively clean. He acknowledges that fertilizer is creating dead zones in places like the Gulf of Mexico and the Black Sea but says the disruptions are worth it when compared to the improved crop yields. [22]

"Our oceans have not been defiled . . . and although the nutrient influx has increased in many coastal waters like the Gulf of Mexico," he continues. "This does not constitute a major problem — in fact, the benefits generally outweigh the costs." [23]

Critics accuse Lomborg of cherry-picking facts that support his arguments and ignoring evidence to the contrary. For instance, Lomborg's book fails to address the crisis in the fisheries, the decline of the coral reefs, the problems caused by alien species and other issues.

And even some fishermen don't share Lomborg's view. "The combination of modern electronics with large fishing ves-

sels has created a technology too powerful for fish stocks to withstand," said Ted Ames, a fisherman from Stonington, Maine, who won a McArthur Genius Grant for his research into the decline of Gulf of Maine fish stocks. [24]

Is ecosystem-based management the solution?

The destruction of life in the oceans presents humans with perhaps the greatest marine policy challenge in history: figuring out how to manage human activities so they don't damage marine biodiversity, critical habitat and overall ecosystem function. Known as "ecosys-

tem-based management," the approach has wide support, including both USCOP and the Pew Commission.

"You're not going to have any fish to catch — or healthy fishing communities — unless there is a healthy marine ecosystem to provide the fish," says Jane Lubchenco, a professor of marine biology at Oregon State University, a former Pew Commissioner and past president of the American Association for the Advancement of Science. "We need mechanisms to better understand how ocean ecosystems work and how we're changing them if we are going to do a better job managing them."

The Black Sea's Cautionary Tale

Ecosystem collapse shows signs of recovery.

From ancient times, humans have been drawn to the Black Sea, a kidney-shaped basin the size of California nestled between Eastern Europe and Asia Minor. Its anchovy and sturgeon stocks sustained Ancient Greece, medieval Byzantium, the Ottoman Empire and Imperial Russia.

In the 20th century, millions of tourists flocked each summer to its beaches in Turkey and on the "Communist Riviera," which stretched from Bulgaria and Romania to Soviet Russia. They swam, feasted on fish and basked in the sunshine, recuperating from winter months in the factories of Budapest and Birmingham. [1]

Then, with astonishing suddenness, the ecosystem collapsed in the early 1990s due to a combination of fertilizer and sewage pollution runoff, destruction of wetlands and the introduction of an aggressive, non-native, jellyfish-like species. Given the plethora of critical ocean ecosystems now in jeopardy, the Black Sea collapse provides a cautionary tale about the fragility of marine ecosystems, say marine scientists.

"The Black Sea is a microcosm of the environmental problems of the planet," warns Janet Lubchenco, a professor of marine biology at Oregon State University. "Solutions to the Black Sea crisis may enlighten, inform and inspire our global challenges." [2]

Few saw it coming. The Black Sea had been subject to pollution for decades: Industrial wastes, oil spills and radiation from the 1986 Chernobyl nuclear accident had been carried to the sea by its tributaries — apparently without dramatic effect. But it was the buildup of raw sewage and fertilizer runoff — coupled with the accidental introduction of an alien, plankton-devouring species, the comb jelly — that triggered the near-death of the sea.

The sea's largest tributary, the Danube River, drains half the European continent during its 2,000-mile journey from the

Black Forest of Switzerland to Romania's Black Sea delta. The last half of that journey winds through Eastern Europe, where for decades every village, town and city flushed its untreated sewage into the river. Starting in the late 1960s, state-owned farms used huge quantities of subsidized chemical fertilizers on their fields, and much of it ran off into the streams feeding the Danube. Hydroelectric projects and navigational canals also damaged or bypassed wetlands that once acted as the river's natural filtering system. And Romania's dictator, Nicolae Ceausescu, waged all-out war on the Danube delta — Europe's greatest wetland — in an ill-conceived attempt to convert it to rice production. [3]

As a result, concentrations of nitrogen and phosphorous nutrients in the Black Sea's ecologically critical northwestern shelf dramatically increased between 1960 and 1980. [4] The nutrients fueled enormous algae blooms in the late 1980s, smothering bottom life by using up the oxygen it needed. As the microscopic plants decomposed, they consumed still more oxygen in vast stretches of the sea, suffocating most other creatures. [5]

Then in 1982, *Mnemiopsis leidyi*, an inch-long comb jelly native to North America, was introduced to the sea in the ballast water of a passing ship. The creature established itself amid the gathering chaos and proceeded to graze the waters clean of survivors. With no natural predators, it ultimately achieved a biomass of 1 billion tons — 10 times the weight of all the fish caught by all the world's fishermen in a year. [6]

"The biomass of other zooplankton dropped sharply, and the catches of commercial fish sharply decreased," noted Yuvenaly P. Zaitsev chief scientist at the Odessa office of the Ukrainian National Academy of Science. "*Mnemiopsis . . .* is usually held responsible for much of what happened." [7]

By the early 1990s, total fish landings had fallen to one-seventh of their previous level, and the signature anchovy catch fell by 95 percent. Slicks of ugly, stinking slime drove tourists from the beaches and prompted long closures at the height of summer. Hundreds of bathers became ill and several died from cholera and other infectious diseases that thrived in the algae-choked environment. In 1999 the World Bank estimated the economic damage to the fisheries sector at $300 million a year and $400 million to tourism. [8]

The past five years have seen considerable progress, however, as the European Union — with its strict environmental regulations — expanded to include 10 former communist countries, including Slovakia, Hungary, Romania, Bulgaria and other nations in the Danube's middle and lower basin. [9]

"When these countries joined the EU, they had to adopt new environmental policies and regulations, which has had the benefit of improving the overall water quality situation in the Danube basin," notes Ivan Zavadsky, program director of the Danube/Black Sea Regional Program in Vienna, a joint project of the United Nations and World Bank, which has pumped $70 million into cleanup projects in the region.

New sewage treatment plants have been built in recent years, and many of the most polluting factories and agricultural enterprises collapsed in the early 1990s. As a result, Zavadsky notes, concentrations of phosphorus and nitrogen — the nutrients that ravaged the Black Sea — have dropped 50 percent and 20 percent, respectively, since 1989. Meanwhile, the *Mnemiopsis* population dropped precipitously after the arrival of the Beroe, another invading comb jelly that feeds exclusively on *Mnemiopsis*. Once the Beroe had eaten all the *Mnemiopsis*, its food source was depleted, so Black Sea populations of both species have now been decimated. [10]

"We're witnessing the first signs of a recovery of the Black Sea ecosystem," says Zavadsky, citing reduced algae blooms and an increase in some bottom plants and animals. "But the situation remains on a knife's edge."

But Janos Zlinszky, the government and public affairs manager of the Regional Environmental Center for Central and Eastern Europe in Szentendre, Hungary, is concerned that many of the gains could be lost if the region's economic recovery outpaces its environmental investments. "Romania and Bulgaria have just joined the EU," he says. "If they decide to focus on intensive agriculture rather than the organic market, we could see great increases in fertilizer and pesticide use."

"There's an extraordinary window of opportunity to take action," says Laurence Mee, director of the Marine Institute at the School of Earth, Ocean and Environmental Sciences at the University of Plymouth, in England. "But it can easily be lost."

[1] Colin Woodard, *Ocean's End: Travels Through Endangered Seas* (2000), pp. 1-25.

[2] From a speech in Trabzon, Turkey, Sept. 20, 1997.

[3] Woodard, *op. cit.*, pp. 13-23.

[4] Amhet Kideys, "Fall and Rise of the Black Sea Ecosystem," *Science*, Aug. 30, 2002, p. 1482.

[5] "Pollution and Problems of the Black Sea," a speech by Radu Mihnea, Romanian Research Institute, in Batumi, Georgia, Sept. 21, 1997.

[6] Kideys, *op. cit.*, p. 1482; Woodard, *op. cit.*, p. 22.

[7] "The Black Sea: Status and Challenges," a speech by Yuvenaly Zaitsev in Novorossiysk, Russia, Sept. 23, 1997.

[8] "Black Sea Transboundary Diagnostic Analysis," Global Environment Facility, August 1997, pp. ii, 15, 123, 125; Woodard, *op. cit.*, p. 22; Emilia Battaglini, "The GEF Strategic Partnership for the Danube/Black Sea," Presentation to World Bank, Bucharest, February 2007.

[9] For background, see Brian Beary, "The New Europe," *CQ Global Researcher*, August 2007.

[10] Kideys, *op. cit.*

And many scientists say they don't yet have that understanding.

For the past two years, dozens of scientists in New England and Canada's maritime provinces have been working to develop enough knowledge to undertake ecosystem-based management in the Gulf of Maine by 2010. Scientists working on the Gulf of Maine Census of Marine Life — part of the world's first pilot project for this type of management — are fanning out across the ecosystem examining sea life, ocean currents and the relationship between habitat, predators and prey. A series of ocean buoys is collecting long-term oceanographic data, and other researchers are using sonar technology to map the ocean bottom in unprecedented detail.

"We need to know the big picture of how it happens," says Gerhard Pohle, acting executive director of the Huntsman Marine Science Centre in St. Andrews, New Brunswick, Canada. "If we take one rivet out of the airplane, will it crash? If so, which rivet?" [25]

"We're just beginning to understand how to do the biology on both the super-tiny and the super-large scales," says Lubchenco, who is studying the California Current ecosystem off the U.S. West Coast. "You have to marry oceanography and ecology and genetics and microchemistry in a very interdisciplinary fashion to better understand the processes driving these ecosystems."

But other scientists — many of them government fisheries managers — say sufficient knowledge already exists to start ecosystem-wide management. "Make no mistake, we currently have sufficient scientific information to move forward with an ecosystem-based approach to management," said Andrew Rosenberg, dean of the College of Life Sciences at the University of New Hampshire, a

USCOP commissioner and former deputy director of the U.S. National Marine Fisheries Service. "The nation's ocean policy should recognize these principles and seek to integrate management within regional ecosystems." [26]

Some scientists and environmentalists challenge the notion that humans can or should try to manage ecosystems. In the long term, they say, how could one manage a constantly changing ecosystem, when ideas about what is "healthy" or "desirable" are often based on present conditions or a theoretical, idyllic state. Others point out that an ecosystem means different things to different people, making them difficult to adequately define. [27]

Proponents are careful to point out that ecosystem-based management does not seek to manage the ecosystem — which would be scientific hubris in their opinion — but rather human activity affecting the ecosystem. "One of the reasons ecosystem-based management was pooh-poohed for many years was that there was this naïve assumption [that] you just learn everything you need to know about the ecosystem and then you manage it," says Lew Incze, director of the Aquatic Systems Group at the University of Southern Maine in Portland. "Now we know that we will never know everything — you can even have 100 years of good data, but the ocean is always changing.

"The question is," he continues, "what type of knowledge would allow you to pursue the idea?"

The Gulf of Maine Census of Marine Life is attempting to develop a basic framework by testing ideas about what factors control the number of upper-level predators like Atlantic cod or humpback whales; what indicators would best track the health and diversity of the entire system and how currents, tides and natural oceanographic cycles shape life there. "We're just now developing the tools to come to grips with this," says Pohle.

David Benton, executive director of the Marine Conservation Alliance, a coalition of Alaskan fishing interests, says ecosystem-based management is potentially very good for commercial fishermen.

"It's very clear to most of our membership that it is in their long-term interest to make sure that we have healthy oceans and fish stocks and that all the associated components of the ecosystem around those stocks are in good shape," he says. "A lot of these companies are looking at how these fisheries are going to support their business — not two years from now but a decade from now or longer."

Should ocean-floor trawling be restricted?

Concerns about the damage caused by shellfish dredges and trawl nets have prompted many scientists, environmentalists and governmental agencies to call for a ban on trawling where it is likely to cause lasting harm.

A 2002 study by the National Academy of Sciences' National Research Council recommended that the U.S. government reduce the impact and extent of bottom trawling to reduce its impact on undersea life. Trawls scrape away cold-water corals, sponges, plants, sea anemones, starfish and other creatures, the report found, and repeated passes can cause a 93 percent reduction in these and other bottom-dwelling animals.

"The more we understand about the ecology of fishes, the more we find that for the animals that live right above the seafloor, the integrity of these seafloors is critical to their survival," said Peter Auster, science director of the National Undersea Research Center at the University of Connecticut and a co-author of the report. [28]

Mark Butler, policy director at the Ecology Action Center in Halifax, Nova Scotia, argues there are other fishing methods that don't damage bottom habitats. "We are talking about protecting the ocean floor, which is often the nursery for young fish. You damage that in a major way, then you perhaps start to impair the health of the fishery itself." [29]

Of particular concern are the effects of trawling on seamounts — underwater mountain havens for life in the deep ocean. A 2006 U.N. study found that many seamount fisheries had been quickly depleted and estimated that some 95 percent of the ecological damage found on the undersea mountains was due to bottom-trawling. The study prompted New Zealand, the United States and other countries to push for a worldwide ban on bottom trawling on the high seas — international waters located more than 200 nautical miles from dry land. The measure was defeated when Canada, Iceland, Russia and China refused to back it. [30]

The Madrid-based environmental organization Oceana advocates limiting bottom trawlers and dredgers to areas where they already fish, excluding them from areas containing deep sea coral and sponge habitat, two species that recover poorly from trawling disruptions. Although the U.S. Commission on Ocean Policy didn't address the trawling issue specifically, the Pew Commission said it should be excluded wherever it will reduce biodiversity or alter or destroy "a significant amount of habitat." "Sensitive habitats as well as areas not currently trawled or dredged should be closed to such use immediately," the report said. [31]

Federal fisheries managers in the United States banned bottom trawling from 300,000 square miles of such habitat off the U.S. West Coast. New Zealand, generally regarded as a leader in oceans policy, banned bottom trawling in 2006 in 30 percent of its waters. [32] Rosenberg of the University of New Hampshire contends trawling doesn't need to be universally banned. "You have to manage where trawling occurs and what level of impact we can sustain without reducing the resource productivity." [33]

Meanwhile, the European Union has been criticized for the "cash for access" deals it made with 12 West African states. Under the agreements — which include few, if any, conservation provisions — hundreds of European trawlers are operating in a "fundamentally unsustainable" manner, according to Milan Ilnyckyj, a doctoral student at Oxford University who has studied the problem. [34]

Si'd Ahmed Ould Abeid, president of Mauritania's National Fisheries Federation, said the fisheries agreements with the EU have been "a catastrophe for the fishermen whose catches are down and for the future of the fish in our waters. . . . The fish are just taken from our water, our fishermen lose their lives and we don't gain anything." [35]

Justin Brashares, an assistant professor of environmental science policy at the University of California-Berkeley, argues that the quickest way to increase the production and sustainability of West African domestic fisheries would be to limit the access of the foreign trawlers.

But EU Fisheries Commissioner Joe Borg argues deals such as the one the EU struck with Mauritania would benefit both parties in terms of "jobs, strengthened monitoring and control, conservation of resources in compliance with scientific assessment and environmental protection." [36]

Paul Molyneaux, author of *The Doryman's Reflection*, about his career as a commercial fisherman in Alaska, New Jersey and Maine, says the damage wrought by industrial-scale fishing is so great that it will be years before trawling will be ecologically appropriate in many areas. "In the short term, I think it should be banned," he says. "In many places they have destroyed the ecosystem foundations by dragging, taking too many fish, and disrupting the stock structure. They put everybody else out of business, and then they went out of business themselves."

But trawler owners argue that, far from damaging the environment,

Marine Conservation Biology Institute/Elliot A. Norse

More than 7 million metric tons of unwanted sea life — bycatch — is caught in the huge nets of commercial fishermen and discarded, including seabirds, stingrays, juvenile fish, sea turtles, dolphins, sharks, crabs, starfish, anemones, sponges and other creatures. Bycatch often dies before it gets thrown back into the sea. In the tropics, bycatch represents 56 percent of the haul.

their gear actually improves marine productivity by plowing the seafloor and churning up nutrients. "A lot of fishermen feel that they are freshening the bottom, sort of turning over the soil, tending a garden, and that this helps certain species," says Bonnie Brady, executive director of the Long Island Commercial Fishermen's Association.

James Kendall, executive director of the New Bedford (Mass.) Seafood Coalition, argues nobody knows whether scallop dredges harm or hurt the fishery. "Does this activity oxygenate the sediment, release buried nutrients, damage herring-egg beds or adversely affect juvenile fish and scallops?" he asks, adding that more study is needed to answer those questions. [37]

The Seattle-based factory trawler Northern Eagle can harvest 50-60 metric tons of pollock per day in the Bering Sea. The decimation of global fisheries is blamed largely on powerful new technologies — including radar, fish finders and satellite tracking and navigation systems — as well as onboard processing plants, flash freezers and nets that stretch for miles.

"I am not suggesting that the act of scalloping on the ocean floor does not create an impact upon it, but that it may not be the adverse one so easily assumed," he continues, noting that the best scallop grounds have remained constant for decades, despite intensive dragging.

Some studies from the intensively trawled North Sea support the fishermen's contentions, at least in part. Scientists have observed 35 times as many fish gathered in areas that had recently been trawled compared to adjacent unfished areas, suggesting the fish were attracted to the disturbed bottom to feed. Bottom trawling appears at least partially responsible for increased growth rates in sole and plaice, two North Sea flat fish, presumably because the disturbances promote the growth of small invertebrates they like to eat. A third study suggested that two other North Sea species — gurnards and whiting — were drawn to trawl tracks to feed on tube worms dredged up by the fishing gear. [38]

Furlong, of the Mid-Atlantic Fishery Management Council, contends dragging and dredging can be appropri-ate in sandy, muddy bottoms like those off New Jersey, Delaware and Maryland. "There's nothing to clear cut down there," he said. "You can't clear cut sand." A 1989 bottom trawling assessment of a sandy area by the Massachusetts Division of Marine Fisheries found no damage to bottom-dwelling lobsters and negligible habitat impacts. [39] ∎

BACKGROUND

Run on the Banks

Fishing has been an important activity since prehistoric times. Fishermen using hand nets, hand lines and open boats may have found the seas teeming, but even then — archaeologists have found — early people depleted seafood resources. Ancient garbage heaps show the size of fish and shellfish often became smaller over time. [40]

The advent of industrialized fishing in the 19th century greatly increased human impact on marine ecosystems. Instead of using baited hooks or traps, fishermen developed gear that could pursue and scoop up fish. In the early 1800s, the development of steam-powered ships allowed fishermen to drag larger net bags across the seafloor. At first it was a clumsy proposition: The bag was held open with wooden beams that often hung up on rocks and other obstructions. In the 1890s, however, British fishermen replaced the beam with a pair of small boards, rigged in such a way that when pulled through water they flew apart like kites, holding the mouth of the net open; a heavy chain kept the net bottom dragging on the seafloor to prevent fish from escaping underneath. The so-called otter trawl was incredibly effective and quickly dominated the North Sea fleet. Starting in 1905, it was deployed in North America. [41]

The destructive potential of trawling was clear as early as 1912, when Congress demanded an investigation of fishermen's claims that it "is such an unduly destructive method that if generally adopted . . . the fishing grounds [will be] quickly rendered unproductive." Investigators recommended that trawls be restricted to a few areas in New England, but their advice was not acted upon.

In the 20th century the scale and power of fishing technology increased enormously. Diesel engines, introduced in the 1920s, were cheaper, safer and more reliable than steam; otter trawls were adapted to trap shrimp, clams, oysters and scallops. Processors invented a way to mass-produce fish fil-lets — which could be sold fresh, canned or smoked at processing plants — and commercial-scale flash freezing, which eventually led to fish "sticks." In the late 1920s, processors like General Foods began building their own trawler fleets, completing the industrialization of the industry.

Continued on p. 250

Chronology

1800s-Early 1900s

Fish stocks decline as fishing becomes more mechanized and new trawling methods are developed.

1905

The steam-powered otter trawl is deployed in North America, making industrial fishing possible in previously unfished areas.

1920s-1940s

Diesel engines replace steam, triggering an expansion of offshore trawling. U.S. processors develop flash freezing. Norwegian fishermen begin dragging for shrimp.

1936

New England halibut catch falls to 2 million pounds from 13.5 million in 1902; haddock falls by two-thirds.

1946

International Convention for the Regulation of Whaling is adopted.

1950s

The first factory trawlers come into use, mainly in North Atlantic. Pollution runoff from land-based development begins disrupting global ecosystems.

1954

A Scottish firm builds the first freezer-equipped "factory" trawler, the *Fairtry*, which is four times the size of conventional trawlers.

1959

Antarctica Treaty is adopted, sparing the southern continent and surrounding oceans from hunting, industrial activity and fishing pressure.

1970s-1980s

Countries extend their territorial seas to prevent foreign overexploitation. Pollution disrupts ocean life. Fisheries decline sharply. Iceland fights "cod war" to extend its territorial waters.

1972

Congress passes Coastal Zone Management, Clean Water and Marine Mammal Protection acts.

1975

U.N. adopts Convention on International Trade in Endangered Species, which helps reduce cross-border trade in marine animal products.

1977

U.S., Canada, others adopt 200-mile territorial limits. Trawlers move to Africa, Asia and the Caribbean.

1982

U.N. Convention on the Law of the Sea becomes the first international agreement to regulate ocean use. International Whaling Commission bans most commercial whaling beginning in 1986.

1989

Norway restricts cod fishing to protect declining stocks; cod recovers by 1992.

1990s-2000s

Black Sea ecosystem collapses. Coral reefs decline worldwide and climate change, pollution and chronic overfishing appear to be driving oceans into crisis.

1990

Pollution, overfishing and the introduction of a non-native comb jelly species devastate the Black Sea.

1991

British North Sea cod stocks have declined by more than two-thirds in 10 years.

1992

Canada closes Grand Banks cod fishery; stock stands at 1 percent of 1965 levels.

1994

U.S. closes key New England fisheries; slow recovery begins.

1998

Bleaching destroys 16 percent of the world's coral reefs. Cause is unknown.

2002

European Union (EU) recommends member states adopt integrated coastal zone management.

2003

Pew Oceans Commission urges immediate action to protect the oceans; Canadian study says stocks of large predatory fish have fallen by 90 percent worldwide since 1950.

2004

U.S. Commission on Ocean Policy calls for ecosystem-based response to protecting oceans and coasts.

2006

Iceland breaks whaling ban. . . . Most EU members have adopted integrated coastal zone management plans. . . . Pro-whaling bloc gets whaling ban declared no longer necessary.

2007

Hundreds of marine species — including corals for the first time — are added to the World Conservation Union's "red list" of species facing the risk of extinction.

Continued from p. 248

But fish stocks could not stand up to the technology. In New England, the halibut catch fell from 13.5 million pounds in 1902 to 2 million in 1936; the haddock catch fell by two-thirds between 1929 and 1936, while winter flounder became so scarce fishermen began targeting the previously spurned yellowtail flounder. By World War II, the yellowtail had been depleted, and near-shore stocks of cod and haddock were driven into commercial extinction. [42]

In 1954, a Scottish whaling firm built the first factory trawler, modeled on the big processing ships that had wiped out Antarctica's whale stocks. *The Fairtry* weighed 2,600 tons, more than four times the size of the conventional trawlers of the day, and was equipped with an onboard processing and freezing plant and nets large enough to swallow the Statue of Liberty. It could operate 24 hours a day for weeks on end, allowing the ships to travel to fisheries thousands of miles from home. The design revolutionized distant-water fishing.

By the 1970s, the Soviet Union was operating more than 700 freezer trawlers, and the two Germanys, Poland, Spain, France and Japan each had dozens more. [43] So-called mid-water trawls were developed to target mackerel, herring and anchovies that lived in medium depths. The ships massed on fishing grounds off Newfoundland, New England, Antarctica and the Bering Sea and mined one fish stock after another into near-oblivion.

Whaling Nations Want Hunting Ban Lifted

Japan, Iceland and Norway are leading the charge.

Great whales are up against a lot these days: Their food supply has diminished and is laced with PCBs and heavy metals, noise from sonar devices plagues their habitat and entangling fishing nets and passing ships are a constant threat. It's no surprise that most varieties — including blue, fin, humpback, sperm and right — remain on the endangered species list.

But in October 2006 Iceland announced it would resume hunting great whales, breaking the International Whaling Commission's (IWC) 21-year-old global moratorium on commercial whaling. *

Iceland's commercial whalers were permitted to kill nine endangered fin whales — the second-largest species after blue whales — and 30 smaller, more abundant minkes during the year ending Aug. 31, 2007, ostensibly for scientific research. By last November, however, they had already killed seven of the endangered fins, setting off a storm of international criticism. [1]

"It's outside all international norms to hunt an endangered species," says Susan Lieberman, director of the World Wildlife Fund's Global Species Program in Rome. "There is a commercial whaling moratorium in effect, so we're not saying it's fine and dandy to be hunting [the non-endangered] minkes. But targeting fin whales is a far more confrontational and aggressive act."

Whale meat is a delicacy in Iceland, Norway and Japan, particularly with older generations, and whaling nations say hunting and consuming whales is part of their cultural heritage. But critics point out that all three countries have difficulty disposing of whale blubber and other non-meat byproducts, leading conservationists to question why they continue the hunt.

Last November 25 countries — including the United States, Great Britain and Australia — demanded a halt to Iceland's hunt. Critics say it undermines the IWC by directly challenging its moratorium. [2]

"They are testing what the international reaction would be, and I think they've found it has been pretty harsh," says Sue Fisher, trade expert at Britain's Whale and Dolphin Conservation Society. "What they're doing is a violation of the ban."

In fact, the moratorium has been unraveling for years, largely because of Japan's diplomatic maneuvering. It used aid and trade measures to convince a small army of previously disinterested Caribbean and Pacific nations to join the IWC and vote with Japan. As a result, in 2006, the pro-whaling bloc achieved a simple majority and passed a symbolic measure declaring the ban no longer necessary. But the measure fell short of the three-quarters majority needed.

In retaliation, the United Kingdom and other anti-whaling nations have recently recruited five disinterested proxies of their own — Croatia, Cyprus, Ecuador, Greece and Slovenia — to help stave off the pro-whaling bloc. [3]

Norway, which is allowed to legally hunt whales commercially, focused exclusively on minkes, which are about an eighth the mass of fin whales. Operating from small vessels, Norwegian fishermen kill between 600 and 800 minkes each year out of an estimated North Atlantic population of over 170,000.

Norway's leading environmental groups support the hunt, arguing it is a sustainable fishery that produces organic meat with fewer inputs than a corporate beef or pork farm. "We use small fishing vessels that consume few inputs and cause almost no pollution — it's very friendly eco-production," says Marius Holm, co-chairman of the Bellona Foundation environmental

---

* *The 1986 moratorium allowed nations to legally continue whale hunting if they filed official reservations prior to the adoption of the ban. Norway made such a reservation. Iceland and Japan did not, though Iceland tried to claim one after the fact, when it first resumed hunting minkes for "scientific purposes" in 2003. Japan conducts a "scientific" research hunt — also allowed under IWC rules — taking about 900 minkes in the Southern Ocean. Non-whaling nations consider the hunt a violation of the spirit of the treaty. Aboriginal hunters are exempt and permitted to hunt a limited number of whales for cultural and religious purposes.*

organization in Oslo. "Our principle is that we should harvest what nature provides, but in a sustainable way regarding the ecosystem as a whole and the specific stocks."

"The hunt we have had along our coast has always been sustainable," says Halvard Johansen, deputy director general of the Norwegian Ministry of Fisheries and Coastal Affairs. "We've been whaling on this coast since the 9th century, and we don't see that big a difference between aboriginal whaling in Alaska, Russia and Greenland and what we do here."

Stefan Asmundsson, Iceland's whaling commissioner, claims his country's hunts are also sustainable, despite targeting an endangered species. "The fin whale stocks being targeted by Iceland are not in any way endangered," he says. "There is no lineage between the stocks in the North Atlantic, which are abundant, and those in the Southern Hemisphere" that were decimated by factory whaling fleets in mid-20th century.

Indeed, many in the West fear a return to those dark days when Norway, Japan and other whaling nations drove many great whales to the brink of extinction to procure industrial oil and pet food. Recovery has been slow. In 2003, a study by Stanford University geneticist Stephen Palumbi suggested that the pre-whaling populations of North Atlantic humpback, fin and minke whales were far larger than previously thought and won't return to exploitable levels for many decades. [4]

Others say Japan and Iceland are defending whaling out of a sense of national pride rather than economic necessity, since their domestic markets have been unable to absorb the meat from their scientific minke hunts. As whale meat piles up in freezers — 4,400 tons according to Greenpeace — Japan has resorted to introducing it in school lunch programs.

"Almost all those who like whale meat are middle-aged and older," admits Kouji Shingru, owner of the only whale-meat retail shop in Tokyo. "Young people have no experience with eating whale. In fact, my shop is one of the only places where young people have a chance to eat it." [5]

Japanese whalers butcher a Baird's beaked whale, a species not in danger of extinction. Despite a 1986 ban on commercial whaling, the International Whaling Commission allows Japan and Norway to hunt whales, and Iceland announced in October 2006 it would resume hunting great whales.

"This is not driven by economics, it's just political, which makes it far more egregious," says Lieberman.

[1] Krista Mahr, "Defying global ban, Iceland to resume commercial whaling after almost 2 decades," The Associated Press, Oct. 17, 2006; Colin Woodard, "Thar She Blows," *E Magazine*, January 2007.

[2] Lewis Smith, "Iceland's whaling sinks tourism: Two dozen nations protest hunting," *Calgary Herald*, Nov. 2, 2006.

[3] Micheal McCarthy, "Pro-hunting Japanese seize control of whaling commission," *The Independent* (London), April 17, 2006, p. 2; Richard Lloyd Perry, "Japan may go it alone after defeat over whaling ban," *The Times* (London), June 2, 2007.

[4] Stephen Palumbi and Joe Roman, "Whales before whaling in the North Atlantic," *Science*, July 25, 2003, p. 508.

[5] Greenpeace International press release, Jan. 30, 2007; Leo Lewis, "Giant of the sea used as petfood," *The Times* (London), Feb. 10, 2006.

In 1977, following Iceland's lead, Canada, the United States and other nations extended their territorial waters from 12 to 200 nautical miles in order to push foreign factory trawlers off their banks and save the fish for themselves. Fish stocks had been damaged, but with proper regulation it was assumed they would recover.

Stocks Crash

After kicking foreign fleets out, the United States, Canada and other na-tions built their own modern trawler fleets. Though the vessels were much smaller, they carried the latest fish-finding gadgetry and were capable of fishing the offshore banks. Encouraged by government incentives, New England trawlers larger than 125 tons grew by 144 percent between 1976 and 1979, while the number of medium-size trawlers nearly doubled. Other governments offered similar incentives, increasing the size of the world's fishing fleet by 322 percent from 1979 to 1989. [44]

In New England and Canada, the domestic fleets quickly wiped out many stocks, including the cod stocks that had attracted Europeans to colonize the region in the first place. Between 1965 and 1999, New England's haddock catch fell by 95 percent, halibut by 92 percent and cod by 40 percent, prompting managers to close many fishing grounds. On Newfoundland's Grand Banks, the greatest cod fishery in the world was reduced by 98.9 percent in the 30 years leading up to its 1992 closure.

The pain of the closures and other fishing restrictions eroded the economic and cultural foundations of many New England and Canadian communi-

A diver in the Great Barrier Reef Marine Park in Australia photographs masses of bleached staghorn coral, which occurs when sea temperatures rise and kill microbes that give coral its bright colors. Rising sea temperatures and other mostly man-made factors are said to be killing coral reefs at an alarming rate.

Great Barrier Reef Marine Park Authority/Paul Marshall

ties, and fueled an exodus of young people from Newfoundland. "We are witnessing an entire generation without hope, enthusiasm and access to meaningful, steady employment," noted Michael Temelini, assistant professor of political science at the Memorial University of Newfoundland. "[Our] 500-year-old civilization is disappearing." [45]

Fifteen years after the closure, Newfoundland's cod appear unable to regain their place in the ecosystem, while depleted fish stocks on Georges Bank and in the Gulf of Maine are recovering slowly. Fish plants have closed throughout eastern North America, and large fishing vessels have vanished from harbors whose residents had fished offshore for centuries. [46]

Similar devastation has occurred from the North Sea to the Gulf of Thailand. "Ten years ago, we could catch anything we wanted," said Sophon Loseresakun, a fisherman in Talumphuk, Thailand. "Now we have almost nothing." [47]

Eventually, the large fishing vessels moved on to the developing world, buying access to fishing grounds that once supported small-scale local fishermen. "European and Russian distant-water fishing fleets shrank and their remnants turned south," recalled Carl Safina, president of the Blue Ocean Institute, a marine conservation advocacy group in East Norwich, N.Y., "under-paying their way into the fishing zones of countries too desperate for foreign cash to say no, lest the same bad offer be accepted by a neighboring country and the boats go there instead." [48]

Today, subsistence fishermen in canoes in Mauritania and other African nations compete with hundreds of government-subsidized Spanish, Russian and Chinese trawlers. West African fish stocks have declined by 50 percent over the past 30 years, and thousands of fishermen have been put out of work. [49] Likewise, South Pacific nations have sold tuna fishing rights to Russian, Chinese and Taiwanese companies, even though the valuable fish is one of their few natural resources.

Other vessels have turned to the high seas, beyond the reach of government, to fish for orange roughy, Patagonian toothfish and other long-lived, slow-to-reproduce species. According to a University of British Columbia study, high-seas bottom trawlers receive $152 million a year in subsidies worldwide.

"Eliminating government subsidies would render the fleet economically unviable," said lead author Rashid Sumaila. [50]

"We are vacuuming the ocean of its content," said French environmentalist Jean-Michel Cousteau, son of the late ocean explorer Jacques Cousteau. "If this continues, there will be nothing left." [51]

Shifting Baselines

Scientists now realize that the damage to the oceans is far worse than previously estimated. New forensic research using fishing logbooks, archaeological evidence and genetic tests reveals that ocean life was far more abundant in the pre-industrial era than anyone had assumed.

For instance, after witnessing the destruction of many of Jamaica's coral reefs, Jackson of the Scripps Institution used historical records to determine how the Caribbean might have looked in 1492. "Think about the wildebeests and lions and all that on the plains of Africa," he says. "Well, there was a world in which the biomass of big animals among the reefs was greater than the biomass of the big mammals of the Serengeti plains." [52]

Based on hunting records, adult green sea turtles — now rare — once numbered at least 35 million and may have exceeded 550 million. "Think about that: 35 million 220-pound turtles grazing on crustaceans, sea grass, starfish and mollusks," he says. "The productivity of those reefs must have been fantastic. The whole mind-set of scientists about what is a 'pristine' reef is completely wrong." His research suggests that by wiping out sea turtles — used for food in the 19th century — humans probably triggered the collapse of sea grass beds, which suffer infections when they are not grazed. [53]

Similarly, W. Jeffrey Bolster a professor of maritime history at the University of New Hampshire, and colleagues used 19th-century logbooks to reconstruct the scale of the cod catch and the average size of the fish. The results stunned fisheries experts. In 1861 alone, small-boat fishermen from several Maine towns — using small sailboats and baited hooks — caught more fish than all the U.S. and Canadian fishing fleets combined caught in the Gulf of Maine between 1996 and 1999. Today, there are virtually no cod in the area.

"Ask yourself, 'What were all those cod eating?' " Bolster says. "When you think about the copepods and krill, all the way up to the alewives and mackerel that had to be present in the inshore area to feed them, it's flabbergasting.

"The world we have today," he points out, was created by humans trying to "manage" the exploitation of fisheries resources. "In terms of engineered outcomes, it's been a disaster."

Climate Change

Global warming is already affecting polar marine ecosystems. On the Antarctic Peninsula, ice-dependent species like Adelie penguins and Weddell seals have moved southwards, replaced by the Gentoo penguins and elephant seals that prefer open water. Krill, the small marine crustaceans that form the basis of the Antarctic food chain, feed on algae that grow on the underside of winter sea ice. Experts fear less sea ice could mean less krill and, thus, less food to go around.

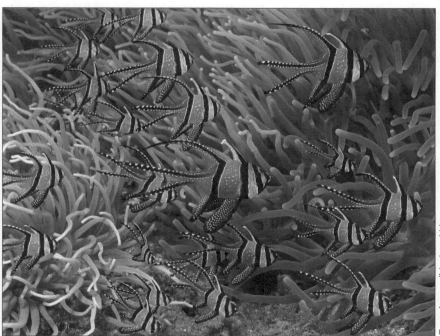

The Banggai cardinalfish, native to Indonesia, is among hundreds of marine animals and corals added to the World Conservation Union's 2007 "red list" of species in jeopardy or facing high risk of extinction. The group blames excessive and destructive fishing activities for the loss of ocean biodiversity.

Getty Images/Norbert Wu

In the Arctic, reduced ice cover is causing starvation and reproductive failure among polar bears, prompting the United States to propose listing them as a threatened species. [54] Walruses and ringed seals also depend on floating ice as habitat.

Fishermen in Greenland have witnessed considerable changes in the composition of the continent's marine species in recent years. Cold-loving shrimp are becoming rarer on the south and west-central coasts, while cod are becoming more numerous. The fishing season is longer in areas where sea ice is failing to form, but polar bear and seal hunters cannot risk going out on the ice with their sled dogs.

"Hunters and fishermen have passed down detailed information about their environment for generations," says Lene Kielsen Holm, director for environment and sustainable development at the Greenland office of the Inuit Circumpolar Council, which represents the interests of the 160,000 Inuit, the indigenous people of the high arctic. "Now they tell us things are changing so quickly everything they have been taught by their elders is no longer accurate."

Scientists also worry that melting polar glaciers, ice caps and ice sheets could slow ocean circulation. Ocean currents normally act as a conveyor belt, moving warm surface waters toward the poles and cold bottom water from polar areas toward the equator. The release of fresh meltwater alters seawater density, which may slow or even stop circulation, with potentially devastating consequences for human and marine life.

A 2005 study indicated that the Gulf Stream, which keeps northern Europe's climate mild, may have slowed by 30 percent since 1992. "We don't want to say the circulation will shut down, but we are very nervous about our findings," said Harry L. Bryden, a specialist in the role of ocean heat and freshwater currents at the School of Ocean and Earth Science at the National Oceanography Centre in

Southampton, England. "They have come as quite a surprise." [55]

Meanwhile, the increase in atmospheric carbon dioxide (CO_2) — one of the major causes of climate change — is making the ocean more acidic, with potentially catastrophic consequences. About 2 billion tons of atmospheric CO_2 ends up in the oceans each year — 10 times the natural rate — and the oceans have become 30 percent more acidic since the Industrial Revolution. That figure is expected to rise to 100 percent to 150 percent by the end of the century.

Increased acidity disrupts the ability of corals and other sea animals to build shells and skeletons. "There's a whole category of organisms that have been around for hundreds of millions of years that are at risk of extinction," says Ken Caldeira, a chemical oceanographer at the Carnegie Institution's Department of Global Ecology at Stanford University. The likely casualties include a range of microscopic creatures at the base of the food chain, including coccolithophores and pteropods; the polyps that build coral reefs; and starfish and sea cucumbers — popular food items for larger creatures. Oysters, scallops, mussels, barnacles and many other creatures may also be affected.

"This is a matter of the utmost importance," said reef expert Ove Hoegh-Guldberg of the University of Queensland in Australia. "I can't really stress it in words enough. It is a do-or-die situation." [56]

Governing the Seas

Historically, the seas were unregulated beyond three nautical miles from land, the effective range of shore-based cannon. But in the 20th century, fishery-dependent Iceland — in an effort to protect its fish from foreign fleets — extended its territorial waters first to four then to 12, 50 and, finally, to 200 nautical miles.

Initially, the world protested. Britain dispatched a fleet of warships to protect its trawlers, triggering three bloodless "cod wars" with Iceland's coast guard. Shots were fired, ships rammed and nets cut. Iceland barred British warplanes from landing at the NATO air base in Keflavik. Finally, in 1976, the world backed down, and soon nations around the world had declared their own 200-mile zones. [57]

The 200-mile limit was codified in the 1982 U.N. Convention on the Law of the Sea (UNCLOS), which has since been ratified by 155 nations. The United States — which had extended its own territorial waters to 200 miles in 1977 — has not yet ratified the treaty, but President George W. Bush has said it should be ratified.

Under UNCLOS, governments can regulate fishing and other economic activities within their 200-mile "exclusive economic zones" (EEZ) and are expected to take a precautionary approach in utilizing their EEZs "according to their capabilities." Legal scholars have said the conservation language is so weak as to be useless.

"Just as with bringing up children, a permissive approach to the law of the sea guarantees spoiling," writes John Charles Kunich, associate professor of law at the Appalachian School of Law in Grundy, Va., and author of *Killing Our Oceans: Dealing with the Mass Extinction of Marine Life*. "It is all too predictable that nations often discover that other pressing needs prevent them . . . from doing anything to protect biodiversity in the oceans." [58]

Nations have cooperated in trying to regulate international exploitation of marine creatures that regularly cross borders through treaty organizations like the International Whaling Commission, the Northwest Atlantic Fisheries Organization and the International Commission for the Conservation of Atlantic Tuna. But the track record of most of these organizations is poor. In most cases, the populations of species

they are supposed to protect have declined due to predation by humans. Even when member nations agree to tougher enforcement, unscrupulous vessel owners can simply re-register their ships in a "flag-of-convenience" nation that does not observe the relevant rules.

International cooperation has succeeded, however, in banning drift-net fishing — which indiscriminately kills any fish or mammals that come in contact with the massive nets — and the 1975 Convention on International Trade in Endangered Species helped reduce cross-border trade in hawksbill turtles, Caspian sturgeon roe and many whale products. The 1959 Antarctica Treaty has spared the southern continent and much of the surrounding ocean from continued hunting, industrial activity and fishing pressure.

But with most of the more ecologically productive parts of the ocean located within nations' EEZs, it will likely fall on national governments to protect them. ∎

CURRENT SITUATION

Government Responses

World governments have been slow to address the ocean crisis, often stepping in only after a fishery has collapsed. And governments that have closed fishing grounds have had mixed results. While cod stocks on Canada's Grand Banks have failed to recover, for instance, New England's haddock, flounder and other species are slowly recovering. Scientists say New England stocks can recover completely if policy makers withstand industry pressure to allow more fishing.

Internationally, China, Iceland, Russia and other deep-sea fishing nations blocked a 2006 U.N. effort to ban high-seas bottom trawling. "There were several countries that really didn't want any controls at all," said U.S. Assistant Secretary of State for Oceans, Environment and Science Claudia McMurray. "We're very disappointed." [59]

Many countries have considered privatizing fisheries by creating individual transferable quotas (ITQs). Under an ITQ system, scientists set quotas on the total allowable catch for a given season, species and fishing ground; shares of the quotas are bought and sold by private entities, and only those holding shares are allowed to fish.

Proponents say that creating "owners" of uncaught fish will encourage responsible stewardship. Opponents say an ITQ system represents a massive transfer of public resources to corporate control.

"Since quotas are bought and sold to the highest bidder, local fishermen — who can't compete with deep-pocketed corporations — are almost inevitably squeezed out," write Pietro Parravano, past president of the Pacific Coast Federation of Fishermen's Associations, and Lee Crockett, formerly of the Marine Fish Conservation Network and now federal fisheries policy director at the Pew Environment Group. "Moreover, because quotas are [initially] set on the number of fish historically caught by individual fishermen, those who tried to allow stocks to replenish by fishing responsibly [are] penalized, while those who fish rapaciously are rewarded." [60]

Others say ITQ rules can be written to prevent the concentration of ownership. Alaska's halibut and sable-fish fisheries adopted ITQs in 1995, but the rules limited absentee ownership and consolidation of shares and allowed communities to buy blocks of shares to divvy up among individuals. A decade later, the fishery was safer and greener, according to Linda Behnken, executive director of the Alaska Longline Fishermen's Association.

Wastewater runoff from coastal developments is one of many threats to marine habitats. Each summer fertilizer runoff from 31 states in the United States flows into the Gulf of Mexico via the Mississippi River, creating a huge dead zone south of New Orleans where few species survive.

U.N. Environmental Program/Topham

"From a resource perspective, it's been an unqualified success," she said. [61]

Zoning the Sea

Some governments have tried to zone the ocean bottom just as cities use zoning rules to separate incompatible land uses and to establish parks and preserves. Environmentalists have long advocated a similar approach, as has the World Bank, the U.S. Commission on Ocean Policy and the Pew Oceans Commission. Sea-floor zoning — and establishing marine reserves where certain human activities are prohibited — is also a crucial part of ecosystem-based management.

"It's inevitable that once we developed methods to reach further and further into the sea, we would have to extend the regulatory framework we have on land to the sea," says Callum Roberts, an expert on marine zoning at the University of York in England.

To date, however, less than 2 percent of the oceans are within marine protect-ed areas, and less than 0.2 percent are classified as no-take marine reserves where no disruptive activities are allowed. [62]

The Central American nation of Belize is leading the way on marine protected areas. With the help of the Global Environment Facility — a joint UN/World Bank environmental grant-making agency based in Washington — and scientists from the United States and Britain, Belize has created a surprisingly comprehensive array of fully protected parks, mixed-use reserves and specialized wildlife sanctuaries aimed at protecting biodiversity while enhancing the country's most important industries: fishing and eco-tourism. [63]

One of the reserves, Glover's Reef, a remote atoll 30 miles offshore, has been fully protected and carefully patrolled for nearly a decade. As a result, depleted commercial seafood species have rapidly rebounded: From 1998 to 2003, queen conch populations jumped by 350 percent and spiny lobster 250 percent. Similar results have been observed in New Zealand, where overall ecological productivity has jumped 50 percent in a reserve founded in 1977. [64]

"Right now, it's like we have a few oases in the desert," says Bill Ballantine, a marine biologist and former director of the University of Auckland's Leigh Marine Laboratory, who says the world needs many more.

To be effective, Belize's reserves must be "big enough so that a reasonable number of species can complete their life cycle within its borders, but small enough so the animals inside will produce larvae that wind up outside its borders and seed other fish populations," says Peter Sale, professor emeritus of biological sciences at the University of Windsor, Ontario. [65]

But there are problems, especially in cash-strapped developing countries like Belize. Running the reserves "is an enormously expensive undertaking," says Janet Gibson, former director of Belize's Coastal Zone Management Authority, "and there are really few ways to raise funds apart from charging entrance fees."

In Indonesia, the U.S.-based Nature Conservancy has provided funding for wardens at Komodo National Park, who have largely driven out the dynamite fishermen who blasted many of the area's reefs to rubble. Researchers have helped the reefs reestablish themselves by piling sandstone and limestone boulders on the sea floor. "Places that were just bare rock and rubble now have great coral growth and are surrounded by fish," says the World Wildlife Fund's senior marine conservation biologist Helen E. Fox, who worked on the project. The technique costs only about $5 per square meter, compared to between $550 and $10,000 to repair corals in the Florida Keys, she says.

In 2006, President Bush created the world's largest fully protected reserve, the 138,000-square-mile Papahānaumokuākea Marine National Monument in the northwestern Hawaiian Islands. The new reserve — 100 times the size of Yosemite National Park and larger than all other U.S. national parks combined — prohibits all exploitative activities except for limited ritualistic fishing by native Hawaiians. Environmentalists and ocean policy advocates widely praised the move, called by Fred Krupp, president of Environmental Defense, "as important as the establishment of Yellowstone." [66]

Some fishermen fear large areas could be declared reserves and closed to fishing. But proponents note that creating a reserve doesn't necessarily preclude fishing; it can, however, restrict fishing by method, times or places.

"What you zone for depends on what you are trying to protect," says Anthony Chatwin, director of the South American marine program at The Nature Conservancy. "Some areas will be reserved to protect marine biodiversity, others purely for fisheries management purposes, but they will be comprehensively put together."

Managing Better?

Some experts have identified examples of sustainable fisheries — those that work within the limits of what marine ecosystems can support. Most include community-based management, in which local communities develop their own rules for how, when and by whom the grounds can be fished.

In the Maine lobster fishery, for instance, catches remain at all-time highs despite an enormous increase in the number of fishermen and traps. Traditionally, lobstermen from each harbor controlled their own piece of the seafloor and defended it from intrusion by anyone without the community's permission. By controlling their own lobster pasture, the fishermen had an incentive to enforce or even enhance conservation laws, according to James Acheson, an anthropologist at the University of Maine. "The whole theory of common property resources like lobster assumes they're bound to be overexploited," he says. "That's nonsense." [67]

Numerous other successful fisheries have similar arrangements, from the coral reefs of Micronesia and Polynesia (where villages control fishing rights to nearby reefs) to the community fishing cooperatives that control inshore grounds in Japan's Ryukyu Islands. They have inspired others to advocate giving fishing communities proprietary rights to the resources they have long relied on in Atlantic Canada and elsewhere. [68]

Molyneaux, the author and former fisherman who received a Guggenheim Fellowship to study sustainable fisheries, points to Chile, where the government has blended privatization with community-based management. Small-scale fishermen are allowed to form unions, which are allocated a slot of ocean bottom. The government sets quotas on commercial species, but the union members decide how to manage the area.

"You don't get sole access to the fish but rather to the sedentary resources: shellfish, seaweed, abalone," says Molyneaux. "It's a form of privatization, but it's community based and intended to keep fishermen in their communities and not moving to shantytowns in the cities."

The unions pay for scientific assessments of the health of their stock but are left to work out the economic strategy for harvesting it. "They figured out how to give people the power to control and promote the resource, a reason for promoting sustainability," he says.

Some fishing gear is clearly less harmful than others. Maine lobstermen fish with baited traps that don't harm juvenile lobsters, oversized lobsters, non-target species or the ocean floor. New Brunswick weir fishermen capture herring in hand-built fish traps, with the rights to use a given location handed down in families. Some Icelandic fishermen combine high-tech with small-scale: fishing from small, locally built boats with hand-baited, computer-tended longlines, sophisticated electronic

Continued on p. 258

At Issue:

Should the moratorium on commercial whaling be lifted?

RUNE FRØVIK
CEO, HIGH NORTH ALLIANCE, NORWAY

WRITTEN FOR *CQ GLOBAL RESEARCHER*,
SEPTEMBER 2007

*w*hile fishing continues to enjoy almost universal acceptance as a means of food production, Western urban society has decided unilaterally to shut down whaling with complete disregard for any culture that still practices it.

Each culture has its own culinary idiosyncrasies. For many Asians, dog meat is a delicacy; the French like their frogs, snails and horse meat, and Australians have a taste for kangaroos. And there are just as many taboos — Indians forego the joy of beefsteak, while Jews and Muslims won't touch pork.

Beset with environmental challenges and yet respectful of cultural differences, the world community thankfully embraced Agenda 21's principle of striving for sustainable development — using renewable resources at rates that are within the resources' capacity for renewal.* Yet the West's cultural imperialists would have whales exempted from the sustainable-use principle — placing them above and apart from the animal kingdom to which they obviously belong.

For those who live close to nature, natural resources play vital roles, both nutritional and cultural, in their lives. Thus, coastal people will continue to harvest what nature provides — be it seals, fish, birds . . . or whales. And in the interest of self-preservation, they will strive to do so sustainably.

Sustainable whaling must be managed in accordance with agreed principles, not by launching destructive attacks on those who engage in exactly what we are striving for — sustainable use — just because one's cultural bias finds a particular harvest unpalatable.

In a world where trade depends on the exchange of money, there are commercial aspects to whalers' lives. In Greenland, Iceland, Japan and Norway whale meat is sold in supermarkets, and expensive whale souvenirs are sold to tourists in Alaska. Until whale meat is accepted as currency, whalers must do their shopping the same way as the rest of us — with cash.

Harvesting nature's surplus, including super-abundant whale resources, means biodiversity and habitat do not have to be destroyed and turned into agricultural land. True environmentalists are concerned not with appearances but with practicing the principles that they preach. In so doing, they have either reached the conclusion, or are getting there, that whaling should not only be continued but should even be increased to provide more people with ecological, healthy and nutritious food.

---

* Agenda 21 — adopted by more than 178 governments at the Earth Summit in June 1992 — is a 300-page plan for achieving sustainable development in the 21st century.

PHILIPPA BRAKES
SENIOR BIOLOGIST, WHALE AND DOLPHIN CONSERVATION SOCIETY

WRITTEN FOR *CQ GLOBAL RESEARCHER*,
SEPTEMBER 2007

*t*he unbridled ravages of commercial whaling — which brought several whale populations to the brink of extinction and significantly depleted many others — should serve as a grave warning to the dangers of the poorly regulated exploitation of marine mammals.

The moratorium was intended to allow whale populations to recover to pre-exploitation levels. Since the moratorium was implemented, however, we have learned what we know and what we don't know about whale populations. We have learned that it's difficult to accurately estimate whale populations and that whales now face new threats from noise and chemical pollution, ship strikes, loss of critical habitat, entanglement in fishing gear and, more recently, challenges due to climate change. All of these threats may influence recovery of whale populations.

Even if the threats to whale populations could be adequately mitigated, many question whether commercial whaling could ever, realistically, be well regulated. The lessons of history — and the burgeoning exploitation of the moratorium's loophole for "scientific" whaling — lead us to conclude that this is unlikely.

Moreover, grave concerns remain as to whether whaling could ever be conducted humanely. Since commercial whaling is conducted for profit, it is argued with good reason that whaling should be held accountable to the same standards for humane slaughter as other animals killed for commercial purposes. It is difficult — even, unpleasant — to imagine a situation in which an animal could escape and be lost during slaughter in an abattoir and be left to die of its injuries. Yet, whales that are injured and escape remain a permanent feature of all whaling practices.

In addition, there are also much broader ethical issues at stake: In the 21st century, many people, even some cultures, no longer view whales as a resource to be exploited but as social beings, with complex lives that should be afforded protection of their interests, not because of their potential value to humankind but because of their own value in and of themselves.

Perhaps rather than asking whether the moratorium should be lifted, the global community should now turn its attention to closing the legal loopholes that permit whaling under objection and whaling for "scientific" purposes, and we should instead ask how we can secure a brighter future for our seaborne cousins so they are protected from commercial hunting permanently.

Continued from p. 256

bottom maps and the capability to sell their catch electronically hours before they ever reach the dock.

Others say conventional gear and management can work, pointing to Alaska, where fisheries managers have prevented overfishing, reduced bycatch and protected habitats. By 2005, the North Pacific Fisheries Management Council — which includes fishing interests, scientists and public officials — had banned bottom trawling in nearly 40 percent of Alaska's federal waters and imposed fishing closures to protect the spawning and nursery grounds of herring, rockfish, crab and other commercial species.

Unlike its New England counterpart, the council never authorized catch quotas above those recommended by its scientific advisers. It also restricted fishing near Stellar sea lion rookeries. Federal managers also banned targeted fishing of key forage species like smelt, capelin and sand lance that feed seabirds and commercial fish.

"People in Alaska had been through some tough times in the past, and they created an ethic that puts protecting the resource first," said Benton of the Marine Conservation Alliance. "People around the world acknowledge Alaska as one of those places where you can look for positive lessons learned." [69]

Saving the Shore

Policy makers are paying considerable attention to calls to develop integrated coastal zone management plans that control human activities throughout an entire watershed and its associated coastline and estuaries.

Three-quarters of the ocean's pollution comes from land-based human activity, and the problem is getting worse. Half of the world's population lives within 62 miles of the coast; in Southeast Asia it's two-thirds. In the United States,

53 percent of the population lives in coastal counties that comprise only 17 percent of the country's landmass. Nine of the world's 10 largest cities are on the coast, and coastal population growth is expected to greatly exceed overall trends. [70] Already, 20th-century development activities have destroyed half of all coastal wetlands. [71]

Reengineering rivers compounds the problem with polluted watersheds. Egypt's Aswan High Dam cut off the flow of nutrients to the Nile Delta, causing the sardine catch to plummet from 18,000 to 600 tons in three years. In Louisiana, decades of levee- and channel-building by the Army Corps of Engineers has transformed the Mississippi River into a ditch that shunts fertilizers and sewage directly into the Gulf of Mexico. And by preventing the river from dropping silt in its delta, the levees are causing the Louisiana bayous to disappear into open water at the rate of 25-35 square miles a year. [72]

"The loss of Louisiana's marshes will incrementally destroy the economy, culture, ecology and infrastructure, not to mention the corresponding tax base of this state and this region," said banker King Milling, chairman of the state-appointed Committee on the Future of Coastal Louisiana. [73]

The U.S. Commission on Ocean Policy and the Pew Oceans Commission have called for integrated, watershed-wide planning in the United States. If implemented, the model might again draw on the experience of Belize, which found that protecting coral reefs required addressing problems far inland.

"To protect the reefs — or any other ecosystem for that matter — you have to take . . . an ecosystem approach," said Gibson, the former head of Belize's Coastal Zone Management Authority (CZMA). "You need to look at pollution in the watersheds, at coastal construction and urban expansion, at fisheries and forest loss, at the effects of tourism and air pollution.

"If there isn't coordination between economic sectors, your efforts to conserve an ecological system are not going to be successful," she continued. The CZMA has coordinated policies governing forestry, fisheries and water quality while sponsoring education programs to show people how rivers, mangroves, cays and reefs interact. Other countries have made advances in creating integrated coastal zone management plans. Australia created a national framework to integrate its state, territorial, regional and local government agencies to manage resources on a river-basin-by-river-basin level. [74] In 2002, the European Union recommended that member states move towards integrated management. A 2007 review found "a positive impact in stimulating progress" but acknowledged it would be a "slow and long-term process." Most member states hadn't adopted national strategies until 2006; six had not done so at all. [75] ∎

OUTLOOK

Out of Mind

The oceans have never gotten the political attention they need, and after the Sept. 11, 2001, terrorist attacks, they seemed to sink even lower on the world's priority list.

"We are moving forward extremely slowly, and in fact we are actually retreating from some of the movement we had toward more concerted international action," says Mee of the University of Plymouth. "There's been concentration on other issues: terrorism, in particular, and security."

Ironically, the urgent attention being focused on climate change has further eclipsed the oceans' problems. "Action by celebrity figures like [former Vice

President] Al Gore has managed to put climate change on the agenda and kept it there very effectively," says Mee. "We don't really have many champions for the marine environment. The oceans don't make it onto people's agenda, resulting in the feeling that action can be postponed."

The United States is a case in point. The chairmen of the Pew Oceans Commission and the U.S. Commission on Ocean Policy joined forces to press for implementation of their recommendations, including issuing "report cards" grading governments' progress The U.S. government got a D+ in 2005 and a C- in 2006, reflecting modest progress.

"The improvement is largely attributable to state action and a few notable federal accomplishments," explained Admiral Watkins and Panetta. [75]

States and regional organizations received an A- after 18 took steps to develop comprehensive strategies for protecting marine systems. But there was little progress in most areas. Even modest programs to establish ocean observation systems (which collect basic oceanographic information) or to monitor and protect the endangered North Atlantic right whale have seen drastic funding cuts in the past year. Getting attention for oceans issues will remain challenging due to the lack of public awareness of the problems, and other congressional priorities, such as the Iraq War.

Internationally, even problems that aren't that difficult to solve have been put on the backburner. For instance, many of the disruptions caused by alien species could be prevented if the shipping industry adopted a ballast water-exchange program in which ballast water is pumped to and from sterilized sources rather than into local harbors.

"It costs people money," notes Mee. "It's a tiny marginal cost on transportation, but it requires a greater sense of purpose to push it through."

In the future, the world community is expected to focus more attention on the negative impact China's burgeoning industrialization is having on the world's oceans. According to the World Wildlife Fund, China is now the largest polluter of the Pacific Ocean. Each year China releases about 2.8 billion tons of contaminated water into the Bo Hai — a sea along China's northern coast — and the heavy metal content of Bo Hai bottom mud is now 2,000 times as high as China's official safety standard. In 2006, heavily industrialized Guangdong and Fujian provinces discharged nearly 8.3 billion tons of untreated sewage into the ocean — up 60 percent since 2001. More than 80 percent of the East China Sea — one of the world's largest fisheries — is now rated unsuitable for fishing, and the Chinese prawn catch has plunged 90 percent over the past 15 years. [77]

Bill Wareham, acting director of the Marine Conservation Program at Canada's David Suzuki Foundation, is pessimistic international agencies will cooperate, citing the refusal in 2006 by China, Russia and other deep-sea fishing nations to support the U.N.'s measure to ban trawling.

If the world cannot depend on the U.N. to mandate the protection of fish habitat and to prevent the ongoing decline of high seas fish stocks, "then we are in a very sad state," he said. "The global governance system has failed to move past the stage of denial." [78] ∎

Notes

[1] John W. Miller, "Global fishing trade depletes African waters," *The Wall Street Journal,* July 18, 2007, p. A1.

[2] "State of World Fisheries and Aquaculture 2006," Food and Agriculture Organization, 2007.

[3] Ransom A. Myers and Bruce Worm, "Rapid Worldwide Depletion of Predatory Fish Communities," *Nature,* May 15, 2003, p. 280.

[4] Kenneth R. Weiss, "A primeval tide of toxins," *Los Angeles Times,* July 30, 2006, p. 1.

[5] See Kieran Kelleher, "Discards in the world's marine fisheries: An update," Technical Paper 470, Food and Agriculture Organization, 2005, pp. xvi, 38.

[6] Eleanor M. Dorsey and Judith Pederson, *Effects of Fishing Gear on the Sea Floor of New England,* Conservation Law Foundation (1998), pp. 1-6.

[7] Clive Wilkinson (ed.), *Status of Coral Reefs of the World: 2004, Vol. 1,* Australian Institute for Marine Science, 2004, p. 7.

[8] Colin Woodard, *Ocean's End: Travels Through Endangered Seas* (2000), pp. 144-160.

[9] *Ibid.,* pp. 44-45, 156; "Research reveals virus link in deaths of reefs," *Evening Herald* [Plymouth, England], May 2, 2007, p. 13.

[10] Anne Platt McGinn, "Safeguarding the Health of the Oceans," Worldwatch Institute, 1999, pp. 26-27.

[11] For background, see Marcia Clemmitt, "Saving the Oceans," *CQ Researcher,* Nov. 4, 2005, pp. 933-956.

[12] For a detailed discussion see Woodard, *op. cit.,* pp. 97-129.

[13] For background, see David Hosansky, "Invasive Species," *CQ Researcher,* Oct. 5, 2001, pp. 785-816.

[14] Colin Woodard, "Battling killer seaweed, *The Chronicle of Higher Education,* Aug. 2, 2002, p. 14.

[15] Kenneth R. Weiss, "Dark tides, ill winds," *Los Angeles Times,* Aug. 1, 2006, p. 1; Weiss, *op. cit.*; Jeremy B. C. Jackson, "Habitat destruction and ecological extinction of marine invertebrates," 2006, http://cbc.amnh.org/symposia/archives/expandingthearc/speakers/transcripts/jackson-text.html; see also Frances M. Van Dolah, "Marine Algal Toxins: Origins, Health Effects, and Their Increased Occurrence," *Environmental Health Perspectives Supplements,* March 2000, www.ehponline.org/members/2000/suppl-1/133-141vandolah/vandolah-full.html.

[16] "Major study predicts bleak future for Europe's seas," University of Plymouth press release, June 7, 2007.

[17] "Current trends project collapse of currently fished seafoods by 2050," National Science Foundation press release, Nov. 2, 2006.

[18] Clemmitt, *op. cit.*

[19] Testimony of Admiral James D. Watkins before the U.S. Senate Committee on Commerce, Science and Transportation, April 22, 2004, p. xi, http://govinfo.library.unt.edu/oceancommission/newsnotices/prelim_testimony.html.

20 "1,600+ scientists warn that the sea is in peril, call for action now," Marine Conservation Biology Institute press release, Jan. 6, 1998, www.gdrc.org/oceans/troubled.html.

21 Quoted in Clemmitt, *op. cit.* Furlong was citing the "Fish Stocks Sustainability Index" of the National Oceanographic and Atmospheric Administration's Fisheries Service, April 1-June 30, 2007, www.nmfs.noaa.gov/sfa/domes_fish/StatusoFisheries/2006/3rdQuarter/Q3-2006-FSSIDescription.pdf.

22 Bjorn Lomborg, *The Skeptical Environmentalist* (2001), pp. 189, 201, 329.

23 *Ibid.*

24 Edie Clark, "Ted Ames and the Recovery of Maine Fisheries," *Yankee Magazine*, November 2006.

25 Colin Woodard, "Saving Maine," *On Earth*, Summer 2003.

26 Testimony before U.S. House Natural Resources Subcommittee on Fisheries, Wildlife and Oceans, April 26, 2007, www.jointoceancommission.org/images/Rosenberg_testimony_H.R.21_04_26_07.pdf.

27 Wayne A. Morrisey, "Science Policy and Federal Ecosystem-based Management," *Ecological Applications*, August 2006, pp. 717-720; D. S. Slocombe, "Implementing ecosystem-based management," Bioscience, Vol. 43 (1993), pp. 612-622; U.S. Department of Commerce, *The Ecosystem Approach: Vol. I*, June 1995.

28 Kenneth R. Weiss, "Study urges trawling ban in fragile marine habitats," *Los Angeles Times*, March 19, 2002, p. 12; Joe Haberstroh, "The bottom of the trawling matter," *Newsday*, March 31, 2002, p. 31; Jeremy Collie, *et al.*, Effects of Trawling and Dredging on Sea Floor Habitat, National Research Council, March 2002.

29 Glen Whiffen, "Dragging technology hurts: advocate," *St. John's Telegram* [Newfoundland], May 4, 2005, p. A4.

30 U.N. General Assembly, "The Impacts of Fishing on Vulnerable Marine Ecosystems," July 14, 2006, www.un.org/Depts/los/general_assembly/documents/impact_of_fishing.pdf; John Heilprin, "UN ban on bottom trawling fails," *Houston Chronicle*, Nov. 25, 2006, p. 2.

31 "America's Living Oceans: Charting a course for sea change," Pew Oceans Commission, May 2003, p. 47, www.pewtrusts.org/uploaded-Files/wwwpewtrustsorg/Reports/Protecting_ocean_life/env_pew_oceans_final_report.pdf.

32 "NZ to close 30pc of waters to trawling," New Zealand Press Association, Feb. 14, 2006.

33 Jeff Barnard, "Seafloors fished by trawlers hold fewer kinds of fish," The Associated Press, April 18, 2007.

34 Milan Ilnyckyj, "The legality and sustainability of European Union fisheries policy in West Africa," *MIT International Review*, Spring 2007, pp. 33-41, http://web.mit.edu/mitir/2007/spring/fisheries.html.

35 Kim Willsheer, "Mauritanians rue EU fish deal with a catch," *The Guardian*, Feb. 9, 2001.

36 Stephen Castle, "EU Trawlers get fishing rights off Africa for pounds 350m," *The Independent*, July 24, 2006.

37 Haberstroh, *op. cit.*, p. A23; James Kendall, "Scallop dredge fishing," in Dorsey and Pederson, *op. cit.*, pp. 90-93.

38 M. J. Kaiser and B. E. Spencer, "The effects of beam-trawl disturbance on infaunal communities in different habitats," *Journal of Animal Ecology, Vol. 65* (1996), pp. 348-358; A.D. Rijnsdorp and P. I. Van Leewen, "Changes in growth of North Sea plaice since 1950 in relation to density, eutrophication, beam trawl effort, and temperature," *ICES Journal Marine Science*, Vol. 53 (1996), pp. 1199-1213; R. S. Millner and C. L. Whiting, "Long-term changes in growth and population abundance of sole in North Sea from 1940 to present," *ICES Journal Marine Science*, Vol. 53 (1996), pp. 1185-1195.

39 Haberstroh, *op. cit.*, p. A23. Also see "The impact of bottom trawling on American lobsters off Duxbury Beach, MA," Massachusetts Division of Marine Fisheries, Oct. 1, 1989.

40 For background see Clemmitt, *op. cit.*, p. 946.

41 Colin Woodard, *The Lobster Coast* (2004), pp. 201-207; William Warner, *Distant Water* (1977), pp. 50-53.

42 Woodard, *ibid.*, p. 204.

43 Warner, *op. cit.*, pp. 52-53.

44 Marcus Gee, "Here's a fine kettle of . . . ," *Globe & Mail* [Toronto], May 16, 2003, p. A23; Woodard, *ibid.*, pp. 223-231.

45 Michael Temelini, "The Rock's newfound nationalism," [Toronto] *Globe & Mail*, June 29, 2007, p. A17.

46 For background, see Michael Harris, *Lament for an Ocean* (1998); Woodard, 2004, *op. cit.*, pp. 223-231.

47 "Fishers fear tough catch curbs again after cod-stocks claim," *Aberdeen Press and Journal*, Oct. 23, 2001, p. 19; John McQuaid, "Overfished waters running on empty," *New-Orleans Times-Picayune*, March 24, 1996, p. A39.

48 Carl Safina, "Fishing off the deep end — and back," *Multinational Monitor*, Sept. 1, 2003, p. 8.

49 Anne Platt McGinn, "Rocking the Boat," Worldwatch Institute, June 1998, pp. 43-44.

50 Margaret Munro, "Fuel subsidies keep trawlers 'strip-mining' sea," *Vancouver Sun*, Nov. 17, 2006, p. A3; see also U. R. Sumaila and D. Pauly, (eds.), "Catching more bait: a bottom-up re-estimation of global fisheries subsidies," Fisheries Centre Research Reports 14(6), p. 2. University of British Columbia, www.fisheries.ubc.ca/members/dpauly/chaptersInBooksReports/2006/ExecutiveSummaryCatchingMoreBait.pdf.

51 Patricia J. James, "Cousteau says lack of care impacts sea," *Telegram & Gazette* (Mass.), May 12, 2006, p. B1.

52 The Serengeti — a vast plain stretching from Kenya to northern Tanzania — is famous for its extensive wildlife.

53 Jeremy B. C. Jackson, "What was natural in the coastal oceans?" Proceedings of the National Academy of Sciences, May 8, 2001, pp. 5412-3; Woodard, 2000, *op. cit.*, pp. 160-161.

54 Woodard, *ibid.*, pp. 208-215; Juliet Eilperin, "US wants polar bears listed as threatened," *The Washington Post*, Dec. 27, 2006, p. 1.

55 Fred Pearce, "Failing ocean circulation raises fears of mini ice-age," NewScientist.com news service, Nov. 30, 2005, http://media.newscientist.com/article.ns?id=dn8398.

56 Elizabeth Kolbert, "The darkening sea," *The New Yorker*, Nov. 20, 2006, p. 67; Usha Lee McFarling, "A chemical imbalance," *Los Angeles Times*, Aug. 3, 2006, p. 1.

57 Woodard, 2004, *op. cit.*, pp. 212-213.

58 John Charles Kucinich, *Killing Our Oceans:*

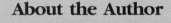

About the Author

Colin Woodard has reported from more than 40 foreign countries on six continents and lived in Eastern Europe for more than four years. He is the author of *Ocean's End: Travels Through Endangered Seas*, a narrative, nonfiction account of the deterioration of the world's oceans. He also writes for *The Christian Science Monitor* and *The Chronicle of Higher Education*. His previous *CQ Global Researcher* was on climate change.

Dealing with the Mass Extinction of Marine Life (2006), pp. 56-57.

[59] "World Digest," *The Capital* (Annapolis, Md.), Nov. 25, 2006, p. A2.

[60] Petro Parravano and Lee Crockett, "Who should own the oceans?" *San Francisco Chronicle*, Sept. 25, 2000.

[61] See Clemmitt, *op. cit.*, p. 941.

[62] Jane Lubchenco, "Global changes for life in oceans," Conference presentation, *M/S Fram* off Greenland, Sept. 8, 2007; Jon Nevill "Marine no-take areas: How large should marine protected area networks be?" white paper, Sept. 4, 2006, available from www.onlyoneplanet.com.au.

[63] For background, see Rachel S. Cox, "Eco-tourism," *CQ Researcher*, Oct. 20, 2006, pp. 865-888.

[64] For background see Colin Woodard, "Belizean bonanza," *The Chronicle of Higher Education*, July 2, 2004, http://chronicle.com/subscribe/login?url=/weekly/v50/i43/43a01301.htm.

[65] *Ibid.*

[66] "U.S Ocean Policy Report Card," Joint Ocean Commission Initiative, February 2007, www.jointoceancommission.org/images/report-card-06.pdf; Colin Woodard, "Faraway, natural and beautiful and it will stay that way," *Trust*, Fall 2006, pp. 2-11.

[67] For a full discussion, see Woodard, *The Lobster Coast, op. cit.*, pp. 267-273.

[68] John Cordell, ed., *Sea of Small Boats, Cultural Survival* (1989), pp. 337-367; R.E. Johannes, "Traditional Law of the Sea in Micronesia," *Micronesica*, December 1977, pp. 121-127; Janice Harvey and David Coon, "Beyond the Crisis in the Fisheries: A proposal for community-based ecological fisheries management," Conservation Council of New Brunswick, 1997.

[69] Brad Warren, "Conserving Alaska's Oceans," Marine Conservation Alliance, 2006, www.marineconservationalliance.org/news/1359_MCA_Report_for_download.pdf.

[70] "Population Trends Along the Coastal United States 1980-2003, National Oceanographic and Atmospheric Administration, March 2005.

[71] Peter Weber, "Abandoned Seas: Reversing the Decline of the World's Oceans," Worldwatch Institute, November 1993, pp. 17-24, www.worldwatch.org/node/874; Woodard, 2000, *op. cit.*, p. 45.

[72] Woodard, *ibid.*, p. 116.

[73] Weber, *op. cit.*, p. 20; Pew Oceans Commission, *op. cit.*, p. 54.

[74] "National Cooperative Approach to Integrated Coastal Zone Management," National Resource Management Ministerial Council, 2006, www.environment.gov.au/coasts/publications/framework/pubs/framework.pdf.

[75] "Report to the European Parliament and Council: An evaluation of ICZM in Europe," European Commission, June 7, 2007, http://eurlex.europa.eu/LexUriServ/LexUriServ.do?uri=CELEX:52007DC0308:EN:NOT.

[76] Leon Pannetta and James Watkins, "State's map for saving the oceans," *The Washington Post*, Feb. 3, 2007, p. A15.

[77] See Elizabeth C. Economy, "The Great Leap Backward?" *Foreign Affairs*, September/October 2007, pp. 38-59.

[78] James Vassallo, "Canada becoming a 'pariah' over trawling," *Prince Rupert Daily News* (BC), Nov. 28, 2006, p. 2.

FOR MORE INFORMATION

Belize Coastal Zone Management Institute, P.O. Box 1884; Belize City, Belize; Central America; +1-501-223-0719; www.coastalzonebelize.org. Coordinates policies that will affect the health of Belize's barrier reef system.

Black Sea Ecosystem Recovery Project, www.bserp.org. A network of scientists and policy experts sponsored by the U.N. Development Program.

Canadian Department of Fisheries and Oceans, 200 Kent St., 13th Floor, Station 13228; Ottawa, Ontario, Canada K1A 0E6; 613-993-0999; www.dfo.ca. Manages Canada's fisheries, oceans and marine research.

International Whaling Commission, The Red House, 135 Station Road, Impington, Cambridge, CB24 9NP, United Kingdom; +44-1223-233-971; www.iwcoffice.org. Regulates whaling and the conservation of whales.

Joint Ocean Commission Initiative, c/o Meridian Institute,1920 L St., N.W., Suite 500, Washington, DC 20036-5037; (202) 354-6444; www.jointoceancommission.org. A vehicle for the chairs of the Pew Oceans Commission and U.S. Commission on Ocean Policy to push implementation of their recommendations.

National Fisheries Institute, 7918 Jones Branch Dr., Suite 700, McLean, VA 22102; (703) 752-8880; www.nfi.org. The main lobbying and advocacy association for commercial fishing and seafood processing industries in the United States.

National Marine Fisheries Service, 1315 East-West Highway, Silver Spring, MD 20910; (301) 713-2239; www.nmfs.noaa.gov. U.S. federal agency responsible for the management of fisheries and international fishing agreements.

Oceana, 2501 M Street, N.W., Suite 300, Washington, DC 20037-1311; (202) 833-3900; www.oceana.org. The world's largest oceans-based environmental group, with branch offices in Europe, South America and the U.S. West Coast.

Ocean Conservancy, 1300 19th St., N.W., Washington, DC 20036; (202) 429-5609; www.oceanconservancy.org. Nonprofit organization promoting healthy and diverse ocean ecosystems and opposing practices that threaten ocean life.

Pew Institute for Ocean Science, 126 East 56th St., New York, NY 10022; (212) 756-0042; www.pewoceanscience.org. A Pew-funded body that conducts, supports, and disseminates scientific information on protecting the world's oceans.

Seaweb, 8401 Colesville Road, Suite 500, Silver Spring, MD 20910; (301) 495-9570; www.seaweb.org. Environmental advocacy group promoting ocean conservation, with branch offices in London and Paris.

The Shark Alliance, Rue Montoyer 39, 1000 Brussels, Belgium; www.sharkalliance.org. A continent-wide coalition of nongovernment organizations working to save Europe's sharks.

University of British Columbia Fisheries Centre, 2202 Main Mall, University of British Columbia, Vancouver, B.C.; Canada V6T 1Z4; 604-822-2731; www.fisheries.ubc.ca. Leading fisheries research institution and home to the scientists who predict the world is facing "seas of slime" and may end up "fishing down the marine food webs."

Bibliography

Selected Sources

Books

Acheson, James M., *Capturing the Commons: Devising Institutions to Manage the Maine Lobster Industry*, **University Press of New England, 2003.**
An anthropologist examines the successes of Maine's lobster fishery and state efforts to incorporate traditional lobster-fishing practices into law.

Barker, Rodney, *And the Waters Turned to Blood*, **Simon & Schuster, 1997.**
An investigative journalist chronicles of the rise of a dangerous marine microorganism in the Mid-Atlantic region of the United States and efforts to contain it.

Ellis, Richard, *The Empty Ocean*, **Island Press, 2004.**
The author of more than 10 books on the oceans and a research associate at the American Museum of Natural History in New York uses history, anecdote and stunning facts to chronicle humanity's predation on the oceans.

Fujita, Rodney M., *Heal the Ocean: Solutions for Saving Our Seas*, **New Society Publishers, 2003.**
A marine ecologist describes successful efforts to confront the problems in the world's oceans.

Meinesz, Alexandre, *Killer Algae: The True Tale of a Biological Invasion*, **University of Chicago Press, 1999.**
The scientist who discovered the problem of *Caulerpa taxifolia*, the "killer seaweed" that is taking over the Mediterranean, gives a first-hand account of how bad an invading species can be.

Molyneaux, Paul, *The Doryman's Reflection: A Fisherman's Life*, **Avalon, 2005.**
A former commercial fisherman gives a personal account of the collapse of U.S. fisheries, with critical insights into the mindset and values that destroyed a beloved culture.

Warner, William, *Distant Water: The Fate of the North Atlantic Fishermen*, **Little, Brown, 1977.**
The classic account of the rise of factory trawlers and the damage they did prior to the advent of 200-mile limits.

Woodard, Colin, *Ocean's End: Travels through Endangered Seas*, **Basic Books, 2000.**
The author, a freelance journalist specializing in ocean issues, describes the degradation of the world's oceans.

Articles

Garrison, Virginia, *et al.*, **"African and Asian dust: from desert soils to coral reefs,"** *Bioscience*, **May 1, 2003.**
Scientists hypothesize that the world's coral reefs may be dying because of dust from African desertification.

Kolbert, Elizabeth, **"The Darkening Sea,"** *The New Yorker*, **Nov. 20, 2006.**
The author provides a detailed and readable account of what carbon emissions are doing to the ocean through acidification.

Kunzig, Robert, **"Twilight of the Cod,"** *Discover*, **April 1995.**
Kunzig chronicles the collapse of the New England and Newfoundland cod fisheries in the late 1980s.

Weiss, Kenneth R., *et al.*, **"Altered Oceans,"** *Los Angeles Times*, **July 30-Aug. 3, 2006.**
In a Pulitzer Prize-winning series, Weiss and others describe the oceans' demise, with emphasis on the "rise of slime."

Reports and Studies

Dorsey, Eleanor, and Judith Pederson (eds.), *Effects of Fishing Gear on the Sea Floor of New England*, **Conservation Law Foundation, 1998.**
This report catalogs the damage caused by bottom trawlers, as viewed by scientists, environmentalists and fishermen.

Harvey, Janice, and David Coon, "Beyond the Crisis in the Fisheries: A proposal for community-based ecological fisheries management," Conservation Council of New Brunswick, 1997.
A groundbreaking report by the Canadian conservation council argues for giving proprietary ownership of fishing resources to the local communities that have relied on them.

Pew Oceans Commission, "America's Living Oceans: Charting a Course for Sea Change," May 2003.
The commission describes the critical state of the world's oceans and recommends an ecosystem-based approach to future ocean management.

United Nations Food and Agriculture Organization, "State of the World's Fisheries and Aquaculture 2006," 2007, www.fao.org/docrep/009/A0699e/A0699e00.htm.
The FAO's latest official report on the state of world fisheries says a quarter of the commercial fish stocks have been over-exploited or depleted and about half are fully exploited.

U.S. Commission on Ocean Policy, "An Ocean Blueprint for the 21st Century," 2004.
The recommendations of a panel established by Congress echo those of the Pew Oceans Commission, while extending analysis to energy, environmental education and the issues afflicting the Great Lakes.

The Next Step:

Additional Articles from Current Periodicals

Commercial Whaling

"Commission Must Compromise for Sake of World's Whales," *Canberra Times*, May 28, 2007.

The 2007 meeting of the International Whaling Commission serves as a reminder that commercial whaling poses political and financial questions as much as environmental and ethical ones.

"Iceland Breaks Ban on Commercial Whaling With First Fin Whale Kill in 20 Years," The Associated Press Worldstream, Oct. 23, 2006.

Ignoring a worldwide moratorium that went into effect in 1986, Iceland announced the harpooning of an endangered fin whale — the first such killing in 20 years.

"Too Much Blubber," *The Economist*, June 17, 2006.

The international moratorium on whaling seeks to conserve whales in the context of a commercial fishery, not to preserve whales for their own sake; therefore, it is reasonable for countries to reexamine the moratorium.

Ecosystem-Based Management

"Canada's New Government Announces a New Approach to Canadian Fisheries," CCN-Matthews News Agency (Canada), April 12, 2007.

The Canadian Minister of Fisheries and Oceans has announced a three-year, $61 million investment in ecosystem-based science and the use of the data in fisheries-management decisions.

Crowder, Larry, "The Oceans Need a Hand," *Los Angeles Times*, Aug. 6, 2006, p. A7.

Ecosystem-based management would allow scientists to diagnose all the problems affecting the oceans and not just those of individual systems.

Richardson, John, "Excellence Is Key to Integrated Maritime Policy," *Lloyd's List* (England), Nov. 30, 2006.

Ecosystem-based planning can only work if coastal communities accept the necessary responsibilities.

Human-Caused Destruction

Corder, Mike, "Sewage Is Growing Problem for Coastal Populations, Environments, UN Report Says," The Associated Press Worldstream, Oct. 4, 2006.

Untreated sewage pouring into the world's oceans is endangering people and animals, according to the U.N. Environment Programme.

Perlman, David, "Greenhouse Gas Turning Oceans Acidic," *The San Francisco Chronicle*, July 6, 2006, p. A4.

A team of government-sponsored scientists has concluded that carbon dioxide is threatening destruction of shell-building organisms that form the base of the entire marine food web.

von Radowitz, John, "Crisis in the Oceans May Put Fish Off Menu by 2048," *The Western Mail* (Wales), Nov. 3, 2006.

The current rate of pollution, habitat destruction and climate change threatens to kill off nearly all ocean species by 2048.

Whitty, Julia, "The Fate of the Ocean," *Mother Jones*, March/April 2006, p. 32.

Ocean problems that were once encountered on a local scale have become pandemic, and even worse.

Ocean-Floor Trawling

"Greenpeace Urges Pacific Nations to Oppose Bottom Trawling" Asia Pulse News Agency, Sept. 27, 2006.

The environmental group Greenpeace is urging Pacific island nations to call on the U.N. to establish a moratorium on unregulated high-seas bottom trawling.

"Ministers Welcome Progress on Bottom-Trawling Regulation" New Zealand Press Association, Nov. 24, 2006.

Despite the U.N. General Assembly's failure to institute a ban on bottom trawling, the New Zealand minister of fisheries praises progress made toward regulating the practice.

Christensen, Jon, "Unlikely Partners Create Plan to Save Ocean Habitat Along With Fishing," *The New York Times*, Aug. 8, 2006, p. F3.

The Nature Conservancy has agreed to buy fishing permits and boats from trawling fisherman in exchange for the implementation of three "no-trawl zones" off the California coast.

CITING CQ GLOBAL RESEARCHER

Sample formats for citing these reports in a bibliography include the ones listed below. Preferred styles and formats vary, so please check with your instructor or professor.

MLA STYLE

Flamini, Roland. "Nuclear Proliferation." CQ Global Researcher 1 Apr. 2007: 1-24.

APA STYLE

Flamini, R. (2007, April 1). Nuclear proliferation. *CQ Global Researcher*, 1, 1-24.

CHICAGO STYLE

Flamini, Roland. "Nuclear Proliferation." *CQ Global Researcher*, April 1, 2007, 1-24.

Voices From Abroad:

FERNANDO CURCEIO
Director, Fisheries Resources Department, Spain

We're trawling on sand

"The EU and Spain maintain that this type of [trawling] measure would provoke illegal fishing activities throughout the world and there would be a proliferation of boats with flags of convenience. . . . We are not trawling over ecosystems that are in danger. But we're not prepared to tell our boats to stop fishing when there's nothing to protect."

AP Worldstream, November 2006

YVONNE SADOVY
Biologist, University of Hong Kong

The wealthier, the fancier

"The taste for fancy, novel, coral reef fish is spreading as wealth is spreading in mainland China. Countries must limit export quotas, create protected areas and encourage consumers to select less threatened species."

National Geographic, April 2007

ASBJÖRN BJÖRGVINSSON
Chairman, Icelandic Whale Watching Association

Whaling provides no social benefits

"There is no way whale hunting can be defined as sustainable industry. There are no economic benefits from whaling as there are no markets for the products.

And whaling is without a doubt damaging to Iceland's international image as a nature destination. It is hard to find any positive social effects from whaling."

The Irish Times, October 2006

ALEX ROGERS
Senior Research Fellow, Zoological Society of London

Trawling burden must shift to governments

"Our research actively demonstrates the vulnerability of deep-sea corals and their associated biodiversity to trawling across seamounts. . . . It is essential that the burden of proof shifts to governments and fisheries when deciding whether it is appropriate to exploit these irreplaceable ecosystems."

The Independent (England), November 2006

PETE DAWSON
CEO, New Zealand Confederation of Commercial Fishermen

Trawling helps explore resources

"Not only do we have a high exchange rate, high fuel costs and a depressed fish market — but now we've got the burden of reduced access to fisheries as well. [Trawling restrictions] deprive a nation of investigating and exploring its continental shelf for further resources."

New Zealand Herald, May 2007

VALENTIN ILYASHENKO
Member, International Whaling Commission, Russia

Japan's whaling dilemma

"Japan has big problems. Providing their traditional food is prohibited by international organizations, and I have no doubts that coastal whaling has analogies with aboriginal whaling."

Kyodo News Service (Japan), May 2007

JULIAN CRIBB
Professor, University of Technology, Sydney, Australia

Put it in the hands of consumers

"While the experts may argue about what is and isn't a sustainable fish stock, the idea of transferring the onus for protecting marine species to consumers, rather than fishers, is a good one. Fishers can hardly be blamed for trying to make a living, but consumers can always purchase and consume more

wisely, sending fishers the correct market signals."

The Australian, February 2007

XAVIER PASTOR
Executive Director, Oceana Europe

Overfishing's domino effect

"European Union vessels search for new fishing grounds further and further in the world's oceans because European waters are widely overfished. . . . We must prevent the same thing from happening in the Pacific."

http://europe.oceana.org

USSIF RASHID SUMAILA
Researcher, University of British Columbia Fisheries Centre

Investing in fish brings returns

"Rebuilding fish [stocks] is like putting money in an interest-bearing savings account instead of spending down the balance. If you grow the principal, you can live off the interest."

PR Newswire, October 2005

DISAPPEARING SPECIES

BY TOM ARRANDALE

Excerpted from the CQ Researcher. Tom Arrandale. (November 30, 2007). "Disappearing Species." *CQ Researcher*, 985-1008.

Disappearing Species

BY TOM ARRANDALE

THE ISSUES

From tree frogs to African gorillas to Mediterranean sharks, some of Earth's most intriguing — and beloved — wild creatures face uncertain futures. Now biologists are predicting that two-thirds of the Northern Hemisphere's estimated 20,000 polar bears will disappear in the next 50 years. The cause: Warming temperatures are rapidly melting the Arctic ice, where they spend most of the year hunting seals and other mammals.

"As the sea ice goes, so goes the polar bear," said Steven Amstrup, a biologist with the U.S. Geological Survey. [1] The agency worries the bears could be extinct in three Arctic ecosystems within 75 years, with a smaller population barely hanging on along Canada's northern coast.

Throughout the world, thousands of other distinctive animals and plants may also be hurtling toward oblivion. India has only 1,500 tigers left, down 50 percent in the last six years, and poaching and habitat loss caused by growing human populations over the next 20 years could condemn tigers to "dwindle to the point of 'ecological extinction,' in which their numbers are too few to play their role as top predator in the ecosystem," said researchers for the World Wildlife Fund and other conservation organizations. [2]

Even inside wildlife sanctuaries, illegal hunting for food and animal parts is taking a rising toll on tigers as well as Africa's gorillas, humans' closest primate relatives. The International Union for Conservation of Nature and Natural Resources (IUCN), also known as

Greenpeace activists in Rio de Janeiro, Brazil, call for an end to timbering in the Amazon rain forest. Environmentalists say highway construction, timbering, mining and agricultural-colonization schemes are accelerating deforestation, river contamination and the demise of indigenous people. Twenty percent of the Amazon has been cut in the past 40 years.

the World Conservation Union, warns the combined toll from the Ebola virus and demand for "bush meat" has reduced the remnant population to critically endangered status, or put it on a downward spiral toward total annihilation in the wild.

The IUCN warned in an October report that nearly a third of the world's 394 known species of apes, monkeys, lemurs and other primates are now in danger of extinction. The study focused on 25 species imperiled by hunting, illegal wildlife trade and logging in the tropical forests that provide them with dwindling strongholds. In Ghana and Ivory Coast, a rare red colobus may already be gone, while remnants of Vietnam's golden-headed langur and China's Hainan gibbon number in the dozens.

"You could fit all the surviving members of these 25 species in a single football stadium; that's how few of them remain on Earth today," said Conservation International President Russell A. Mittermeier, who chairs the IUCN's primate panel. [3]

Meanwhile, the IUCN recently added 188 more species to its "Red List" of threatened wildlife. In addition to the gorilla, new listings included the Sumatran orangutan, Indian crocodile, Asian red-headed vulture, two Galapagos Islands corals, Mexico's Santa Catalina Island rattlesnake and 10 seaweed species. A Malaysian herb, the woody-stalked Begonia, was listed as officially extinct, and China's Yangtze River dolphin was listed as possibly extinct. [4]

But the most staggering species losses are likely among insects, amphibians and millions of even smaller organisms that inhabit remote tropical forests or deep ocean waters.

"Each species is a small piece of [the global threat], but it all adds up," says renowned tropical biologist Thomas E. Lovejoy, president of the Washington-based Heinz Center for Science, Economics, and the Environment. Lovejoy, who coined the term "biological diversity" in 1980, concludes that "we're in the first stages of a mass extinction."

At current rates, "one-third to one-half of all species on Earth are predicted to be extinguished in the next century," adds Duke University conservation biologist Stuart Pimm. Biologists so far have described a total of 1.9 million species of plants and animals, but the

'Hot Spots' List Includes 34 Regions Worldwide

Thirty-four regions around the world are "hot spots," or vulnerable to destruction, according to Conservation International. The regions include 22 tropical forests, mostly in tropical Africa, South America and Southeast Asia. Together, they once covered nearly 16 percent of the globe but today only 14 percent of their original area remains. Hot spots result from population growth and efforts by Third World nations to boost food production and stimulate lagging economies, often by timbering and development efforts in tropical forests and other undeveloped lands.

Source: Conservation International, February 2005

IUCN has completed formal studies on the status of fewer than 3 percent. [5] The total number of plants and animals on Earth, however, may be exponentially larger. E. O. Wilson, the famed Harvard University entomologist and best-selling author, reports that estimates of the actual number of species "range, according to the method used, from 3.6 million to 100 million or more." [6]

In the last 50 years, wildlife has come under increasing pressure as population exploded and Third World nations began developing tropical forests and other undeveloped lands to boost food production and stimulate lagging economies. In South America and Indonesia, settlers continue cutting and burning tropical forests that comprise

the planet's most biologically rich ecosystems. Tropical rain forests cover only 6 percent of Earth's land surface, yet their terrestrial and aquatic habitats hold more than half of the world's known species. But the forests are being cut down at a devastating clip. Twenty percent of South America's Amazon rain forest has been lost in the last 40 years, and Africa's 800,000-square-mile Congo Basin rain forest also is rapidly being logged and converted to agricultural use.

Twenty years ago, Duke University biologist Norman Myers identified 25 vulnerable areas, or "hot spots," that house 40 percent of the planet's species. [7] Myers and other Conservation International scientists now have expanded the list to 34 regions, in-

cluding 22 tropical forests. Most are in Africa, South America and Southeast Asia, but they also include parts of Australia, Japan, the Caucasus temperate forests, China, the Caribbean and the California Coast from Mexico to Oregon.

"We have cut much of the heart out of biodiversity," Wilson wrote. [8] Many known insects, amphibians, microbes, fungi and rare plants are at severe risk, even as biologists continue to identify previously unknown species. Just in the last year, two new lizard species were found in Brazil. Also, scientific surveys in the Guinean forest of West Africa recorded three previously unknown bat species and found 16 more specimens of an endangered horseshoe bat that had been thought down to nine individuals. [9]

Gorillas and Indian Crocs Added to 'Threatened' List'

More than 16,000 species are included on the 2007 Red List of Threatened Species, including 12 percent of the birds, 20 percent of the mammals and 29 percent of the frogs and other amphibians that biologists have studied so far. Recent additions to the list include the African gorilla, Sumatran orangutan, Indian crocodile, Asian red-headed vulture, two Galapagos Islands corals and Mexico's Santa Catalina Island rattlesnake. China's Yangtze River dolphin is listed as possibly extinct. The list is compiled by the International Union for Conservation of Nature and Natural Resources (IUCN).

	Number of described species	Number of species evaluated by 2007	Number of threatened species in 1996/98	Number of threatened species in 2007	Number threatened in 2007, as % of species evaluated
Vertebrates					
Mammals	5,416	4,863	1,096	1,094	22%
Birds	9,956	9,956	1,107	1,217	12%
Reptiles	8,240	1,385	253	422	30%
Amphibians	6,199	5,915	124	1,808	31%
Fishes	30,000	3,119	734	1,201	39%
Subtotal	**59,811**	**25,238**	**3,314**	**5,742**	**23%**
Invertebrates					
Insects	950,000	1,255	537	623	50%
Molluscs	81,000	2,212	920	978	44%
Crustaceans	40,000	553	407	460	83%
Corals	2,175	13	N/A	5	38%
Others	130,000	83	27	42	51%
Subtotal	**1,203,375**	**4,116**	**1,891**	**2,108**	**51%**
Plants					
Mosses	15,000	92	N/A	79	86%
Ferns and allies	13,025	211	N/A	139	66%
Gymnosperms	980	909	142	321	35%
Dicotyledons	199,350	9,622	4,929	7,121	74%
Monocotyledons	59,300	1,149	257	778	68%
Green algae	3,715	2	N/A	0	0%
Red algae	5,956	58	N/A	9	16%
Subtotal	**297,326**	**12,043**	**5,328**	**8,447**	**70%**
Others					
Lichens	10,000	2	N/A	2	100%
Mushrooms	16,000	1	N/A	1	100%
Brown Algae	2,849	15	N/A	6	40%
Subtotal	**28,849**	**18**	**N/A**	**9**	**50%**
TOTAL	**1,589,361**	**41,415**	**10,533**	**16,306**	**39%**

Source: International Union for Conservation of Nature and Natural Resources

As habitat destruction continues, however, biologists fear countless species are vanishing without ever being discovered and studied.

Big, visible animals with limited ranges and slow reproductive rates that live high on the food chain generally have been most vulnerable, such as carnivores and primates. Many mammals, birds, reptiles and fish are increasingly beset by polluted water, competition from exotic competitors and well-armed poachers and industrial fishing fleets. Now, climbing temperatures are an increasing factor. In addition to putting polar bears at risk by melting Arctic ice, global warming also is bleaching coral reefs and raising the acid content in seawater, killing off plankton and other crustaceans at the base of the marine food chain.

Some biologists and observers contend that evidence is too sketchy to conclude that a mass extinction has started. But others see evidence that cumulative human impacts are launching an ecological catastrophe that could prove more devastating than climate change by itself.

"To scientists, this is an unparalleled calamity, far more severe than global warming," University of Chicago biologist Jerry A. Coyne and Harvard University Professor Hopi E. Hoekstra recently warned. "Life as we know it would be impossible if ecosystems collapsed," they contend. "Yet that is where we're headed if species extinction continues at its current pace." [10]

The Earth's accumulating loss of biodiversity "is huge and it's irreversible, and global warming isn't irreversible," adds Duke University's Pimm. "Once you drive a species to extinction, it's gone." Pimm and Duke colleague Myers also describe biodiversity loss as "today's most significant environmental problem. We can clean up pollution, but we cannot re-create species. 'Jurassic Park' is a [movie] fantasy." [11]

Yet, Pimm and others see some hope that humans will eventually recognize that preventing extinction of species is in their own best interest. Pimm and Myers note, for example, that "biodiversity conservation offers a host of spin-off benefits, like protecting watersheds that are essential for drinking water and fisheries." Besides, Pimm adds, "who would want to tell their kids that they witnessed the demise of lions and tigers and bears?"

As scientists study the viability of Earth's species and threatened habitats, here are key questions they are asking:

Is the mass extinction of wild species imminent?

Species have evolved and disappeared since life on Earth began nearly 4 billion years ago. Fossil records reveal at least five mass extinctions — the result of natural cataclysms — that eliminated most species, the most recent 65 million years ago. Over time, the world's biodiversity has always bounced back. But now some biologists fear human development of global resources is setting off a sixth mass extinction, one that could permanently impoverish global biological resources.

Up to 30,000 plant and animal species are now disappearing every year because of human activity, biologists estimate. The United Nations Environment Programme warned recently that "changes to biodiversity currently underway . . . are more rapid than at any time in human history." [12] The IUCN calculates that species are now being lost at between 1,000 to as much as 10,000 times higher than the "background" rate at which natural evolutionary forces would cause them to vanish.

The IUCN's most recent "Red List" identifies more than 16,000 species as potentially threatened with extinction, including 12 percent of the birds, 20 percent of the mammals and 29 percent of the frogs, turtles and other amphibians that biologists have studied so far.

So far, roughly three-quarters of known extinctions have occurred on islands, where distinctive species that evolved separately from the rest of the world have been extremely vulnerable when humans began moving into their habitat, according to the U.N.'s *Millennium Ecosystem Assessment*. [13] In the last two decades, however, biologists have found half of the extinctions on continental mainland, with tropical rain forests especially vulnerable.

Biodiversity losses are compounded when native species are displaced by the competitors that thrive in the urban settings or farms that replace forests and other native ecosystems. A panel of *Millennium Assessment* biologists found that bird extinctions so far have been concentrated on oceanic islands, especially Hawaii and New Zealand. Sri Lanka has the highest number of recorded amphibian extinctions, but losses are accelerating in South American mountains and forests, Caribbean islands and Australia. Mammals are vanishing most rapidly in Australia and the Caribbean. [14] Some species like the American bald eagle and peregrine falcon have rebounded, but scientists aren't optimistic that other declining plants and animals

can be rescued as remaining wild habitat is chopped up. "Nature is resilient, but it's not resilient enough to overcome the changes we're now experiencing," the University of Chicago's Coyne says. "We're taking away all the space that it needs to come back."

But skeptics contend that scientific data is too sketchy to conclude a biological holocaust is looming. Amy L. Kaleita, an expert on soil and water conservation at Iowa State University, agrees that "many ecosystems are quite delicate, and removing certain species or communities has consequences." However, "there are very few quantifiable measures of biodiversity," she adds. "Once you get to larger ecosystems, it becomes very difficult" to assess the extent of species losses.

Kaleita co-authored a 2007 assessment of environmental threats published by the conservative American Enterprise Institute and Pacific Research Institute arguing that whether or not global biodiversity "should be considered in 'crisis' depends on which guesstimate of the magnitude of the problem one finds most plausible. As usual, the more alarmist projections receive the most media notice." [15]

In a recent *BioScience* article, several biologists concurred that "the information is so meager and poor that the evaluations in many cases are merely intelligent guesses." [16] Predicting future species losses "is a very tricky thing, because everybody's overstating the case on every side of the debate," says co-author Daniel B. Botkin, a University of California-Santa Barbara biologist. Botkin himself works on saving whales and other endangered wildlife, but he worries that focusing debates on exaggerated warnings about mass extinctions based on questionable scientific assumptions are "moving people away from what they ought to be doing about saving species."

In his controversial 2001 book *The Skeptical Environmentalist*, Bjorn Lomborg, a Danish professor of statistics, contended that a commonly used forecast of losing 40,000 species every year "has become part of our environmental litany." Yet that commonly accepted figure is based on assumptions that if 90 percent of a habitat is removed, half the species found there will vanish, he says. But Lomborg notes that the species/area relationship has not held up: When biologists attempted to count species numbers in the transformed habitats of Eastern North America, Brazil's Atlantic Coast forests and Puerto Rico, the rate of extinction turned out to be much less dramatic, he says. Lomborg notes that more cautious models agree that extinctions are climbing above natural levels — but at a rate of only about 0.7 percent every 50 years, much smaller than the most alarming projections. Losing that many "over a limited time span is not a catastrophe but a problem — one of many that mankind still needs to solve," he contends [17]

In fact, Seymour Garte, a professor of environmental and occupational health at the University of Pittsburgh, says worst-case predictions of mass extinctions overlook real progress in preserving habitat and restoring endangered species in North America and other developed areas. But Garte concurs that habitat destruction is taking a heavy toll on biodiversity in the Amazon and Southeast Asian rain forests. "They've got it half right," Garte says. "We really don't know what's going on."

Will global warming increase species losses?

Wildlife biologists are detecting troubling signs that rising atmospheric temperatures have begun compounding the habitat losses and other ecological pressures already devastating many plant and animal species. In April 2007, the Intergovernmental Panel on Climate Change (IPCC) reported "approximately 20-30 percent of plant and animal species assessed so far . . . are likely to be at increasingly high risk of extinction as global mean temperatures exceed a warming of 2-3 degrees C above

Continued on p. 993

Four in Danger

Threatened species on the "Red List" compiled by the International Union for Conservation of Nature and Natural Resources (IUCN) include (clockwise, from top left): the Western lowland gorilla, Sumatran orangutan, Egyptian vulture and Fiji crested iguana. A total of 41,415 species are listed, including 16,306 facing extinction.

NHPA/Photoshot/Martin Harvey

NHPA/Photoshot/Daniel Heuclin

NHPA/Photoshot/Henry Ausloos

NHPA/Photoshot/A.N.T. Photo Library

Calculating the Value of Ecosystem 'Services'

Scientists try to show conservation makes financial sense.

Ecologists and economists have been debating for years about the economic value — if any — of wild species. In recent years, however, there has been growing recognition that natural landscapes, including the plants and animals they support, provide critical "ecosystem services" to human communities worth $33 trillion a year.

After Hurricane Katrina raged across the U.S. Gulf Coast in 2005, biologists contended that the damage was so severe because development had eliminated much of the Gulf of Mexico's natural coastal wetlands that normally would have absorbed some of the storm's fury. A recent study, moreover, catalogs the storm's unprecedented destruction of 320 million trees in the region.[1] Similarly, scientists say coastal damage from a 2004 tsunami that killed more than 200,000 people in Southeast Asia was greatest along coastlines where mangrove swamps that once buffered storms have been replaced by shrimp farming ponds. Around the globe, the loss of natural biological diversity similarly may be eroding the benefits of nature that once were taken for granted, such as clean water and air, flood control, disease prevention and healthful foods.

"Ecosystem services are essential to human existence and operate on such a grand scale and in such intricate and little-explored ways, that most could not be replaced by technology," Stanford University ecologist Gretchen C. Daily wrote. "Yet, escalating impacts of human activities on natural ecosystems imperil their delivery."[2]

In 2005, the *Millennium Ecosystem Assessment* study commissioned by the United Nations found that economic development around the world has begun to curtail the Earth's ability to provide for future generations. Drawing on the findings of nearly 1,400 experts, the project assessed 24 benefits that provide materials, regulate environmental conditions or enhance human spiritual and aesthetic values. It concluded that only four functions — farming, livestock grazing, fish farming and sequestering carbon dioxide — have improved in the last half-century. Fifteen are in decline. In addition, ecosystem damage has degraded nature's ability to purify water and air, control floods, prevent soil erosion and pollinate crops.

The assessment's board declared that a potential massive wave of species extinctions is "threatening our own well-being. The loss of services derived from ecosystems is a significant barrier to the achievement of (U.N.) goals to reduce poverty, hunger and disease."[3]

The problem, Daily says, is that "until now there has been little incentive to measure or manage natural capital: It has been treated as essentially inexhaustible." A decade ago, a team of scientists came up with an estimate that ecosystems supply services worth $33 trillion a year to the global economy. In one effort that showed how natural services benefit even the largest cities, New York City saved $10 billion on a new water filtration plant by investing $1.5 billion in protecting its Catskill Mountains watershed against development. Costa Rica in 1996 began taxing fossil fuels to pay landowners to leave tropical forest intact, while an electric utility, water companies and a brewery now pick up the cost of protecting the watershed of Quito, Ecuador.[4] The Chicago Climate Exchange now brokers carbon trades that reward landowners for maintaining forests and rangelands that absorb and store carbon dioxide from the atmosphere.

Economists, meanwhile, are trying to sort through how to define and measure ecological services and account for them in a "green" gross national product. "The benefits of nature are too important and too large to be 'left off the table' of national accounting," writes James Boyd, a senior fellow at Resources for the Future.[5]

Some biologists and environmental groups caution that calculating ecological services can never replace conserving nature for its own sake. Indeed, they doubt financial markets will ever be capable of keeping up with technological change or protecting predators that kill people or livestock and other species. For example, scientists calculated that native bees provided $60,000 a year in pollination services to a Costa Rican coffee plantation. After coffee prices dipped, however, the owners shifted to growing pineapples that don't require pollination. "If we oversell the message that ecosystems are important because they provide services, we will have effectively sold out nature," wrote biologist Douglas J. McCauley, a Stanford graduate student.[6]

But ecologists are now working on other methods to quantify the economic contributions of ecosystems and devise market-based mechanisms to conserve them. In 2006, Stanford teamed with The Nature Conservancy and World Wildlife Fund to undertake the Natural Capital Project, a 10-year venture to demonstrate how the value of ecological services can encourage conservation in such diverse habitats as China's Yangtze River Basin, Tanzania's mountain rain forests and California's Sierra Nevada Mountains.

Calculating the value of ecosystems, scientists hope, will encourage governments, industries and native communities to keep them intact. But the calculations will be arduous, and meanwhile development in many regions is accelerating.

"I worry that we're not going to have enough time," Daily says.[7]

---

[1] Henry Fountain, "Katrina's Damage to Trees May Alter Carbon Balance," *The New York Times*, Nov. 20, 2007, p. F3.

[2] Gretchen C. Daily, "Developing a Scientific Basis for Managing Earth's Life Support Systems," *Conservation Ecology* 3, 1999, www.consecol.org.

[3] *Millennium Ecosystem Assessment, Living Beyond Our Means*, Natural Assets and Human Well-Being, 2005, p. 3.

[4] David Wolman, "How to Get Wall Street to Hug a Tree," *Los Angeles Times*, Feb. 11, 2007.

[5] James Boyd, *The Nonmarket Benefits of Nature: What Should Be Counted in Green GDP?* Resources for the Future, May 2006.

[6] Douglas J. McCauley, "Selling out on nature," *Nature*, Sept. 7, 2006.

[7] Quoted in Peter Aldhous and Bob Holmes, "Radical rethink required to save rare species; Conservationists are trying a host of new strategies in a bid to avert a modern-day extinction crisis," *New Scientist*, Sept. 15, 2007.

Continued from p. 991

pre-industrial levels." The IPCC scientists noted that a European heat wave in 2003 sparked wildfires and drought that damaged forests and freshwater ecosystems, potentially foreshadowing "very likely progressive impairment of ecosystem composition and function if such events increase in frequency." [18]

Researchers already have seen evidence in some regions: Sea levels have begun rising, snow cover has dropped, glaciers and Arctic ice have melted and precipitation patterns are changing. The number of glaciers in Glacier National Park has fallen to 26 from 150; and the rest could melt in another 25 years. Saltwater has begun intruding on Florida Keys marshlands, while bark beetle outbreaks have begun killing spruce trees on a million acres of Kenai Peninsula forest in Alaska. Drought and invasive grasses are making the Southwest's native saguaro cacti and Joshua trees more susceptible to exotic insects and devastating wildfire.

"This is actually something we see from pole to pole, and from sea level to the highest mountains in the world," said Lara Hansen, the World Wildlife Fund's chief climate scientist. [19]

Climate change "may not lead to extinctions per se, but it will alter how ecosystems function," accelerating biodiversity losses, says Dennis Ojima, a Heinz Center global warming expert. Beginning in the 1990s, entomologists found evidence that rare butterflies have been forced northward as temperatures grew warmer, and British researchers from Earthwatch reported badgers have responded to shifting climate conditions by bearing young earlier in the year. Populations of small mammals like voles and wood mice have declined as diminishing vegetative cover made them more vulnerable to foxes, weasels and owls. [20]

Like polar bears, walruses may be losing access to prime ocean feeding grounds as ice recedes in the Arctic. Yellowstone National Park's rebounding grizzly bear population could again collapse if hotter temperatures combine with disease to eliminate white bark pine trees that supply bears with nutritious nuts when they prepare to den in winter. Rocky Mountain pikas, distinctive small rodents, have been forced to higher elevations by warming daytime temperatures. Federal and state fisheries managers have been restocking cold-water Rocky Mountain streams with native cutthroat trout, but they worry that restored populations will be cut off from each other by rising water temperatures in lower-elevation rivers.

Climbing atmospheric carbon dioxide levels are making the oceans more acidic, possibly disrupting the marine food web by dissolving the exoskeletons of microscopic plankton. Rising seawater temperatures also are killing coastal coral reefs, leaving them bleached and colorless. "In the shallow seas of the Caribbean, a number of reef systems are under threat," Ojima says. "Those reef systems are critical nursery areas for important marine species."

In 2006, the federal government listed two Caribbean coral species as threatened under the Endangered Species Act. In January 2008 the U.S. government will decide whether to list the polar bear as a federally protected endangered species. "If we don't proceed to address climate change really quickly and systematically, then all of the best conservation efforts will go for naught," the Heinz Center's Lovejoy says.

However, other biologists still doubt the link between global warming and extinction. In March, Botkin and 18 international colleagues challenged the accuracy of recent predictions of the damage from climate change. They make note of what they call "the Quaternary conundrum," revealed by fossils from the last 2.5 million years (the Quaternary period), when temperatures fluctuated as glaciers covered the Northern Hemisphere and then retreated.

"While current empirical and theoretical ecological forecasts suggest that many species could be at risk from global warming, during the recent ice ages few extinctions are documented," they noted.

During previous rapid climate shifts, 68 percent of European trees — but only one North American species — disappeared, they report. North American mammoths, ground sloths and other large mammals also were eliminated, but human hunters migrating across the Bering Sea land bridge probably helped exterminate them. Currently, Botkin adds, "the evidence that global warming will have serious effects on life is thin. Most evidence suggests the contrary." [21]

Botkin agrees that climate change is real and supports replacing fossil fuels with carbon-free alternatives. But he argues that focusing on climate change could divert attention from dealing with more demonstrable threats to species. Orangutans, for instance, are clearly imperiled by loss of their forest homelands, Botkin says. "In our fear of global warming, it would be sad if we fail to find funds to purchase those forests before they are destroyed, and thus let this species go extinct."

Despite the wide consensus about the global warming threat, some pro-business groups, including the American Enterprise Institute and Competitive Enterprise Institute, still have doubts. Some conservative religious thinkers also say global warming has not been sufficiently demonstrated to justify limiting economic development that benefits humans. Countering evangelicals who back action to control carbon dioxide emissions, Professor E. Calvin Beisner of Knox Theological Seminary, in Fort Lauderdale, Fla., told lawmakers last year that 19,000 scientists have signed a petition arguing that higher CO_2 concentrations actually "enhance plant growth and so contribute to feeding the human population and all other species." [22]

Will losing species impair human well-being?

Many experts say humans will pay a price if too many species are lost. As species disappear, the ecosystems they live in are also collapsing or disappearing, taking with them sources of pure water, food and other necessities. "Human activities have taken the planet to the edge of a massive wave of species extinctions, further threatening our own well-being," the 2005 *Millennium Ecosystem Assessment* declared. [23]

Experts disagree, however, over whether keeping entire ecosystems intact will be as essential for human communities as it is for wild plants and animals. In a widely reported estimate a decade ago, economists and environmental scientists calculated that ecosystems contribute $33 trillion or more a year in materials, energy and other services that benefit human communities. [24]

Disappearing honeybees demonstrate the potential for calamity if key species vanish. American bee colony numbers have fallen 40 percent in the last 60 years, and last winter about a quarter of American beekeepers began reporting heavy losses of bees that failed to return to hives. Some blamed the puzzling phenomenon, dubbed "colony collapse disorder," on global warming. Preliminary findings, however, pointed to a paralyzing virus, possibly carried by bees imported from Australia.

Nonetheless, the honeybee losses demonstrated how the sophisticated agricultural industry still relies on wild pollinators. In many other ways, the World Resources Institute declared in a 2005 *Millennium Ecosystem Assessment* that "biodiversity and human well-being are inextricably linked." [25] The Earth's natural ecological bounty supplies clean air, water, food and other "ecosystem services" human communities need to survive. Coastal mangroves and wetlands protect humans by absorbing the fury of tropical storms and hurricanes, for instance. Forests and grasslands store water, retard floods, and purify the air — keeping even the most urbanized cities livable.

In the *Millennium Assessment*, more than 1,300 experts concluded that meet-

ing human requirements for food, water, fiber and energy is putting unprecedented pressure on productive ecosystems. While the resulting changes have helped to improve the lives of billions, they wrote, "at the same time they weakened nature's ability to deliver other key services such as purification of air and water, protection from disasters, and the provision of medicines. . . . [T]he ability of the planet's ecosystems to sustain future generations can no longer be taken for granted." [26]

In a notable example, commercial fishing catches peaked in the 1980s, but at least a quarter of valuable marine fish are still being harvested at rates that can't be sustained. [27] Overfishing caused the Newfoundland cod fishery to collapse in 1992, costing 20,000 jobs. In the Pacific Northwest, an estimated 72,000 jobs have vanished because salmon runs have been depleted by dams and habitat losses.

Declining biodiversity also may encourage the spread to humans of serious wildlife-born diseases like Lyme disease, avian influenza and West Nile virus. For generations, native societies have converted wild plants and animals into medicines and herbal remedies. A 2002 Harvard Medical School report found that commonly used drugs for humans were being made from 119 chemical compounds derived from 90 plant species. [28] The bark of the Pacific yew tree, for instance, yields taxol, a drug effective in treating breast and ovarian cancer. The Madagascar periwinkle's sap produces another anti-cancer drug, while tree bark is used to make aspirin and quinine. Scientists says that only a small portion of plants and animals have been screened for potential life-saving pharmaceuticals, and "the search for natural medicinals is a race between science and extinction," entomologist Wilson warns. [29]

"If we value the life sciences, we shouldn't wreck the fundamental library on which they're built," says the Heinz Center's Lovejoy. "We're losing the books without even reading them."

Free-enterprise advocates acknowledge that some visible species are in trouble, but they caution against overreacting to dire predictions. Sally Satel, a physician and American Enterprise Institute scholar, says wild plants and animals are growing less important to drug researchers, who she says are now focusing on screening synthetic molecules for leads on new medications. Satel calls "bioprospecting" for useful wild plants and animals a "high risk and very low yield" venture" that amounts to "a small and shrinking percentage of the portfolio of major drug companies." [30]

Even some biologists say that human benefits from protecting species are difficult to demonstrate. "In truth, ecologists and conservationists have struggled to demonstrate the increased material benefits to humans of 'intact' wild systems" over lands now being used for farming and other purposes, contends Martin Jenkins, a U.N. Environment Programme biologist. "In terms of the most direct benefits, the reverse is indeed obviously the case; this is the logic that has driven us to convert some 1.5 billion hectares of land area to highly productive, managed and generally low-diversity systems under agriculture." [31]

Even considering general ecological services, Jenkins says developed landscapes managed and maintained by humans can provide what communities require. "Where increased benefits of natural systems have been shown, they are usually marginal and local," he adds. ∎

BACKGROUND

Astonishing Variety

New species are constantly being discovered, further revealing Earth's complexity. "The totality of life, known as the biosphere to scientists and

Continued on p. 996

Chronology

1900-1970
U.S. laws and treaties protect threatened wildlife.

1900
Lacey Act prohibits interstate commerce of animals killed in violation of state game laws.

1914
Passenger pigeons go extinct with the death of the last specimen in the Cincinnati Zoo.

1918
Migratory Bird Treaty Act protects certain birds crossing national boundaries.

1958
International Union for the Conservation of Nature and Natural Resources (IUCN) established.

1962
Rachel Carson's best-seller *Silent Spring* galvanizes the U.S. environmental movement by linking the decline of wild species to pesticides.

1966
Endangered Species Preservation Act requires conservation of threatened species and government acquisition of habitat.

1968
Former Secretary of Defense Robert S. McNamara becomes president of the World Bank, expanding its development lending to Third World nations for dams, roads and improved "green revolution" agricultural techniques.

———— • ————

1970s-1980s
U.S. strengthens wildlife protection as concern grows about loss of species in developing

nations. Conservationists push World Bank, U.N. and national governments to protect species in Third World.

1972
Marine Mammal Protection Act protects seals, dolphins and other sea mammals. Clean Water Act requires cleanup of sewage and industrial discharges that harm aquatic species.

1973
Convention on International Trade in Endangered Species outlaws trade in endangered wild species. President Richard M. Nixon signs Endangered Species Act.

1980
American biologist Thomas E. Lovejoy III coins term "biological diversity." Smithsonian Institution biologist Terry Erwin estimates Earth contains 30 million insect species.

1984
Lovejoy proposes debt-for-nature swaps. . . . Conservation groups begin assuming poor nations' debt in return for protecting threatened habitat. . . . By the early 1990s, 20 agreements had been reached forgiving $110 million in debt.

1987
U.N. Brundtland Commission calls for protecting natural ecosystems and sustainable economic development.

1988
British ecologist Norman Myers identifies 10 threatened tropical forest "hot spots" with exceptional plant and animal diversity. Myers and Conservation International subsequently expand the list to 34 hot spots.

———— • ————

1990-2007
International agencies attempt to im-

plement international biodiversity-protection agreements.

1992
More than 100 nations at the Earth Summit in Rio De Janeiro, Brazil, approve Convention on Biodiversity. President Bill Clinton signs the treaty in 2003, but the United States never ratifies it. . . . World Bank establishes new position: vice president for environmentally sustainable development.

1995
World Bank lending for environmental-protection projects approaches $1 billion annually; bank continues setting standards for environmental protection for nations receiving loans.

2002
Parties to the Convention on Biodiversity set a target of significantly reducing the rate of biodiversity losses by 2010. Heads of state endorse the goal at World Summit on Sustainable Development in South Africa.

2005
U.N.'s *Millennium Ecosystem Assessment* concludes that human activities have degraded 60 percent of 24 ecological services that natural systems provide for human communities. . . . Murder in Brazil of American-born nun Dorothy Stang, a rain forest activist, draws attention to controversies over international efforts to prevent tropical deforestation.

2007
IUCN's updated Red List adds 188 species to a critically endangered category, declares Malaysian herb extinct and reports China's Yangtze River dolphin may also have vanished. . . . Separate IUCN report finds that a third of the world's primate species are on the path to extinction.

Mystery of the Vanishing Pollinators

Honeybees are disappearing worldwide.

The mystery began in the fall of 2006, when beekeepers in North America began noticing that honeybees were disappearing at an unusually rapid rate.

In the following months, beekeepers in at least 35 states discovered that 30-90 percent of bees were not returning to their apparently healthy hives. Up to half of all the managed bee colonies in the United States may have been lost, prompting the Department of Agriculture to launch an emergency probe of "colony collapse disorder" (CCD). [1]

Moreover, biologists who study pollinators fear that "if honeybee numbers continue to decline at the rates documented from 1989 to 1996, managed honey bees will cease to exist by 2035," University of Illinois entomology Professor May R. Berenbaum told a 2007 congressional hearing. [2]

North America's managed honeybees have been in decline for several decades, plagued by parasitic mites, microbes, beetles and aggressive, invasive Africanized bees. They also had been increasingly stressed by declines in pollen, overcrowding, contaminated water and exposure to pesticides. But scientists have not isolated a specific cause for CCD. In fall 2007, however, a team of agricultural researchers reported they had detected a link to a bee virus first detected by Israeli scientists that possibly was carried into the United States by Australian bees imported to supplement the country's dwindling honeybee hives. [3] Meanwhile, CCD has been blamed, without scientific proof, on factors ranging from climate change to cell phone signals. [4]

The economic stakes of a bee collapse are huge — for consumers as well as the nation's agriculture industry. Some 135,000 U.S. beekeepers maintain 2.4 million hives, leasing most to farmers when it's time to pollinate more than 90 crops; some hives also are used to produce honey for consumption in the United States. The annual value of crops that honeybees pollinate has been estimated at up to $19 billion, including most of the nation's high-nutrition fruits, vegetables and nuts, as well as hay for livestock. California's $2 billion almond crop alone requires 1.3 million bee colonies. [5]

Wild pollinators may be just as valuable. Organic farmers in California's productive Central Valley rely on native wild bees to pollinate tomatoes, melons and other high-value crops, inviting wild pollinators to visit by preserving adjacent nesting sites and planting mixed patches of crops. A 2006 study by two entomologists calculated that wild bee species and other native insects provided pollinating services worth more than $3 billion a year to U.S. farmers. [6] Unlike "domesticated" honeybees, which don't vibrate to shake pollen loose, native species "buzzpollinate" plants, vibrating to shake pollen grains free. Research has determined that wild bees play essential roles in producing watermelons, blueberries and cranberries.

Along with managed honeybees, wild bees and other species also pollinate home gardens as well as flowering plants that provide vital wildlife habitat. Globally, roughly 75 percent of crops that are grown for food, fiber, spices, condiments, beverages and medicines are pollinated by animals, according to the National Biological Infrastructure Pollinators Project, a multipartner effort led by the U.S. Geological Survey. [7] Pollinators are essential for growing coffee beans, cocoa for chocolate and

Continued from p. 994

Creation to theologians, is a membrane of organisms wrapped around the Earth so thin that it cannot be seen edgewise from the space shuttle, yet so internally complex that most species composing it remain undiscovered," writes Harvard's Wilson. "From Everest's peak to the floor of the Mariana Trench, creatures of one kind or another inhabit virtually every square inch of the planetary surface." [32]

New plants and animals have been evolving into new species since singlecelled organisms emerged roughly 3.5 billion years ago. Species have appeared and then vanished ever since, and an estimated 99 percent of those that ever existed are now extinct. Their disappearance opens places for new forms to take hold that are better adapted to changing climate and habitat conditions.

Over the last 250 years, biologists have identified and named between 1.5 million and 1.8 million species — merely the tip of the species iceberg. Estimates of the total number of species in existence today range from 3.6 million to 100 million or more. Recent advances in molecular biology, genetics and mathematics are giving scientists powerful new tools for detecting and studying smaller and smaller organisms. Oceans cover about 70 percent of the planet, and they teem with life — from bacteria to giant fish and whales.

In addition, scientists have identified 10,000 ant species, but further study in tropical regions could double the number; and 16 million more fungi species could eventually be added to the 69,000 already found. Each year 2,000 new flowering plants are described, including 60 new species in the United States and Canada. Between 1985 and 2001, the list of known amphibian species expanded by one-third, from 4,003 to 5,282, and Wilson sees "little doubt that in time it will pass 6,000."

Roughly 10,000 bird species are known, but during the 1990s sophisticated fieldstudy techniques — including recordings of differences in mating songs — began detecting genetically isolated populations that possibly could double the number. Also in the 1990s, four previously unknown land mammals were discovered along the Laos-Vietnam border, includa number of oilseeds that provide dietary staples.

Three-quarters of Earth's flowering plants reproduce from pollen carried to female plants by insects, birds, bats and some other far-ranging mammals. In fact, an estimated 400,000 mobile species that are drawn to flowers for their nectar carry pollen from plant to plant. The mainstays are mostly bees, ants, wasps and other insects, but also bats, monkeys and a threatened Madagascar lemur that carries pollen to a rare palm tree.

Like North America's honeybees, wild bumblebees and other native pollinators are also in trouble across much of the globe. A number of bats and insects important in pollinating have been in steep decline, and some wild bumblebees and butterflies may already have vanished. Scientists blame the decline on pesticides sprayed on lawns and fields as well as changes to forests and grasslands, which have been fragmented or converted to single-crop farmland. No global data have been compiled, but scientists report declining populations for such key pollinating species as bats and lemurs, hummingbirds and sunbirds, European butterflies and British and German bumblebees.

In response, France has banned a class of chemical pesticides that in high concentrations can be acutely toxic to bees

The loss of bee colonies in at least 22 states in 2006 prompted a Department of Agriculture probe of "colony collapse disorder" (CCD).

and in lower doses impair bees' ability to navigate back to their hives.

The world's staple crops of wheat, rice, barley and other grains either pollinate themselves or rely on wind currents. But bees and other species provide a crucial service in supplying humans with adequate diets. "Were pollinators to vanish, we wouldn't starve," Stanford University ecologist Gretchen C. Daily and journalist Katherine Ellison have written. "But our menus would most likely become more boring, more costly, and less healthful." [8]

[1] Renee Johnson, "Recent Honey Bee Colony Declines," Congressional Research Service, Aug. 14, 2007.

[2] Testimony before House Subcommittee on Horticulture and Organic Agriculture, March 29, 2007.

[3] Diana L. Cox-Foster, *et al.*, "A Metagenomic Survey of Microbes in Honey Bee Colony Collapse Disorder," *Science*, Oct. 12, 2007, pp. 283-287.

[4] "Colony Collapse Disorder Action Plan," Colony Collapse Disorder Steering Committee, U. S. Department of Agriculture, June 20, 2007.

[5] For background, see "Status of Pollinators in North America," National Research Council, National Academies of Science, 2007.

[6] John E. Losey and Mace Vaughan, "The Economic Value of Ecological Services Provided by Insects," *BioScience*, April 2006, pp. 311-323.

[7] See http://pollinators.nbii.gov.

[8] Gretchen C. Daily and Katherine Ellison, *The New Economy of Nature* (2002), p. 209.

ing two barking deer, a striped hare and the cowlike saola or "spindlehorn."

The Extinction Record

In normal times, perhaps one species in a million goes extinct every year. At least five times in Earth's history, however, mass extinctions wiped out at least one-fourth of all species. Roughly 250 million years ago, a catastrophic event — perhaps a climate shift, volcanic eruptions or change in ocean salinity — may have eliminated more than three-fourths of terrestrial vertebrate families and 96 percent of marine species. Most recently, dinosaurs and flying lizards vanished 65 million years ago, possibly because

a large asteroid crashed into Earth, sparking cataclysmic climate shifts that altered the Earth's atmosphere and wiped out 85 percent of all species. [33]

That event cleared the way for the rise of mammals, including Homo sapiens, about 100,000 years ago in Africa. After early humans spread across the continents, some large mammals and flightless birds vanished, including giant mammoths, ground sloths, saber-toothed tigers and North American camels. Some paleontologists think primitive hunters, rather than a cataclysmic event, decimated slow and lumbering beasts after developing weapons.

"As a rule around the world, wherever people entered a virgin environment, most of the megafauna soon vanished," Harvard's Wilson writes. That's a possible explanation for the loss of mammoths after Asian people began migrating to North America when dropping sea levels opened a temporary Bering Sea land bridge. Madagascar, an island off East Africa, evolved distinct, large forms of life after it broke off from India, including the 10-foot-tall "elephant bird" (species *Aepyornis*), huge tortoises, pygmy hippopotamuses and giant lemurs. With the exception of the Nile crocodile, those species vanished when human agricultural villages appeared around the 11th century. New Zealand and Hawaii also developed distinctive fauna and flora, and similar die-offs occurred in Polynesia when settlers arrived. [34]

The loss of unique biological re-

Is Amazon Rain Forest Nearing 'Tipping Point'?

Despite gains, conservationists fear deforestation rate.

Biologist Thomas E. Lovejoy III began working in Brazil's immense and still largely unexplored Amazon rain forest in 1965.

The vast region — roughly the size of the continental United States — has given rise to about a tenth of the world's distinct species of birds, 3,500 fish, 150 bats, hundreds of other unique mammals and — conceivably — a staggering 20 million insects. The hot, humid Amazon forest even creates its own hydrological cycle, as trees release half the moisture they take in back to the atmosphere.

In 1980, the Amazon's teeming natural life inspired Lovejoy to coin the term "biological diversity" in recognition of the complex ecological webs that its plants and animals created as they've evolved. Lovejoy still revels in visiting a region he describes as "this place of perpetual biological surprise."

Indeed, just this fall, a Dutch biologist discovered a new species of Amazon peccary double the size of the region's other wild pigs.[1] Yet Lovejoy now fears that "the Amazon forest is very close to the tipping point" — the point where rapid human disturbance will terminally disrupt its hydrological cycle and doom what may be the Earth's richest concentrations of plants and animals.

In the last 40 years, roughly one-fifth of Brazil's 1.6 million square miles of rain forests have been cut down as the eight South American nations that share the Amazon Basin pushed for economic growth.[2] In the 1970s, for example, Brazil's military government implemented an "occupy it or risk losing it"

policy encouraging destitute settlers from overcrowded cities to move into the Amazon over new highways providing access deep into the forests. Since then, loggers have built more than 105,000 miles of illegal roads to harvest mahogany and other hardwoods for international markets.[3]

Afterwards, much of the cleared land was used for cattle grazing, making Brazil the world's largest beef exporter. Soybean production also has soared in the last decade, with the intensive agriculture operations producing damaging fertilizer and pesticide runoff into the region's rivers. "Ten or 15 years ago, they were literally giving away chain saws" to settlers moving into the forests, Lovejoy says.

Land fraud and violence were rife, culminating in the 2005 murder of environmental activist Dorothy Stang, an American-born nun who had incurred the wrath of loggers and landowners for her outspoken efforts on behalf of the poor and the environment. Now the World Bank, World Wildlife Fund and several other international organizations have begun working with the government of Brazilian President Luiz Inacio Lula da Silva to establish a network of national parks and preserves protecting 190,000 square miles of the Amazon.

In 2006, Cargill International, a U.S.-based food conglomerate, and other Brazilian soy traders agreed to a moratorium on buying soy from logged Amazon farmlands. In the last two years, Brazil reported that logging in Amazon forests had been reduced substantially from previous levels.

sources accelerated after Europeans began colonizing remote islands and interior regions. Since 1500, the IUCN calculates humans have exterminated 785 species, mostly by hunting, destroying natural habitats or by releasing exotic competitors into the wild. Another 65 decimated species hang on in zoos and research facilities. Most of the birds that went extinct were from New Zealand, Hawaii and other oceanic islands; most of the extinct mammals were from the Caribbean and Australia.[35]

Continuing Threats

After World War II, American plant geneticist Norman Borlaug developed new strains of high-yield wheat and launched a "Green Revolution" that

improved crop production in populous areas such as India, Africa and Mexico. In the 1970s, the World Bank began giving loans and other financial aid to expand food production, build dams and highways and otherwise expand the economies of former European colonies.

Green Revolution methods have helped feed rising populations through intensified farming methods, including expanded use of irrigation and chemical fertilizers and pesticides. Those methods may have headed off the conversion to cropland of nearly 200 million acres of forest and grasslands, but they also taxed water supplies and created polluted runoff that also degrades ecosystems and impacts wildlife.[36]

Bank-sponsored projects in Indonesia since then have forced native minorities off their lands and destroyed vast areas

of wetlands and tropical forests, notes Michael Goldman, a University of Minnesota professor of sociology who studies World Bank policies for developing Third World nations. In the Brazilian Amazon, he writes, "Bank-financed highways and timber, mining and agricultural-colonization schemes sped up deforestation, river contamination and the death of forest-dwelling indigenous people."[37]

Around the globe, biologists now warn that untold millions of species could be pushed to the brink of extinction. Habitat destruction takes the most serious toll. More than four-fifths of threatened birds, mammals and amphibians depend on habitat that's disappearing as human development spreads across their homelands. Freshwater ecosystems damaged by dams, pollution and stream-bank development host the highest percentages

"It is possible to grow while preserving the environment," President da Silva said in an August 2007 radio address. "The challenge we face is how to use the forest and environmental preservation to improve the lives of people." [4]

Despite such efforts, conservation groups are not sure the deforestation can be stopped. They are particularly concerned that Cargill is balking at closing a new soybean export port, and the national government has proposed spending $800 million to pave the last 500 miles of a controversial 1,100-mile-long highway deep into the Amazon. International interest also is growing in Brazil's sugarcane-based ethanol industry as an alternative to burning fossil fuels.

"Brazil has reduced its deforestation rate, but now it turns out it's going back up again," Lovejoy says. In Brazil, nationalist political sentiment objects to outside conservation groups trying to influence how the Amazon is developed. Brazil contains two-thirds of the Amazon forests, and another seven national governments also hold sovereignty. "It is, ultimately, the geographic territory of those countries, and how the rest of the world engages in the Amazon is obviously a contentious issue," Lovejoy says.

Deforestation scars Amazon rain forest in northern Brazil. More than 105,000 miles of illegal roads are used to harvest mahogany and other Amazon hardwoods for international markets.

AFP/Getty Images/Antonio Scorza

Yet the Amazon's fate has global implications. So much carbon dioxide is released to the atmosphere by the vast logging and burning operations in the Amazon that Brazil, along with Indonesia, ranks just behind the heavily industrialized United States and China in its contributions to global greenhouse-gas emissions.

Meanwhile, shifting ocean conditions in recent years have created drought in the Amazon, and Lovejoy fears the resulting loss of vegetation may be getting close to disrupting the unique rainfall cycle, with global environmental consequences.

"If we pass that tipping point, we're going to have lost that contribution to life on Earth," he says.

[1] "New Species of Peccary — Pig-like Animal — Discovered in Amazon Region," *Science News*, Nov. 9, 2007, www.sciencedaily.com/releases/2007/11/071105153607.htm.

[2] Michael Goulding, "Tomorrow's Amazonia; Biodiversity in Jeopardy," *The American Prospect*, September 2007.

[3] Scott Wallace, "Last of the Amazon," *National Geographic*, Jan. 1, 2007, p. 40.

[4] Quoted in Michael Astor, "Brazil's president says preserving environment does not impede economic growth," The Associated Press, Aug. 13, 2007.

of endangered species; in 20 countries with the most complete Red List assessments, 17 percent of freshwater fish are threatened.

Salmon and other anadromous species — those that move between salt and fresh waters — have been hit particularly hard by dams, farming and urban development, creating bottlenecks along spawning streams. About half the Earth's forests have already been cut down, while vast expanses of biologically diverse grasslands and deserts have been converted to farmlands growing just a few crops.

Human encroachment has cut off long-distance migrations by elephants in India and Kenya, springbok and wildebeest in Namibia and South Africa, and the saiga antelope on Central Asian steppes. By one estimate, bison, pronghorn antelope, elk and deer in Yellowstone National Park have lost 75 percent of their migration routes to lower winter habitat. [38] In forested regions of the tropics, destructive development is quickly consuming some the world's richest, most diverse ecosystems with uncounted numbers of unique and vulnerable plants and animals. Logging and conversion of tropical forests to agriculture is combining with drought to threaten to disrupt natural hydrological cycles that make Africa's Congo River Basin and South America's Amazon Basin possibly the world's richest biological reserves. [39]

Even where logging leaves some patches of trees standing, cutting up habitat leaves some distinctive plants and animals confined. Building dams and flood-control structures has similarly carved up river habitat. While some species can thrive by moving into transformed lands or waters, smaller fragments can only support diminished populations that are unable to migrate to find enough habitat. Starting in 1979, the Heinz Center's Lovejoy and his colleagues have been studying how species have fared in patches of Brazilian rain forest surrounded by cleared agricultural holdings. They've found that fragments smaller than one square kilometer lost half their bird species in less than 15 years. "As expected, many species — large mammals, primates, under-story birds and certain beetles, ants, bees, termites and butterflies — are highly sensitive to [habitat fragmentation], and some have disappeared from even the largest fragments," they found. [40]

Other threats to plant and animal species include:

• **Alien species** — Next to changing habitats, invasions by exotic fish and other organisms account for the heaviest decline of freshwater species. Hawaii has no native ants, for instance, but two species from South America and Africa that traveled to the islands with humans eradicated many native insects and contributed to the disappearance of insect-consuming birds. Hawaii originally hosted just two mammals, the monk seal and a native hoary bat, but now feral pigs and house cats, along with immigrant rats and mongooses, have devastated forest plants and contributed to the extinction of all but 35 of at least 125 birds unique to the islands. [41]

In New Zealand, an average of 11 exotic plants have been introduced every year since Europeans settled in 1840. On Guam, poisonous brown tree snakes from New Guinea eliminated all 10 native forest bird species. In the Black Sea, the comb jelly, introduced by ships from the U.S. East Coast, wiped out 26 commercial fisheries. Introduced in 1904, an Asian fungus by the mid-1900s had killed off all but a hundred or so of the American chestnut trees that once dominated Northeastern U.S. forests. Asian carp, red fire ants, toxic weeds, tamarisk trees and Kudzu vines have overrun native ecosystems throughout North America. And the Baltic Sea now contains a hundred exotic creatures, a third of them native to North America's Great Lakes, while 170 species have invaded the Great Lakes, including a third from the Baltic. [42]

• **Hunting and fishing** — Poaching takes a toll on tigers, rhinoceroses and a number of other imperiled species. India's beleaguered tigers are still hunted for animal parts used in traditional Asian medicine; similarly, the fate of the endangered Sumatran rhinoceros may be sealed by poachers seeking valuable rhino horns. African hunters illegally take nearly 2 million tons of wild bush meat annually from the Congo River Basin, or six times the rate that biologists believe can be sustained without exterminating widely hunted species. Ocean fishing fleets have expanded and begun exploiting deeper waters, but catches peaked in the 1980s,

and today 20 percent of fishing stocks have crashed, according to the U.N.'s 2007 global biodiversity study. [43]

• **Pollution and disease** — Nutrients from fertilizers used on Midwestern farms are washing into the Mississippi River, creating a 5,000-8,000 square mile "dead zone" in the Gulf of Mexico where organisms are starved for oxygen. Although manufacturers dispute the findings, some scientists see evidence that spraying the pesticide atrazine on Midwestern corn fields is deforming frogs and amphibians in downstream rivers and lakes. Scientists also are finding evidence that metropolitan sewage system discharges contain birth control compounds and other chemicals excreted by large urban populations. Those contaminants have begun "feminizing" male fish in Lake Mead, the Potomac River and other U.S. waters, disrupting normal reproduction.

According to the American Bird Conservancy, more than 670 million birds are exposed to pesticides sprayed on U.S. farms a year, and 10 percent die. Avian flu, Lyme disease, plague and other afflictions can also hit wildlife populations hard. In November, Minnesota duck hunters found 3,000 dead scaup, a blue-billed diving duck, and several hundred coot on the shore of Lake Winnibigoshish. Laboratory tests identified the killer as an intestinal parasite carried by invasive snails the animals had ingested. North America's scaup population has fallen by 44 percent in the last 30 years, from 6.3 million to 3.5 million.

Conservation Efforts

Charles Darwin described how species evolve in his groundbreaking *Origin of Species* in 1859, while 19th-century writers like Henry David Thoreau and naturalists like John Muir began to advocate preserving wilderness. By the early 1900s, sportsmen, including President Theodore Roosevelt, were pushing for the creation of na-

tional parks and refuges for wildlife. The Lacey Act of 1900 barred commercial hunting of migratory birds for hat plumes and also implemented a series of treaties with Canada, Mexico, Japan and the Soviet Union to protect birds that cross international borders. Meanwhile, state governments established agencies to restore depleted wildlife by enforcing hunting and fishing seasons, restocking streams and purchasing key habitat.

In less-developed regions, however, uncontrolled hunting continued. After World War II, global concern for wild species increased. In 1948, more than a hundred conservation groups joined with 18 national governments to create the IUCN. Today its members include 82 nations, 122 environmental agencies and more than 800 non-government organizations. [44] Concern about environmental issues spread after Rachel Carson's 1962 book *Silent Spring* warned that DDT and other pesticides were decimating wild birds and causing other environmental damage. Congress responded by adopting stringent laws, including the 1973 Endangered Species Act.

The same year, the United States led international negotiations by 80 countries to create the Convention on International Trade in Endangered Species of Wild Fauna and Flora. Now signed by 173 nations, the treaty established international rules that afford some protection for 30,000 animal and plant species by barring or regulating trade in live specimens and animal parts and pelts. The International Whaling Commission in 1986 declared a moratorium on hunting whales for commercial purposes, although Japan's whaling fleet continues taking 1,400 whales a year under an exemption for scientific research. In 2007, Japan drew widespread criticism when it announced its fleet would take 50 endangered humpback whales in the Antarctic whale sanctuary.

In Europe and North America, wildlife-protection efforts, such as pollution-control standards and expansion of protected habitats, largely have paid off.

Continued on p. 1002

At Issue:

Should religious people care more about mass extinction?

E. O. WILSON
HARVARD UNIVERSITY PROFESSOR OF ENTOMOLOGY

FROM AN INTERVIEW ON *RELIGION & ETHICS NEWSWEEKLY*, PUBLIC BROADCASTING SYSTEM, NOV. 16, 2006

*t*he Creation — Living Nature — is in deep trouble. At current rates of loss, half the plant and animal species on Earth could be either gone or at least fated for early extinction by the end of the century. . . . Surely we can agree that each species, however inconspicuous and humble it may seem to us at this moment, is a masterpiece of biology, and well worth saving.

The great challenge of the 21st century is to raise people everywhere to a decent standard of living while preserving as much of the rest of life as possible.

If we fail to prevent mass extinctions, what we will lose would have otherwise been of incalculable value to future generations — and to our generation as well — in scientific knowledge and in potential new products. . . . It's been estimated that the wild creatures of this world, and the ecosystems they form, provide roughly $30 trillion worth of services scot-free to humanity every year.

If massive extinctions occur, the long-term loss would be the pauperization of the Earth. We'd get along, but it would just be a far poorer, less stable Earth.

What difference would mass extinction make to people? There are many reasons that are fundamental to the human weal. Unimaginably vast sources of scientific information and biological wealth would be destroyed. . . . Gone forever will be undiscovered medicines, crops, timber, fibers, soil-restoring vegetation, petroleum substitutes and other products and amenities.

The good news is that if we use our science, if we use common sense, we can actually increase the productivity of the world while saving all, or most of the remaining species. And we can do it in part by studying and making use of new food crops and new genes that can improve existing crop productivity and by restoring for agricultural purposes parts of the world that have become wasteland.

The Bible preaches stewardship — that is, it gives us responsibility. It gives us the rule of the Earth, the dominion of the Earth, but it also gives us responsibility. It does not tell us, really, to transform everything into a cornfield, you know, to make everything just produce more people and more products to feed more people.

If extinctions continue, we will have severe spiritual loss — a bad conscience — for having allowed the world's biodiversity — and I will also call it "the Creation" — to disappear. . . . Human beings depend on balanced, harmonious ecosystems for our lives.

RICHARD LAND
PRESIDENT, ETHICS AND RELIGIOUS LIBERTY COMMISSION, SOUTHERN BAPTIST CONVENTION

FROM AN INTERVIEW ON *RELIGION & ETHICS NEWSWEEKLY*, PUBLIC BROADCASTING SYSTEM, NOV. 17, 2006

*a*s Christians, we have an obligation and a responsibility for "creation care." Since we believe that God created everything for a purpose, we ought to, as an act of stewardship, try to keep some of everything . . . alive until we discover God's purpose for it and use it for that purpose.

For example, the rosy periwinkle in the Amazon was on the verge of extinction, and now an enzyme that can be extracted from the plant provides a leukemia treatment.

Under Genesis, Chapter 1, in the Bible, human beings have dominion, but under Chapter 2 man is put into the garden to till it and to keep it. That means to bring forth its fruit, to plow it, to cultivate it. We're not just to worship nature in its pristine form. We have a divinely mandated responsibility to both develop the Earth for human betterment and to protect it and guard it.

E. O. Wilson defines nature exclusive of humanity. But we believe that God created the creation for humankind. While we are to respect all life, we must treat human life with reverence. In Christian theology, there is a hierarchy of species, and there is a firebreak between human beings and the rest of creation. It is to human beings that God gave a soul.

Humans don't have a right to destroy other species. We do have a right to make value judgments about whether human beings are more important than other species. For example, God created spotted owls, and we ought to keep some alive. But if the choice is between keeping all the spotted owls alive and causing 10,000 families the loss of their livelihood, I say keep some spotted owls alive, not all of them.

Dr. Wilson is far more ready than I am to protect species with policies that would cause a radical reduction in living standards. . . . I think human beings have preeminence in the creation over other species.

Wilson and other biologists advocate controlling population growth to protect biodiversity, but I think if you unleash human ingenuity we can preserve the environment, and this Earth can sustain a lot more people than we have now.

Wilson describes human beings as an alien species doing damage to nature in the same way that red ants are an alien species to North America doing damage to the habitat of North America. We fundamentally disagree on that.

Continued from p. 1000

Forests have begun expanding after farm fields have been abandoned and wilderness protected from logging. In the United States, the federal, state and local governments have enlarged protected areas, and conservation groups have been buying private holdings and working with landowners to establish conservation easements that rule out development.

Under Endangered Species Act protection, Yellowstone's grizzly bears recovered from a low ebb 30 years ago. Wolves, exterminated from the Rocky Mountains in the 1920s, were reintroduced to Yellowstone and to the Idaho wilderness in 1995, and packs begun spreading rapidly. In the 1990s, the Clinton administration brokered a deal with conservation groups and the timber industry that curtailed old-growth-forest logging in the Pacific Northwest to save the last habitat for the northern spotted owl. Last year, federal, state and local governments in Washington state worked with Native American tribes and local watershed groups to develop a comprehensive plan to restore Puget Sound's endangered salmon runs. [45]

Meanwhile, under pressure from environmental groups and the U.S. government, the World Bank revised lending standards in the 1990s to force developing nations to create environmental agencies and look for sustainable economic development that minimizes ecological damage.

In 1987, the World Commission on Environment and Development focused attention on global environmental problems, including the continued loss of biological resources, and called for sustainable approaches to economic development. Five years later, the 1992 Earth Summit in Rio de Janeiro, Brazil, created the Convention on Biological Diversity. Now adopted by 178 countries, it launched nation-by-nation surveys of fauna and flora and called for stronger protection for endangered species, including establishing parks and reserves to preserve habitat.

President Bill Clinton signed the biodiversity treaty in 2003, but since then a Republican-led Congress and the George W. Bush administration have not pushed for Senate ratification. When the U.N. convened a 2002 World Summit on Sustainable Development, Secretary General Kofi Anan put biodiversity and ecosystems on the agenda along with health, agriculture, energy and water and sanitation. Participating nations set a goal of beginning to at least stop the increase in biodiversity losses by 2010. In June 2007, leaders of the Group of Eight industrialized nations endorsed the 2010 objective.

The United Nations has designated 2010 as the International Year of Biodiversity. But although 178 nations signed on, U.N. agencies and national governments are just beginning to assess the status of species and ecosystems. The United States has ignored the initiative. ■

CURRENT SITUATION

Tenuous Progress

Only a few marine ecosystems have significant protection. Australia protects its Great Barrier Reef, and President Bush in 2006 created a marine national monument off the northwestern Hawaiian Islands. On land, many of the largest remaining forests in Southeast Asia and South America's vast Amazon Basin are being cut down or fragmented. China's timber imports have tripled since 1993 as its economy expanded rapidly, and Chinese logging companies have begun cutting heavily in old-growth forests in Cambodia and Myanmar. [46]

Congress, meanwhile, has backed away from major changes to limit Endangered Species Act protections proposed by Re-

publicans before Democrats took control of the House and Senate in the 2006 elections. But state and local governments are still debating how much land to preserve as roads, suburbs and resorts crowd onto some of North America's best remaining habitat for the Florida panther, several songbirds and other threatened species. In the Pacific Northwest, the spotted owl's survival may again be in doubt, as competing barred owls invade their habitat and the Bush administration reconsiders the Clinton administration policy that put most old-growth forests off-limits to logging.

At the same time, federal agencies and Western congressional delegations still resist removing four Snake River dams that biologists and wildlife groups contend prevent endangered salmon populations from swimming upriver to spawn in high-altitude streams and rivers. Climate change could transform ecosystems in national parks, just as logging, livestock grazing and subdivision development have begun to cut off corridors that wildlife might need to reach new habitat. Yellowstone's grizzlies, now off the federal endangered species list, could once again be endangered if subdivisions are built in their habitat, imperiling their food sources. Oil and gas drilling in Wyoming is crimping pronghorn antelope migration from Grand Teton National Park, while Yellowstone's pronghorn are in decline, isolated from nearby herds by ranches, hay fields and new homes.

Migratory wildlife herds throughout Western states "continue to lose connectivity," says John Varley, Yellowstone's former natural resource director. "Yellowstone is becoming increasingly an island." For now, grizzlies' comeback and wolf reintroduction have restored a nearly complete natural ecosystem to Yellowstone and adjacent national forests. Indeed, both are being removed from federal protection, and state wildlife managers in several states are planning to reopen wolf and grizzly hunting seasons. Wildlife groups are planning legal challenges to the decisions.

Adding Up the Losses

The IUCN's Red List continues to paint a grim picture for many other species. Just in the last decade, the IUCN has added 2,520 species to wildlife classified as vulnerable, endangered and/or critically endangered. Since 1996, the number of threatened reptiles has climbed from 253 to 341, including 73 at critical risk. The 2007 update reported that one in four mammals, one in eight birds and one in three amphibians are in jeopardy among the species studied.

Among other findings, the IUCN notes that six of the world's eight bear species and nearly half of the sharks and rays in the Mediterranean Sea are threatened.

"Life on Earth is disappearing fast and will continue to do so unless urgent action is taken," IUCN officials declared. [47]

The IUCN has been working for 60 years to document species truly at risk of declining beyond the point where recovery will be possible. In 1996, the organization began following rigorous procedures for tracking the status of imperiled species for annual Red List updates. So far, the group's scientific advisers have assembled data on roughly 41,500 species that may be at risk. They expect the number to increase to 60,000 plants and animals in 2008. Large gaps remain, however, in scientists' knowledge of 300,000 plants as well as most species that live in arid landscapes or freshwater and ocean habitats. [48]

Other reports are equally ominous. Nearly half of all tortoises and freshwater turtles are considered to be in trouble. Around the globe, scientists have noted that many frogs and salamanders have gone into steep declines, and Australia's unique northern gastric breeding frog and Costa Rica's golden toad vanished altogether after 1989. Biologists say amphibian species are particularly vulnerable to fungal disease, pollution, drought and the continued loss of water habitat.

The United Nations' *Millennium Ecosystem Assessment* found "across a range of taxonomic groups, the population size or range (or both) of the majority of species is declining. . . . Studies of amphibians globally, African mammals, birds in agricultural lands, British butterflies, Caribbean and IndoPacific corals and commonly harvested fish species show declines in population of the majority of species." [49]

Wild tigers, Asia's largest predators, have vanished from Bali and Java and are now confined to isolated pockets comprising 7 percent of their historic range. India has set aside 28 tiger reserves, but more than 380,000 people live in those forested areas, grazing livestock and gathering firewood. Tribes have resisted government efforts to move them to other regions, and with enforcement poorly funded, poaching is rampant. Inside Congo's Virunga National Park, poachers have been killing gorillas for local consumption and also a growing international trade in bush meat.

Cutting Deals

Pressured by conservation groups and others, national governments by 2000 had set aside more than 44,000 national parks and other areas that protect habitat on roughly 10 percent of the Earth's land surface — about the size of India and China combined. [50] International agencies and conservation groups are also helping indigenous communities turn their wild areas to economic advantage. Under a 2003 agreement with the World Bank and the World Wildlife Fund, Brazil has established a network of national parks and reserves in the Amazon. Governments and conservation organizations now work with local residents to encourage eco-tourism and resource development without devastating tropical forests.

In Costa Rica, for instance, ecotourism focused on its tropical forests has replaced banana production as the economic mainstay. Working with the Wildlife Conservation Society, Costa Rica has set aside 17 percent of its lands in parks, refuges

and privately owned reserves and is actively protecting jaguars, sea turtles and other native species. Conservation International has partnered with indigenous communities to build an eco-lodge in Bolivia's Madidi National Park and worked with Ghana's wildlife agency on a Kakum National Park visitor center.

Now Madagascar, once headed for ecological disaster after cutting 90 percent of its forests, is preserving what's left of natural habitat and reconnecting it by reforesting wildlife-movement corridors as part of a plan to build up "green" tourism in rural regions. The World Bank and non-governmental organizations are pursuing "debt-for-nature" swaps that ease developing nations' heavy loan burdens to free up money for preserving landscapes from logging and other development.

In October, the United States forgave $26 million in Costa Rica's debt in exchange for its commitment to filling critical gaps in its protection of rain forests and other ecosystems.

The Brazilian government, meanwhile, reports that tougher enforcement last year slowed logging in the Amazon, but the country is also planning to finish paving the last 500 miles of a highway into the remote region that could set off a disastrous rush to raze more lush tropical stands to graze cattle and grow soybeans. [51] ∎

OUTLOOK

Population Pressure

Earth's population has doubled since 1960, surpassing 6 billion people, compounding the pressure on ecosystems. Three-quarters of the world's poor live in rural regions, many on incomes averaging $600 a year. While population growth is slowing in developed countries, the global total is projected to jump to 8-10 billion by mid-century, with growth concentrated in the crowded and impoverished cities of the Middle East, South Asia and sub-Saharan Africa. [52]

Population growth will force developing nations to expand agriculture and develop natural resources in regions often rich with species. As it is, nearly 2 billion people live in the 34 regions Conservation International identifies as biodiversity hot spots, including more than 312 million within about six miles of protected habitat areas. The group estimates it would cost as much as $100 billion to expand protected landscapes within the 34 hot spots to include the most vulnerable habitat, plus $5.5 billion annually to police poaching and prevent other damaging intrusions.

Harvard entomologist Wilson has called for an urgent multinational campaign to do nothing less than:

- Protect the remaining hot spots;
- Protect the Amazon, Congo and other rain forests;
- Halt logging in old-growth forests;
- Protect coral reefs and define other marine hot spots;
- Clean up freshwater ecosystems;
- Promote ecotourism and tighten security in wildlife preserves;
- Promote captive breeding of vanishing species; and
- Complete the inventory of the world's biological diversity.

With those concerted efforts, human standards of living could continue being raised through mid-century and "the great majority of ecosys-tems and species still surviving can also be protected, he maintains. "One key element, the protection and management of the world's existing natural reserves, could be financed by a one-cent-per-cup tax on coffee." [53]

So far, prospects for accomplishing Wilson's ambitious agenda aren't promising. By one estimate, spending on conserving biodiversity in developing nations has already exceeded $10 billion. [54] Still, only 5 percent of Earth's tropical forests have effective protection, while the degradation accelerates. The seven major industrial nations in 1992 pledged $1.2 billion to preserve the Amazon, but 10 years later had committed only $350 million to the effort. "The international response to accelerating deforestation has been anemic," the Heinz Center's Lovejoy wrote in 2004. "As a result, yesterday's nightmare scenarios are becoming today's reality." [55]

But some developing nations consider it unfair for wealthy countries that long ago decimated their own natural landscapes to ask less fortunate lands to put tropical forests off-limits. In Brazil, for instance, business and political leaders see rain forest reserves as stalling the country's hopes for dams, highways, ports and other developments that could give their countrymen better lives. "[T]his is a new form of colonialism," said Lorenzo Carrasco, a Brazilian critic of Amazon conservation programs." [56]

But biologists counter that preserving ecosystems is an essential step to keeping human societies going. Right now, says the University of Chicago's Coyne, "the message isn't getting through, and it won't until people feel an individual loss" when species disappear.

Preventing mass extinction and supplying human needs at the same time will remain a difficult challenge. Many species may already be doomed, but some scientists nonetheless find reason to hope that enough biological resources can still be saved. "Not all of this is inevitable," says Lovejoy. "That's the point of ranting and raving about it. It's not a matter of whether we can prevent extinction entirely; the issue is whether we can still minimize it." ∎

Notes

[1] Quoted in John Roach, "Most Polar Bears Gone by 2050, Studies Say," *National Geographic News*, www.nationalgeographic.com, Sept. 10, 2007.

[2] Eric Dinerstein, *et al.*, "The Fate of Wild Tigers," *BioScience*, June 2007, p. 513.

[3] Russell A. Mittermeier, *et al.*, "Primates in Peril: The World's 25 Most Endangered Primates," *Primate Conservation 2007*, pp. 1-40.

[4] See 2007 IUCN "Red List of Threatened Species," www.iucn.org/themes/ssc/redlist.htm.

[5] International Union for the Conservation of Nature and Natural Resources, "A Global Species Assessment," Executive Summary, 2004.

[6] Edward O. Wilson, *The Future of Life* (2002), p. 14.

[7] Norman Myers, "Threatened biotas: 'Hot spots' in tropical forests," *The Environmentalist*, 1988, pp. 1-20.

[8] Wilson, *op. cit.*, p. 102.

[9] Conservation International, *Hot Spots E-News*, spring 2007.

[10] Jerry A. Coyne and Hopi E. Hoekstra, "The Greatest Dying, A fate worse than global warming," *The New Republic*, Sept. 24, 2007, p. 7.

[11] Norman Myers and Stuart Pimm, "The Last Extinction?" *Foreign Policy*, March 1, 2003, p. 28.

[12] United Nations Environmental Program, "Global Environment Outlook: Environment for Development (GEO4)," 2007, p. 157.

[13] "United Nations Millennium Ecosystem Assessment, Current State and Trends Assessment," 2005, p. 79.

[14] *Ibid.*, pp. 99-111.

About the Author

Tom Arrandale freelances from Livingston, Mont., on environmental issues. He is a columnist for Congressional Quarterly's *Governing* magazine and has written for *Planning Magazine, High Country News* and *Yellowstone Journal*. He authored *The Battle for Natural Resources* (CQ Press, 1983). He visits Yellowstone National Park regularly to hike, snowshoe and photograph wildlife. He graduated from Dartmouth College with a history degree and from the University of Missouri with a master's degree in journalism.

[15] Steven F. Hayward and Amy Kaleita, Index of Leading Environmental Indicators 2007, Pacific Research Institute and American Enterprise Institute for Public Policy Research, 2007, p. 59.

[16] Daniel B. Botkin, *et al.*, "Forecasting the Effects of Global Warming on Biodiversity," *BioScience*, March 2007, p. 229. For background see Colin Woodard, "Curbing Climate Change," *CQ Global Researcher*, February 2007, pp. 25-48; and Marcia Clemmitt, "Climate Change," *CQ Researcher*, Jan. 27, 2006, pp. 73-96.

[17] Bjorn Lomborg, *The Skeptical Environmentalist* (2001), p. 257.

[18] Intergovernmental Panel on Climate Change, IPCC Fourth Assessment Report, Working Group II Report, "Impacts, Adaptation and Vulnerability," April 2007, pp. 211-272.

[19] Quoted in David A. Fahrenthold, "Climate Change Brings Risk of More Extinctions," *The Washington Post*, Sept. 17, 2007, p. A7.

[20] "Climate Change: the impact on biodiversity," Earthwatch Institute, www.earthwatch.org.

[21] Daniel B. Botkin, "Global Warming Delusions," *The Wall Street Journal*, Oct. 17, 2007, p. A19.

[22] Testimony before Senate Environment and Public Works Committee, Oct. 20, 2006.

[23] Millennium Ecosystem Assessment, *Living Beyond Our Means, Natural Assets and Human Well-Being* (2005), p. 3.

[24] Robert Costanza, *et al.*, "The value of the world's ecosystem services and natural capital," *Nature*, May 15, 1997, pp. 253-260.

[25] World Resources Institute, "Ecosystems and Human Well-being: Biodiversity Synthesis, Millennium Ecosystem Assessment," 2005, p. iii.

[26] *Millennium Ecosystem Assessment, op. cit.*, p. 3.

[27] *Ibid.*, p. 10.

[28] Eric Chivian, ed., *Biodiversity: Its Importance to Human Health*, Harvard Medical School Center for Health and the Global Environment (2002).

[29] Wilson, *op. cit.*, p. 123.

[30] Sally Satel, "Diminishing Biodiverse Returns," *Tech Central Station*, Feb. 16, 2005.

[31] Martin Jenkins, "Prospects for Biodiversity: State of the Planet," *Science*, Nov. 14 2003, pp. 1175-1177.

[32] Information in this section is from Wilson, *op. cit.*, unless otherwise noted.

[33] Unless otherwise noted, material in this section is from David Hosansky, "Mass Extinction," *CQ Researcher*, Sept. 15, 2000, pp. 713-744.

[34] Wilson, *op. cit.*, pp. 79-102.

[35] Jonathan E. M. Baillie, Craig Hilton-Taylor and Simon N. Stuart, "A Global Species Assessment," IUCN Species Survival Commission, 2004.

[36] The World Bank, "World Development Report 2008: Agriculture for Development," 2007.

[37] Michael Goldman, *Imperial Nature* (2005), pp. 35-36.

[38] Joel Berger, "The Last Mile: How to Sustain Long-Distance Migration in Mammals," *Conservation Biology*, April 2004, pp. 320-331.

[39] Eugene Linden, Thomas Lovejoy, and J. Daniel Phillips, "Seeing the Forest," *Foreign Affairs*, July-August 2004, p. 8.

[40] William Laurance, *et al.*, "The Biological Dynamics of Forest Fragments Project: 25 Years of Research in the Brazilian Amazon," Tropinet, September 2004, www.atbio.org.

[41] Wilson, *op. cit.*, p. 49. For background, see David Hosansky, "Invasive Species," *CQ Researcher*, Oct. 5, 2001, pp. 785-816.

[42] Millennium Ecosystem Assessment, *op. cit.*, p. 12.

[43] United Nations Environment Program, *Global Environment Outlook* (2007), pp. 157-192.

[44] Nicholas A. Robinson, "IUCN as catalyst for a law of the biosphere: acting globally and locally," *Environmental Law*, March 22, 2005, p. 249.

[45] See Tom Arrandale, "Confluence of Interest," *Governing*, September 2007, p. 54.

[46] See Elizabeth C. Economy, "The Great Leap Backward?" *Foreign Affairs*, September/October 2007, p. 86.

[47] IUCN press release, Sept. 12, 2007.

[48] See "An Overview of the IUCN Red List," www.incnredlist.org/info/programme#introduction.

[49] Millennium Ecosystem Assessment, *Ecosystems and Human Well-Being, Biodiversity Synthesis* (2003), p. 3.

[50] IUCN, "Protected Areas," www.iucn.org.

[51] Jack Chang, "Highway Project Pits Progress Against Environment," McClatchy-Tribune News Service, Aug. 14, 2007.

[52] Millennium Ecosystem Assessment, *Current State & Trends Assessment* (2005), p. 74.

[53] Wilson, *op. cit.*, p. 164.

[54] R. David Simpson, *Conserving Biodiversity Through Markets: A Better Approach*, Property and Environment Research Center (2004).

[55] Linden, *et al.*, *op. cit.*

[56] Quoted in Larry Rohter, "In the Amazon: Conservation or Colonialism?" *The New York Times*, July 27, 2007, p. A4.

FOR MORE INFORMATION

Conservation International, 2011 Crystal Dr., Suite 500, Arlington, VA 22202; (703) 341-2400; www.conservation.org. Promotes biological research and works with national governments and businesses to protect 34 biodiversity hot spots worldwide.

John H. Heinz Center for Science, Economics and the Environment, 900 17th St., N.W., Suite 700, Washington, DC 20006; (202) 737-6307; www.heinzctr.org. Fosters collaboration among industry, environmental groups, academics and government to gather data and find solutions to environmental threats.

IUCN, International Union for the Conservation of Nature and Natural Resources, Gland, Switzerland; www.iucn.org. Compiles Red List of threatened plant and animal species and tracks biodiversity-conservation efforts around the globe.

Natural Capital Project, 371 Serra Mall, Department of Biological Sciences Stanford University, Stanford, CA 94305-5020; www.naturalcapitalproject.org. Joint venture by Stanford University, Nature Conservancy and World Wildlife Fund that explores ecosystem services and economic incentives for preserving biodiversity.

The Nature Conservancy, 4245 North Fairfax Dr., Suite 100, Arlington, VA 22203-1606; (703) 841-5300; www.nature.org. Supports land conservation to protect wildlife habitat in the U. S. and more than 30 nations.

United Nations Environment Programme, P.O. Box 30552, Nairobi, Kenya; +254 20 7621 234; www.unep.org. The worldwide United Nations program charged with overseeing the Convention on Biological Diversity as well as the Convention on Climate Change.

World Resources Institute, 10 G St., N.E., Suite 800, Washington, DC 20002; (202) 729-7600; www.wri.org. Conducts research on ecosystem threats and works with indigenous communities to balance human and wildlife needs.

World Wide Fund for Nature (WWF), Av. Du Mont-Blanc 1196, Gland, Switzerland; +41 22 364 91 11; www.panda.org. Funds conservation in more than 100 countries.

Bibliography

Selected Sources

Books

Daily, Gretchen C., and Katherine Ellison, *The New Economy of Nature, The Quest to Make Conservation Profitable*, Island Press, 2002.
A Stanford University professor and a journalist discuss how governments, nonprofit organizations and entrepreneurs are trying economic incentives to conserve wildlife and natural ecosystems.

Garte, Seymour, *Where We Stand, A Surprising Look at the Real State of Our Planet*, AMACON Books, 2007.
A University of Pittsburgh environmental health professor argues that popular media overlook real progress in correcting environmental problems and protecting endangered species, although he also calls for immediately accelerating protection for the Amazon rain forest.

Gunter, Michael M. Jr., *Building the Next Ark*, Dartmouth College Press, 2004.
The director of the Rollins College International Relations Program outlines the growing role of non-governmental organizations like The Nature Conservancy, Conservation International and the World Wildlife Fund in international efforts to preserve biodiversity.

Lomborg, Bjorn, *The Skeptical Environmentalist, Measuring the Real State of the World*, Cambridge University Press, 2001.
A controversial professor of statistics at the University of Aarhus in Denmark contends that politicians and conservationists exaggerate environmental threats, including the risk of mass extinction of species.

Wilson, Edward O., *The Creation, An Appeal to Save Life on Earth*, W.W. Norton, 2006.
In a letter to an imaginary Baptist preacher, Wilson challenges religious leaders' inattention to a looming mass extinction.

Wilson, Edward O., *The Future of Life*, Vintage Books, 2002.
A prolific author and acclaimed entomologist at the Harvard University Museum of Comparative Biology discusses the plight of endangered species.

Articles

Botkin, Daniel B., *et al.*, "Forecasting the Effects of Global Warming on Biodiversity," *Bio Science*, March 2007.
A University of California at Santa Barbara professor emeritus and international colleagues challenge how biologists make estimates of massive species losses.

Botkin, Daniel B., "Global Warming Delusions," *The Wall Street Journal*, Oct. 17, 2007.
Botkin argues that overreacting to the threat that climate change poses to wildlife could divert attention from habitat destruction that more seriously threatens rare species.

Coyne, Jerry A., and Hopi E. Hoekstra, "The Greatest Dying, a fate worse than global warming," *The New Republic*, Sept. 24, 2007.
A University of Chicago ecology professor and a Harvard biologist contend a looming mass extinction of wild species would have dire consequences for humans.

Fahrenthold, David A., "Climate Change Brings More Risk of Extinctions," *The Washington Post*, Sept. 17, 2007.
Farhrenthold reports on current studies indicating that global warming has begun to harm wildlife.

Linden, Eugene, *et al.*, "Seeing the Forest, Conservation on a Continental Scale," *Foreign Affairs*, July-August 2004.
The authors, who include a tropical biologist and a former diplomat, make a scholarly case for a market-based plan to help impoverished African nations protect remaining tropical forests.

Shoumatoff, Alex, "The Gasping Forest," *Vanity Fair*, May 2007.
A journalist travels through the Amazon rain forest with his son and reports on how logging and drought imperil the world's most biologically diverse region.

Reports and Studies

International Union for the Conservation of Nature and Natural Resources, 2007, www.iucn.org.
IUCN Red List of Threatened Species. The organization's Web site provides the most recent listings, case studies, photos and statistics on 41,415 species the organization has studied.

Millennium Ecosystem Assessment, *Living Beyond Our Means, Natural Assets and Human Well-Being*, 2005.
A four-year study examines natural ecosystem functions and the value of the services they provide.

United Nations Environment Programme, *Global Environment Outlook 4*, Oct. 25, 2007.
Published on the 20th anniversary of the 1987 Brundtland Commission report "Our Common Future," this UN report updates the state of global environmental conditions.

World Resources Institute, *Restoring Nature's Capital*, 2007.
The global conservation organization outlines findings from the Millennium Ecosystem Assessment project along with an agenda for initiatives to preserve and restore natural ecosystem services.

The Next Step:

Additional Articles from Current Periodicals

Ecosystems

"Scientists Setting Dollar Value for Ecosystems," Agence France-Presse, Nov. 1, 2006.

American and Canadian researchers have developed a scientific model to measure the costs and benefits of ecosystems.

Casey, Michael, "Tsunami Spurs Renewed Interest in Protecting Asia's Battered Coastlines," The Associated Press, Sept. 7, 2006.

Many countries are funding restoration projects for ecosystems destroyed during the 2004 tsunami.

Wolman, David, "How to Get Wall Street to Hug a Tree," Los Angeles Times Magazine, Feb. 11, 2007, p. 12.

Environmentalists are working with investment bankers on putting price tags on nature and its resources.

Global Warming

"Climate Change: Fossil Record Points to Future Mass Extinctions," Agence France-Presse, Oct. 23, 2007.

A new study shows a correlation between rising temperatures and mass extinctions over the past 500 million years.

Eilperin, Juliet, "Warming Tied to Extinction of Frog Species," The Washington Post, Jan. 12, 2006, p. A1.

Global warming is responsible for pushing dozens of frog species to the brink of extinction over the past three decades, according to a team of Latin American and U.S. scientists.

Kay, Jane, "Where Are All the Birds?" The San Francisco Chronicle, July 4, 2006, p. A2.

About 12 percent of all bird species are expected to be endangered or extinct by 2100, largely due to anticipated habitat loss linked to global warming.

Kolber, Elizabeth, "Butterfly Lessons," The New Yorker, Jan. 9, 2006, p. 32.

The full extent of the transformation of life on Earth by global warming is a matter of speculation, but the process has already begun.

Mass Extinction

Biemer, John, "Scientists Sound Alarm for World's Amphibians," Chicago Tribune, July 7, 2006, p. A4.

Fifty international experts have warned that a new fungal disease has the potential to wipe out half of all exposed amphibian species.

Goodman, Brenda, "To Stem Widespread Extinction, Scientists Airlift Frogs in Carry-On Bags," The New York Times, June 6, 2006, p. F3.

Two Atlanta conservationists requested permission from airlines to fly home with frogs from Panama in order to save the species from extinction.

Velasquez-Manoff, Moises, "One Way to Help Species Facing Habitat Loss: 'Escape Routes,' " The Christian Science Monitor, June 21, 2007, p. 25.

Preserving large tracts of wilderness to establish healthy populations will provide "biological corridors" that would help species adapt to changing conditions and avoid mass extinction.

Whitty, Julia, "By the End of the Century Half of All Species Will Be Gone. Who Will Survive?" Mother Jones, May-June 2007, p. 36.

The number of factors contributing to the extinction of species has reached an alarming level over the past decade alone.

Pollinators

Higgins, Adrian, "Saving Earth From the Ground Up," The Washington Post, June 30, 2007, p. C1.

Pulitzer Prize-winning biologist E. O. Wilson outlines the consequences of a world without pollinators and other bugs.

Kolbert, Elizabeth, "Where Have All the Bees Gone?" The New Yorker, Aug. 6, 2007, p. 56.

The number of bee hives across the country has dropped by almost half since the early 1980s.

Schmid, Randolph E., "Scientists Worry It May Be Bye-Bye to the Birds and the Bees," The Associated Press, Oct. 18, 2006.

The National Research Council warns that the endangerment of pollinators may spell trouble for many crops.

CITING CQ RESEARCHER

Sample formats for citing these reports in a bibliography include the ones listed below. Preferred styles and formats vary, so please check with your instructor or professor.

MLA STYLE

Jost, Kenneth. "Rethinking the Death Penalty." CQ Researcher 16 Nov. 2001: 945-68.

APA STYLE

Jost, K. (2001, November 16). Rethinking the death penalty. CQ Researcher, 11, 945-968.

CHICAGO STYLE

Jost, Kenneth. "Rethinking the Death Penalty." CQ Researcher, November 16, 2001, 945-968.

CARBON TRADING

BY JENNIFER WEEKS

Excerpted from the CQ Global Researcher. Jennifer Weeks. (November 2008). "Carbon Trading." *CQ Global Researcher*, 295-320.

Carbon Trading

BY JENNIFER WEEKS

THE ISSUES

It's little wonder that Tirumala temple in Tirupati, in the south Indian state of Andhra Pradesh, prepares 30,000 meals for visiting Hindu pilgrims daily. The shrine is among the busiest religious pilgrimage sites in the world. In years past, cooks fired up pollution-spewing diesel generators to power their stoves to boil water in massive cauldrons. But today there's a new, clean energy source: the sun. Curved solar collectors heat water up to 280 degrees Centigrade, creating steam to cook foods such as rice, lentils and vegetables. [1]

"With most businesses, the first question is of economics," says engineer Deepak Gadhia, whose company built the system. "But spiritual organizations look at larger issues. They want energy that is spiritually positive." [2]

In fact, the temple does quite well financially, too, and so do many other temples, schools and government offices throughout India that use energy-saving systems built by Gadhia and his wife. The energy they save enables them to amass credits that can be used in a process called "carbon trading" — buying and selling rights to emit greenhouse gases.

Two years ago, the energy-saving systems at those sites were approved as carbon credit sources under the Kyoto Protocol. [3] The international agreement is designed to stem global warming, and — among other things — allows some developing countries to profit from projects that reduce emissions of greenhouse gases (GHGs) that cause

A worker pours chemicals into a vat of molasses used to make ethanol in Simbhacli, Uttar Pradeshi, India. Replacing gasoline with ethanol in cars can reduce emissions of carbon-based "greenhouse gases" (GHGs), created by burning fossil fuels, which contribute to climate change. Projects in developing countries that produce such alternative fuels are part of an international carbon trading scheme that allows polluters in industrialized countries to "offset" some of their GHG emissions by buying pollution credits from companies in developing countries.

AP Photo/Mustafa Quraishi

climate change. (*See sidebar, p. 302.*)

Under the protocol, most of the world's wealthy countries agreed to reduce their GHG emissions by fixed percentages between 2008 and 2012, mainly by reducing energy use and switching to low-carbon fuels. But if they can't reach the required reductions, rich nations can also "offset" some of their GHG emissions by buying credits from energy-saving projects — like the Gadhia solar cookers — in developing countries.

If U.N. officials certify that those projects reduce GHG emissions beyond levels that would have occurred otherwise, they can sell "certified emission reductions," each representing one avoided metric ton of carbon dioxide (CO_2). Companies in industrialized nations buy these credits to help reach their GHG reduction targets.

Virtually all scientists agree that human use of carbon-based fossil fuels such as oil, coal and natural gas is raising concentrations of heat-trapping gases in the atmosphere to the highest levels in at least 650,000 years. [4] The gases are called "greenhouse" gases because their heat-trapping properties warm the Earth's surface, much as the glass walls of a greenhouse retain the sun's heat. Unless countries sharply reduce their GHG emissions by mid-century, the buildup of greenhouse gases — often referred to as "carbon" emissions since carbon dioxide (CO_2) is by far the most abundant GHG in Earth's atmosphere — could cause dramatic planetary warming. Climate scientists predict that higher temperatures will cause melting of the polar ice caps, rising sea levels and more intense droughts, floods and hurricanes. [5]

The Kyoto Protocol, which was signed in 1997 and went into effect in 2005, requires major industrialized countries (except for the United States, which failed to ratify the agreement) to reduce their GHG emissions, on average, by 5.2 percent below 1990 levels. [6] Members of the European Union vowed to reduce their emissions even farther — to 8 percent below 1990

Which Countries Emit the Most Carbon Dioxide?

Australia and major oil producing countries like the United States, Norway, Russia, Canada, Saudi Arabia, Kuwait and other Gulf states emit the most carbon dioxide (CO$_2$) per capita. Carbon dioxide is the most abundant greenhouse gas — one of several blamed for causing global warming.

Carbon Dioxide Emissions Per Capita, 2004
(in metric tons*)

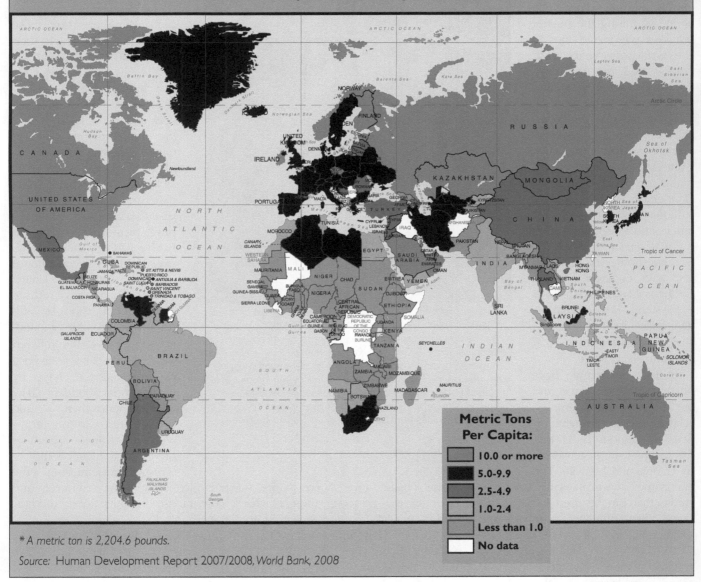

Metric Tons Per Capita:

- 10.0 or more
- 5.0-9.9
- 2.5-4.9
- 1.0-2.4
- Less than 1.0
- No data

* A metric ton is 2,204.6 pounds.

Source: Human Development Report 2007/2008, World Bank, 2008

levels by 2012. At the same time, the EU launched the world's largest mandatory carbon emissions trading system, in which governments cap national emissions and allow polluters to buy and sell permits to emit carbon dioxide. Australia, Canada and Japan are developing their own emission reduction systems, which will likely include some form of carbon trading.

Global interest in carbon trading is part of a gradual movement toward market-based environmental policies — strategies that give polluters economic incentives to clean up instead of simply telling them how much pollution they can release and what kinds of controls to install. The approach makes sense because climate change is what scholars refer to as a "commons problem" — in which a resource (in this case, Earth's atmosphere) is held in common by everyone. Individual polluters profit more by using and degrading a common resource than by cleaning it up while their competitors continue polluting.

"The rational man finds that his share of the cost of the wastes he discharges into the commons is less than the cost of purifying his wastes before releasing them," wrote biologist Garrett Hardin in a famous 1968 essay that identified commons problems as a central challenge for modern societies. "Since this is true for everyone, we are locked into a system of 'fouling our own nest,' so long as we behave only as independent, rational, free-enterprisers." [7]

Climate experts agree that one of the best ways around the commons problem is to "put a price on carbon" by making factories, power plants and other large GHG sources pay for their emissions. Hitting them in the pocketbook gives them more incentive to clean up — for example, by imposing a tax so that every source pays for its own GHG emissions at some set rate per ton.

However, an alternative approach — trading emission allotments — has become increasingly popular in recent decades. It is usually enacted through so-called cap-and-trade policies, in which regulators set an overall cap on emissions and then issue quotas that limit how much pollution each company can release. If a company wants or needs to emit more than its allowance, it must buy permits from cleaner companies that don't need all their allotments. Over time, regulators can lower a country's cap to further reduce total pollution.

Advocates say carbon emissions trading encourages companies to use clean fuels and technologies because

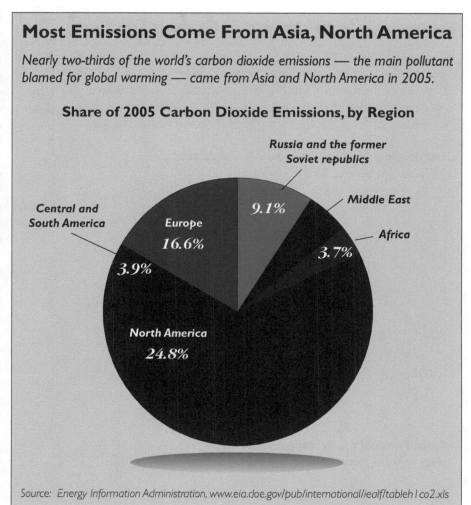

Most Emissions Come From Asia, North America

Nearly two-thirds of the world's carbon dioxide emissions — the main pollutant blamed for global warming — came from Asia and North America in 2005.

Share of 2005 Carbon Dioxide Emissions, by Region

Russia and the former Soviet republics 9.1%

Middle East

Central and South America 3.9%

Europe 16.6%

Africa 3.7%

North America 24.8%

Source: *Energy Information Administration, www.eia.doe.gov/pub/international/iealf/tableh1co2.xls*

firms that reduce their own emissions can then sell their unneeded allowances. "The carbon market gives companies an incentive to reduce emissions so they can make money," says Henrik Hasselknippe, global carbon services director for Point Carbon, an international market research firm in Oslo, Norway. Moreover, he predicts, since carbon trading tells companies to limit their emissions but lets them decide how, it will stimulate research and development into a wide range of new, clean technologies. "It puts an infrastructure in place that releases capital for long-term investments," Hasselknippe explains.

Global carbon markets have grown quickly since the Kyoto Protocol entered into force in 2005. The total

value of international carbon trades increased more than 80 percent between 2006 and 2007, from €22 billion ($33 billion) to €40 billion ($60 billion). [8] The market is expected to grow still larger as Europe lowers its cap on GHG emissions, and new trading systems gear up in some U.S. states and in other countries. [9]

Moreover, public support is growing for the U.S. government to act on climate change. President George W. Bush rejected the Kyoto treaty shortly after taking office in 2001, claiming that capping GHG emissions would harm the U.S. economy. But since then 23 states have joined regional carbon trading schemes, and the United States is widely expected to participate in a post-Kyoto agreement to limit GHG

Europe Leads the World in Carbon Trading

The European Union accounted for 70 percent of the €40 billion ($60 billion) spent worldwide to buy carbon emission allowances in 2007. The Clean Development Mechanism, which allows companies in industrialized countries to buy emission credits from companies in the developing world, accounted for 29 percent.

Distribution of Carbon Trading Contracts, by Financial Value, 2007

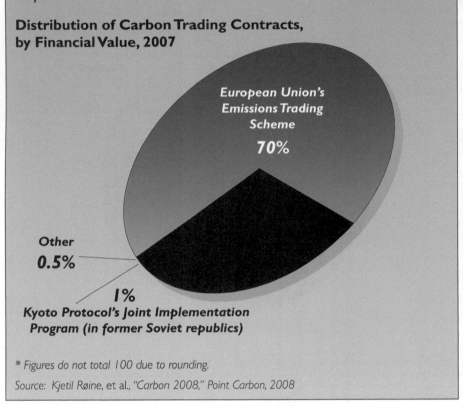

European Union's Emissions Trading Scheme

70%

Other
0.5%

1%
Kyoto Protocol's Joint Implementation Program (in former Soviet republics)

* Figures do not total 100 due to rounding.

Source: Kjetil Røine, et al., "Carbon 2008," Point Carbon, 2008

be done more cheaply.

Supporters say offsets are primarily designed to lower the cost of meeting Kyoto targets, and that it shouldn't matter where reductions take place because a ton of CO_2 causes the same amount of warming whether it's released in Germany or Malaysia. But others worry that if rich countries rely on offsets too heavily, they will have little incentive to reduce fossil fuel use or develop cleaner technologies. Ultimately, they argue, developing countries will refuse to make deep cuts in their own emissions if they see little change in rich countries.

"The developed world is responsible for the majority of greenhouse gas emissions," the World Wildlife Fund warned in 2007. "If the [European Union] is to maintain its status as a major player in global climate change negotiations, then it must put its own back yard in order first and ensure that Europe is placed firmly on a path towards a low carbon economy." [12]

Moreover, say critics, offset projects sometimes credit "anyway tons" — reductions from projects that would have gone forward anyway, even without extra revenue from selling emission reductions. Reductions are supposed to be "additional" to business as usual, but that concept can be hard to prove.

As governments, corporations and advocacy groups weigh the pros and cons of carbon trading, here are some issues they are debating:

Are current trading systems working?

Global carbon markets are booming, but some experts question whether carbon trading systems are making emission reductions affordable or reducing GHG emissions at all.

Two markets dominated world carbon trading through 2007. The European Union's Emissions Trading Scheme (EU ETS) accounted for 70 percent of trades by value, followed by the Clean

emissions after 2012. [10] Many U.S. political leaders, including both major presidential candidates, say the United States should create a cap-and-trade system similar to Europe's to cut GHG emissions in the United States far below 1990 levels by 2050. [11]

Ironically, several market-based elements were included in the Kyoto agreement at U.S. insistence in hope of convincing the United States to sign on to the treaty. They included two programs that let companies in industrialized countries offset some emissions by investing in carbon reduction projects elsewhere. The Clean Development Mechanism (CDM) paves the way for projects in devel-

oping countries, such as Gadhia's steam cookers at the temples in India, while Joint Implementation (JI) supports projects in other industrialized countries, mainly former Soviet satellite countries that are transitioning to market economies.

However, offset projects are controversial for several reasons. First, companies in industrialized countries can emit more carbon than is allowed under their countries' total allowable levels under the Kyoto Protocol by buying credits from developing countries, which have no emission caps. In effect, offsets allow industrialized countries to outsource reductions to places where they can

Development Mechanism, which accounted for 29 percent. Joint Implementation projects and all other carbon trading forums generated less than 2 percent.

During its trial phase from 2005 through 2007, the EU ETS produced mixed results. Carbon allowances initially traded for €20-30 (about $30-$45) per ton of CO_2, but in April 2006 the Czech Republic, Estonia, the Netherlands, Switzerland and France announced that their 2005 GHG emissions had been lower than expected. Demand for allowances fell sharply. Share prices plunged to €10-15 ($15-23) within a few days. And prices for allowances that were valid only for the trial period — and hence not usable after 2007 — fell to almost zero in early 2007. Allowances then stabilized at €15-25 ($23-38) for the second trading period. [13]

Some observers called the price gyrations a sign that the ETS was failing. Open Europe, a London think tank, charged that ETS had failed to provide either a "workable market in carbon" or reduced emissions. [14] Others said market volatility was not surprising for the trial phase of a new system without historical data to guide it.

"Since companies had not previously been required to track and disclose emissions, there were no hard numbers on which to base allocations," wrote Annie Petsonk, an attorney for the New York-based Environmental Defense Fund. "So companies were asked how much they'd need to emit, and naturally they said, 'A lot!' When emissions data became available and companies saw that cutting emissions was easier than they anticipated, the price of allowances plummeted." [15]

In a detailed assessment, Massachusetts Institute of Technology (MIT) economists A. Denny Ellerman and Paul Joskow pointed out that ETS was not intended to deliver big emissions

Carbon Trading Tripled

Nearly 3 billion metric tons worth of greenhouse gas emission allowances were traded in 2007, more than triple the amount sold in 2005.

Annual Volume of Carbon Emission Contracts, 2005-2007
(in billions of metric tons)

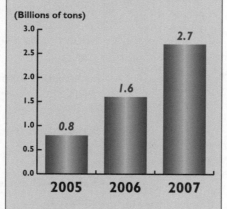

(Billions of tons)

* A metric ton is 2,204.6 pounds.

Source: Kjetil Røine, et al., "Carbon 2008," Point Carbon, March 11, 2008

cuts during its trial run, and that estimating emissions for any given year is difficult because weather patterns and fuel prices affect fossil fuel use. Similar fluctuations occurred when the United States launched a trading program for sulfur dioxide (SO_2) allowances in the late 1990s, they noted, and as in the SO_2 program, ETS allowance prices settled down once policy makers had some real emissions data to work with. [16]

In its second trading period, which runs from 2008 through 2012, the EU's total emissions cap is 6.5 percent below the 2005 level. "Leaders learned their lesson after they over-allocated allowances in Phase I, and the cap is more stringent now. They have definitely done a better job in Phase 2," says Anja Kollmuss, an analyst at the Stockholm Environment Institute (SEI).

EU leaders now are grappling with new challenges for Phase 3, which starts in 2013, including bringing more emitters under the pollution cap. Currently the ETS only covers six sectors — energy, iron and steel, cement, glass, ceramics and pulp and paper — which produce about 45 percent of EU emissions. The European Parliament voted in July 2008 to include aviation emissions, beginning in 2012, and EU government ministers formally approved the policy in October over industry resistance. [17] Airlines assert that their industry has been hit hard by high oil prices and that the EU does not have legal authority to regulate emissions from flights, regardless of where airlines are based. (See "At Issue," p. 313.) European leaders also propose to include emissions from petrochemicals, aluminum and ammonia production in Phase 3.

Another critique points up flaws in both the ETS and CDM systems. In a 2007 report, the World Wildlife Fund warned that many EU countries might allow emitters to use offset credits from CDM and JI projects to meet most or all of their Phase 2 EU emission limits. Because they are not buying allowances from other EU sources, that would mean they aren't really cutting carbon among EU emitters. [18]

That prospect raises two problems, says Kollmuss. First, the Kyoto Protocol and EU directives say offsets should be "supplemental" to direct reductions. "When emitters can use a high fraction of offset credits, some sectors may not have to actually cut their emissions at all," she says.

Second, some offsets fail the "additionality" test, critics say, which occurs when the GHG reductions they produce are not additional to what would have happened anyway. For example, if local law already requires landfill owners to collect methane emissions instead of venting them into the air, they should not be able to market that action as a CDM project

How Greenhouse Gases Are Measured

When discussing greenhouse gas (GHG) emissions, businesses and government agencies often use shorthand terms, like "carbon" or "carbon dioxide," to refer to the various gases emitted when carbon-based fuels are burned.

The Kyoto Protocol and other schemes to regulate greenhouse gases cover six major types of emissions that remain in the atmosphere for a significant time, trapping heat that is reflected back to Earth, which warms the planet's surface. Most are caused by various human activities.

Climate scientists have assigned each gas a global warming potential (GWP), based on its heat-trapping properties. A GWP value measures the impact a gas has on the climate over a given time period (usually 100 years) compared to the heat-producing impact of a ton of carbon dioxide (CO_2) — the most abundant greenhouse gas. For example, methane's GWP value is 25, which means that a ton of methane released into the atmosphere will cause as much warming as 25 tons of CO_2 over a 100-year period. [1] Thus, the higher the GWP, the more global warming the gas causes.

Carbon trading schemes allow emitters to trade allowances to release some or all of the six types of gases, whichever are covered by a particular system. For example, under the Kyoto Protocol, so-called Clean Development Mechanism (CDM) projects in developing countries can generate credits that they can then sell abroad by reducing their own emissions from any of the six GHG categories. Each credit certifies that the project has reduced greenhouse gas emissions by the equivalent of one metric ton (2,205 pounds) of carbon dioxide per year.

Under the European Union's Emissions Trading System (EU ETS), an electric power company in Italy might buy credits to cover excess CO_2 emissions created by its coal- or oil-fired power plants. These credits could come from CDM projects that reduced other GHG emissions through such actions as collecting methane emissions from landfills or reducing hydrofluorocarbon leakage at aluminum-smelting plants. Using the GWP values for these gases, project owners can calculate how many tons of CO_2 equivalent the project releases or avoids, and then sell the reduction credits easily across international borders.

Types of Greenhouse Gases

GHG Categories	GWP Value*	Major Sources
Carbon dioxide (CO_2)	1	Fossil fuel combustion, deforestation
Methane (CH_4)	25	Landfills, rice paddies, digestive tracts of cattle and sheep
Nitrous oxide (N_2O)	298	Fertilizer, animal waste
Hydrofluorocarbons (HFCs)	Varies (up to 14,800)	Semiconductor manufacturing and other industrial processes
Perfluorocarbons (PFCs)	Varies (up to 12,200)	Same as HFCs, plus aluminum smelting
Sulfur hexafluoride (SF_6)	22,800	Electrical transmission systems, magnesium and aluminum production

* Global warming potential

Source: U.S. Environmental Protection Agency

[1] U.S. Environmental Protection Agency, "Inventory of U.S. Greenhouse Gas Emissions and Sinks: Fast Facts," April 2008.

---

and sell the emission credits to a company in an industrialized country. Conversely, they say, if there is no clear financial reason to carry out a project unless it can produce CDM credits that can be sold, then the project is probably additional.

"Additionality is a simple concept, but it often comes down to subjective decisions," says Kollmuss. "And it's very easily fudged."

As one example, Stanford University law professors Michael Wara and David Victor pointed out in a 2008 paper that nearly all new renewable and gas-fired power plants in China are applying for CDM credits, even though China's energy sector is growing rapidly and the Chinese government has asked companies to invest in non-coal energy sources. Given these trends, they contend, China would probably be moving toward lower-carbon fuels even without CDM credits for new power plants. "[I]n practice, much of the current CDM market does not reflect actual reductions in emissions, and that trend is poised to get worse," the authors argued. [19]

Such controversies have spurred development of an entirely new industry of consultants and third-party certifiers who screen and verify claims from "green" development projects and help buyers find high-quality offset sources. (*See sidebar, p. 308.*)

U.N. officials acknowledge that additionality is a key challenge but argue that the CDM has effective rules for measuring it. They also point out that that the CDM has generated three times more funding for climate-friendly technology transfers to developing countries than direct foreign aid programs.

"Has the Kyoto Protocol's Clean Development Mechanism met the goal for which it was designed?" Yvo de Boer, executive secretary of the U.N. Framework Convention on Climate Change, asked in October. "In my view, the answer is yes." [20]

The EU has barred using reforestation projects in developing countries as offsets because regulators say reductions from these projects are hard to measure and can be quickly reversed (for example, if a forest plantation burns down). Ironically, developing countries without large industrial sectors would have a better chance of earning money through the CDM if the EU accepted forestry credits, since farming and forestry projects are among their best options for slowing climate change.

Carbon marketers generally see the CDM as an important tool despite its flaws. "CDM has the strictest review and approval process for emission reduction projects in the world." says Point Carbon's Hasselknippe. "Some offset projects in North America [where companies are experimenting with emission reductions and trading] are even more questionable than CDM projects. Without a regulated market, anything goes."

Are there better ways to cut emissions?

Creating carbon markets and trading carbon emission allowances is the best way to speed the transition to a low-carbon world, say proponents, because it puts a limit on carbon pollution and creates big profit incentives for cutting emissions. But critics see it as a complicated scheme that isn't guaranteed to deliver innovative energy solutions. Instead, some say, carbon taxes would be a simpler and more direct way to slow climate change.

Both approaches make polluters pay for carbon emissions, which spurs investments in cleaner technologies — with one important difference. In cap-and-trade schemes regulators specify how much pollution can be emitted, but they can't predict exactly how much allowances will cost once trading starts. Many factors, including weather, economic conditions and the discovery of new technologies influence fossil fuel use, which can drive demand for carbon allowances either up or down.

Economists can model what allowance prices may look like, but experience can be quite different from predictions, as the U.S. acid rain trading program of the 1990s (*see p. 310*) and the trial phase of EU ETS both showed.

Carbon taxes, on the other hand, charge polluters a set rate for each ton of greenhouse gases released, so there are no surprises about compliance costs. Regulators can't be sure, however, how taxes will affect pollution levels because they don't know how businesses will handle those costs. Some companies may pay taxes on their emissions and pass the expense on to consumers, while others clean up their operations to avoid the extra charge. In other words, carbon taxes offer more certainty for businesses, but cap-and-trade systems provide more certainty that the environment will improve.

"A tax doesn't put any legal limits on how much pollution can be released, so it's like a blind bet," says Fred Krupp, president of the Environmental Defense Fund (EDF). "You know what the ante is, but not what the payoff will be. Only a cap guarantees results."

Norway has achieved mixed results since it imposed a $65-per-ton carbon tax on oil and gas companies in 1991. The tax prompted StatoilHydro, one of Norway's largest energy companies, to sharply cut its carbon emissions, largely by pumping them into an undersea reservoir. Today the firm is one of the world's few companies doing large-scale geologic storage of CO_2 emissions. [21]

But StatoilHydro also has expanded drilling operations since the tax was levied. So, even though the company is more carbon efficient than many other big energy producers, its net emissions have increased as world demand for oil has grown. Today Norway's total GHG emissions are 15 percent higher than in 1991. Norway still has the tax in place, but it also has joined the EU ETS, even though it is not an EU member.

Cap-and-trade supporters also argue that carbon trading generates larger investments in new technologies than taxes do, because polluters can turn emissions into income by cleaning them up and selling their unneeded allowances. "A tax creates no such market and, so, fails to enlist the full range of human potential in a struggle where every bit of creativity is needed," writes Krupp. [22] But many energy experts say a whole suite of measures is needed to commercialize new energy technologies and that the process shouldn't be left up to market forces. Rather, they argue, a combination of big governmental investments and other measures like tax credits and clean energy targets are needed to help ensure that clean technologies are put to use.

"Emissions trading won't do much to stimulate investment in research and development of technologies that may be able to deliver deep cuts in emissions in the future," says Chris Riedy, research director at the CAP Institute for Sustainable Futures at the University of Technology in Sydney, Australia. "Markets are very good at meeting short-term goals but not so good at looking many years ahead."

Australia is developing a national carbon trading plan, Riedy notes, but it also has established a national target to generate 20 percent of its energy from renewable fuels by 2020. "That will ensure that renewable energy is developed over time until it can establish itself in the market," says

Riedy. "We need to give the industry some long-term certainty."

The challenge is even larger in fast-growing countries like China, India and Brazil, which are just now industrializing and have not yet accepted binding caps on GHG emissions. As those countries raise their living standards over the next several decades, they will account for a rising share of world

ket, and China is the biggest source, trading can help China make more cuts because businesses will see value in carbon."

But several things must happen before carbon trading becomes a useful tool for cutting Chinese GHG emissions, Yang continues. First, Beijing must make a political commitment to reducing emissions. Then the

share the burdens fairly. Because developed nations got rich from fossil-fueled growth and produced most of the human-driven warming that has occurred to date, the framers of the Kyoto Protocol decided that developed countries should make the deepest GHG emissions cuts. However, large developing countries like China and India are rapidly becoming the world's biggest carbon sources, so it is also crucial to limit their emissions while allowing their citizens to enjoy rising standards of living, say climate experts.

"[W]e need to provide resources to see that the developing countries don't get hooked onto the same path of development that we have," said Rajendra K. Pachauri, chairman of the Intergovernmental Panel on Climate Change (IPCC), which advises governments on climate science. [23]

The Clean Development Mechanism was designed as a first step to help poor countries grow while reducing their emissions. But critics argue that CDM projects primarily benefit the rich nations that sponsor them and that some actually damage the environment in the host countries.

For instance, the environmental advocacy group International Rivers charged in a 2007 report that awarding carbon reduction credits to numerous hydropower projects resulted in "blindly subsidizing the destruction of rivers, while the dams it supports are helping destroy the environmental integrity of the CDM." The study contended that the CDM has few standards to block projects that harm nearby ecosystems and that many hydropower projects applying for CDM credit would clearly be built in any case. As examples it cited a 60-megawatt dam in Kenya that started construction in 1999 (before the CDM was established) and an 880-megawatt dam in Brazil that applied for CDM validation six months after it began generating power in May 2007. [24]

China's booming growth has made it one of the world's top emitters of carbon dioxide, the most abundant greenhouse gas (GHG). Advocates of carbon trading say that if China were to set formal limits on its GHG emissions, polluters would have an incentive to cut emissions in order to trade their allowances for cash.

AFP/Getty Images/Frederic J. Brown

energy consumption. It is crucial to help those countries move onto clean energy pathways in order to slow climate change.

For instance, carbon trading could become an important option for China at some point, says Yang Fuqiang, chief representative in Beijing for the U.S.-based Energy Foundation. "China is now the top CO_2 emitter in the world, and we expect that its emissions will be much larger by 2030, perhaps as much as 20 percent of world emissions," he says. "If carbon becomes a commodity that is traded in the mar-

Chinese government must fund development of clean energy sources. Carbon trading will not work, however, without a strong legal system to ensure trades are protected and penalties enforced if partners violate the rules.

"China's legal systems aren't strong enough for carbon trading yet," says Yang.

Does carbon trading help developing countries?

Global climate change policy has been complicated by the need to create strategies that enable countries to

Funding is not the only yardstick, replies U.N. spokesman David Abbass. "A company might have the ability to undertake an emission-reduction improvement, but not the incentive," he says. "If CDM was a motivating factor, then the project could potentially qualify, regardless of when construction was begun. In most hydro projects, CDM is providing incentives for efficiency improvements such as installing more efficient turbines. Such a decision could be undertaken after dam construction has begun or even after the dam has entered operation."

Forest carbon credits are also controversial. Under the CDM program, carbon credits can be granted for planting trees on formerly forested land that is either being reforested or used for other purposes. Many early CDM forestry projects were commercial tree plantations that were popular because planting swathes of fast-growing tree species absorbs large quantities of carbon. But opponents complained that such projects sometimes ended up clearing large areas of native forest, expelling local populations and damaging the environment.

"The fact that eucalyptus absorbs carbon dioxide to grow . . . can never be used to justify the environmental, social, economic and cultural damage that has occurred in places where large-scale monoculture tree plantations have been implemented in our country," wrote 53 unions and nonprofits in 2003 opposing a tree plantation proposed by a company called Plantar in the Brazilian state of Minas Gerais. The project ultimately was approved by the CDM board after three tries, not for absorbing carbon into the trees but for using a low-carbon process to turn those trees into charcoal. [25]

"The CDM is riven with fraud, just like other government-to-government aid programs, and it doesn't save any carbon," says Michael Northcott, a divinity professor at Scotland's Edinburgh University who views carbon trading as a route by which governments can avoid imposing hard limits on GHG emissions. Citing projects like the Plantar venture, Northcott writes, "The new global carbon market is not incentivizing real reductions in emissions. But it has created tremendous, new trading opportunities and new opportunities for fraud and injustice." [26]

Now, however, awarding credits for forest protection is gaining new support from tropical countries and conservation experts, who say forests can soak up carbon emissions, protect biodiversity and provide economic benefits to developing nations. Advocates are proposing some new approaches to make this method more rigorous. For example, avoided emissions would be measured at the national level instead of project by project, so it would be harder for a host country to claim credit for saving one forest while it cut down others. [27]

Advocates say the new approach would reward countries that preserve their forests instead of cutting them down and then seeking carbon credits for new tree plantations. "Central African countries consider that their efforts made in managing forests deserve to be recognized and supported, because they are positive for climate," the 15-member Coalition of Rainforest Nations contended in 2007. [28] More than 300 national leaders, research institutes and conservation groups have signed a policy statement urging governments to include tropical forests in global carbon markets. [29]

As negotiations on a post-Kyoto climate treaty proceed, CDM officials say the program needs to be scaled up. "Carbon markets and market-based mechanisms, like the [CDM], are essential for achieving the large shifts in investment required . . . to put the world on a clean path to development," said the U.N.'s de Boer. [30]

For the long term, some experts are thinking beyond the CDM model. "The CDM only lets developing countries trade credits if they prove additionality project-by-project, which is a nightmare. It's cumbersome, it leads to endless arguments and small countries have been squeezed out by big projects in China, India and Brazil," says EDF's Krupp. "We should . . . offer all developing nations technical assistance and more generous emissions targets if they agree to cap their emissions quickly. We need a global system where everyone agrees to a cap that's fair, given their level of development."

Even CDM advocates agree that benefits have been spread unequally up to now. About three-quarters of all CDM projects to date are located in China, Brazil, India and South Korea. [31] Many poor regions like sub-Saharan Africa, which are extremely vulnerable to the negative impacts of climate change, have seen little benefit from carbon trading.

"So far, the poorest developing countries have been bypassed — and there have been limited benefits for broad-based sustainable development" from carbon trading, the U.N. Development Programme observed in its 2007/2008 *Human Development Report*. "Marginal women farmers in Burkina Faso or Ethiopia are not well placed to negotiate with carbon brokers in the City of London."

However, the report noted, new approaches, such as "bundling" many small, rural projects together for CDM credit, could help poor countries participate. [32] Under a 2006 initiative called the Nairobi Framework, the U.N. is working to channel CDM projects to countries in sub-Saharan Africa. In 2008 the U.N. Environment Programme estimated that CDM projects in Africa could generate nearly $1 billion worth of credits by 2012. [33] ∎

BACKGROUND

Who Pays for Pollution?

The fledgling global carbon trading industry represents the intersection of two complex debates that stretch back for more than a century. Scientists have worked since the early 1800s to understand how Earth's climate systems function and whether human actions affect them. And for nearly as long, economists who study the environment have sought cost-effective ways to control pollution.

Climate science has been international from its earliest days. In 1859 Irish physicist John Tyndall showed that certain gases in the atmosphere absorbed heat. Svante Arrhenius, a Swedish chemist, built on this idea with his calculation in 1896 that doubling the quantity of CO_2 in the atmosphere would raise Earth's average by 5 to 6 degrees Centigrade. Other researchers have shown that natural processes also influence climate cycles. For example, in 1860, Scottish physicist James Croll theorized that regular variations in Earth's orbit could trigger ice ages. Eighty years later Milutin Milankovic, a Ser-

bian geophysicist, calculated these variations more precisely and developed a theory of glacial periods, now known as Milankovic cycles.

Other environmental issues were more urgent in the early 1900s. Air and water in industrialized countries were already heavily polluted from factory operations and urban growth, but governments had little power to respond. In Britain and the United States the nuisance doctrine — an historic concept of English common law — held that people should not use their property in ways that infringed heavily on their neighbors and that injured parties could sue those responsible for noise, odors and toxic discharges. Noxious facilities such as metal smelters were frequent early targets for nuisance lawsuits in the United States.

However, nuisance law was ineffective at controlling harmful discharges and emissions from large-scale industrial production. With pollution coming from many sources, it was hard to prove direct connections between discharges and impacts. Moreover, by the early 1900s, U.S. courts had come to view pollution as an unavoidable result of economic activity. Rather than shutting down dirty factories, they generally weighed harms against benefits and compensated plaintiffs for serious damages while allowing polluters to keep operating. [34]

Governments then developed new approaches, like zoning, which established rules for using large areas of land. City and state agencies enforced a growing body of public health laws barring practices such as dumping untreated waste into waterways. In 1920, British economist Arthur Pigou proposed a new option: pollution taxes. Pollution, he argued, was a "negative externality" — a production cost that polluters did not have to pay for. If manufacturers were taxed for their pollution they would

Continued on p. 308

Indonesian environmental activists at the U.N. climate change conference in Bali, Indonesia, last December demonstrate against a proposal to award carbon credits to tropical countries that join the Reducing Emissions From Deforestation and Degradation (REDD) program. The protesters say the Indonesian government can't handle the delicate and complicated carbon trading scheme and that the program will benefit developed countries and large corporations at the expense of indigenous communities. Delegates agreed to include forest conservation in future discussions on a new global warming treaty.

AP Photo/Firdia Lisnawati

Chronology

1900s-1960s
As scientists study Earth's climate, experts debate controlling pollution efficiently.

1920
British economist Arthur Pigou suggests taxing polluters for the indirect costs of their emissions.

1945
Researchers start developing models to test atmospheric behavior.

1957
American geochemist Charles David Keeling begins measuring atmospheric carbon dioxide (CO_2) levels in Hawaii, where readings are not skewed by pollution.

1960
British economist Ronald Coase proposes tradable emission allowances.

1970s-1980s
Scientists warn that humans may be causing global warming. Stricter pollution controls are enacted.

1970
Congress creates Environmental Protection Agency, expands Clean Air Act.

1972
First major global environmental conference — held in Stockholm, Sweden — spurs creation of United Nations Environment Programme.

1976
Scientists identify deforestation as a major cause of climate change.

1980
U.S. President Ronald Reagan's election signals a backlash against technology-specific regulations.

1987
Montreal Protocol sets international limits on ozone-destroying gases.

1988
U.N. creates Intergovernmental Panel on Climate Change (IPCC) to provide expert views on global warming.

1990
IPCC says global temperatures are rising and likely to keep increasing. . . . U.S. adopts emissions trading to reduce acid rain.

1990s
Governments pledge to tackle climate change, but worry about costs.

1992
The United States and over 150 nations pledge to cut greenhouse gas (GHG) emissions below 1990 levels by 2000.

1995
IPCC finds that global warming has a "human-driven" signature.

1997
Kyoto Protocol is adopted after intense negotiations, requiring developed countries to cut GHG emissions 5.2 percent, on average, below 1990 levels by 2012. U.S. Senate refuses to ratify it until developing nations also are required to make cuts.

2000s-Present
Carbon emissions trading begins, primarily in Europe. Support grows in United States for action to reduce GHGs.

2001
IPCC says major global warming is "very likely."

2002
Clean Development Mechanism (CDM) — which allows industrialized countries to partly fulfill their carbon-reduction commitments by purchasing GHG reductions in developing countries — begins.

2003
Chicago Climate Exchange launches voluntary GHG trading system for selected U.S. companies and nonprofits.

2005
Kyoto Protocol enters into force with only the United States and Australia as non-participating developed countries. . . . EU Emissions Trading Scheme begins trials. . . . Seven Northeastern states agree to form GHG cap-and-trade system for electric power plants.

2006
EU carbon allowance prices plummet after emissions are lower than expected. . . . Global carbon trading reaches $30 billion, triple the previous year's level. . . . California promises to cut CO_2 emissions 25 percent by 2020 and to start trading emissions in 2012.

2007
IPCC says climate warming is mostly due to human activities. . . . Australia joins Kyoto Protocol. . . . Three more states join Northeastern cap-and-trade system. . . . 180 countries agree to negotiate a post-Kyoto climate change treaty.

2008
EU emissions trading scheme enters Phase 2, with tighter caps. . . . U.N. proposes stricter standards for CDM projects.

Continued from p. 306

have an incentive to pollute less, according to the theory.

Economists generally agreed with Pigou's approach, but environmental regulation did not gain a serious foothold until after World War II. Economic growth expanded worldwide in the 1950s and '60s, first in the United States and then in post-war Western Europe and Japan. Governments began to limit industrial pollution, but instead of taxing it they applied so-called command-and-control standards, which told polluters how much pollution they could release and often specified what kind of technologies had to be used to clean up their operations. The same standards applied to all producers, whether their operations were relatively clean or dirty. As a result, these laws imposed much larger costs on some sources than on others.

In 1960 University of Chicago economist Ronald Coase proposed a way to control pollution with lower total costs to society. If rights to pollute could be bought and sold, he argued, polluters could bargain and find an efficient way to distribute those rights. Other economists took up his idea and called for government regulators to limit total quantities of pollutants and then create markets for pollution rights.

"[N]o person, or agency, has to set the price — it is set by the competition among buyers and sellers of rights," wrote Canadian economist John Dales in 1968. [35] This approach was more effective, proponents contended, because producers (who knew more about their own costs and production methods than regulators) could decide who would clean up and find the best ways to do it.

International Cooperation

By the 1960s, protecting the environment was a national political issue in many industrialized countries. Social Democrat Willy Brandt campaigned for chancellor in West German in 1961 with a promise to clean up air pollution. Japanese activists began suing large polluters in the mid-1960s, pressuring regulators and industrialists into adopting tighter con-

Nonprofit Auditors Keep Projects 'Honest'

Gold Standard projects provide jobs, help the environment.

Power outages and voltage fluctuations once plagued the Honduran city of La Esperanza, and many rural residents in the surrounding countryside had no electricity at all.

Now a small hydroelectric project on the nearby Intibuca River reliably produces 13.5 megawatts of electricity — enough to power 11,000 households for a year — while avoiding 37,000 tons of annual carbon dioxide emissions from diesel generators previously used to produce electricity. And because it is a so-called run-of-river project, it generates electricity without damming the river.

The La Esperanza Hydroelectric Project is the first to be certified as reducing greenhouse gas (GHG) emissions under the Kyoto Protocol's Clean Development Mechanism (CDM). The CDM allows Third World developers whose projects reduce carbon emissions to sell "emission credits" — equal to the emissions they avoid — to polluting companies in industrialized countries.

The project also will provide a variety of other sustainable benefits in the community, such as reducing local residents' use of carbon-consuming trees for fuel, encouraging reforestation and providing reliable jobs and technical skills for the dam construction, maintenance and operating staffs, providing running water for households near the project and engaging more women in work and community life.

How can La Esperanza's developers prove their facility will provide all those benefits? The project is being evaluated by Gold Standard, an independent, nonprofit organization in Basel, Switzerland. Founded by the World Wildlife Fund and other non-governmental organizations and funded by public and private donors, Gold Standard accredits high-quality CDM projects that benefit the local community and cut carbon emissions. Gold Standard approval gives carbon-credit buyers extra assurance that the carbon credits they are purchasing come from measurable GHG reductions that have clearly benefited the host countries where they were carried out.

Nonprofits like Gold Standard have emerged to provide extra certification for carbon offset projects because of concerns that private verification companies, which are paid by project developers, have a financial incentive to certify that the projects they are auditing reduce carbon emissions just to get them approved. And the CDM Executive Board, which reviews CDM applications, does not have enough staff to verify all of the information submitted by auditors.

"Right now, good auditors get their projects approved, but that shouldn't be the only incentive," says Stanford University law professor Michael Wara. The CDM board has "done the best it can, but it's in an untenable situation," he contends, because it is understaffed and facing a growing demand for offset credits.

Some critics have claimed that as a result of these conflicts of interest and other problems, carbon markets, in effect, are generating "rights to pollute."

"We require project developers to make positive contributions to local communities in two out of three categories — economic, social and environmental — and our screening process gives them numbers they can use to rate what they're delivering in each area," says Caitlin Sparks, U.S. marketing director for Gold Standard. "We monitor those promises through the full life of the project. U.N.-accredited auditors validate and verify all of the documents, and the information is re-verified after the project starts."

For instance, all CDM projects are supposed to promote "sustainable development," but it's usually left up to the host country to define what that means. However, Gold Standard makes its own judgment.

"They are doing the sorts of things that should be applied wholesale to CDM," says Wara. 'They dig in and do better verification, which costs more and makes the process more time-consuming, but that needs to happen. We need more scrutiny of these projects."

Gold Standard projects have three key features: They must focus on renewable energy or energy efficiency to help promote a transition to a clean-energy economy; developers must prove that the carbon reductions will be "additional" to business as usual (this test is optional when projects go through CDM review but is required by Gold Standard); and they must show that their projects will make measurable economic contributions to sustainable development in host communities.

"A free market for credits will tend to focus on quantities of tons," says Sparks. "The Gold Standard is meant to focus on quality" of emission reductions.

Gold Standard projects in India and South Africa reflect the program's diversity and focus on quality:

- The Shri Chamundi biomass co-generation power plant in Karnataka, India, will generate 16 megawatts of electricity from biomass fuels such as eucalyptus branches, coconut fronds, rice husks and cashew shells. It will also use waste heat to produce steam for manufacturing, replacing boilers that run on heavy fuel oil. The plant will create more than 800 jobs, including collecting and preparing biomass, converting previously useless crop residues into fuel. It also will reduce open burning of crop wastes in fields, which pollutes the atmosphere and local water supplies.
- In Cape Town, South Africa, the Kuyasa housing service upgrade project installed ceiling insulation, solar hot water heaters and energy-efficient lighting in a low-income housing development and will install similar improvements in future developments. Making homes more energy-efficient will reduce CO_2 emissions, local air pollution and the danger of household fires. [1]

---

[1] Information about these projects comes from validation reports in the Gold Standard Registry, http://goldstandard.apx.com; and "Reducing the Carbon Footprint of the UN: High-Level Event on Climate Change," U.N. Headquarters, Sept. 24, 2007, www.un.org/climatechange/2007highlevel/climate-friendly.shtml.

---

trols. In 1970 two versions of Earth Day were launched: an international celebration on the date of the spring equinox, formally endorsed by the United Nations, and a U.S. observance on April 22 that drew millions of Americans to rallies and teach-ins.

National governments began setting standards for air and water quality, waste management and land conservation. Then a 1972 international conference on the environment, held in Stockholm, set lofty goals for international cooperation and led to the formation of the United Nations Environment Programme. The conferees declared that most environmental problems in the developing world were caused by poverty and underdevelopment, and that rich countries should try to reduce the gap between rich and poor countries. [35]

Meanwhile, international cooperation was growing in the field of climate science. In the 1950s and '60s, international research groups in the United States, England, Mexico and elsewhere developed circulation models to simulate climate processes and began testing theories about how the system might change in response to natural or manmade events. French, Danish, Swiss, Russian and U.S. scientists drilled into ice sheets in Greenland and Antarctica and analyzed air bubbles trapped thousands of years earlier to determine how the atmosphere's composition had changed over time. A growing body of climate studies showed that many processes shaped global climate patterns, and that human actions could disrupt the system.

In 1976, frustrated with the slow pace of pollution reductions under the Clean Air Act, U.S. policymakers began experimenting with market-based measures. As a first step, companies were permitted to build new factories in polluted areas if they bought credits from nearby sources that had reduced emissions below legal limits. In 1977 Congress amended the act to allow policies like banking credits (saving them for use or sale in the future). In 1982 the Environmental Protection Agency (EPA) used a trading program to phase out lead from gasoline. Refiners were issued tradable lead credits that they could sell if they were already blending unleaded gasoline or use while retooling their plants. Lead, which had been outlawed from U.S. gasoline, was finally eliminated by 1987.

Other nations also tried market-based environmental policies, primarily pollution taxes. Many European countries — including West Germany, the Netherlands, Czechoslovakia and Hungary — taxed water pollution discharges to help fund sewage treatment and bring water quality up to healthy standards. [37] France and Japan imposed charges for air pollution emissions. China also adopted air and water pollution taxes in the early 1980s, although these levies were quite low, and a large share of the funds were distributed back to pollution sources as subsidies. [38]

Then, in an important milestone for global environmental cooperation, 23 nations signed the Montreal Protocol in 1987, agreeing to restrict production and use of industrial chemicals that were damaging Earth's protective ozone layer. Over the next decade, as

climate and that policy makers needed to act. In 1988 the United Nations established the Intergovernmental Panel on Climate Change (IPCC) to advise national governments about climate science and potential impacts from global warming. But critics, including large corporations and President George H. W. Bush, argued that the scientific evidence was uncertain and that reducing GHG emissions would seriously harm economic growth.

ministration was worried about costs. A U.S. carbon tax was not an option: The administration had suffered an embarrassing defeat in 1993 when it proposed a BTU tax (a levy on the energy content of fuels), only to be blocked by fellow Democrats in Congress. Instead, U.S. negotiators at Kyoto pushed to include emissions trading and credits for funding offset projects in developing countries and former Eastern Bloc nations.

Although the final agreement included these policies and the Clinton administration signed the treaty, the Senate voted 95-0 for a resolution against ratifying it unless developing countries also had to make binding reduction pledges. President George W. Bush, who had promised during his campaign to limit carbon dioxide emissions, repudiated the Kyoto agreement shortly after taking office, arguing that mandatory GHG reductions (even through market-based mechanisms) would harm the U.S. economy.

> **"Marginal women farmers in Burkina Faso or Ethiopia are not well placed to negotiate with carbon brokers in the City of London."**
>
> — *U.N. Human Development Report 2007/2008*

science showed that damage was still occurring, more nations joined, and members amended the agreement to eliminate the substances completely. Several nations used allowance trading systems to phase out domestic use of ozone-depleting chemicals, including the United States, Canada, Mexico and Singapore.

The protocol established some other important precedents: It relied on expert advice from scientists, forced governments to act in time to prevent serious environmental harm and required developed nations to help developing countries adjust to the ban without harming their living standards. [39]

Confronting the Evidence

By the late 1980s many environmentalists and scientists believed human activities were affecting Earth's

Other nations, led by Western European countries with strong Green parties, wanted a binding agreement to limit greenhouse emissions. The Framework Convention on Climate Change (FCCC), signed at the 1992 Earth Summit in Rio de Janeiro, Brazil, amounted to a compromise: It called only for voluntary reductions in greenhouse gases to 1990 levels but laid out a path for further action. Some countries — including the Netherlands, Sweden, Finland, Norway and Denmark — passed domestic carbon taxes to reduce their emissions. But total GHG emissions from industrialized nations kept rising, making it clear that mandatory targets and timetables would be needed.

In 1997 FCCC members adopted the Kyoto Protocol, which required signatories to make specific reductions (averaging 5.2 percent below 1990 levels) by 2012. U.S. President Bill Clinton supported the goal, but his ad-

Nonetheless, President Bush embraced the idea of emissions trading to address domestic air pollution issues and sought to build on a successful program initiated a decade earlier under the first Bush administration. In 1990 Congress had amended the Clean Air Act to create an emissions trading program for sulfur dioxide (SO_2) and nitrogen oxide (NO_x), two pollutants from fossil fuel-fired power plants. These emissions formed acids in the atmosphere that fell back to Earth in rain and snow, damaging forests, soils and buildings. The so-called acid rain trading program, which began in 1995, capped emissions of SO_2 (with looser limits for NO_x) and set up a trading market for emission allowances.

The program was widely viewed as a success. EPA reported in 2004 that a decade of emissions trading had reduced the power sector's SO_2 and NO_x emissions 34 and 38 percent, respectively, below 1990 levels. [40] Economists estimated trading had saved $1 billion or more per year over command-and-control approaches. [41] Touting these results, President Bush proposed emissions trading initiatives to cut U.S. SO_2 and NO_x emissions even further and suggested using a trading scheme to control mercury emissions. But congressional critics argued that these measures did not cut far or fast enough and that emissions trading was the wrong way to reduce toxic pollutants like mercury. [42]

As the Bush administration continued to oppose cutting GHG emissions, other U.S. leaders grew increasingly worried about climate change. Sens. John McCain, R-Ariz., and Joseph Lieberman, D-Conn., offered carbon cap-and-trade legislation in 2003 and 2005 and reintroduced the bill in 2007. Seeing the political handwriting on the wall, large U.S. corporations began to endorse carbon controls.

"We know enough to act on climate change," said the U.S. Climate Action Partnership, an alliance of major corporations including Alcoa, Dupont and General Motors. The group called on Congress to pass mandatory GHG limits and create a cap-and-trade system to attain them. [43] ■

CURRENT SITUATION

A New Player?

As Americans increasingly worry about climate change, many observers expect the United States to limit its greenhouse gas emissions and create a domestic carbon trading system after the 2008 elections. Multiple cap-and-trade bills were introduced in both houses of Congress in 2007 and 2008, including several with bipartisan support. [44] And the two major-party presidential candidates, Sens. McCain and Barack Obama, D-Ill., both pledged to set up a cap-and-trade system and to pursue deep cuts in U.S. GHG emissions.

Since the United States is one of the world's largest GHG emitters, U.S. entry into carbon trading would dramatically expand global carbon markets. New Carbon Finance, a market research firm in London, estimated in October that the total value of world carbon trading would reach $550 bil-

River waters crash into a Buddhist temple during high tide on the outskirts of Bangkok, Thailand. Climatologists say higher global temperatures are causing polar ice caps to melt, raising sea and river levels in low-lying coastal areas. Carbon trading schemes are the world's current answer to the question of how to control global warming.

lion by 2012 and just over $2 trillion by 2020, even without U.S. participation. If the United States introduces a federal cap-and-trade system, however, those figures would increase to $680 billion by 2012 and more than $3 trillion by 2020. [45] By way of comparison, $3 trillion is roughly the size of the combined world markets for oil, coal, natural gas and electricity today. [46]

Two legislative proposals — one debated by the Senate in mid-2008 and a House Energy and Commerce committee proposal released on Oct. 7 — offer some indication of what national cap-and-trade legislation might look like. Both bills would cap U.S. greenhouse gas emissions and set up a trading system to reduce them. The House bill would require a 6 percent cut below 2005 levels by 2020, and the Senate bill calls for a 19 percent cut. By 2050, however, the House measure would reduce emissions by 80 percent below 2005 levels, compared to 71 percent under the Senate bill.

Along with public concern and growing scientific evidence that human activities are warming the planet, another factor pushing U.S. policy makers to act is a 2007 Supreme Court ruling which held — contrary to the Bush administration's position — that carbon dioxide was a pollutant under the Clean Air Act and that the EPA had authority to regulate it. [47] "CO_2 controls are clearly coming. The only remaining questions are when and who is going to do the controlling," said Rep. Rick Boucher, D-Va., chair of the

House Energy and Commerce Committee's Subcommittee on Energy and Air Quality, in late 2008. A coauthor of the committee's cap-and-trade proposal, Boucher said he thought Congress rather than the EPA should lead on regulating carbon and that he planned to hold hearings on cap-and-trade legislation early in 2009. [48]

If Congress does pass such legislation, its effectiveness will depend

bon cap-and-trade system. RGGI is designed to reduce GHG emissions from electric power plants 10 percent below current levels by 2018. Unlike systems that have given polluters emission allowances for free, RGGI auctioned off its first batch of allowances and will invest the proceeds — $38.5 million, at a final price of $3.07 per ton of CO_2 — in energy efficiency and renewable energy programs.

to 15 percent below 2005 levels by 2020. Trading is scheduled to start in 2012, with a second phase beginning in 2015. [50]

"The Western Climate Initiative is increasingly the system that many observers see as a possible precursor to a U.S. federal system because of its size and design features. They've received input from some key experts who were involved in setting up the EU system," says Hasselknippe of the Point Carbon research firm. However, if Congress enacts national GHG controls, that system would almost certainly replace regional cap-and-trade programs.

Thousands of planes will be required to cut their carbon emissions now that the European Union has decided that airliners should be included in EU carbon emission caps under Phase 3 of the Kyoto Protocol climate change treaty, beginning in 2012. Airlines are resisting, saying that their industry has been hit hard by high oil prices and that the EU does not have legal authority to regulate emissions from flights that originate in other countries. Above, planes in Glasgow, Scotland.

Getty Images/Jeff J. Mitchell

Beyond Kyoto

Global negotiators are working on a follow-on agreement to the Kyoto Protocol, which only limits signatories' emissions through 2012, although some countries have made longer-term commitments. For example, in 2007 EU countries pledged to cut their total GHG emissions 20 percent by 2020 and to increase this target to 30 percent if other nations sign a post-Kyoto treaty.

At a contentious international conference in 2007 in Bali, Indonesia, negotiators agreed on basic principles for crafting a post-Kyoto agreement. The plan calls for finalizing a new treaty in 2009 (to take effect in 2013) that includes deep cuts in developed countries' greenhouse emissions and unspecified "mitigation actions" by developing countries. It also pledges to develop policies that reward tropical countries for protecting their forests and creates a fund using a surcharge on CDM projects to help poor countries adapt to climate change impacts. [51]

Many developed countries wanted emissions cuts of 25 to 40 percent in rich countries by 2020, but the United States refused to approve an agenda

on which sectors it covers, how quickly it cuts emissions and whether it compensates businesses and consumers for higher costs. Carbon marketers will watch closely to see how strictly the U.S. limits the use of offset credits from foreign sources such as CDM projects.

Some states are launching regional cap-and-trade schemes to show the approach can work and to build support for national action. In September, 10 Northeastern states, stretching from Maryland to Maine, launched the Regional Greenhouse Gas Initiative (RGGI) — the first mandatory U.S. car-

State officials called the first RGGI auction a success. "Demand was high, and fears of low-ball bidding did not come to pass," said Democratic New York Gov. David Paterson. "Instead, RGGI has used market forces to set a price on carbon." [49]

At nearly the same time, seven Western states and four Canadian provinces agreed on the basics of a broader regional cap-and-trade program that would cover emissions from electricity generation, industry, transportation and residential and commercial energy use. The initiative would cut members' GHG emissions

Continued on p. 314

At Issue:

Should the European Union cap aviation carbon emissions?

JOAO VIEIRA
POLICY OFFICER, EUROPEAN FEDERATION FOR TRANSPORT AND ENVIRONMENT

FROM *T&E BULLETIN*, JULY 22, 2008

*a*fter years of us and others highlighting the environmental damage caused by aviation, the [European Union] has finally done something to try and counteract its impact. It has shown courage, in particular, in standing up to threats from the USA and against a background of abysmal inaction from the International Civil Aviation Organisation, the body charged with regulating emissions from aircraft under the Kyoto Protocol. . . .

So why are we at *T&E* so reluctant to be happy about this? There are two reasons. The terms on which aviation has entered the ETS [Emissions Trading Scheme] will mean very limited reductions in emissions from aircraft [which] might create the illusion that other measures that would do much more to reduce emissions . . . are no longer needed. And . . . the ETS might now be seen as a "silver bullet" solution for emissions from transport. . . .

Airlines will be allowed to buy permits from other sectors without restrictions, so their emissions will continue to grow. Instead of changing to greener technologies and operations, the aviation sector is likely to limit its climate efforts to buying permits in the carbon market. In addition, this directive only addresses CO_2 [carbon dioxide] emissions, ignoring the fact that NO_x [nitrogen oxides] is emitted from aircraft . . . and aviation-induced clouds also have climatic impact. It will mean aviation remains the least-efficient and most climate-intensive mode of transport.

The limitations of a cap-and-trade system's ability to effectively reduce emissions from transport should be a lesson for EU decision-makers, some of whom seem tempted by the idea of emissions trading for road transport.

The ETS is . . . for large, fixed-emission facilities. Transport . . . has numerous operators of mobile emissions sources, which do not face international competition [since] transport is a geographically bound activity.

T&E has said all along that including aviation in the ETS can only be a first step. If the transport sector is to reduce its emissions, other measures to address the climatic impacts of all modes of transport will be needed.

Without the courage to apply fuel taxation, fair and efficient infrastructure charging and strict emission standards, applying emissions trading to transport will simply allow transport to keep growing its emissions. . . . That is unfair to [other] industries, and irresponsible to future generations.

GIOVANNI BISIGNANI
DIRECTOR GENERAL AND CEO, INTERNATIONAL AIR TRANSPORT ASSOCIATION (IATA)

FROM REMARKS AT THE FARNBOROUGH [ENGLAND] INTERNATIONAL AIR SHOW, JULY 18, 2008

*t*oday, airlines are in crisis. Oil is above $140. Jet fuel is over $180. In five years fuel went from 14 percent of operating costs to over 34 percent. If oil averages $135 for the rest of the year, the industry bill will be $190 billion. And next year it could be over $250 billion. . . .

IATA's environment leadership is delivering results. We worked with our members to implement best practices in fuel management. In 2007 this saved 6.7 million tonnes of CO_2 and $1.3 billion in cost.

We also worked with governments and air navigation service providers. Optimising 395 routes and procedures in 81 airports saved 3.8 million tonnes of CO_2 and $831 million in costs.

We could save up to 73 million tonnes of CO_2 with better air traffic management, but, while painting themselves green to win votes, governments are slow to deliver results. . . .

IATA supports emissions trading, but it must be global, fair and effective. Europe's approach could not be more wrong.

First, it's not an effective incentive. Developed when oil was $55 per barrel, it was meant to be an economic stick to force airlines to become more fuel-efficient. Europe's politicians had not foreseen the giant club of $140 oil.

It has beaten the life out of 25 airlines already this year, and we expect many more to follow into bankruptcy protection if they can afford it or straight into liquidation if they cannot. To survive, airlines are doing everything possible to reduce fuel burn. The [Emissions Trading Scheme] will add costs but will not improve the results. . . .

Second, the timing is wrong. Why make long-range policy decisions in the moment of a crisis when the future is completely uncertain — even five years out. And why make fuel more expensive when it is at its highest level ever — an 87 percent increase in the last year? Clearly, green politics has got in the way of good policy. . . .

How can Europe expect to charge an Australian airline for emissions over the Middle East on a flight from Asia to Europe? This will be challenged at [the International Civil Aviation Organisation] and in the International Court of Justice. And a responsible industry could easily be caught in a trade war of a layering of punitive economic measures.

Instead of cleaning up the environment, Europe is creating an international legal mess.

Growers burned down a dense forest in Sumber, Kalimantan, Indonesia, to make way for a palm oil plantation. Deforestation accounts for about 20 percent of human-generated greenhouse gas emissions worldwide. Environmentalists point out that forest preservation is one of the most cost-effective ways to address climate change.

AP Photo/Ed Wray

Continued from p. 312

with specific targets. U.S. representatives were booed during the talks, and at one point Papua New Guinea's representative was cheered when he told them, "If you're not going to lead, get out of the way." Ultimately, however, the U.S. supported the principles — the first time that the Bush administration had agreed to negotiate climate targets with other nations. [52]

It is not yet clear what shape a post-Kyoto agreement may take. It could set binding national emissions targets, like the Kyoto treaty, or build on pledges by individual countries or groups of countries. Some nations have already made significant commitments outside the Kyoto framework. The European Parliament, for example, is already setting emissions caps and planning to auction some carbon allowances in the third phase of EU ETS, to start in 2013. [53]

Some developing countries have also pledged to reduce their contribution to climate change. China's current five-year plan, which runs through 2010, calls for reducing the energy intensity of gross domestic product (the amount of energy used to produce each unit of income) 20 percent below 2005 levels by 2010. Beijing is also working to generate 10 percent of national energy demand with renewable sources by 2010 and 15 percent by 2020; by contrast, the U.S. currently gets about 7 percent of its energy from renewables. And Costa Rica has pledged to become carbon-neutral, as have New Zealand, Monaco, Norway and Iceland. [54]

Beyond these steps, however, experts warn that unless large developing countries like China, India, Indonesia and Brazil accept binding carbon caps soon, it will be impossible to avoid disastrous climate change. "If China and India keep doing what they're doing, their emissions will be tremendous," says Kollmuss of the Stockholm Environment Institute. "At the same time, these countries need to develop, so we need to find a just and equitable climate solution that will get them to buy in."

The U.N. Development Programme seconded this view in its 2007/2008 *Human Development Report*, which urged large developing countries to accept emissions targets proportional to what they could accomplish. "Any multilateral agreement without quantitative commitments from developing countries will lack credibility," the report asserted. However, it also argued that it would be impossible to negotiate such an agreement unless wealthy countries provided money and technology to help poorer nations adopt low-carbon strategies. [55]

Some advocates in developing countries worry that they will be asked to take on GHG reduction commitments when many rich nations have not cut their emissions significantly (or, in the case of the United States, at all).

"The message from Bali is that the fight against climate change will be brutal and selfish," says Sunita Narain, director of the Centre for Science and Environment in New Delhi. She agrees that India is "devastatingly vulnerable" to climate change impacts like floods and heat waves. By signing an action plan in Bali without hard reduction targets or timetables, she argues, "The world powers have reneged on all of us. Now developing countries will be even more reluctant to engage. Hardliners will say, 'We told you so.' "

In September U.N. Secretary-General Ban Ki-moon announced a cooperative program to test ways of managing tropical forests to keep them healthy and store large amounts of carbon. Norway donated $35 million for the first phase, which will involve at least nine countries in Africa, Asia and Latin America. The program seeks to pave the way for including forest conservation in a post-Kyoto treaty.

"This initiative will not only demonstrate how forests can have an important role as part of a post-2012 climate regime," said Ban, "it will also help build much needed confidence that the world community is ready to support the implementation of an inclusive, ambitious and comprehensive climate regime, once it is ratified." [56] ■

OUTLOOK

Cost of Inaction

As world leaders struggle to address this fall's global financial meltdown, some policy makers say now is the wrong time to impose further limits on greenhouse gas emissions. Putting a price on carbon, they worry, will raise energy costs when economies are already sputtering.

In October, for example, some East European countries tried unsuccessfully to delay the auctioning of EU ETS emission allowances, and conservative U.S. legislators questioned whether the economy could handle the added impact of cap-and-trade legislation. [57] If the world goes through a prolonged recession, energy prices are likely to fall, which would ease the financial crunch somewhat but would also reduce some of the imperative to shift away from fossil fuels.

Indeed, controlling carbon emissions won't be cheap. The total cost of controlling global warming could cost 1-2 percent of world gross domestic product — or roughly $350 to $700 billion — per year over the next few decades, according to several prominent economists, including Nicholas Stern of Great Britain and Jeffery Sachs of the United States. [58]

But advocates say it's more urgent than ever to act on climate change. Since renewable fuels like wind, solar and geothermal energy are free or low-cost, investing in them now will not only reduce GHG emissions but also make nations less dependent on oil and gas. And, they argue, green technologies can generate thousands of new, high-paying jobs.

Supporting this view, a 2008 study by David Roland-Holst, an economist at the University of California, calculated that energy efficiency policies in California from 1976 through 2006 had saved households some $56 billion and created about 1.5 million jobs. [59]

"The longer we wait to cap our emissions, the farther we fall behind in the remaking of a $6 trillion economy," says Environmental Defense Fund President Krupp.

Moreover, the cost of inaction is likely to be much higher than those of cutting emissions. Climate change will have major impacts worldwide, especially in poor countries that have few resources to protect people or move them out of harm's way. Global policy experts warn that recent progress against poverty in developing countries could be wiped out by climate change impacts like crop failures, water shortages and catastrophic flooding in river deltas that could leave millions hungry and homeless.

"If we are to avoid the catastrophic reversals in human development that will follow in the wake of climate change, we need to more than halve emissions of greenhouse gases," wrote Kevin Watkins, lead author of the U.N.'s *Human Development Report*, during the Bali climate conference. "That will not happen without a global accord that decarbonises growth and extends access to affordable energy in the developing world: a shake-up in energy policy backed by a programme similar to the post-Second World War Marshall Plan." [60] Under that initiative, the United States spent about $13 billion from 1947 through 1951 to rebuild war-torn Western Europe. The price tag for a program on the same scale, measured in 2007 dollars, would be roughly $740 billion. [61]

Rising concerns about costs make it increasingly likely that carbon trading will be a central part of the climate change solution, since it offers the opportunity to make cuts where they are most affordable. But cap-and-trade programs alone will not be enough. Government also must fund energy research and development; tighten energy efficiency standards and create markets for new technologies by setting national renewable energy targets. The overall goal, says IPCC Chair Pachauri, is to create a cleaner, less resource-intensive development path.

Pachauri often recalls Mahatma Gandhi's quip when asked whether India's people should have the same standard of living as the British. "It took Britain half the resources of the planet to achieve this prosperity," Gandhi replied. "How many planets will a country like India require?" [62] ∎

Notes

[1] Mamuni Das, "Germany To Buy Carbon Credits From TTD Solar Kitchen," *The Hindu Business Line.com*, Aug. 24, 2005, www.the-hindubusinessline.com/2005/08/24/stories/2005082402960100.htm; "Solar Amenities Way Above Sea Level," *The Statesman*, Oct. 15, 2006, www.thestatesman.net/page.arcview.php?clid=30&id=161337&usrsess=1; Madhur Singh, "India's Temples Go Green," *Time*, July 7, 2008, www.time.com/time/world/article/0,8599,1820844,00.html.

[2] Singh, *ibid.*

[3] http://cdm.unfccc.int/UserManagement/FileStorage/4WZXEVUUTRCJDV4AC6SY7VSL0KBFC5.

[4] David Adam, "World Carbon Dioxide Levels Highest for 650,000 Years, Says U.S. Report," *The Guardian*, May 13, 2008, www.guardian.co.uk/environment/2008/may/13/carbonemissions.climatechange.

[5] For background, see Colin Woodard, "Curbing Climate Change," *CQ Global Researcher*, February 2007, pp. 27-52.

[6] For background, see Mary H. Cooper, "Global Warming Treaty," *CQ Researcher*, Jan. 26, 2001, pp. 41-64.

[7] Garrett Hardin, "The Tragedy of the Commons," *Science*, Dec. 13, 1968, pp. 1243-1248.

[8] "Carbon 2008" Point Carbon, March 11, 2008, p. 3.

[9] Fiona Harvey, "World Carbon Trading Value Doubles," *Financial Times*, May 7, 2008, http://us.ft.com/ftgateway/superpage.ft?news_id=fto050720082214562909.

[10] "Regional Initiatives," Pew Center on Global Climate Change, www.pewclimate.org/what_s_being_done/in_the_states/regional_initiatives.cfm.

[11] Sen. Barack Obama (D-Ill.) endorsed cutting U.S. emissions 80 percent below 1990 levels by 2050, while Sen. John McCain (R-Ariz.) called for reducing at least 60 percent below 1990 levels on the same timetable. "Science Debate 2008," www.sciencedebate2008.com.

[12] "Emission Impossible: Access to JI/CDM Credits in Phase II of the EU Emissions Trading Scheme," World Wildlife Fund-UK, June 2007, p. 10, http://assets.panda.org/downloads/emission_impossible_final_.pdf.

[13] A. Denny Ellerman and Paul Joskow, "The European Union's Emissions Trading System in Perspective," Pew Climate Center, May 2008, figure 1, p. 13, www.pewclimate.org/docUploads/EU-ETS-In-Perspective-Report.pdf.

[14] "Europe's Dirty Secret: Why the EU Emissions Trading Scheme Isn't Working," Open Europe, 2007, p. 16, www.openeurope.org.uk/research/etsp2.pdf.

[15] "What's Really Going On in the European Carbon Market," Environmental Defense Fund, June 27, 2007, http://blogs.edf.org/climate411/2007/06/27/eu_carbon_market/.

[16] Ellerman and Joskow, op. cit., pp. 12-15.

[17] James Kanter, "Europe Forcing Airlines to Buy Emissions Permits," The New York Times, Oct. 25, 2008, p. B2.

[18] "Emission Impossible . . . ," op. cit., pp. 3-4.

[19] Michael W. Wara and David G. Victor, "A Realistic Policy on International Carbon Offsets," Working Paper #74, Program on Energy and Sustainable Development, Stanford University, April 2008, p. 5, http://pesd.stanford.edu/publications/a_realistic_policy_on_international_carbon_offsets/.

[20] Yvo de Boer, "Prepared Remarks for Public Debate on the Kyoto Mechanisms," New York, Oct. 9, 2008.

[21] Leila Abboud, "An Exhausting War On Emissions," The Wall Street Journal, Sept. 30, 2008, p. A15.

[22] Fred Krupp and Miriam Horn, Earth: The Sequel: The Race to Reinvent Energy and Stop Global Warming (2008), p. 247.

[23] "A Conversation with Nobel Prize Winner Rajendra Pachauri," Yale Environment 360, June 3, 2008, http://e360.yale.edu/content/print.msp?id=2006.

[24] Barbara Haya, "Failed Mechanism: How the CDM is Subsidizing Hydro Developers and Harming the Kyoto Protocol," International Rivers, November 2007, http://internationalrivers.org/files/Failed_Mechanism_3.pdf.

[25] Oliver Balch, "Forests: A Carbon Trader's Gold Mine?" ClimateChangeCorp.com, May 7, 2008, www.climatechangecorp.com/content.asp?ContentID=5305; for project details and review documents, see "Project 1051: Mitigation of Methane Emissions in the Charcoal Production of Plantar, Brazil," United Nations Framework Convention on Climate Change, http://cdm.unfccc.int/Projects/DB/DNV-CUK1175235824.92/view.

[26] Michael S. Northcott, A Moral Climate: The Ethics of Global Warming (2007), p. 136.

[27] William F. Laurance, "A New Initiative to Use Carbon Trading for Tropical Forest Conservation," Biotropica, vol. 39, no. 1 (2007), pp. 20-24, www.globalcanopy.org/themedia/NewCarbonTrading.pdf.

[28] Keya Acharya, "Rainforest Coalition Proposes Rewards for 'Avoided Deforestation,' " Environmental News Network, Aug. 15, 2007, www.enn.com/ecosystems/article/21854.

[29] "Forests in the Fight Against Climate Change," www.forestsnow.org.

[30] De Boer, op. cit.

[31] "CDM Experiences and Lessons" (presentation), slide 5, U.N. Development Programme, April 1, 2008, http://unfccc.meta-fusion.com/kongresse/AWG_08/downl/0401_1500_p2/Krause%20UNDP%20JI_CDM1.pdf.

[32] "Fighting Climate Change: Human Solidarity in a Developed World, Human Development Report 2007/2008 (2008), United Nations Development Programme, p. 155.

[33] " 'Global Green Deal' — Environmentally-Focused Investment Historic Opportunity for 21st Century Prosperity and Job Generation," United Nations Environment Programme, press release, Oct. 22, 2008.

[34] Richard N. L. Andrews, Managing the Environment, Managing Ourselves: A History of American Environmental Policy (1999), pp. 127-128.

[35] J. H. Dales, Pollution, Property and Prices (1968), p. 801.

[36] The final conference declaration is online at www.unep.org/Documents.Multilingual/Default.asp?DocumentID=97&ArticleID=1503.

[37] See Thomas H. Tietenberg, Environmental and Natural Resource Economics, 5th ed. (2000), pp. 454-455.

[38] Randall A. Bluffstone, "Environmental Taxes in Developing and Transition Economies," Public Finance and Management, vol. 3, no. 1 (2003), pp. 152-55.

[39] Richard Elliot Benedick, Ozone Diplomacy: New Directions in Safeguarding the Planet (1998), pp. 314-320.

[40] "Acid Rain Trading Program, 2004 Progress Report," U.S. Environmental Protection Agency, October 2005, pp. 2, 10, www.epa.gov/airmarkt/progress/docs/2004report.pdf.

[41] Robert N. Stavins, "Lessons Learned from SO2 Allowance Trading," Choices, 2005, p. 53, www.choicesmagazine.org/2005-1/environment/2005-1-11.htm; Nathaniel O. Keohane and Sheila M. Olmstead, Markets and the Environment (2007), p. 184.

[42] For background see Jennifer Weeks, "Coal's Comeback," CQ Researcher, Oct. 5, 2007, pp. 817-840. The Bush administration then issued regulations through EPA to promote emissions trading, but in 2007 the D.C. Circuit Court held that the EPA did not have authority under the Clean Air Act to develop such broad trading programs.

[43] "A Call for Action," Jan. 22, 2007, U.S. Climate Action Partnership, p. 2, www.us-cap.org/ClimateReport.pdf.

[44] For a summary of bills pending in September 2008, see "Comparison of Legislative Climate Change Targets," World Resources Institute, Sept. 9, 2008, www.wri.org/publication/usclimatetargets.

[45] "Carbon Market Round-Up Q3 2008," New

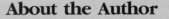

About the Author

Jennifer Weeks is a *CQ Researcher* contributing writer in Watertown, Mass., who specializes in energy and environmental issues. She has written for *The Washington Post*, *The Boston Globe Magazine* and other publications, and has 15 years' experience as a public-policy analyst, lobbyist and congressional staffer. She has an A.B. degree from Williams College and master's degrees from the University of North Carolina and Harvard.

Carbon Finance, Oct. 10, 2008; www.new-carbonfinance.com/download.php?n=2008-10-10_PR_Carbon_Markets_Q3_20082.pdf&f=fileName&t=NCF_downloads.

[46] Simon Kennedy, " 'Carbon Trading' Enriches the World's Energy Desks," Marketwatch, May 16, 2007.

[47] Massachusetts v. Environmental Protection Agency, 549 U.S. 497 (2007).

[48] Rep. Rick Boucher, remarks at the Society of Environmental Journalists annual conference, Roanoke, Va., Oct. 17, 2008.

[49] "Governor Paterson Hails Nation's First Global Warming Cap and Trade Auction A Success," Sept. 29, 2008, www.ny.gov/governor/press/press_0929083.html.

[50] For details see www.westernclimateinitiative.org/.

[51] Robert N. Stavins and Joseph Aldy, "Bali Climate Change Conference: Key Takeaways," Harvard Project on International Climate Agreements, Dec. 18, 2007, http://belfercenter.ksg.harvard.edu/publication/17781/bali_climate_change_conference.html.

[52] Daniel Howden and Geoffrey Lean, "Bali Conference: World Unity Forces U.S. to Back Climate Deal," The Independent, Dec. 16, 2007, www.independent.co.uk/environment/climate-change/bali-conference-world-unity-forces-us-to-back-climate-deal-765583.html; Gary LaMoshi, "Bumpy Ride Ahead for Bali Road Map," Asia Times, Dec. 18, 2007, www.atimes.com/atimes/Southeast_Asia/IL18Ae01.html.

[53] Ian Traynor and David Gow, "EU Promises 20% Reduction in Carbon Emissions by 2020," The Guardian, Feb. 21, 2007, www.guardian.co.uk/environment/2007/feb/21/climatechange.climatechangeenvironment Pete Harrison and Gerard Wynn, "EU Lawmakers Watch Credit Crisis in Climate Fight," Reuters, Oct. 7, 2008.

[54] Stefan Lovgren, "Costa Rica Aims to Be 1st Carbon-Neutral Country," National Geographic News, March 7, 2008, http://news.nationalgeographic.com/news/2008/03/080307-costa-rica.html.

[55] "Fighting Climate Change, . . ." op cit., pp. 27-28.

[56] " 'Redd'-Letter Day for Forests: United Nations, Norway United to Combat Climate Change from Deforestation, Spearheading New Programme," U.N. press release, Sept. 24, 2008.

[57] Pete Harrison and Gerard Wynn, "EU Lawmakers Watch Credit Crisis in Climate Fight," Reuters, Oct. 7, 2008, www.reuters.com/article/rbssIndustryMaterialsUtilitiesNews/idUSL711408420081007?sp=true; Dina Cappiello, "Economic Woes Chill Effort to Stop Global Warming," The Associated Press, Oct. 12, 2008, http://ap.google.com/article/ALeqM5jFaQmoLWbpKq8HH1AAQ5GoGZjz0gD93OTVC00; James Kanter, "Europe's Leadership in Carbon Control at Risk in Credit Crisis," The New York Times, Oct. 21, 2008, p. B10.

[58] Juliette Jowit and Patrick Wintour, "Cost of Tackling Global Climate Change Has Doubled, Warns Stern," The Guardian, June 26, 2008, www.guardian.co.uk/environment/2008/jun/26/climatechange.scienceofclimatechange; Jeffrey D. Sachs, Common Wealth: Economics for a Crowded Planet (2008), pp. 308-311.

[59] David Roland-Holst, "Energy Efficiency, Innovation, and Job Creation in California," Center for Energy, Resources and Economic Sustainability, University of California, Berkeley, October 2008.

[60] Kevin Watkins, "Bali's Double Standards," The Guardian, Dec. 14, 2007, www.guardian.co.uk/commentisfree/2007/dec/14/comment.bali.

[61] Niall Ferguson, "Dollar Diplomacy: How Much Did the Marshall Plan Really Matter?" The New Yorker, Aug. 27, 2007, p. 81.

[62] "A Conversation with Nobel Prize Winner Rajendra Pachauri," op. cit.

FOR MORE INFORMATION

Centre for Science and Environment, 41 Tughlakabad Institutional Area, New Delhi, India; (+91)-11-29955124; www.cseindia.org. An independent public interest organization that works to increase awareness of science, technology, environment and development issues.

China Sustainable Energy Program, The Energy Foundation, CITIC Building, Room 2403, No. 19, Jianguomenwai Dajie, Beijing, 100004, P.R. China; (+86)-10-8526-2422; www.efchina.org. A joint initiative funded by U.S. foundations to support China's policy efforts to promote energy efficiency and renewable energy.

The Gold Standard, 22 Baumleingasse, CH-4051, Basel, Switzerland; (+41)-0-61-283-0916; www.cdmgoldstandard.org. A nonprofit that screens carbon offset projects and certifies initiatives that provide measurable economic, environmental and social benefits.

Institute for Sustainable Futures, University of Technology, L11, 235 Jones St., Broadway, Sydney, Australia; (+61)-2-9514-4590; www.isf.uts.edu.au. Research institute that works with Australian businesses and communities to promote sustainable environmental and design policies.

Intergovernmental Panel on Climate Change, 7bis Ave. de la Paix, C.P. 2300, CH-1211 Geneva 2, Switzerland; (+41)-22-730-8208; www.ipcc.ch. A U.N.-sponsored organization created to advise national governments on climate change science.

Regional Greenhouse Gas Initiative, 90 Church St., 4th Floor, New York, NY 10007; (212) 417-7327; www.rggi.org. A joint venture launched in 2008 by 10 Northeastern states to reduce greenhouse gas emissions from the electric power sector through carbon emissions trading.

U.N. Development Programme, One United Nations Plaza, New York, NY 10017; (212) 906-5000; www.undp.org. Works to cut poverty and use aid effectively.

World Wildlife Fund — UK, Panda House, Weyside Park, Godalming, Surrey GU7 1XR, United Kingdom; (+01)-483-426444; www.wwf.org.uk. The British arm of an international conservation organization.

Bibliography

Selected Sources

Books

Krupp, Fred, and Miriam Horn, *Earth: The Sequel*, Norton, 2008.

The president and senior staff writer, respectively, at the U.S.-based Environmental Defense Fund describe innovators who are leading a clean-energy revolution and argue that the United States should adopt a carbon cap-and-trade system to boost investments in innovative energy technologies.

Northcott, Michael S., *A Moral Climate: The Ethics of Global Warming*, Orbis, 2007.

An Episcopal priest and divinity professor at the University of Edinburgh views climate change as an ethical issue and criticizes carbon trading as biased toward rich countries and large greenhouse gas emitters.

Tietenberg, Thomas H., *Emissions Trading: Principles and Practice*, 2nd edition, Resources for the Future, 2006.

An environmental economist shows how emissions trading became popular as an alternative to command-and-control regulation and assesses successes, failures and lessons learned in 25 years of application.

Zedillo, Ernesto, ed., *Global Warming: Looking Beyond Kyoto*, Brookings, 2008.

Authors from around the globe examine how to structure a post-Kyoto climate change agreement that can reduce emissions quickly enough to avert disastrous warming.

Articles

"C is for Unclean," *Down to Earth*, Dec. 15, 2007.

A critique of the Clean Development Mechanism (CDM) by India's Centre for Science and Environment argues that the program has been taken over by carbon entrepreneurs and turned into a financial tool instead of fighting climate change.

"First Africa Carbon Forum Fosters Clean Climate Projects," *Environment News Service*, Sept. 4, 2008.

Only a fraction of CDM projects are in Africa, but African leaders and international development officials want to increase the continent's share.

Arrandale, Tom, "Carbon Goes to Market," *Governing*, September 2008, p. 26.

As Congress debates cap-and-trade policies, nearly half the states are working on their own carbon trading schemes.

Scott, Mark, "Giant Steps for Carbon Trading in Europe," *Business Week*, Jan. 23, 2008.

The EU Emissions Trading Scheme is setting stringent, new targets, which will make carbon credits more valuable.

Szabo, Michael, "Problems Plague Canada's Emissions Trading Plans," *Reuters*, May 8, 2008.

Canada wants to start carbon trading, but some of its provinces have already adopted their own schemes, and emissions from the Canadian oil industry are rising.

Turner, Chris, "The Carbon Cleansers," *Canadian Geographic Magazine*, October 2008, p. 3.

Norway's carbon tax on the oil and gas industry, adopted in 1992, has spurred research into cleaner energy technologies, as well as carbon capture and storage.

Reports and Studies

"Carbon 2008," Point Carbon, March 11, 2008, www.pointcarbon.com/polopoly_fs/1.912721!Carbon_2008_dfgrt.pdf.

An international market research firm focusing on carbon markets provides an overview of global carbon trading and major carbon policy trends.

"Fighting Climate Change: Human Solidarity in a Developed World," *Human Development Report 2007/2008*, 2008, United Nations Development Programme, http://hdr.undp.org/en/media/hdr_20072008_en_complete.pdf.

Climate change is a major threat to human development and is already undercutting global efforts to reduce poverty in some parts of the world. This report calls for urgent action on a post-Kyoto agreement and policies to help poor countries adapt to unavoidable climate change impacts.

Ellerman, A. Denny, and Paul Joskow, "The European Union's Emissions Trading System in Perspective," Pew Center on Global Climate Change, May 2008, www.pew-climate.org/docUploads/EU-ETS-In-Perspective-Report.pdf.

Two economists from the Massachusetts Institute of Technology conclude that the EU ETS is still a work in progress but has successfully set a European price for carbon emissions and offers important lessons for U.S. leaders as they debate cap-and-trade policies.

Wara, Michael W., and David G. Victor, "A Realistic Policy on International Carbon Offsets," *Working Paper #74*, Program on Energy and Sustainable Development, Stanford University, April 2008, http://iis-db.stanford.edu/pubs/22157/WP74_final_final.pdf.

Two Stanford University law professors recommend major reforms to the Kyoto Protocol's Clean Development Mechanism, which they say awards credits for projects that don't really reduce emissions, and argue the United States should not rely on offsets to lower the cost of reducing carbon emissions.

The Next Step:

Additional Articles from Current Periodicals

Alternatives to Carbon Trading

"Domestic Carbon Tax as Alternative to Global Trading," *Canberra Times* (Australia), June 2, 2007.

Implementing a carbon tax as opposed to carbon trading could create subsidies for non-carbon energy sources and energy-related foreign assistance.

"Greenpeace Offers REDD Alternative," *Jakarta Post* (Indonesia), Nov. 28, 2007.

The environmental group Greenpeace has called for the Indonesian government to support efforts to finance deforestation prevention in lieu of a carbon trading system.

Hyun-cheol, Kim, "Carbon Tax to Be Introduced in 2010," *Korea Times*, Aug. 22. 2008.

South Korea is likely to introduce a carbon tax by 2010.

Prasad, Monica, "On Carbon, Tax and Don't Spend," *The New York Times*, March 25, 2008, p. A27.

Carbon taxes have led to a decrease in emissions in Denmark, where per capita emissions were 15 percent lower in 2005 than in 1990.

Developing Countries

"Carbon Trading to Benefit Farmers," Chinadaily.com.cn, Oct. 24, 2007.

The Clean Development Mechanism encourages developed countries to invest in carbon emission reduction projects in the developing world.

Akosile, Abimbola, "Continent Benefits Least from Carbon Market," *This Day* (Nigeria), Nov. 21, 2006.

A report by the World Bank suggests Africa is the continent hardest hit by climate change and benefits the least from the carbon market.

Vaish, Nandini, "The Greener Side of Carbon," *India Today*, Nov. 19, 2007.

The Kyoto Protocol's regulation of carbon emissions trading has created a huge opportunity for large developing countries such as India and China.

Emission Levels

"Carbon Dioxide Emissions Rise 3%," The Associated Press, Sept. 26, 2008.

Man-made carbon dioxide emissions rose 3 percent worldwide in 2007, according to a group of international climatologists.

Bradsher, Keith, "China to Pass U.S. in 2009 in Emissions," *The New York Times*, Nov. 7, 2006, p. C1.

China is scheduled to surpass the United States in 2009 —

about a decade ahead of previous predictions — as being the largest emitter of carbon dioxide.

Shourie, Dharam, "China's CO2 Emissions Much Above Previous Estimates," *Press Trust of India*, March 11, 2008.

The growth in China's carbon dioxide emissions is outpacing earlier estimates, making the stabilization of greenhouse gases much more difficult than anticipated.

Trading Schemes

"Achieving Zero," *Bangkok Post* (Thailand), Sept. 25, 2007.

European proposals to include aviation in their emissions trading scheme have put economic measures at the center of a political debate.

"Emission Trading Scheme May Result in Energy Cost Blowout," *Asia Pulse* (Australia), March 20, 2008.

Any emissions trading scheme intended to combat global warming would inevitably raise electricity and gas prices.

Dorsey, Michael K., "License to Pollute," *Los Angeles Times*, April 1, 2007, p. A4.

Climate change may be wreaking havoc before regulators realize that cap-and-trade systems are not effectively reducing carbon emissions.

Velasquez-Manoff, Moises, "Do Carbon Offsets Live Up to Their Promise?" *The Christian Science Monitor*, Jan. 10, 2007, p. 13.

There is currently no proven method to measure the effectiveness of current carbon trading schemes, and no easy way to determine whether offsetting companies are doing what they promise.

CITING CQ GLOBAL RESEARCHER

Sample formats for citing these reports in a bibliography include the ones listed below. Preferred styles and formats vary, so please check with your instructor or professor.

MLA STYLE

Flamini, Roland. "Nuclear Proliferation." CQ Global Researcher 1 Apr. 2007: 1-24.

APA STYLE

Flamini, R. (2007, April 1). Nuclear proliferation. *CQ Global Researcher*, 1, 1-24.

CHICAGO STYLE

Flamini, Roland. "Nuclear Proliferation." *CQ Global Researcher*, April 1, 2007, 1-24.

Voices From Abroad:

ED MILIBAND
Energy and Climate Change Secretary United Kingdom

Everyone has a role

"It's easy to assume that carbon trading is for big governments to talk about at global summits, but the truth is that everyone has a role to play in tackling climate change, and it is great to see this exciting activity being carried out by business at a local level."

Yorkshire Post (England), October 2008

MARK DEMBITZ
Vice President, Carbon Capital, China

Countries should spend accordingly

"It would not be fair to Uganda asking them to spend a lot of money reducing greenhouse gases when this legitimate money is needed for development. This same standard cannot be applied to the United States, a developed country which should be spending money reducing greenhouse gases."

South China Morning Post, August 2007

ANDREI MARCU
President and CEO, International Emissions Trading Association

Carbon isn't free anymore

"Carbon trading points to a price of a ton of carbon, so people tend to treat it with more respect and make sure it is carefully used. People are coming to realize that carbon is not a free commodity anymore and are coming up with ideas to monetize it."

BusinessWorld (Philippines), July 2007

STEVE RAYNER
Professor of Science and Civilization, Saïd Business School, Oxford University

Pollution credits clear the conscience

"These companies may be operating with the best will in the world, but they are doing so in settings where it's not really clear you can monitor and enforce their projects over time. What these companies are allowing people to do is carry on with their current behavior with a clear conscience."

The New York Times, February 2007

ANTHONY ABRAHAM
Executive Director, Macquarie Bank, Australia

Banks are taking notice

"This happy banker is here because of the opportunities there, and he is looking to the upside . . . and we see why he is smiling . . . the World Bank tells us that in 2006 we had three times the volume of carbon trading in the market that we did the year before. . . . This is a growing market, and a very large market."

AAP News (Australia), May 2007

JIANG WEIXIN
Vice Minister, National Development and Reform Commission, China

Opportunities for all

"Developed countries get opportunities to emit greenhouses gases at a relatively low economic cost and achieve their emission reduction targets, while developing countries obtain benefits such as funding and technology transfer, which will boost their efforts to pursue sustainable development."

Inter Press Service (Italy), November 2006

RUTH LEA
Director, Centre for Policy Studies, England

Why a carbon tax is better

"If there have to be green measures for 'tackling climate change', taxes are . . . preferable. . . . The cost of carbon is known with taxes, and transparent, whereas prices can be very volatile and uncertain under trading schemes. . . . The imposition of carbon taxes can be used to lower taxes elsewhere in the economy to maintain overall competitiveness of business and/or people's real disposable income. . . . Finally, tax collection is administratively straightforward and can cover all emitters — not the case with trading schemes."

Daily Telegraph (England), November 2006

SIMON POWELL
Head of Regional Research — Power, Gas and Utilities, CLSA Ltd., Malaysia

Everyone benefits

"Whether it is a developed state or a developing country, all markets hold a card in the emissions games, and no matter who wins the trading game, the outcome, which is to reduce pollution, will benefit everyone. Expansions of the EU scheme to new sectors, and the likely establishment of trading mechanisms elsewhere, will lead to substantial carbon-market growth."

Business Times (Malaysia), April 2007

6

LOOMING WATER CRISIS

BY PETER BEHR

Excerpted from the CQ Global Researcher. Peter Behr. (February 2008). "Looming Water Crisis." *CQ Global Researcher*, 27-56.

Looming Water Crisis

BY PETER BEHR

THE ISSUES

As 2007 came to a close, the steady drumbeat of headlines about China's worst drought in a half-century affirmed Prime Minister Wen Jiabao's earlier warning that the crisis threatens "the survival of the Chinese nation." [1]

The alarming developments included:

- The drying up of 133 reservoirs in burgeoning Guangdong Province, leaving a quarter of a million people facing water shortages. [2]

- The lowest levels since 1866 on portions of the Yangtze River, restricting barge and ship traffic and reducing hydroelectric output on China's largest river, even as pollution from 9,000 industrial plants along its course jeopardizes drinking water supplies. [3]

- Near-record low levels in vast Lake Poyang, restricting water supplies for 100,000 people. [4]

"My house used to be by the side of the lake," villager Yu Wenchang told the Xinhua News Agency. "Now I have to go over a dozen kilometers away to get to the lake water." [5]

Similar woes are being reported across the globe, as one of the worst decades of drought on record afflicts rich and poor nations alike. While scientists hedge their conclusions about whether long-term climate change is causing the dry spell, many warn that Earth's gradual warming trend unquestionably poses a growing threat to water supplies and food production in arid regions. Already, population growth and economic expansion are straining water supplies in

Kenyan children play in raw sewage in Kibera, Nairobi's largest slum. Each year 1.8 million children — 5,000 a day — die from waterborne illnesses due to a lack of access to sanitation and clean water.

AFP/Getty Images/Simon Maina

many places, particularly in the poorest nations. But despite an unending series of international water conferences — attended by thousands of experts — no consensus has emerged on how to make adequate clean water available to all people in affordable, environmentally sustainable ways. [6]

A fifth of the world's population — 1.2 billion people — live in areas experiencing "physical water scarcity," or insufficient supplies for everyone's demands, according to a 2006 study by the International Water Management Institute that draws on the work of 700 scientists and experts. Another 1 billion face "economic scarcity," in which "human capacity or financial resources" cannot provide adequate water, the report found. [7]

While drought and expanding populations visibly affect the world's lakes and rivers, a less-visible problem also threatens water supplies. Accelerated pumping of groundwater for irrigation is depleting underground aquifers faster than they can be refreshed in densely populated areas of North China, India and Mexico. And land and water resources there and beyond are being degraded through erosion, pollution, salination, nutrient depletion and seawater intrusion, according to the institute.

A United Nations task force on water predicted that by 2025, 3 billion people will face "water stress" conditions, lacking enough water to meet all human and environmental needs. [8] By that time, there will be 63 major river basins with populations of at least 10 million, of which 47 are either already water-stressed, will become stressed or will experience a significant deterioration in water supply, according to a separate study by the World Resources Institute incorporating the U.N. data. [9] (*See map, p. 30.*) *

As water depletion accelerates, drought is undermining nature's capacity to replenish this essential resource, punishing the planet's midsection — from eastern Australia and northern China through the Middle East and sub-Saharan Africa to the U.S. Sun Belt, the Great Plains and northern Mexico.

---

* "Water stress" occurs when less than 1,700 cubic meters (448,000 gallons) per person of new fresh water is available annually from rainfall or aquifers for human use, making populations vulnerable to frequent interruptions in water supply.

Serious Shortages Projected for Many Regions

Water shortages are expected to afflict much of the Earth by 2025, as growing populations use vastly more water for daily life and farming. Areas likely to be hardest hit include China, Western Europe, the United States, Mexico and a wide swath of the globe's midsection from India to North Africa. In the most severe cases, humans are expected to use up to 40 percent of the available water, compared with the current average withdrawal, or use, rate, of 10 percent.

Projected Water-Use Rates, 2025 *

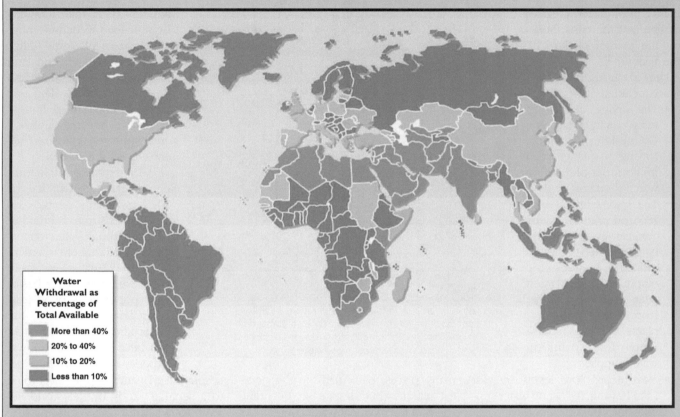

Water Withdrawal as Percentage of Total Available

- More than 40%
- 20% to 40%
- 10% to 20%
- Less than 10%

* Based on data from 1996-2000

Source: World Meteorological Organisation, Global Environment Outlook, U.N. Environment Programme, Earthscan, www.unep.org/dewa/assessments/ecosystems/water/vitalwater/21.htm#21b

In the United States, chronic alarms over depleted water resources in the Southwestern states have spread to the Southeast. The water level in giant Lake Sidney Lanier outside Atlanta has dropped about a dozen feet in this decade, causing an intense struggle among Georgia and neighboring Alabama and Florida over rights to the lake's diminished flows. [10]

And drought conditions worldwide are likely to worsen as the effects of climate change are felt, many scientists warn. [11] Climate change is expected to expand and intensify drought in traditionally dry regions and disrupt water flows from the world's mountain snowcaps and glaciers.

Finally, a new threat to global water supplies has emerged: terrorism. "The chance that terrorists will strike at water systems is real," said Peter H. Gleick, president of the Pacific Institute for Studies in Development, Environment and Security in Oakland, Calif. [12] Modern public water systems are designed to protect users from biological agents and toxins, but deliberate contamination by terrorists could kill or sicken thousands, he said. Since the Sept. 11, 2001, terrorist attacks most major U.S. cities have sent the federal government confidential reports on the vulnerability of local water supplies, and the Environmental Protection Agency's (EPA) Water Sentinel Initiative is designing a water-contamination warning system. [13]

Perhaps the grimmest long-range prediction on water availability was issued by the Met Office Hadley Centre for Climate Prediction and Research in London. Using supercomputer modeling, the center projected that if current trends continue, by this century's end drought will have spread across half the Earth's land surface due to climate change, threatening millions of lives. Moreover, "extreme drought" — which makes traditional agriculture virtually impossible — will affect about a third of the planet, according to the group's November 2006 report.

"Even though (globally) total rainfall will increase as the climate warms, the proportion of land in drought is projected to rise throughout the 21st century," the report said. [14]

"There's almost no aspect of life in the developing countries that these predictions don't undermine — the ability to grow food, the ability to have a safe sanitation system, the availability of water," said Andrew Simms, policy director of the liberal London-based New Economics Foundation. [15] The consequences will be most dire for the planet's poorest inhabitants, he added. "For hundreds of millions of people for whom getting through the day is already a struggle, this is going to push them over the precipice."

Access to safe, fresh water separates the well-off — who can treat water as if it were air — from the world's poorest, who hoard it like gold. In the United States, the average consumer uses nearly 160 gallons of water per day, summoned by the twist of a faucet. In much of Africa, women often trudge for hours to and from wells, carrying the two to five gallons per person used by the typical person in sub-Saharan Africa. (*See graph, above.*) [16]

But the lack of clean water is not only inconvenient. It can also be deadly. Each year 1.8 million children — 5,000 per day — die from waterborne illnesses such as diarrhea, according

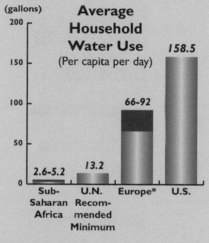

World Water Consumption Varies

The average American uses nearly 160 gallons of water per day for drinking, cooking, bathing and sanitation — more than any other nationality and more than twice the amount used by many Europeans. People in sub-Saharan Africa use only about a quarter of the 13 gallons the United Nations sets as a minimum basic standard.

Average Household Water Use
(Per capita per day)

(gallons)

- Sub-Saharan Africa: 2.6–5.2
- U.N. Recommended Minimum: 13.2
- Europe*: 66–92
- U.S.: 158.5

* *Consumption among European countries ranges from 66-92 gallons*

Source: World Water Council

to the United Nations. "That's equivalent to 12 full jumbo jets crashing every day," said U.N. water expert Brian Appleton. "If 12 full jumbo jets were crashing every day, the world would want to do something about it — they would want to find out why it was happening." [17]

Policymakers are trying various ways to solve the global water challenge, including contracting with private firms to operate urban water and sanitary systems, adopting new conservation technologies, enacting multination pacts to manage regional watersheds and increasing funds for water projects in the world's poorest regions. Water experts advocate "environmental flow" policies — the release of enough water from dams to sustain the environment of rivers, wetlands and underground aquifers. [18]

And their efforts seem to be paying off — at least in some areas. Between 1990 and 2002, more than 1 billion people in the developing world gained access to fresh water and basic sanitation. But because of population growth, the total number of people still lacking safe water remained more than a billion, and there was no change in the number lacking basic sanitation. [19]

In 2003, the U.N. General Assembly designated the period from 2005 to 2015 as the International Decade for Action on "Water for Life." And the U.N.'s new Millennium Development Goals include a campaign to cut in half by 2015 the proportion of people without sustainable access to safe drinking water and basic sanitation — at a cost of more than $10 billion per year. [20] Currently, governments and international agencies like the U.N. and World Bank provide only $4 billion a year in aid for water and sanitation projects. [21]

"We will see these issues play out silently: dry rivers, dead deltas, destocked fisheries, depleted springs and wells," wrote Margaret Carley-Carlson, chairwoman of the Global Water Partnership in Stockholm, and M. S. Swaminathan, president of the Pugwash Conferences on Science and World Affairs in Chennai, India. [22] "We will also see famine; increased and sometimes violent competition for water, especially within states; more migration; and environmental devastation with fires, dust, and new plagues and blights."

Averting that future will require fundamental changes in governmental policies and human practices governing the use, conservation and value of water, experts agree.

As water experts and policymakers discuss how to conserve and protect future water supplies, here are some of the questions they are debating:

Poorest Lag Far Behind in Access

Although progress has been made since 1990, only 37 percent of the residents in sub-Saharan Africa and South Asia had access to sanitation services in 2004. Sub-Saharan Africa lags behind the rest of the world in access to reliable sources of clean water. Meanwhile, more than 90 percent of those living in the industrialized countries, Central and Eastern Europe, Latin America, the Caribbean and the former Soviet republics had access to water in 2004.

Percentage of Population with Access to:
(1990-2004)

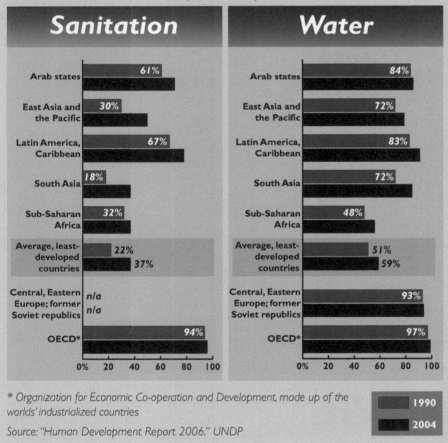

* *Organization for Economic Co-operation and Development, made up of the worlds' industrialized countries*

Source: "Human Development Report 2006," UNDP

Are we running out of water?

Amid the growing alarm about water shortages, water expert Frank Rijsberman offers a contrarian perspective. "The world is far from running out of water," he says. "There is land and human resources and water enough to grow food and provide drinking water for everyone." [23]

The issue is how efficiently water is used, says Rijsberman, former director of the International Water Management Institute in Colombo, Sri Lanka. Every year, about 110,000 cubic kilometers * of rain falls on Earth's surface, of which humans withdraw just over 3 percent — about 3,700 cubic kilometers — from rivers and groundwater to use in cities, industries and farming. About 40,000

---

* 1 cubic kilometer would cover an area of about 810,000 acres with one foot of water.

cubic kilometers flows into rivers and is absorbed into groundwater, and the rest evaporates.

Much of the water used by humans is returned to watersheds as wastewater, farm runoff or discharges from energy and industrial plants, with only a small fraction used for drinking and cooking. [24] Irrigation claims 70 percent of total water withdrawals, 22 percent is used by industry and the rest goes for homes, personal and municipal uses. [25]

Water isn't running out everywhere, said Canadian journalist Marq de Villiers, author of *Water: The Fate of Our Most Precious Resource*. "It's only running out in places where it's needed most. It's an allocation, supply and management problem." [26]

It's also a demand problem: Over the past half-century, millions of people have migrated from colder, wetter, northern climates to warmer, drier, southern locales such as the American Southwest or southern France, putting new pressure on those expanding "Sun Belt" communities to build irrigation systems, tap into groundwater supplies or rechannel large amounts of river water.

Experts agree that the world should not be facing an overall water-scarcity crisis. But water supplies in much of Africa, parts of China, southern Europe, northern Mexico and the American Southwest and high plains aren't meeting demand, and climate change may be accelerating the problem, the experts say. [27] The issues add up to what the World Commission on Water calls the "gloomy arithmetic of water." [28]

In addition, man has transformed most of the world's great rivers. For example, the Danube — Central Europe's "lifeline" — has been dredged, deepened, straightened, channelized and obstructed by dams and fishing weirs. It is now "a manufactured waterway," says de Villiers, with more than a third of its volume withdrawn for human use, compared to an average of about 10 percent for other rivers. [29]

Pollution is also reducing the world's supply of potable water. In Asia, many rivers "are dead or dying," according to Rijsberman. The Musi River near India's Hyderabad technology center has become "a dwindling black wastewater stream," he writes. "[Y]et the cows that produce the curd and the dairy products for Hyderabad are bathing in that black and stinking water." [30] In China, 265 billion gallons of raw sewage is dumped into the Yangtze River every year. [31]

The depletion and despoiling of the world's reservoirs, rivers and watersheds also contribute to the problem. During the 20th century, more than half the wetlands in parts of Australia, Europe, New Zealand and North America were destroyed by population growth and development. The loss of wetlands increases water runoff, which exacerbates flooding, reduces the replenishment of aquifers and leaves rivers and lakes more vulnerable to pollution. [32]

Aquifers — the immense storehouses of water found beneath the Earth's surface — are the largest and fastest-growing source of irrigation water. Depleting those underground rivers will have deleterious effects on the 40 percent of the planet's agricultural output that relies on irrigation from groundwater. [33] Experts say some of that water — which dates back to past ice ages — would take eons to refresh but is being consumed in less than a century.

"Large areas of China, South Asia and the Middle East are now maintaining irrigation through unsustainable mining of groundwater or over-extraction from rivers," said the U.N. "Human Development Report 2006." [34] The problem is widespread in Mexico, India and Russia, as well, although precise data are not available for many countries. [35]

In his seminal 1986 book *Cadillac Desert*, the late Marc Reisner warned about the long-term effects of water policies in the Western United States, including the depletion of the giant Ogallala Aquifer, which runs south-

Dead plants in a parched field attest to the ravages of the worldwide drought in Liujiang, a county of southwest China's Guangxi Zhuang Autonomous Region, in November 2006. Nearly all of the region's 84 counties were affected by the drought, which was caused by unseasonably warm temperatures, according to the Xinhua News Agency.

AP Photo/Xinhua, He Fenglun

ward from South Dakota to Texas. It has two distinctions, he wrote, "one of being the largest discrete aquifer in the world, the other of being the fastest-disappearing aquifer in the world." [36]

In the 1930s a farmer on the Great Plains could raise a few gallons per minute from the Ogallala, using a windmill-driven pump. After the New Deal brought electricity to the region and oil and gas discoveries provided plenty of cheap fuel, electric pumps raised 800 gallons per minute.

"All of a sudden, irrigation became very energy- and labor-efficient. You turn on the switch and let it run," says Robert M. Hirsch, associate director for water at the U.S. Geological Survey.

"There was an explosion of irrigated agriculture, particularly on the high plains, and in California."

In 1937, West Texas had 1,116 irrigation wells. Thirty years later it had 27,983. By 1977, Texas was withdrawing 11 billion gallons of groundwater a day to grow corn, cotton and other crops in what once had been part of the Great American Desert, Reisner wrote. [37]

Now, experts say the Ogallala — a resource that could have lasted hundreds of years — will be virtually depleted within the lifetimes of today's farmers.

Yet during the optimism and opportunism that characterized development of the modern American West,

Continued on p. 35

Is Access to Clean Water a Human Right?

The question is at the heart of a global debate.

Should all humans have guaranteed access to clean water, or is water an increasingly scarce commodity that should be priced according to its value?

The question stands at the center of a global debate over threats to the world's water resources, as competition for water increases among industry, farming and households. It is also critical in efforts to protect the long-term environmental viability of rivers, lakes and aquifers.

The debate goes back at least to 1992, when an international commission on water and the environment meeting in Ireland issued the "Dublin Principles," which were later adopted by a U.N. panel. The commission concluded: "Water has an economic value . . . and should be recognized as an economic good." Only by recognizing that economic value can water "be properly conserved and allocated to its most important uses." [1]

But the principle also declared it a "basic right of all human beings to have access to clean water and sanitation at an affordable price." Poor households cannot compete with industry for scarce water supplies. Nor could most farmers, who typically receive subsidized prices for irrigation water.

The U.N. Committee on Economic, Social, and Cultural Rights declared in 2002 that all people are entitled to an essential minimum amount of clean water. "Water is fundamental for life and health," it said. "The human right to water is indispensable for leading a healthy life in human dignity." [2]

Canadian activist Maude Barlow says, "You can't really charge for a human right; you can't trade it or deny it to someone because they don't have money." Barlow is co-author of *Blue Gold: The Battle Against Corporate Theft of the World's Water.* [3]

The other side in the debate argues that until water is priced and valued as a scarce resource it will be wasted and billions of dollars required annually to extend water service to the poor and fix leaking water systems will not be forthcoming. Two years before the U.N. declared clean water a human right, the World Water Council — which reflects the views of international lenders and the water-supply industry — called for "full pricing" of water to reflect its "economic, social, environmental and cultural values." [4]

Farmers in dry regions throughout the world get water at preferential rates — or at no charge at all — as a matter of government policy. But if farmers were required to pay the full price for water, they could not compete with industry, which would be willing and able to pay market price.

In industrial countries, 60 percent of the water withdrawn from freshwater sources is used by industry, mainly to generate electricity. The developing world is moving rapidly in the same direction. China's industrial water use, for example, is projected to grow fivefold by 2030. [5]

"As urban centers and industry increase their demand for water, agriculture is losing out," said the U.N. "Human Development Report 2006." [6]

And the world's poor cannot compete with either farmers or business for water at market prices, said the report. About a third of those without access to clean water live on less than $1 a day. Twice that many live on less than $2 a day. "These figures imply that 660 million people lacking access to [safe] water have, at best, a limited capacity to pay more than a small amount for a connection to water service," the report said. "People might lack water because they are poor, or they might be poor because they lack water." The end result is the same: a limited ability to pay for water. [7]

A man protesting the privatization of water in Cochabamba, Bolivia, waves a sign saying "what is ours is ours and it cannot be taken away."

American water expert Peter H. Gleick calls for a truce in the water rights dispute in favor of problem-solving. Workshops on privatization standards and principles for implementing a human right to water "would be far more likely to produce progress," he writes. [8]

---

1 "Dublin Statements and Principles," Global Water Partnership; www.gwpforum.org/servlet/PSP?iNodeID=1345.

2 Meena Palaniappan, *et al.*, "Environmental Justice and Water," *The World's Water 2006-2007*, p. 117.

3 Jeff Fleischer, "Interview with Maude Barlow," *Mother Jones*, Jan. 14, 2005, www.motherjones.com/news/qa/2005/01/maude_barlow.html.

4 "Ministerial Declaration of The Hague," World Water Forum, March 22, 2000.

5 Palaniappan, *op. cit.*, p. 125.

6 See "Summary, Human Development Report 2006: Beyond scarcity: Power, poverty and the global water crisis," United Nations Development Programme, p. 17, http://hdr.undp.org/en/media/hdr2006_english_summary.pdf.

7 *Ibid.*, pp. 49-52.

8 Peter Gleick, "Time to Rethink Large International Water Meetings," *The World's Water 2006-2007*, p. 182.

Continued from p. 33

worries about future water supplies evaporated. "What are you going to do with all that water?" the late Felix Sparks, former head of the Colorado Water Conservation Board, asked in the mid-1980s. "When we use it up, we'll just have to get water from somewhere else." But today, "somewhere else" is not an answer, say authors Robin Clarke, editor of climate publications for the United Nations and World Meteorological Organization, and environmental author Jannet King. The co-authors of *The Water Atlas* insist water must be considered a finite resource. [38]

Should water be privatized?

In 2000, street fighting broke out between government forces and political activists, rural cocoa farmers and residents of shantytowns on the hilly outskirts of Cochabamba — Bolivia's third-largest city. The dispute was over privatization of the city's water supplies.

The year before, Cochabamba had turned its water and sanitation system over to Aguas del Tunari, a coalition of multinational and Bolivian water and engineering corporations whose biggest stakeholder was Bechtel Corp., based in San Francisco. [39] This was the high-water mark of a global, pro-market movement toward deregulation and privatization of state-owned monopolies in water, electricity and other services. [40] The World Bank and other international lenders had been supporting privatization strategies in hopes that investments and better management by private industry would help bring water and sanitation to more than a billion poor people whose governments couldn't or wouldn't do the job.

But Cochabamba's privatization included a costly dam and pipeline to import more water, which required sharp rate increases starting at 35 percent. Some customers' water bills doubled. Farmers outside the city, who had enjoyed free water, suddenly had to pay. The city erupted in protest,

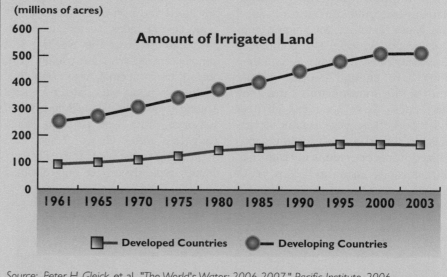

Irrigation Doubled in Developing Nations

The amount of irrigated land more than doubled in developing countries in the past four decades, increasing faster than in the developed world. But the rate of increase for both has slowed in recent years because of heavy draws on groundwater aquifers and competition from industry for water.

Amount of Irrigated Land

(millions of acres)

Years: 1961 1965 1970 1975 1980 1985 1990 1995 2000 2003

■— Developed Countries   ●— Developing Countries

Source: Peter H. Gleick, et al., "The World's Water: 2006-2007," Pacific Institute, 2006

the water company's officials fled and their contract was rescinded. [41] The government reclaimed the water operations, and Cochabamba became a rallying cry against privatization and globalization for the political left.

Elsewhere, however, corporate involvement in water and sanitation system operations has not ceased. Veolia Water, a subsidiary of the French firm Veolia Environment SA — the world's largest water-services firm — signed a $3.8 billion, 30-year, contract in 2007 to supply drinking water to 3 million residents of the Chinese river port city of Tianjin. Since 1997, Veolia has signed more than 20 water and sanitation contracts in China, and supplies more than 110 million people in 57 countries worldwide. [42]

These projects, and smaller-scale versions in poorer nations, suggest that while the inflamed debate over water privatization continues, threats of water scarcity and climate change may help

accelerate the search for private-sector support.

The percentage of the world's population served at some level by private firms has grown from 5 percent in 1999 to 11 percent — or 707 million people — in 2007, according to *Pinsent Masons Water Yearbook*, a widely consulted summary of private-sector water projects. [43]

Opponents of privatization argue that safe drinking water and adequate sanitation are essential human rights, obligating governments to provide them at affordable rates or free if necessary. "If it's a human need, it can be delivered by the private sector on a for-profit basis. If it's a human right, that's different," says Canadian anti-globalization activist Maude Barlow, co-author of *Blue Gold: The Battle Against Corporate Theft of the World's Water.* "You can't really charge for a human right; you can't trade it or deny it to someone because they don't have money." [44]

Bringing in private firms to run water and sewer operations does not make the services more efficient or affordable, opponents also argue, but forces the poor to pay for corporate profits, shareholder dividends and high executive salaries. "The efficiencies don't happen," asserts Wenonah Hauter, executive director of Food and Water Watch, a Washington anti-globalization group. "The companies simply lay off staff members until they don't have enough people to take care of the infrastructure. And they raise rates. We've seen this all over the world." Last year Hauter's organization issued a study claiming privatized water operations in California, Illinois, Wisconsin and New York

pears to have improved water quality in urban areas but not in rural communities, said the study. After privatization began, water bills for the poor rose about 10 percent but declined for the wealthy, reflecting a scaling back in government subsidies to poorer consumers. Similar shifts occurred in both privatized and non-privatized cities. [48]

In central cities, water-rate subsidies tend to favor the wealthy and middle classes, who are usually connected to municipal water systems, while the poor often are not, says American journalist Diane Raines Ward, author of *Water Wars: Drought, Flood, Folly, and the Politics of Thirst.* And by keeping water rates artificially low, utilities typically collect only about a

privatization has declined in Latin America and sub-Saharan Africa but increased in Europe and Asia, according to the *Water Yearbook.* [50] The average contract size also has diminished since the 1990s, it said, due to a trend away from mega-contracts with multinational water companies in favor of "local and possibly less contentious contracts." [51]

A U.N.-sponsored analysis cites Chile and parts of Colombia among the successful examples of collaborative water and sanitation services. In Cartagena, Colombia's fifth-largest city, the local government retains control of the pipes and facilities and raises investment capital, but a private firm runs the service. Today, nearly all the city's residents have water in their homes, up from only one-quarter in 1995. [52] Chile's water program offers subsidies to the poorest households, guaranteeing an essential minimum of supply of up to 4,000 gallons per month. Deliveries are monitored to limit cheating, and every household must have a water meter to verify usage. [53]

Experts say the political problems of water privatization cannot be managed without effective government regulation and consumer involvement at all levels. Both elements were missing in Cochabamba but are present in Chile, the U.N. report says. [54]

The outlook is bleak for rancher Andrew Higham's parched land in Gunnedah, in northwestern New South Wales, Australia, in October 2006. Scarce winter rains caused drought across much of the continent last fall and led to severely reduced wheat and barley harvests.

AP Photo/Peter Lorimer

Will water scarcity lead to conflicts?

In 1995 Ismail Serageldin, a World Bank vice president, predicted that "the wars of the next century will be over water." [55]

The reality has been different thus far. "Water resources are rarely the sole source of conflict, and indeed, water is frequently a source of cooperation," writes Gleick, of the Pacific Institute for Studies in Development, Environment and Security, in the new edition of *The World's Water 2006-2007.* [56] The survey of reported conflicts over water in the past 50 years, compiled by Oregon

charged more for water than comparable publicly owned systems. [45]

Privatization advocates dispute Hauter's claims, and facts to settle the issue are illusive. A 2005 survey by the AEI-Brookings Center for Regulatory Studies found "no systematic empirical evidence comparing public and private water systems in the United States." [46]

A study by the Inter-American Development Bank of water rates in Colombia said prices charged by privatized systems were not significantly different from those charged by public systems. [47] And privatization ap-

third of their actual costs, so they don't raise enough money to expand pipelines to unserved poor neighborhoods, she says. [49]

The rural poor or those living in urban slums often must haul water home from public wells or buy it from independent merchants — delivered by truck or burro — at much higher prices. In Cairo, Egypt, for instance, the poor pay 40 times the real cost of delivery; in Karachi, Pakistan, the figure is 83 times; and in parts of Haiti, 100 times, Ward says.

In the years since Cochabamba galvanized the left against privatizing water,

State University researchers, found 37 cases of violence between nations, all but seven in the Middle East. [57]

In 1964, Israel opened its massive National Water Carrier canal to carry water from the Sea of Galilee and the Jordan River to its farms and cities. Syria retaliated to maintain its access to the Jordan by starting two canals to divert Jordan flows for its uses. Skirmishes by military units and raids by the newly established al-Fatah forces escalated until Israeli air strikes halted the diversion projects. By then, Israel and the Arab League were on the road to the Six-Day War of 1967. [58]

"The attacks by Syria, Egypt, and Jordan that eventually followed had many causes, but water remained a priority for both sides," says author Ward. [59]

Still, more than 200 water treaties have been negotiated peacefully over the past half-century. The Partition of India in 1947, for instance, could have led to war between India and newly created Pakistan over control of the mighty Indus River basin. Instead, the two nations were brought together with World Bank support over a perilous decade of negotiations, signing the Indus Water Treaty in 1960. Three rivers were given to Pakistan, and three to India, with a stream of international financial support for dams and canals in both countries. Even when war raged between the two nations in later years, they never attacked water infrastructure. [60]

"Most peoples and even nations are hesitant to deny life's most basic necessity to others," Ward wrote. Two modern exceptions occurred during the Bosnian War (1992-1996), when Serbs "lay waiting to shoot men, women and children arriving at riverbanks or taps around Sarajevo carrying buckets or bottles," and during Saddam Hussein's regime in Iraq, when he diverted the lower waters of the Tigris and Euphrates rivers to destroy the homes and livelihood of the Marsh Arabs. [61]

Mountain Snowpack Is Shrinking

The amount of snow covering the globe's highest mountains has been shrinking over the past half-century, upsetting crucial seasonal water flows that restock rivers, lakes, reservoirs and aquifers. Scientists think short-term climate conditions like El Niño and long-term warming caused by climate change are to blame. By century's end, only 16 percent of New Zealand's current snowpack will remain.

Snowpack Now and Projected in 2100

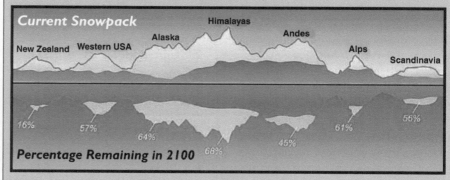

Source: Pacific Northwest National Laboratory, http://picturethis.pnl.gov/picturet.nsf/by+id/AMER-6PWV.V?opendocument.

Except for such instances, cooperation over water resources is common today, even if sometimes grudging and incomplete, says Undala Alam, a professor and specialist in water diplomacy at Britain's Cranfield University. "Turkey was releasing water for Syria and Iraq; the Nile countries are preparing projects jointly to develop the river; the Niger countries have a shared vision for the basin's development, and the Zambezi countries are working within the Southern African Development Community," she notes. [62]

But analysts warn that growing stress on water supplies, coupled with the impact of climate change, will create combustible conditions in the coming years that will undermine collaboration over water.

There is plenty of precedence for the concern, notes Gleick, who describes the history of violence over fresh water as "long and distressing." [63] The latest volume of *The World's Water* lists 22 pages of historical water conflicts —

beginning in about 1700 B.C. with the Sumerians' efforts to dam the Tigris River to block retreating rebels.

In the future, climate change is expected to extend and intensify drought in Earth's driest regions and disrupt normal water flows from mountain snowcaps in Europe, North America and Central Asia. "Climate change has the potential to exacerbate tensions over water as precipitation patterns change, declining by as much as 60 percent in some areas," warned a recent report by a panel of retired U.S. generals and admirals convened by CNA, a think tank with longstanding ties to the military. "The potential for escalating tensions, economic disruption and armed conflict is great," said the report, "National Security and the Threat of Climate Change." [64]

On the simplest level, the report said, climate change "has the potential to create sustained natural and humanitarian disasters on a scale far beyond those we see today." Already, it said, Darfur, Ethiopia, Eritrea, Somalia, Angola, Nigeria, Cameroon and Western Sahara have all been hit hard by tensions that can be traced in part to environmental causes. [65] If the drought continues, the report said, more people will leave their homelands, increasing migration pressures within Africa and into Europe. [66]

The impact will be especially acute in the Middle East, where about two-thirds of the inhabitants depend on water sources outside their borders. Water remains a potential flashpoint between the Israelis and Palestinians, who lack established rights to the Jordan River and receive only about 10 percent of the water used by Israel's West Bank settlers. [67] "Only Egypt, Iran and Turkey have abundant fresh water resources," the CNA report said.

The military advisers urged the United States to take a stronger national and international role in stabilizing climate change and to create global partnerships to help less-developed nations confront climate impacts. [68]

Currently, there is only a weak international foundation for water collaboration, according to the U.N. Human Development report. While a 1997 U.N. convention lays out principles for cooperation, only 14 nations have signed it, and it has no workable enforcement mechanism. In 55 years, the International Court of Justice has decided only one case involving international rivers. [69]

It is possible, however, that as the awareness of climate impacts on water supplies deepens, so will the urgency for governments to respond. "Unlike the challenges that we are used to dealing with, these will come upon us extremely slowly, but come they will, and they will be grinding and inexorable," said former Vice Adm. Richard H. Truly, a former astronaut who headed the U.S. National Aeronautics and Space Administration (NASA) and served as a CNA consultant. [70] "They will affect every nation, and all simultaneously." ■

China Leads World in Dam Building

China has three times as many dams as the United States and more than all the next 11 countries combined. Dams typically generate electricity, control flooding and provide water for irrigation.

Countries with the Most Dams

1.	China	22,000
2.	United States	6,575
3.	India	4,291
4.	Japan	2,675
5.	Spain	1,196
6.	Canada	793
7.	South Korea	765
8.	Turkey	625
9.	Brazil	594
10.	France	569
11.	South Africa	539
12.	Mexico	537

Source: Peter H. Gleick, et al., "The World's Water: 2002-2003," Pacific Institute, 2002

BACKGROUND

Taming Water

The ruins of irrigation canals 8,000 years old have been found in Mesopotamia. [71] Remains of water-storage dams 5,000 years old survive in Egypt and Jordan. Humans have been using waterwheels for milling and threshing since the 1st century B.C., and by 1291 China had completed its Grand Canal running nearly 1,800 kilometers between Beijing and Hangzhou. [72] Water power drove mining, metal drilling, textile and milling industries at the dawn of the Industrial Age. [73] In the United

States, the opening of the Niagara Falls hydroelectric power station in 1896 — built by American entrepreneur and inventor George Westinghouse and backed by financier J. P. Morgan and others — inaugurated water's use to generate electric power. [74]

But these early accomplishments were dwarfed in the 20th century by nearly 100 years of massive dam projects that have transformed most of the world's major rivers. Between 1950 and 2000, the number of dams higher than 50 feet increased sevenfold — to more than 41,000 structures impounding 14 percent of the world's average river runoff. [75] By 2000, large dams were supplying nearly a fifth of all electrical power worldwide.

Dams also have been critical in the rapid expansion of irrigated farming. "Half of the world's large dams are built exclusively for irrigation, supporting about 12 to 16 percent of world food production, according to the World Commission on Dams." [76]

About 12 percent of large dams were constructed specifically to provide drinking water and sanitation (and a similar percentage were built to control flooding), and many have multiple uses. Whether used for energy, community water, flood control or agriculture, dams have been a key tool of economic growth, according to the commission. Typically, however, the full consequences of such giant projects were not taken into consideration, the commission said. In the past century, the world built, on average, one large dam per day without asking whether it was getting a fair return from the $2 trillion investment, said South African Minister of Education Kader Asmal, who chaired the commission. [77]

The dam-building blitz was enabled by political bias in favor of dams. From the Aswan High Dam in Egypt to the Hoover Dam on the Colorado River, big dams have stood as pre-

Continued on p. 40

Chronology

19th Century
The industrial age and urbanization create critical need for municipal water treatment and sanitation services.

1848
The Public Health Act, followed in 1852 by the Metropolitan Water Act, lead to investments in water treatment and sanitation that dramatically reduce waterborne illnesses in Britain by the end of the century.

1876
Berlin city planner James Hobrecht starts work on drainage system and waterworks that channels sewage to fields as fertilizer. He designs similar systems for Moscow, Cairo and Tokyo.

1900-1980
Governments around the globe launch major dam construction.

1902
The Aswan Dam on the Nile River in Egypt is completed to control flooding and regulate water flow for agriculture.

1910
Chlorination begins in the United States. Typhoid fever from polluted drinking water falls from 25 deaths per 100,000 people to almost zero.

1936
Hoover Dam on the Colorado River is opened, fulfilling an agreement among Southwestern states and cities to share the river's flow.

1945
Large-scale groundwater irrigation begins expanding in the Western United States, aided by rural electrification and innovations in pumping and irrigating technology.

1986
Major dam construction has largely stopped in the U.S. but continues in Asia.

1980s
Britain's public utilities are sold to private firms in the late 1980s, triggering a wave of water-system privatization worldwide.

1994
Construction begins on the Three Gorges Dam on China's Yangtze River, designed to be the largest in the world. More than 1 million people will have to be relocated as the river rises.

2000s
Drought spreads worldwide, heightening concern about climate change's impact on water scarcity. Privatization strategies shift.

2000
Violent public protests against higher water rates force cancellation of a water-privatization plan in Cochabamba, Bolivia. But privatization continues in China, India and other parts of the world. . . . U.N. adopts the Millennium Development Goals calling for halving the number of people without access to safe water and adequate sanitation by 2015.

2002
China approves massive South-to-North Water Diversion Project, which will eventually link the country's four major rivers to bring water from the south to the arid north.

2005
Group of Eight industrialized nations pledge to double their aid for water, sanitation and other development projects in poorer nations by 2010.

2006
Waterborne-disease epidemics strike Karachi, Lahore and other Pakistani cities, caused by the leakage of sewage and industrial wastes into damaged water-distribution pipelines. . . . U.N. Human Development Report warns that without a major increase in investment, improvements in water and sanitation services will fall far short of the Millennium Development Goals.

2007
Multi-year drought afflicts China, the Horn of Africa, Turkey, Australia, Spain and the U.S. Sun Belt. . . . A director of the Three Gorges Project warns that rising waters behind the dam could cause "water pollution, landslides and other geological disasters," but other officials later say environmental problems will be ameliorated. . . . The Intergovernmental Panel on Climate Change predicts that freshwater resources will decrease in large river basins over the next decade due to drought. . . . Drought prompts several Australian cities to commission new desalination plants. Algeria also has a major plant in construction.

2008
The scarcity of water in key agricultural areas has contributed to soaring world prices for wheat, soybeans, corn, rice and poultry, and the trend is likely to continue this year, say agricultural forecasters.

Continued from p. 38

eminent symbols of governments' engineering prowess and the use of state power to control devastating floodwaters and feed economic expansion. "Colossal engineering works bestow big contracts and big benefits, divide up waters, hold them fast, channel them away from some and give them to others," says author Ward. "It has always been politics that start the bulldozers moving." [78]

While dams helped expand supplies of drinking water, hydropower and irrigation water, they also attract population expansion that eventually strains the new resources. And until recently, the commission said, policymakers have not fairly considered the damaging impact dams have on downstream rivers and aquifers and the populations that are forced to move to make room for reservoirs.

Resistance to major new dam projects emerged with the rise of the environmental movement in the 1970s, particularly in Europe and the United States, as advocates pointed to the harm caused by dam construction. In the industrial world, "it is now more likely that a dam will be torn down than a new one will go up," says Ward.

But in China, South Asia and South America, dams are "multiplying like mushrooms," she writes. [79] If China's new Three Gorges Dam — the world's largest hydroelectric project — had been built midway through the past century, it might have been considered one of the world's great engineering feats. The main wall of the massive, 60-story structure spanning the Yangtze River was

Empowering Women May Quench Thirst

'Should girls be kept home from school to collect water?'

For millennia, the nomadic Tuareg people have lived by herding, migrating across vast rangelands south of the Sahara Desert to water and graze their animals. But decades of drought are destroying the traditional Tuareg way of life — killing herds, drying up grazing lands and forcing many to settle in villages.

"We used to saddle the camel and put all the nice things on its back and put on our nice clothes and go," Tuareg chief Mohamed Ag Mata told a reporter last year. "We were afraid of nothing." [1]

But the changes, paradoxically, offer hope for a better existence to women and children in the male-dominated Tuareg culture — giving them access to the employment, education and social rights that will give them a greater say in community water policy.

Increasingly, educating and empowering women is seen as an effective way to expand access to clean water across the developing world. Virtually every major international organization dealing with water scarcity is calling for change in women's decision-making roles, including The World Bank, the U.N. Human Development Programme, the World Health Organization, the World Water Forum, the World Commission on Dams and the Stockholm-based Global Water Partnership.

In the developing world, the job of hauling water rests, literally, almost entirely on women's shoulders. A UNICEF study in 23 sub-Saharan countries found that a quarter of women spent 30 minutes to an hour each day collecting and carrying water, and 19 percent spent an hour or more. In Mile Gully, an impoverished rural area of Jamaica, hauling the family's water can take a woman two to five hours a day. [2]

The high incidence of waterborne disease caused by the lack of clean water further burdens women in poor, rural communities, because they are the primary caregivers for the sick.

"Should a woman care for a sick child or spend two hours collecting water?" asks the latest U.N. human development report. "Should girls be kept home from school to collect water, freeing time for mothers to grow food or generate income? Or should they be sent to school to gain the skills and assets to escape poverty?" [3]

But despite having to bear the greatest burden caused by a lack of water, women "play no role in the decision making for their communities," said Margaret Mwangi, a specialist in forestry and environmental issues who has worked for UNESCO. [4]

Women's lack of property rights prevents them from having a say in how water is distributed in their provinces. Women own less than 15 percent of the world's land and in many countries cannot legally own property separately from their husbands. "Lacking rights to land, millions of women in South Asia and sub-Saharan Africa are denied formal membership rights to participate in water-user association meetings," according to the U.N. "Human Development Report 2006." [5] And even those who are welcomed at irrigation-association meetings often cannot find the time. "Meetings are on Friday nights. At that time, after cooking for my husband and the kids, I still have a lot of work to do around the house," said a woman in Ecuador. "Even if I go to the meeting, it's only to hear what the men have to say. Men are the ones who talk and discuss." [6]

Unless both water and gender policies are reformed, water scarcity threatens to worsen women's plight, says the Sri Lanka-based International Water Management Institute. [7] Studies recommend a wide range of strategies to strengthen women's roles in gaining access to water and sanitation services, including micro-credit and micro-insurance programs that target women; training programs in rural irrigation and sanitation processes; creating rural women's councils and broadcasting radio programs on women's issues. [8]

But gender traditions can prove hard to change. Along the Bay of Bengal coast, decision-making and financial control over irrigation systems in the Indian state of Andhra Pradesh, has been decentralized, giving more authority to local communities. Nevertheless, only 4 to 5 percent of the women surveyed in two districts believed they could influence decisions in village meetings. "Women, and particularly poor women, rarely participate," the human development report concluded. [9]

Experts say progress for women requires a push from the bottom and pressure from the top. Legislation in Uganda, for instance, requires that all agencies — from national to village levels — include at least 30 percent female representation.

"Affirmative action may not remove cultural barriers," the U.N. report said, "but it does challenge their legitimacy." [10]

That may even be happening among the Tuaregs, says Hadijatou, describing her new life in a village near Timbuktu in Mali. "Before, everything was given to us by the men. When you are given what you need by other people, you are dependent on them. But when you are producing what you need, you depend on nobody. So life is far better now." [11]

Women and girls in poor countries, like these in Pakistan, bear the greatest burden from a lack of potable water.

[1] Richard Harris, "Drought Forces Desert Nomads to Settle Down," National Public Radio, July 2, 2007.

[2] Polioptro Martinez Austria and Paul van Hofwegen, "Synthesis of the 4th World Water Forum," Mexico City, 2006 p. 4, www.worldwaterforum4.org.mx/files/report/SynthesisoftheForum.pdf.

[3] "Human Development Report 2006 — Beyond Scarcity: Power, Poverty and the Global Water Crisis," U.N. Development Programme, 2006, p. 87; http://hdr.undp.org/en/reports/global/hdr2006/.

[4] Margaret Mwangi, "Gender and Drought Hazards in the Rangelands of the Great Horn of Africa," Women & Environments International Magazine, Spring 2007, p. 21.

[5] "Human Development Report," op. cit., p. 194.

[6] Ibid.

[7] David Molden, ed., "Summary," Water for Food, Water for Life: A Comprehensive Assessment of Water Management in Agriculture, International Water Management Institute, p. 10, www.iwm.cgiar.org/assessment/files_new/synthesis/Summary_SynthesisBook.pdf.

[8] Austria and Hofwegen, op. cit., pp. 43, 63, 96.

[9] "Human Development Report," op. cit., p 193.

[10] Ibid., p. 194.

[11] Harris, op. cit.

completed in 2006, and by the end of 2008 it is expected to deliver up to 18 million kilowatts per hour — nearly a tenth of the electricity needs of China's surging economy. [80] But in today's perspective, the monumental structure symbolizes the threat of environmental destruction caused by major dam construction.

The gargantuan project's human and environmental costs have alarmed opponents. According to the Chinese government, more than a million riverside residents were forced to move as the water rose behind the dam. The rising water levels have triggered some massive landslides on the riverbanks, and a senior government official warned last September that if such ecological and environmental dangers are not dealt with, "the project could lead to a catastrophe." [81] But more recently, Chinese officials have insisted the project will be operated safely. [82]

Some water experts still advocate new, smaller dams in developing countries to control flooding, store irrigation water and generate electricity. For instance, most of India's rainfall occurs in about 100 hours during the monsoon season. While reservoirs capture some of these torrents, most escape to the sea. [83]

The 4th World Water Forum in Mexico City in 2006 cited Norway as a role model for the value of dams. "Electricity from hydropower was the key factor in transforming Norway from one of the poorest countries in Europe a century ago to the industrialized and wealthy nation of today," said Anita Utseth, Norway's deputy minister for petroleum and energy. [84]

China's 60-story Three Gorges Dam — the world's largest hydroelectric project — is expected to deliver up to 18 million kilowatts per hour by the end of 2008, nearly a tenth of China's electricity needs. Critics say the Yangtze River structure symbolizes environmental destruction that can be caused by major dam construction. More than a million riverside residents were forced to move as the water rose behind the dam.

As Ward notes, "In some places, if no reservoirs are built, poor people will be denied the means to improve their lives."[85]

Regulating Water

Who owns water? What rights do water users have? How should conflicts be resolved?

The globe's oldest recorded societies were formed not only for defense but also to try to control the flow of water in rivers that were crucial to farming.

Under Roman law, water resources were the property of the state, which was responsible for their development and protection. Islamic water law in the parched Middle East followed a similar path — irrigation canals and ditches had to be adequately planned and spaced to prevent infringement on others' water sources.[86]

But ancient codes also recognized that landowners did not have to share well water in times of scarcity, and people living closest to rivers and lakes had first claim on their waters. "The cistern nearest to a water channel is filled first, in the interests of peace," said the 12th-century Jewish theologian and philosopher Maimonides.[87]

Eventually, the doctrine of "riparian" rights — giving those living closest to water the first claim on its use — became merged with a "public trust" doctrine, holding that water was a common resource to be managed by the state for the common benefit. One person's use of water could not infringe on a neighbor's reasonable needs. The doctrine was embraced by many countries in Europe — including Britain, France and Spain — which then exported the principle to their colonies abroad.[88]

But in the Western United States, Chile and Mexico, private water rights were recognized, particularly after the 1848 gold rush in the United States. A miner finding a gold seam would claim water from the nearest creek to wash dirt away from precious nuggets. His claim established a "first-in-time, first-in-use" priority allowing him to take as much as he needed.[89] This "prior appropriation" doctrine was a starter's gun for unchecked diversion and exploitation of water resources to create farms and cities in the arid West.[90]

Of course, in most legal debates about water, individual rights are submerged by political elites, rulers and dominant factions. For instance, Senegalese law provides for a democratic distribution of irrigated lands. But in practice, tribal nobles' descendants still claim the lion's share of the land and allocate rights to powerful outsiders, including military leaders, politicians and judges.[91]

In Central Asia during Soviet rule, for instance, Kyrgyzstan, Tajikistan and Uzbekistan — which abut the Syr Darya and Amu Darya river basins — shared the reservoir and hydropower output from Kyrgyzstan's largest reservoir. Now, as separate states, their cooperation has virtually ended, according to a U.N. report. Kyrgyzstan is holding on to more of its reservoir volume in order to increase its hydropower exports, severely reducing irrigation flows in the other two countries. A constructive dialogue "has been conspicuously absent," according to the U.N. Human Development Report. [92]

Two international treaties — the Helsinki Rules of 1966, adopted by the International Law Association, and articles adopted by the U.N. International Law Commission — specify how cross-border water disputes should be resolved. Rivers that divide nations, according to the treaties — must be considered common resources, not under any one nation's control. They also advocate a policy of "no harm" — each riparian nation has the right to "equitable utilization" of a shared water supply, and a nation's water use should not damage its neighbors' water needs. Prior appropriation claims are not allowed, and countries are called on to share accurate information on water resources. [93]

Before fair water-use policies can be expected, governments must consider the needs of the poor and politically weak, whose water needs are usually greatest, says water expert Sandra Postel, director of the Global Water Policy Project in Massachusetts. That means "adding seats around the table," she says. [94] For example, Ghana successfully expanded its water and sanitation services in recent years after water policy was decentralized, and village and district water councils were formed. The result: improved planning and more reasonable priorities for water funding, according to the U.N. [95]

Misusing Water

In Mumbai's Dharavi slum, which lies between the international airport and the city's financial center, an estimated 1 million Indians live in huts and shanties. With only one toilet for every 1,440 people, gutters overflow with waste in the rainy season, turning the streets into open sewers. [96]

A Masai herdsman in Kenya's Rift Valley tends to a dying sheep in January 2006. Today 850 million people live in areas where adequate food supplies are at risk because of drought — two-thirds of them in South Asia and sub-Saharan Africa, where the impact of climate change on food production is expected to be worst.

Such unsanitary conditions in developing countries contribute to a plague of diarrhea that not only kills an estimated 1.8 million children across the globe each year but also repeatedly sickens many times that number, leaving them malnourished and vulnerable to other diseases and keeping them from school. [97]

Trachoma, spread by a fly that breeds in human feces, afflicts nearly 6 million people worldwide, causing widespread blindness. The disease "is a passport to poverty," says the U.N.'s "Human Development Report," because it prevents victims from working. [98]

"At the start of the 21st century one in five people living in the developing world — some 1.1 billion people in all — lack access to clean water," and nearly half the developing world has no access to adequate sanitation. [99]

Some of the same conditions existed in Europe and the United States during most of the 19th century, as farm families migrated to vast urban slums. During the summer in 1858, the stench of untreated sewage in the Thames River (called the "Great Stink" by the *London Times*) forced the Parliament to close temporarily. [100] In the

Water Schemes Range From Monumental to Zany

Many are costly and controversial.

After wildfires devastated parts of Southern California and dry winds parched the Southwest last summer, New Mexico's Gov. Bill Richardson, a Democratic candidate for president, made a startling proposal.

"States like Wisconsin are awash in water," Richardson told the *Las Vegas Sun*, proposing that water from the Great Lakes be piped to his state. With nearly 20 percent of the world's fresh water, the five vast lakes are an irresistible target to promoters, politicians — and potentially — corporate water suppliers.

"You're going to see increasing pressure to gain access to this supply," said Aaron Packman, a professor of civil and environmental engineering at Northwestern University. "Clearly, it's a case of different regional interests competing for this water." [1]

In 1985, Quebec provincial officials floated a plan to sell Canadian river water to the high plains states. The Great Replenishment and Northern Development Canal (GRAND) envisioned pumping river water from Quebec into a reservoir in Ontario. From there the water could travel by aqueduct to the Great Lakes and then by canal to the American plains. Nothing came of it.

Even grander was the North American Water and Power Alliance in the 1960s, which proposed to dam dozens of north-flowing rivers in Canada's western provinces, channeling their waters into a new reservoir 500 miles long — about the distance from Pittsburgh to Chicago — to irrigate the high plains and refill the Colorado River's flows to California. But the half-trillion-dollar price tag sank the proposal. And its environmental cost would have been "ungraspable," says Canadian journalist Marq de Villiers, author of *Water: The Fate of Our Most Precious Resource.* [2]

The idea still seems to stir anxieties in Canada. An off-hand remark by President George W. Bush in 2001 about the benefits of Canadian water exports to the United States caused a brief uproar, and last year Canada's Liberal Party Leader Stephane Dion accused the government in Ottawa of trying to put the matter back on the U.S.-Canada agenda — a claim both governments denied. [3]

Mega-project dreams still survive in China, however, which is planning a series of canals to carry Yangtze River water thousands of miles to the Yellow River and on to the huge cities of Beijing and Tianjin. [4]

And in 2004, Israel agreed to purchase the equivalent of 35 million gallons of freshwater per day from Turkey's Manavgat River, to be shipped aboard tankers. Turkish water exports to Cyprus and other water-short destinations were also under consideration. The agreement was put on hold in 2006 after high-

er oil prices made tanker transportation uneconomical, according to both countries. [5]

And then there are the slightly zany ideas, such as Calgary entrepreneur James Cran's plan to move water by sea inside floating 5,000-ton plastic bags, or towing Arctic icebergs to distant metropolises. [6]

Most monumental water-moving schemes, however, trigger intense opposition. Gov. Richardson's proposal, for instance, is opposed by eight Great Lakes-area states and two Canadian provinces, and drew a terse, unequivocal "No" from Michigan's Democratic Gov. Jennifer Granholm. [7]

The monumental California Aqueduct brings vitally needed water 444 miles from to Southern California farmers and residents.

[1] Tim Jones, "Great Lakes key front in water wars," *The Chicago Tribune*, Oct. 28, 2007, www.chicagotribune.com/news/local/chi-water_bdoct28,1,5145249.story.

[2] Marq de Villiers, *Water: The Fate of Our Most Precious Resource* (1999), p. 260.

[3] Launce Rake, "Canadians fearful of U.S. water grab," *Las Vegas Sun*, May 28, 2006. See also Bruce Campion-Smith and Susan Delacourt, "Secret talks underway, Dion claims," *The Toronto Star*, Aug. 18, 2007, p. 18.

[4] Diane Raines Ward, *Water Wars: Drought, Flood, Folly and the Politics of Thirst* (2002), p. 171.

[5] Josef Federman, "Israel, Turkey put landmark water agreement into deep freeze," The Associated Press, April 5, 2006

[6] de Villiers, *op. cit.*, p. 277.

[7] Jones, *op. cit.*

1890s Britain's infant mortality rate was about the same as Nigeria's today.

Water-treatment legislation in mid-19th-century Britain mandated creation of municipal water companies, followed

by an expansion of sanitation services after 1880. By 1910, the country's rate of infant mortality had fallen by nearly 40 percent. The critical factor for change, according to some experts,

was an extension of voting rights in Britain beyond property-owning classes to those who lacked clean water and sanitation, creating a political constituency for reforms. [101]

But the cost of expanding sanitation services in the developing world today would be immense. A study led by former International Monetary Fund President Michael Camdessus estimates that closing the sanitation services gap could cost $87 billion over the next two decades. [102]

Experts today debate which must come first in the developing world: good governance or solutions to water and sanitation needs. In *The World's Water 2006-2007*, the authors conclude that "eradicating corruption and political interference and ensuring the participation of all stakeholders will be critical to the successful governance of water." [103] The 1,320-km (820 miles) Rhine River — long called "the sewer of Europe" — was cleaned up after a 1986 industrial fire in Basel, Switzerland, allowed more than 30 tons of pesticides, dyes, mercury and other poisons to spill into the river. Millions of fish were killed, and major cities had to close their municipal water intakes. German Chancellor Angela Merkel, a former environmental minister, has called the "rebirth of the Rhine . . . one of the great environmental success stories of the century." [104]

But to the east, the blighted Danube testifies to the damage caused by selfish, beggar-thy-neighbor policies by the nations along its 1,770-mile course. Efforts to reengineer the river began in the 16th century with projects to steer its annual floods into canals or trap them behind huge dams. Historically, vast amounts of untreated human wastes and toxic industrial effluents were also dumped into the river, a situation that worsened during the Cold War. Soviet-era bosses had no scruples about flushing wastes into rivers, and Moscow pushed Hungarian and Czech governments to divert the Danube into man-made waterways in order to speed Russian barge traffic. Parts of the project were begun, but before it went very far, the Soviet Union collapsed.

In the 1990s, a third of the Danube's flow was being taken for human use — an extraction rate that researchers warned would be unsustainable if the vast region served by the river faced continued growth and a prolonged drought. [105]

Today the Danube nations are working toward the river's recovery, and the European Union has pledged $3.3 billion to help. [106]

Ocean Waters

As a last resort, some wealthy nations and cities facing serious shortages of freshwater — including the Persian Gulf states, Israel, Singapore and a handful of cities in California, Florida and Australia — have turned to desalination of the oceans' limitless resources.

More than 10,000 desalting plants were in operation or contracted for construction in January 2005, with a total capacity of 9.6 billion gallons per day. At present, however, desalination plants have the capability to provide just three one-thousandths of daily global freshwater consumption. [107]

Governments usually subsidize up to a third of the consumer cost for desalinated water because it costs many times more than typical urban water service. [108] Improved technology and engineering dropped average desalination costs from $1.60 per 264 gallons in 1990 to about 60 cents in newer plants by 2002. But construction costs have risen recently due to higher steel prices, and operating costs have climbed sharply as energy prices have risen. (Energy costs account for one-third to more than one-half of the expense of desalinating water.) The safe disposal of the brine residue from desalination remains an environmental issue that is still not well researched, according to "The World's Water" report. [109]

Some experts believe desalination will begin to grow at double-digit rates

as water becomes scarce and prices for conventional water rise. [110] For example, California had 20 desalination plants in the construction pipeline in 2006, which could increase California's desalination capacity 100-fold, providing about 7 percent of the freshwater the state used in 2000. ∎

CURRENT SITUATION

Melting Snows

The deep snowpack covering the world's mountains in winter is a renewable gift of nature — melting in spring to restock rivers, lakes, reservoirs, aquifers and eventually flowing back into the ocean. Then evaporating ocean water turns into snowfall in the mountains, repeating the cycle. But the snowpacks have been shrinking — and melting earlier — over the past half-century, upsetting crucial seasonal water flows. [111]

In parts of the Western United States, for example, the snowpack is down to 40 percent of normal. While short-term climate conditions like El Niño are partly to blame, Earth's predicted warming trend is expected to cause dramatic changes in the future. The U.S. Department of Energy's Pacific Northwest National Laboratory has forecast, for instance, that by century's end South America's Andes Mountains will have lost half of their winter snow cover, and ranges in Europe and the U.S. West nearly half. [112] (*See figure, p. 37.*)

"Our main reservoir is snow, and it's going away," says Phillip Mote, a professor of atmospheric sciences at the University of Washington in Seattle. [113]

The early melting is seen as an indication that climate change is already affecting water scarcity. "Some of what's happening with the early snow melts could be due to variations based on ocean circulation," said Gregg Garfin, project manager of the University of Arizona's Institute for the Study of Planet Earth. "But there's a pretty large fraction that can't be explained that way, and we think that's due to increasing temperatures." [114]

Wildfires destroyed hundreds of homes in Malibu, Calif., last October during a severe drought, reigniting the perennial national debate about the folly of building homes in water-scarce areas.

The amount of snow melting into the Colorado River Basin has declined by 10 to 30 percent over the last 30 years, according to Brad Udall, director of the University of Colorado's Western Water Assessment. [115]

Earlier melting has caused unseasonal spring flooding in parts of the West, [116] while a decade of drought has left forests more vulnerable to fires and beetle infestations, said Tom Swetnam, director of the University of Arizona's Laboratory of Tree-Ring Research. "Lots of people think climate change and the ecological responses

are 50 to 100 years away," he said. "But it's not. It's happening now." [117]

In the American Southwest — as in other arid regions of the world — drought is the biggest, most persistent enemy, but scientists are divided over how much of the current drought is due to long-term climate change.

A report by scientists at the Met Office Hadley Centre for Climate Prediction and Research — Britain's weather office — cites evidence that rising emissions of greenhouse gases have exacerbated drought conditions during the past half-century. Greenhouse gases — the carbon dioxide, methane, nitrous oxide and ozone produced from burning fossil fuels — are causing the Earth's temperature to rise by helping to trap the sun's heat, creating a greenhouse effect, according to the Intergovernmental Panel on Climate Change.

"Further research is required," said the Hadley Centre report, not-

ing "the potential seriousness of future climate change impacts if CO_2 emissions continue to increase substantially." [118]

These judgments are still hedged. "It is quite possible that . . . climate change may be having an impact on droughts, not only in the U.S. but around the world," says Michael J. Hayes, director of the National Drought Mitigation Center at the University of Nebraska, Lincoln. But that's not clear yet. "I don't think we can use what is happening today as an argument for climate change. It should open our eyes to the potential impacts that might occur."

Looking further ahead into the century, scientists warn that drought is likely to persist over longer periods, hitting hardest at the world's most-vulnerable arid regions. [119]

A trend toward extreme weather events, ranging from drought and high temperature to violent storms and flooding, is already evident, according to the National Center for Atmospheric Research. [120] Global warming is likely to fuel even more extreme weather, center researchers said. "There's a two-third's chance there will be a disaster," says Nobel laureate Steven Chu, director of the Lawrence Berkeley National Laboratory, "and that's in the best scenario." [121]

The Southwestern United States and other regions appear headed, by mid-century, to a condition of permanent drought caused by global warming, concluded Columbia University's Lamont-Doherty Earth Observatory, after surveying recent studies. "[G]lobal warming not only causes water shortage through early snow melt, which leads to significant water shortage in the summer over the Southwest, but it also aggregates the problem by reducing precipitation," said Mingfang Ting, senior research scientist at Lamont-Doherty and co-author of the survey. [122]

Continued on p. 48

AP Photo/Reed Saxon

At Issue:

Should water be privatized?

TERRY L. ANDERSON
EXECUTIVE DIRECTOR, PERC, SENIOR FELLOW, HOOVER INSTITUTION

WRITTEN FOR *CQ GLOBAL RESEARCHER*
FEBRUARY, 2008

*n*o one washes a rental car" is a truism that suggests that ownership is crucial to stewardship. We also might say, "No one conserves water" for the same reason — too often it's not clear who benefits from conserving water because it's unclear who owns the water. As long as water's cheap, why fix the leaky faucet or switch to an efficient irrigation system?

Making the ownership link is relatively easy, because water is already claimed by someone — either a municipality, individual farmers or a government agency.

In practice, however, claims compete with one another, especially when water is scarce. Miners and farmers on the Western frontier in the 19th century devised the prior-appropriation system, whereby water owners were allowed to resolve conflicts by moving water to higher-valued uses, and trades between farmers have gone on for a century.

The recent drought in the Southeast has raised a red flag about scarcity. The best mechanism for allocating water is to clarify ownership among municipal, agricultural, industrial and environmental users and allow trades. If Atlanta must buy water from lower-valued agricultural users, farmers will have an incentive to save water and sell it, and municipal consumers will face a higher price and thus an incentive to conserve.

Some worry that water markets will put undue burden on the poor while the rich continue enjoying their country club lawns. But the poor could be issued water stamps, akin to food stamps, for buying water. Or suppliers could charge less for minimum amounts of water needed for necessities and increase the price of water for luxuries.

When water rights are allocated through political processes, the poor usually do not get many of the initial rights, forcing them to purchase water if they are to get any. And data from the Chilean water markets suggest that the poor don't fare much better when water is traded on the open market. Perhaps there should be some guaranteed survival quantity of water that is a basic human right.

The problem is not a failure of water markets, but a failure of political allocation, which will not be rectified by preventing water markets from delivering water at a profit to all, regardless of income.

As water scarcity increases in the 21st century, water bureaucracies will bring more conflict, while water markets will foster more cooperation. With this choice, it will be impossible to keep a good water market down

WENONAH HAUTER
EXECUTIVE DIRECTOR
FOOD & WATER WATCH

WRITTEN FOR *CQ GLOBAL RESEARCHER*
FEBRUARY, 2008

*i*n the early 1990s, multinational corporations began to view water services as an important, new profit center — especially in the United States, where 85 percent of water utilities are public. With 1.2 billion people in the developing world lacking access to safe drinking water, the corporations lobbied the World Bank to condition its loans for water services on privatization.

Since then, numerous failed ventures have proven that the cost of privatizing water is too high. It was certainly too high for Tanzania, which terminated a 10-year contract with Biwater after two years of poor management left the government short $3.25 million and the poorest citizens of Dar es Salaam without water. Likewise, massive rate hikes and poor management led Bolivia to end a 40-year contract with Bechtel after only a few months. Similar ventures in Argentina, the Philippines, Indonesia, South Africa and the United Kingdom also have proven unworkable.

In the United States, many municipalities have considered privatization to upgrade their aging systems, but the ventures have been plagued by corruption, high rates, poor service and public outrage. Atlanta terminated a 20-year contract with United Water 16 years early due to bungled emergency responses, boil-water alerts, discolored water and billing difficulties. In 2002, a coalition of citizens' organizations in New Orleans defeated what would have been the largest water-privatization initiative in the United States. Meanwhile, residents of Stockton, Calif., sued the city for failing to perform a proper environmental review of the city's water-privatization contract.

Given this abysmal track record, new solutions are necessary to meet water needs. For example, some U.S. cities have cut costs by improving internal management. Phoenix saved $77 million by working with a labor management team to optimize staffing, organize self-directed work teams and utilize new technology. Similarly, San Diego saved $37 million by developing a more cost-conscious management system.

Safeguarding our water systems is a vital public responsibility. Yet, shockingly, the Environmental Protection Agency estimates that each year we fall $22 billion short of our water infrastructure spending needs. To address this funding gap, we must ensure that public utilities can upgrade and maintain their systems without turning to privatization.

At Food & Water Watch, we support a Clean Water Trust Fund to help ensure that the future of America's water lies in publicly accountable management and secure, clean, affordable water for all.

Continued from p. 46

Wasted Water

Although Atlantans worry that falling levels in Lake Lanier are jeopardizing their drinking water, they might be surprised to learn that up to 18 percent of the city's water is being lost through leaky pipes and wastefulness. [123] In London, the mayor asked the Thames Water company to stanch the 238 million gallons per day being lost through old, leaky water pipes rather than build a costly desalination plant to purify Thames River water. Regulators have

design is to get rid of it as fast as you can: get it off the roof, off the street, into the storm sewers and rush it off into the ocean, never used by anybody." Extravagant water consumption continues in wealthy residential areas and in farming regions where inefficient surface and sprinkler irrigation systems waste up to 25 percent of the water they use. [126] Efficient "drip" irrigation systems — which deliver water directly onto the crops, reducing evaporation to only 5 percent — are used on less than 1 percent of irrigated lands worldwide, largely because of higher equipment costs. [127]

irrigate parks and golf courses. Illinois, Florida, California, Arizona and Ohio reported the largest increases. The U.S. Geological Survey estimated that the amount of graywater used more than doubled between the 1970s and 1995. [129]

Water conservation also is expanding in the construction field. In the United States, Canada, Brazil, India and three-dozen other nations, water-saving green architecture for commercial and government buildings is growing in popularity, but the large-scale use of water-conservation practices by water utilities is still "very rare," says Vickers. [130]

Hirsch says policymakers must recognize that rivers and lakes need sustained flows of water to maintain their long-term environmental viability — and their full range of usefulness. Although this movement is "still in its infancy," at least 70 nations have begun programs to conserve or restore water flows in rivers. [131] For example, a $10 billion project in Florida aims to restore the natural flow of the Kissimmee River, and programs in Australia, Israel, Finland, Thailand, South Africa and Zambia would release flood waters from dams to move sediments downstream and expand plant and animal habitats.

> **"If 12 jumbo jets [full of children] were crashing every day, the world would want to do something about it. They would want to find out why it was happening."**
>
> *— U.N. water expert Brian Appleton, commenting on the deaths each day of 5,000 children due to lack of access to clean water.*

since approved the project because of the urgency of the shortages. [124]

U.S. water systems lose an estimated 15 percent to 25 percent of their water through leakage, and older or poorly maintained networks around the world lose more than 40 percent. However, notes American water conservation consultant Amy L. Vickers, water leakage is "chronically underestimated, ignored, or treated as a tired 'Unsolved Mystery,' " by utilities. [125]

In coastal cities, vast amounts of rainwater and melting snow are "basically lost," flowing into storm drains that flow out to sea, says the U.S. Geological Survey's Hirsch. "The urban

But according to Vickers and other experts, a conservation ethic is beginning to emerge, particularly where water supplies are threatened by drought. For example, aggressive conservation strategies in Boston and Albuquerque, N.M., are reducing systemwide demand by 25 and 18 percent, respectively. "A few other systems, such as New York City, have also realized substantial water savings and wastewater volume reductions that have allowed them to avert major infrastructure expansions," Vickers notes. [128]

In the United States, some states have begun requiring cities to use reclaimed wastewater ("graywater") to

"When given a chance, rivers often heal," write the Global Water Policy Project's Postel and Brian Richter, a staff director at the Nature Conservancy, in *Rivers for Life*. [132] ∎

OUTLOOK

Thirst and Hunger

The world's population, now about 6 billion, is likely to jump by more than a third by 2050, reaching nearly

9 billion, according to the U.N. [133] Such a large increase will cause not only thirst but also hunger, says the International Water Management Institute. The average European uses about 13 gallons of water a day for drinking, cooking and sanitation. But the food an individual consumes in a typical day requires 800 to 900 gallons to grow. [134]

"The world needs roughly 70 times more water to produce food than it needs for cities," says Rijsberman, former director of the International Water Management Institute in Sri Lanka. [135]

Since 1950, water withdrawals for human use have tripled and irrigated cropland doubled. Today, despite important increases in farming productivity, 850 million people live in areas where adequate food supplies are at risk — two-thirds of them in South Asia and sub-Saharan Africa, where the impact of climate change on food production is expected to be worst. [136]

The productivity of irrigated cropland has increased dramatically in the past half-century, according to the World Bank. The production of rice and wheat, for instance, increased 100 percent and 160 percent in that period, respectively, with no increase in water use per bushel. "However, in many (river) basins, water productivity remains startlingly low," the bank reports. Without greater agricultural productivity or major shifts in farming locations, the amount of water needed for farming will jump 70 to 90 percent by 2050, according to the assessment. [137]

As food requirements continue to rise, the increase in irrigation has slowed as underground water levels have begun to recede. Farmers also face growing competition for water from industry. [138]

Rijsberman predicts the average price of water used in agriculture worldwide could increase by two to three times in the coming decades, inflating

AP Photo/Carlos Osorio

National Geographic/Getty Images/James L. Stanfield

Spraying vs. Dripping

Due to evaporation, irrigation systems like this one in Lakefield Township, Mich. (top), waste up to 25 percent of the water they distribute. More expensive drip irrigation systems, like this one in Israel (bottom), lose only about 5 percent to evaporation but are used on less than 1 percent of irrigated lands worldwide.

global food prices. In addition, industries and power producers can outbid farmers for scarce water. If irrigated harvests are cut back through a lack of water or because water is diverted to industrial use, world grain prices will rise even more. [139]

Policymakers still must resolve a major question about water pricing:

Should it be priced competitively, according to its value, like wheat, rice and other food commodities grown with water? A handful of governments have done just that: Chile allows landowners with water on or under their property to trade water rights to the highest bidders. Mexico, several Australian states and California also have water-trading programs. [140] In Texas, the flamboyant oil trader T. Boone Pickens has created a company, Mesa Water, to buy water from landowners above the Ogallala aquifer, to sell to water-short Texas cities. [141]

But many experts think trading water as a commodity is a non-starter for most governments. "There is no movement in the real world, with elected officials," says U.S. water-law expert Robert Glennon. "Water pricing is the third rail of water politics."

Another option, supported by the World Bank and others, is trading "virtual" water. That occurs when a country with scarce water or poor agricultural land concentrates on developing export goods to earn the money needed to import food from water-rich nations with productive, low-cost food producers. Such trades, which would require a lowering of agricultural trade barriers, could bring down food production costs in water-poor countries and help reduce global water consumption, according to a World Bank report. Wheat grown in India, for example, consumes four times more water than wheat grown in France. By importing maize rather than growing it, Egypt reduces its national water consumption by 5 percent. [142]

But importing "virtual" water also has a downside. "In Morocco, for example, one study showed that while the nation as a whole would benefit from agricultural trade liberalization, those benefits would be concentrated on the urban population; farmers — particularly poor farmers — stood to lose," said a World Bank report. [143] For that reason, critics of expanded international trade oppose the "virtual" approach.

Other advocates call for greater reliance on rain-fed farming. Just over half of the world's food, by value, is produced using rainfall, but this sector — dominated by poor rural farmers — has traditionally been ignored by food producers and governments in favor of major irrigation strategies.

"Upgrading rain-fed areas has high potential both for food production and for poverty alleviation," says the International Water Management Institute. Increasing small-scale rainwater storage with supplemental irrigation and better land management could produce quick output gains in these areas. [144]

If farmers continue to depend on irrigation for 40 percent of their water, producing an acceptable diet for 2.4 billion more people in the next 30 years would require another 20 Nile Rivers or 97 Colorado Rivers, says water expert Postel. "It is not at all clear where this water is to come from." [145] ∎

Notes

[1] Charles C. Mann, "The Rise of Big Water," *Vanity Fair*, May 2007; Reuters, "China drought threatens water supply for millions," March 28, 2007.

[2] Xinhua News Agency, "Drought leaves nearly 250,000 short of drinking water in Guangdong," *People's Daily Online*, Dec. 13, 2007; http://english.people.com.cn/90001/90776/6320617.html.

[3] Jonathan Watts, "Dry, Polluted, Plagued by Rats: The Crisis in China's Greatest Yangtze River," *The Guardian* (Britain), Jan. 17, 2008; http://chinaview.wordpress.com/category/environment/drought/.

[4] Xinhua News Agency, "Climate change blamed as drought hits 100,000 at China's largest freshwater lake," *People's Daily Online*, Dec. 14, 2007; http://english.people.com.cn/90001/90776/6321329.html.

[5] Chris O'Brien, "Global Warming Hits China," Forbes.com, Jan. 6, 2008; www.forbes.com/opinions/2008/01/04/poyang-lake-china-oped-cx_cob_0106poyang.html.

[6] Peter H. Gleick, "Time to Rethink Large International Water Meetings," *The World's Water 2006-2007*, Island Press, p. 182; www.world-water.org/.

[7] David Molden, ed.; "Summary," *Water for Food, Water for Life: A Comprehensive Assessment of Water Management in Agriculture*, International Water Management Institute, p. 10; www.iwmi.cgiar.org/assessment/files_new/synthesis/Summary_SynthesisBook.pdf.

[8] See "Summary, Human Development Report 2006: Beyond scarcity: Power, poverty and the global water crisis," United Nations Development Programme, p. 26, http://hdr.undp.org/en/media/hdr2006_english_summary.pdf.

[9] Carmen Revenga, *et al.*, "Executive Summary, Pilot Analysis of Global Ecosystems: Freshwater Systems," World Resources Institute, 2000, pp. 4, 26; www.wri.org/publication/pilot-analysis-global-ecosystems-freshwater-systems.

[10] Stacy Shelton, "Lake Lanier hits lowest point since its construction," *The Atlanta-Journal Constitution*, Nov. 19, 2007; www.ajc.com/metro/content/metro/stories/2007/11/19/lanierlowweb_1120.html?cxntlid=homepage_tab_newstab. For background, see Mary H. Cooper, "Water Shortages," *CQ Researcher*, Aug. 1, 2003, pp. 649-672.

[11] M. Falkenmark, *et al.*, "On the Verge of a New Water Scarcity: A call for good governance and human ingenuity," Stockholm International Water Institute (SIWI) Policy Brief, 2007, p. 17. For background, see Colin Woodard, "Curbing Climate Change," *CQ Global Researcher*, February 2007, pp. 27-50.

[12] "Water and Terrorism," *The World's Water 2006-2007, op. cit.*, p. 1.

[13] Environmental Protection Agency, "Water Sentinel Initiative," www.epa.gov/watersecurity/pubs/water_sentinel_factsheet.pdf.

[14] Met Office Hadley Centre, "Effects of climate change in developing countries," November 2006, pp. 2-3, www.metoffice.gov.uk/research/hadleycentre/pubs/brochures/COP12.pdf; Michael McCarthy, "The Century of Drought," *The Independent* (London), Oct. 4, 2006, p. 1.

[15] McCarthy, *ibid.*, p. 1.

[16] World Water Council.

[17] Quoted in "Billions without clean water," March 14, 2000, BBC, http://news.bbc.co.uk/2/hi/676064.stm.

[18] David Katz, "Going with the Flow," *The World's Water 2006-2007, op. cit.*, pp. 30-39.

19 Data Table 5, Access to Water Supply and Sanitation by Region, *The World's Water 2006-2007, op. cit.,* p. 258.

20 "Synthesis of the 4th World Water Forum," August 2006, p. 23-24, www.worldwaterforum4.org.mx/files/report/SynthesisoftheForum.pdf. For background on Millennium Development Goals, see www.un.org/millenniumgoals and "U.N. Fact Sheet on Water and Sanitation," 2006, www.un.org/waterforlifedecade/factsheet.html.

21 "Human Development Report 2006," *op. cit.,* p. 8; http://hdr.undp.org/en/reports/global/hdr2006/.

22 Australian Broadcasting Corp., "Issues in Science and Technology," transcript, Sept. 22, 2007.

23 Frank Rijsberman, Charlotte Fraiture and David Molden, "Water scarcity: the food factor," *Issues in Science and Technology,* June 22, 2007.

24 *Ibid.*

25 Sharon P. Nappier, Robert S. Lawrence, Kellogg J. Schwab, "Dangerous Waters," *Natural History,* November 2007, p. 48.

26 Marq de Villiers, *Water: The Fate of Our Most Precious Resource* (1999), p. 267.

27 "World hit by water shortage," *Birmingham Post,* Aug. 21, 2006, p. 10; http://icbirmingham.icnetwork.co.uk/birminghampost/news/tm_method=full%26objectid=17597105%26siteid=50002-name_page.html.

28 "Water Resources Sector Strategy, Strategic Directions for World Bank Engagement," World Bank, 2004, p. 5, www-wds.worldbank.org/external/default/WDSContentServer/WDSP/IB/2004/06/01/000090341_20040601150257/Rendered/PDF/28114.pdf.

29 de Villiers, *op. cit.,* pp. 176-177.

30 Frank R. Rijsberman, "1st Asia-Pacific Water Summit," MaximsNews Network, Oct. 8, 2007; ww.abc.net.au/7.30/content/2006/s1716766.htm.

31 Diane Raines Ward, *Water Wars: Drought, Flood, Folly and the Politics of Thirst* (2002), p. 171.

32 Nappier, *et al., op. cit.*

33 "Human Development Report," *op. cit.,* p. 176. Also see Meena Palaniappan, Emily Lee and Andrea Samulon, "Environmental Justice and Water," *The World's Water 2006-2007, op. cit.,* p. 125.

34 "Human Development Report," *ibid.*

35 Palaniappan, *et al., op. cit.*

36 Marc Reisner, *Cadillac Desert, the American West and its Disappearing Water* (1986), p. 10.

37 *Ibid.,* p. 437.

38 *Ibid.,* pp. 10-11; also see Robin Clarke and Jannet King, *The Water Atlas* (2004).

39 "Approaches to Private Participation in Water Services," World Bank, 2006, p. 213.

40 Daniel Yergin and Joseph Stanislaw, *Commanding Heights: The Battle for the World Economy* (2004).

41 Juan Forero, "Multinational Is Ousted, but Local Ills Persist," *The New York Times,* Dec. 15, 2005, p. 1; also, Public Citizen, "Water Privatization Case Study: Cochabamba, Bolivia," pp. 1-2, www.tradewatch.org/documents/Bolivia_(PDF).PDF; and Bechtel Corp. statement, "Cochabamba and the Aquas del Tunari Consortium," www.bechtel.com/assets/files/PDF/Cochabambafacts0305.pdf.

42 "European environment giant Veolia to increase investment in China to $2.5 billion by 2013," Xinhua News Agency, Nov. 1, 2007.

43 *Pinsent Masons Water Yearbook 2007-08,* p. xii, www.pinsentmasons.com/media/1976627452.pdf.

44 Quoted in Jeff Fleischer, "Blue Gold: An Interview with Maude Barlow," *Mother Jones,* Jan. 14, 2005, www.motherjones.com/news/qa/2005/01/maude_barlow.html.

45 "Economic Failures of Private Water Systems," Food & Water Watch, Dec. 2007, www.foodandwaterwatch.org/water/waterprivatization/usa/Public_vs_Private.pdf.

46 Scott Wallsten and Katrina Kosec, "Public or Private Drinking Water?" AEI-Brookings Joint Center for Regulatory Studies, March 2005, pp. 2, 7; www.reg-markets.org/publications/abstract.php?pid=919.

47 Felipe Barrera-Osorio and Mauricio Olivera, "Does Society Win or Lose as a Result of Privatization?" Inter-American Development Bank, Research Network Working Paper #R-525, March 2007, p. 19.

48 *Ibid.,* p. 21.

49 Ward, *op. cit.,* pp. 206-207.

50 *Pinsent Masons Water Yearbook, op. cit.,* p. 3.

51 *Ibid.,* p. 5.

52 Paul Constance, "The Day that Water Ran Uphill," *IDB America,* Inter-American Development Bank, Dec. 9, 2007, www.iadb.org/idbamerica/index.cfm?thisid=3909&lanid=1.

53 "Human Development Report," *op. cit.,* p. 92.

54 *Ibid.,* p. 179; Ward, *op. cit.,* p. 210.

55 Malcolm Scully, "The Politics of Running Out of Water," *The Chronicle of Higher Education,* Nov. 17, 2000.

56 Peter H. Gleick, "Environment and Security," *The World's Water 2006-2007, op. cit.,* p. 189.

57 "Human Development Report," *op. cit.,* p. 221.

58 Benny Morris, *Righteous Victims* (2001), pp. 303-304.

59 Ward, *op. cit.,* p. 174.

60 *Ibid.,* p. 85.

61 *Ibid.,* p. 192.

62 Undala Alam, letter to the *Financial Times,* April 1, 2006, p. 6. Also see www.transboundarywaters.orst.edu/publications/related_research/Alam1998.pdf.

63 Gleick, "Environment and Security," *op. cit.,* p. 189.

64 *Security and the Threat of Climate Change* (2007), CNA Corp., p. 3; http://securityandclimate.cna.org/.

65 *Ibid.,* p. 20.

66 *Ibid.,* p. 22.

67 "Human Development Report," *op. cit.,* p. 216; Clarke and King, *op. cit.,* p. 79.

68 *Ibid.,* p. 47.

69 "Human Development Report," *op. cit.,* p. 218.

70 CNA, *op. cit.,* p. 14.

71 "Dams and Development: A New Framework for Decision Making," World Commission on Dams, Nov. 16, 2000, p. 8; www.dams.org/report/wcd_overview.htm.

72 Xinhua News Agency, "China's Grand Canal Queues for World Heritage Status," July 6, 2004, www.china.org.cn/english/culture/100401.htm.

73 Terry S. Reynolds, *Stronger Than a Hundred Men: A History of the Vertical Water Wheel* (1932), pp. 32, 142.

74 Jill Jonnes, *Empires of Light: Edison, Tesla, Westinghouse, and the Race to Electrify the World* (2004).

75 Revenga, *et al., op. cit.,* p. 12.

76 World Commission on Dams, *op. cit.,* p. 9.

77 *Ibid.,* p. ii.

78 Ward, *op. cit.,* p. 51.

79 *Ibid.,* p. 46.

80 Lin Yang, "China's Three Gorges' Dam Under Fire," *Time,* Oct. 12, 2007, www.time.com/time/world/article/0,8599,1671000,00.html. Also see Bruce Kennedy, "China's Three

Gorges Dam," CNN.com, 2001, www.cnn.com/SPECIALS/1999/china.50/asian.superpower/three.gorges/.

[81] Jonathan Watts, "Three Georges Dam risk to environment, says China," *The Guardian*, Sept. 27, 2007.

[82] Xinhua Financial News, "Chinese Government Fights Back in Defense of Three Gorges Dam," Nov. 27, 2007; Jim Yardley, "China vigorously defends the Three Gorges Dam project," *The International Herald Tribune*, Nov. 28, 2007, p. 3.

[83] Revenga, *et al.*, *op. cit.*, p. 28.

[84] "Synthesis of the 4th World Water Forum," *op. cit.*

[85] Ward, *op. cit.*, p. 47.

[86] *Ibid.*, p. 187.

[87] de Villiers, *op. cit.*, p. 59. Also see "Islamic Water Management and the Dublin Statement," The International Development Research Center, Canada, www.idrc.ca/en/ev-93949-201-1-DO_TOPIC.html.

[88] Katz, *op. cit.*, p. 37.

[89] Robert Glennon, *Water Follies: Groundwater Pumping and the Fate of America's Fresh Waters* (2002), p. 16.

[90] *Ibid.*, p. 14.

[91] "Human Development Report," *op. cit.*, p. 185.

[92] *Ibid.*, p. 214.

[93] Ward, *op. cit.*, p. 188.

[94] Sandra Postel and Brian Richter, *Rivers for Life, Managing Water for People and Nature* (2003), p. 168.

[95] "Human Development Report," *op. cit.*, p. 103.

[96] *Ibid.*, p. 37.

[97] *Ibid.*, p. 42.

[98] *Ibid.*, pp. 45-46.

[99] *Ibid.*, p. 33.

[100] *Ibid.*, p. 29.

[101] *Ibid.*, p. 30, citing Frances Bell and Robert Millward, "Public Health Expenditures and Mortality in England and Wales, 1870-1914," pp. 221-249.

[102] Palaniappan, *et al.*, *op. cit.*, p. 131, citing the World Water Council, "Financing Water for All," Global Water Partnership, March 2004.

[103] *Ibid.*, p. 139.

[104] de Villiers, *op. cit.*, p 171.

[105] *Ibid.*, pp. 172-177.

[106] *Ibid.*, p. 174; "Human Development Report," *op. cit.*, p. 219; "Future Danube Flood Actions Depend On International Cooperation," Commission for the Protection of the Danube, April 21, 2006; www.icpdr.org/icpdr-pages/pr20060421_danube_flood.htm. For background, see Brian Beary, "The New Europe," *CQ Global Researcher*, August, 2007.

[107] Peter H. Gleick, Heather Cooley, Gary Wolff, "With a Grain of Salt: An Update on Seawater Desalination," *The World's Water 2006-2007*, *op. cit.*, p. 55.

[108] *Ibid.*, pp. 68-70.

[109] *Ibid.*, pp. 78-79.

[110] *Ibid.*, p. 161. General Electric Co. projects an annual growth rate of 9 to 14 percent, growing from $4.3 billion in annual desalination expenditures in 2005 to $14 billion in 2014.

[111] P. W. Mote, *et al.*, "Declining Mountain Snowpack in Western North America," *Bulletin of the American Meteorological Society 86*, January 2005, pp. 49-48.

[112] "New Century of Thirst for World's Mountains," Pacific Northwest National Laboratory, May 18, 2006, www.pnl.gov/news/release.asp?id=158.

[113] Eric Bontrager, "West will need to revisit water, land management in light of global warming, experts say," *Land Letter*, Sept. 28. 2006; www.eenews.net/ll/ (subscription required).

[114] Shaun McKinnon, "Southwest Could Become Dust Bowl, Study Warns," *The Arizona Republic*, April 6, 2007, p. 1; www.azcentral.com/arizonarepublic/news/articles/0406climate-report0406.html.

[115] Bontrager, *op. cit.*

[116] *Ibid.*

[117] Shaun McKinnon, "Snow runoff: What's at Stake," *The Arizona Republic*, Nov. 25, 2007, p. 8; and Stephen Saunders and Maureen Maxwell, "Less Snow, Less Water: Climate Disruption in the West," The Rocky Mountain Climate Organization, September 2005, pp. 2, 9, 19, www.rockymountainclimate.org/website%20pictures/Less%20Snow%20Less%20Water.pdf.

[118] "UK Government: Global drought in the 21st century," M2 Presswire, Oct. 26, 2006; www.continuitycentral.com/news02870.htm.

[119] "Fourth Assessment Report — Climate Change 2007: Synthesis Report, Summary for Policymakers," Intergovernmental Panel on Climate Change, Nov. 17, 2007, p. 8; www.ipcc.ch/pdf/assessment-report/ar4/syr/ar4_syr_spm.pdf.

[120] Claudie Tebaldi, Katharine Hayhoe, Julie M. Arblaster and Gerald A. Meehle, "Going to Extremes," Institute for the Study of Society and Environment, National Center for Atmospheric Research, 2006, p. 22; www.cgd.ucar.edu/ccr/publications/tebaldi_extremes.pdf.

[121] Jon Gertner, "The Future is Drying Up," *New York Times Magazine*, Oct. 21, 2007; www.nytimes.com/2007/10/21/magazine/21water-t.html?_r=1&oref=slogin.

[122] "New Study Shows Climate Change Likely to Lead to Periods of Extreme Drought in Southwest North America," Lamont-Doherty Earth Observatory, April 6, 2007, www.ldeo.columbia.edu/news-events/new-study-shows-climate-change-likely-lead-periods-extreme-drought-southwest-north-ameri.

[123] "A Review of Water Conservation Planning for the Atlanta, Georgia Region, August 2006, prepared for the Florida Department of Environmental Protection," Pacific Institute, p. 23, www.pacinst.org/reports/atlanta/atlanta_analysis.pdf.

[124] "Report to the Secretaries of State for Communities and Local Government and Food and Rural Affairs," *The Planning Inspectorate*, Sept. 29, 2006, p. 7-8, www.communities.gov.uk/documents/planningandbuilding/pdf/319931, and www.thameswater.co.uk/UK/region/en_gb/

About the Author

Peter Behr is a Washington freelance writer who has worked for more than 25 years at *The Washington Post*, where he reported on energy issues and served as business editor. A former Nieman Fellow at Harvard University, Behr was a public policy scholar at the Woodrow Wilson International Center for Scholars and is now writing a book about the U.S. electric power grid. His report on "Energy Nationalism" appeared in the July 2007 issue.

content/News/News_001394.jsp?SECT=Section
_Homepage_000431.

[125] Amy L. Vickers, "The Future of Water Conservation: Challenges Ahead," The Universities Council on Water Resources (UCOWR), p. 52, www.ucowr.siu.edu/updates/pdf/V114_A8.pdf. Also see Marcia Clemmitt, "Aging Infrastructure," *CQ Researcher*, Sept. 28, 2007, pp. 793-816.

[126] Clarke and King, *op. cit.*, p. 87.

[127] "Re-engaging in Agricultural Water Management — Challenges and Options," The International Bank for Reconstruction and Development/The World Bank, 2006, p. 3, web.worldbank.org/WBSITE/EXTERNAL/TOPICS/EXTARD/0,,contentMDK:20858509~pagePK:210058~piPK:210062~theSitePK:336682,00.html.

[128] Vickers, *op. cit.*, p. 52.

[129] Harriet Emerson and Mohamed Lahlou, "Conservation: It's the Future of Water," National Drinking Water Clearinghouse, www.nesc.wvu.edu/ndwc/ndwc_conservarticlesetc.htm/harrietarticle.html.

[130] "Fact Sheet," U.S. Green Building Council, www.usgbc.org/DisplayPage.aspx?CMSPageID=222.

[131] Katz, *op. cit.*, p. 32.

[132] Postel and Richter, *op. cit.*, p. 201.

[133] "World Population Prospects," U.N. Department of Economic and Social Affairs, 2006, http://esa.un.org/unpp/.

[134] Molden, *op. cit.*, p. 1; *Birmingham Post*, *op. cit.*, p. 10.

[135] Quoted in Kerry O'Brien, "Water scarcity 'due to agriculture,' " Australian Broadcasting Corp. Transcripts, Aug. 16, 2006; www.abc.net.au/7.30/content/2006/s1716766.htm.

[136] Molden, *op. cit.*, pp. 7-8.

[137] *Ibid.*, p. 14.

[138] "Re-engaging in Agricultural Water Management," *op. cit.*, p. 38.

[139] Rijsberman, *op. cit.*

[140] Postel and Richter, *op. cit.*, pp. 112-117.

[141] Jim Getz, "Kaufman County won't vote on Pickens' freshwater district: But Roberts County calls election on Pickens' pitch for freshwater district," *The Dallas Morning News*, Sept. 5, 2007.

[142] "Re-engaging in Agricultural Water Management," *op. cit.*, p. 102.

[143] *Ibid.*, p 103.

[144] International Water Management Institute, *op. cit.*, p. 10.

[145] de Villiers, *op. cit.*, p 24.

FOR MORE INFORMATION

The CNA Corp., 4825 Mark Center Drive, Alexandria, VA 22311; (703) 824-2000; www.cna.org. A nonprofit research organization that operates the Center for Naval Analyses and the Institute for Public Research, concentrating on security, defense and other government-policy issues.

Food and Water Watch, 1616 P St., N.W., Suite 300, Washington, DC 20036; (202) 683-2500; www.foodandwaterwatch.org. A liberal research and advocacy organization focused on water resources, food security, sanitation and globalization issues.

Intergovernmental Panel on Climate Change, C/O World Meteorological Organization, 7bis Avenue de la Paix, C.P. 2300, CH-1211 Geneva 2, Switzerland; 41-22-730-8208/84; www.ipcc.ch. Intergovernmental research body.

International Water Management Institute, 127, Sunil Mawatha, Pelawatte, Battaramulla, Colombo, Sri Lanka; 94-11 2880000, 2784080; www.iwmi.cgiar.org. Research group supported by 60 governments, private foundations and international organizations.

The Met Office Hadley Centre, Met Office, FitzRoy Road, Exeter, Devon, EX1 3PB, United Kingdom; 44 (0)1392 885680; www.metoffice.gov.uk/research/hadley-centre. Britain's official center for climate-change research.

Pacific Institute, 654 13th St., Preservation Park, Oakland, CA 94612; (510) 251-1600; www.pacinst.org. A nonpartisan think tank studying development, environment and security issues.

Property and Environment Research Center, 2048 Analysis Dr., Suite A, Bozeman, MT 59718; (406) 587-9591; www.perc.org. A pro-market research and advocacy group.

Public Citizen, 1600 20th St., N.W., Washington, DC 20009; (202) 588-1000; www.citizen.org. A liberal consumer-advocacy group.

Stockholm International Water Institute, Drottninggatan 33, SE — 111 51 Stockholm, Sweden; 46 8 522 139 60; www.siwi.org. A research organization affiliated with the Swedish government.

U.N. Human Development Office, 304 E. 45th St., 12th Floor, New York, NY 10017; (212) 906-3661; hdr.undp.org/en/humandev. Publishes an annual report on health, economic and other social conditions.

U.S. Geological Survey, 12201 Sunrise Valley Dr., Reston, VA 20192; (888) 275-8747; www.usgs.gov. The government's mapping agency and research center on water resources, geology, natural hazards and other physical sciences.

World Bank, 1818 H St., N.W., Washington, DC 20433; (202) 473-1000; www.worldbank.org. Provides technical and financial assistance to developing countries.

World Water Council, Espace Gaymard, 2-4 place d'Arvieux, 13002 Marseille, France; 33 491 994100; www.worldwatercouncil.org. An international research and advocacy group of government and international agency officials, academics and corporate executives; sponsors World Water Forum every three years.

Bibliography

Selected Sources

Books

Clarke, Robin, and Jannet King, *The Water Atlas*, The New Press, 2004.
Editors at the World Meteorological Organization present a visual primer on water scarcity, sanitation shortfalls and climate impact on water resources.

de Villiers, Marq, *Water: The Fate of Our Most Precious Resource*, Stoddard Publishing Co., 1999.
A Canadian journalist provides a global overview of challenges confronting the world's water supplies.

Glennon, Robert, *Water Follies: Groundwater Pumping and the Fate of America's Fresh Waters*, Island Press, 2002.
An attorney and water-policy expert advocates new policies to preserve Western U.S. aquifers.

Olivera, Oscar, *Cochabamba! Water Rebellion in Bolivia*, South End Press, 2004.
The leader of the Bolivian protest against water privatization gives his side of the conflict.

Postel, Sandra, and Brian Richter, *Rivers for Life, Managing Water for People and Nature*, Island Press, 2003.
Experts at the Global Water Policy Project in Massachusetts (Postel) and The Nature Conservancy (Richter) chronicle the campaign to restore environmental conditions in threatened rivers.

Reisner, Marc, *Cadillac Desert, the American West and Its Disappearing Water*, Penguin Books, 1986.
This award-winning classic by a former Natural Resources Defense Council expert critiques federal land and irrigation policies and their impact on water use in the West.

Ward, Diane Raines, *Water Wars: Drought, Flood, Folly and the Politics of Thirst*, Riverhead Books, 2002.
An environmental writer reviews controversial global policies affecting dams, water treaties and other water-resource issues.

Articles

Mann, Charles C., "The Rise of Big Water," *Vanity Fair*, May 2007.
A correspondent for *Science* and *The Atlantic Monthly* explores the controversy over privatization programs for water and sanitation worldwide.

Reports and Studies

"Approaches to Private Participation in Water Services: A Tool Kit," The World Bank, 2006, http://publications.world-bank.org/ecommerce/catalog/product?item_id=4085442.
The world's major foreign-aid lender provides lessons learned from water-privatization efforts.

"Beyond scarcity: Power, poverty and the global water crisis," United Nations Development Programme, U.N. Human Development Report 2006, http://hdr.undp.org/en/reports/global/hdr2006.
This detailed review of worldwide water and sanitation challenges includes case studies of successes and failures.

"Comprehensive Assessment of Water Management in Agriculture: Water for Food, Water for Life," International Water Management Institute, 2006, www.fao.org/nr/water/docs/Summary_SynthesisBook.pdf.
A consulting group based in Sri Lanka reports on the impact of water scarcity on global irrigation and food production.

"Dams and Development: A New Framework for Decision Making," World Commission on Dams, 2000, www.dams.org/report.
Water experts, educators and government officials assess issues surrounding major dam construction and operations.

"IPCC Fourth Assessment Report," Intergovernmental Panel on Climate Change, 2007, www.ipcc.ch/ipccre-ports/ar4-syr.htm.
A scientific panel sponsored by the World Meteorological Organization and the United Nations Environment Programme issues its most recent outlook on climate change threats.

"National Security and the Threat of Climate Change," CNA Corp., December 2007, http://securityandclimate.cna.org/report/.
A panel of retired U.S. generals and admirals forecasts security issues that will emerge as a result of climate change.

"Synthesis of the 4th World Water Forum, Mexico City," World Water Council, 2006, www.worldwatercouncil.org/index.php?id=1386.
An international committee presents a summary of its fourth conference on water-resources issues.

Gleick, Peter H., *et al.*, "The World's Water 2006-2007, The Biennial Report on Freshwater Resources," The Pacific Institute for Studies in Development, Environment, and Security, 2006, www.pacinst.org/publications/worlds_water/2006-2007/index.htm.
The institute's latest review of global water-resource issues includes chronologies of water conflicts and analyses of strategies for sustainable freshwater resource management.

The Next Step:

Additional Articles from Current Periodicals

Conflicts

"Climate Change Will Cause New Wars — UN," *Canberra Times*, June 22, 2007.

Changing patterns of rainfall and fights over food production could lead to potential conflicts in Africa's Sahel region and in East Asia, according to U.N. Environment Programme Executive Director Achim Steiner.

"Managing the Water Wars of the Future," *Indian Express*, May 12, 2007.

Overcoming the environmental and political difficulties of water scarcity will help manage the conflicts stemming from a lack of access to clean and safe water.

"Water is Running Out: How Inevitable Are International Conflicts?" Integrated Regional Information Networks (United Nations), Oct. 23, 2006.

Water resources are decreasing amid a growing world population and increasing water consumption, making water-related conflicts inevitable.

Baldauf, Scott, "Climate Change Escalates Darfur Crisis," *The Christian Science Monitor*, July 27, 2007, p. 1.

Competition for water in Sudan's Darfur region has long sparked conflict in the area, and is one of the main causes of the current fighting.

Moving Water

"Bangladesh, India to Hold Talks on Sharing River Water," PTI News Agency (India), July 20, 2007.

India and Bangladesh are expected to discuss water-sharing and diversion issues involving eight major trans-boundary rivers.

"China, Spain Vow to Strengthen Water Resources Utilization Cooperation," Xinhua News Agency (China), Oct. 14, 2007.

Chinese Vice Premier Hui Liangyu has promised to strengthen bilateral cooperation with Spain over water resources.

Jones, Tim, "Great Lakes Key Front in Water Wars," *Chicago Tribune*, Oct. 28, 2007, p. A1.

Eight Great Lakes-area states and two Canadian provinces have proposed a regional water compact that would prevent water diversions from the immediate region.

Privatization

Mitra, Barun, "Water Scarcity in Plenty," *Business Day* (South Africa), March 23, 2006.

Government intervention has contributed significantly to water scarcity, while private initiatives have sustained local markets and improved access for millions.

Sasa, Mabasa, "Privatisation of Water Not the Answer," *The Herald* (Zimbabwe), Dec. 29, 2007.

Privatizing water supplies will mean corporations report to the World Bank rather than to state governments.

Seno, Alexandra A., "Very Liquid Assets," *Newsweek*, Aug. 20, 2007.

Savvy asset-management companies have turned concerns over water shortage into burgeoning investment-fund businesses.

Wallach, Jason, "Privatizing Water and the Criminalization of Protest," NACLA News (New York), July 24, 2007.

Despite corruption in El Salvador's government, citizens want the water system to be under state control rather than in private hands.

Water Supply

Doyle, Alister, "Increasing Agricultural Use a Drain on World Water Supply," *The Washington Post*, Aug. 21, 2006, p. A2.

Rising demand for irrigation in the production of food and biofuels will likely aggravate water scarcity.

Fan, Grace, "Abundant Water Makes Brazil Last Agriculture Frontier, as Scarcity Looms in Rest of World," The Associated Press, Aug. 24, 2006.

Abundant freshwater will help Brazil dominate global food supplies for the next several decades while water scarcity threatens the economies of other countries.

Girma, Biruk, "Poverty, Not Scarcity, to Blame for Lack of Water, Poor Sanitation — Report," *Daily Monitor* (Ethiopia), Nov. 14, 2006.

Millions of people lack access to safe water and sanitation due to inequality and poverty, according to the U.N.

CITING *CQ GLOBAL RESEARCHER*

Sample formats for citing these reports in a bibliography include the ones listed below. Preferred styles and formats vary, so please check with your instructor or professor.

MLA STYLE

Flamini, Roland. "Nuclear Proliferation." CQ Global Researcher 1 Apr. 2007: 1-24.

APA STYLE

Flamini, R. (2007, April 1). Nuclear proliferation. *CQ Global Researcher*, 1, 1-24.

CHICAGO STYLE

Flamini, Roland. "Nuclear Proliferation." *CQ Global Researcher*, April 1, 2007, 1-24.

Voices From Abroad:

LOIC FAUCHON
President, World Water Council

Privatization contributes to exodus

"A lot of poor people are leaving their countries to go to rich countries. Isn't it preferable, isn't it cheaper, to pay so that these people have water, sewage [treatment] and energy . . . [so they can] stay in their own countries?"

The Associated Press, March 2006

FADEL KAWASH
Head, Palestinian Water Authority

Justice provides water to all

"In the Middle East, water is a political issue. Israel's occupation of the West Bank . . . presents a major obstacle to development projects especially in the water sector. There is enough water in the West Bank if there is justice and sharing."

Agence France-Presse, March 2006

BAN KI-MOON
Secretary General, United Nations

Water is crucial to all problems

"Safe drinking water and adequate sanitation are crucial for poverty reduction, crucial for sustainable development, and crucial for achieving any and every one of the Millennium Development Goals."

Speech at opening of U.N. water exhibit, New York, October 2007

KRISTA HANSON
Program Director, Committee in Solidarity with the People of El Salvador

Privatization increases rates

"If we take the electricity sector and telecommunications as guides, privatization has meant higher rates, lower quality, less access and less sovereign control over public services."

NACLA News (New York), July 2007

ABEL MAMANI
Minister of Water, Bolivia

Water: an inalienable right

"Access to fresh drinking water for everyone, in all our countries, is a fundamental right. For us, water is life. Establishing a right to water, therefore is another way of recognising the right to life already enshrined by the United Nations. Recognise water as a human right!"

World Water Assembly, Brussels, March 2007

CAROLINE SAINT-MLEUX
Head, Care International, Iriba, Chad

Water can help Darfur

"You have to have the warring parties talk about a common need, and after that you might have them talk about something else that would start giving other solutions to the conflict."

The Christian Science Monitor, July 2007

MUHAMMAD SAEED AL-KINDI
Minister of Environment and Water, United Arab Emirates

Droughts are shifting the focus

"The concern has recently taken a greater importance with the increased frequency and prolonged periods of drought being experienced worldwide. Consequently, this century will see a greater emphasis being placed on desalinated water and the re-use of treated water, and the extent to which these important resources will contribute to the overall supply of water globally."

Agence France-Presse, April 2007

PRINCE WILLEM-ALEXANDER OF THE NETHERLANDS
Chairman, U.N. Advisory Board on Water and Sanitation

Proper sanitation improves dignity

"Clean water and sanitation are not only about hygiene and disease, they're about dignity, too. Relieving yourself in hazardous places means risking everything from urological disease to harassment and rape. Many examples show that self-esteem begins with having a safe and proper toilet facility."

Speech at U.N. International Year of Sanitation global launch, November 2007

FRANK RIJSBERMAN
Director General, International Water Management Institute

Old models won't work

"The last 50 years of water management practices are no model for the future when it comes to dealing with water scarcity. We need radical change in the institutions and organizations responsible for managing our earth's water supplies and a vastly different way of thinking about water management."

The Associated Press, August 2006

FUTURE PROSPECTS...

EMPTY...

Olle Johansson/Sweden

RAPID URBANIZATION

BY JENNIFER WEEKS

Excerpted from the CQ Global Researcher. Jennifer Weeks. (April 2009). "Rapid Urbanization." *CQ Global Researcher*, 91-118.

Rapid Urbanization

BY JENNIFER WEEKS

THE ISSUES

India's most infamous slum lives up to its reputation. Located in the middle of vast Mumbai, Dharavi is home to as many as 1 million people densely packed into thousands of tiny shacks fashioned from scrap metal, plastic sheeting and other scrounged materials. Narrow, muddy alleys crisscross the 600-acre site, open sewers carry human waste and vacant lots serve as garbage dumps. There is electricity, but running water is available for only an hour or so a day. Amid the squalor, barefoot children sing for money, beg from drivers in nearby traffic or work in garment and leather shops, recycling operations and other lightly regulated businesses.

Moviegoers around the globe got a glimpse of life inside Dharavi in last year's phenomenally popular Oscar-winning film "Slumdog Millionaire," about plucky Jamal Malik, a fictional Dharavi teenager who improbably wins a TV quiz-show jackpot. The no-holds-barred portrayal of slum life may have been shocking to affluent Westerners, but Dharavi is only one of Asia's innumerable slums. In fact, about a billion people worldwide live in urban slums — the ugly underbelly of the rapid and haphazard urbanization that has occurred in many parts of the world in recent decades. And if soaring urban growth rates continue unabated, the world's slum population is expected to double to 2 billion by 2030, according to the U.N. [1]

But all city dwellers don't live in slums. Indeed, other fast-growing cities presented cheerier faces to the

Children scavenge for recyclables amid rubbish in the Dharavi slum in Mumbai, India. About a billion people worldwide live in slums — where sewer, water and garbage-collection services are often nonexistent. If impoverished rural residents continue streaming into cities at current rates, the world's slum population is expected to double to 2 billion within the next two decades, according to the United Nations.

AFP/Getty Images/Rob Elliott

world last year, from Dubai's glittering luxury skyscrapers to Beijing's breathtaking, high-tech pre-Olympic cultural spectacle.

Today, 3.3 billion people live in cities — half the world's population — and urbanites are projected to total nearly 5 billion (out of 8.1 billion) worldwide by 2030. [2] About 95 percent of that growth is occurring in the developing world, especially in Africa and Asia. [3]

These regions are going through the same threefold evolution that transformed Europe and North America over a 200-year period between 1750 and 1950: the industrialization of agriculture, followed by rural migration to cities and declining population growth as life ex-

pectancy improves. But today's developing countries are modernizing much faster — typically in less than 100 years — and their cities are expanding at dizzying rates: On average, 5 million people in developing countries move to cities every month. As urban areas struggle to absorb this growth, the new residents often end up crowded into already teeming slums. For instance, 62 percent of city dwellers in sub-Saharan Africa live in slums, 43 percent in southern Asia, 37 percent in East Asia and 27 percent in Latin America and the Caribbean, according to UN-HABITAT, the United Nations agency for human settlements. [4]

UN-HABITAT defines a slum as an urban area without at least one of the following features:

- Durable housing,
- Adequate living space (no more than three people per room),
- Access to clean drinking water,
- Access to improved sanitation (toilets or latrines that separate human waste from contact with water sources), or
- Secure property rights. [5]

But all slums are not the same. Some lack only one basic necessity, while others lack several. And conditions can be harsh in non-slum neighborhoods as well. Thus, experts say, policies should focus on specific local problems in order to make a difference in the lives of poor city dwellers. [6]

Cities "are potent instruments for national economic and social development. They attract investment and create wealth," said HABITAT Executive Director Anna Tibaijuka last April. But, she warned, cities also concentrate

World Will Have 26 Megacities by 2025

The number of megacities — urban areas with at least 10 million residents — will increase from 19 to 26 worldwide by the year 2025, according to the United Nations. The seven new megacities will be in Asia and sub-Saharan Africa. Most megacities are in coastal areas, making them highly vulnerable to massive loss of life and property damage caused by rising sea levels that experts predict will result from climate change in the 21st century.

● Megacities Existing in 2007
○ New Megacities in 2025

Source: UN-HABITAT

poverty and deprivation, especially in developing countries. "Rapid and chaotic urbanization is being accompanied by increasing inequalities, which pose enormous challenges to human security and safety." [7]

Today, improving urban life is an important international development priority. [8] One of the eight U.N. Millennium Development Goals (MDGs) — broad objectives intended to end poverty worldwide by 2015 — endorsed by world leaders in 2000 was environmental sustainability. Among other things, it aims to cut in half the portion of the world's people without access to safe drinking water and achieve "significant improvement" in the lives of at least 100 million slum dwellers. [9]

Delivering even the most basic city services is an enormous challenge in many of the world's 19 megacities — metropolises with more than 10 million residents. And smaller cities with fewer than 1 million inhabitants are growing even faster in both size and number than larger ones. [10]

Many fast-growing cities struggle with choking air pollution, congested traffic, polluted water supplies and inadequate sanitation services. The lack of services can contribute to larger social and economic problems. For example, slum dwellers without permanent housing or access to mass transit have trouble finding and holding jobs.

And when poverty becomes entrenched it reinforces the gulf between rich and poor, which can promote crime and social unrest.

"A city is a system of systems. It has biological, social and technical parts, and they all interact," says George Bugliarello, president emeritus of Polytechnic University in New York and foreign secretary of the National Academy of Engineering. "It's what engineers call a complex system because it has features that are more than the sum of its parts. You have to understand how all of the components interact to guide them."

Improving life for the urban poor begins with providing shelter, sanitation and basic social services like health care and education. But more is needed to make cities truly inclusive, such as guaranteeing slum dwellers' property rights so they cannot be ejected from their homes. [11]

Access to information and communications technology (ICT) is also crucial. In some developing countries, ICT has been adopted widely, particularly cell phones, but high-speed Internet access and computer use still lag behind levels in rich nations. Technology advocates say this "digital divide" slows economic growth in developing nations and increases income inequality both within and between countries. Others say the problem has been exaggerated and that there is no critical link between ICTs and poverty reduction.

Managing urban growth and preventing the creation of new slums are keys to both improving the quality of life and better protecting cities from natural disasters. Many large cities are in areas at risk from earthquakes, wildfires or floods. Squatter neighborhoods are often built on flood plains, steep slopes or other vulnerable areas, and poor people usually have fewer resources to escape or relocate.

For example, heavy rains in northern Venezuela in 1999 caused mudslides and debris flows that demolished many hillside shantytowns around the capital city of Caracas, killing some 30,000 people. In 2005 Hurricane Katrina killed more people in New Orleans' lower-income neighborhoods,

Tokyo Is by Far the World's Biggest City

With more than 35 million residents, Tokyo is nearly twice as big as the next-biggest metropolises. Tokyo is projected to remain the world's largest city in 2025, when there will be seven new megacities — urban areas with at least 10 million residents. Two Indian cities, Mumbai and Delhi, will overtake Mexico City and New York as the world's second- and third-largest cities. The two largest newcomers in 2025 will be in Africa: Kinshasa and Lagos.

Population of Megacities, 2007 and 2025
(in millions)

2007		2025 (projected)	
Tokyo, Japan	35.68	Tokyo, Japan	36.40
New York, NY/Newark, NJ	19.04	Mumbai, India	26.39
Mexico City, Mexico	19.03	Delhi, India	22.50
Mumbai, India	18.98	Dhaka, Bangladesh	22.02
São Paulo, Brazil	18.85	São Paulo, Brazil	21.43
Delhi, India	15.93	Mexico City, Mexico	21.01
Shanghai, China	14.99	New York, NY/Newark, NJ	20.63
Kolkata, India	14.79	Kolkata, India	20.56
Dhaka, Bangladesh	13.49	Shanghai, China	19.41
Buenos Aires, Argentina	12.80	Karachi, Pakistan	19.10
Los Angeles/Long Beach/ Santa Ana (CA)	12.50	Kinshasa, Dem. Rep. Congo	16.76
Karachi, Pakistan	12.13	Lagos, Nigeria	15.80
Cairo, Egypt	11.89	Cairo, Egypt	15.56
Rio de Janeiro, Brazil	11.75	Manila, Philippines	14.81
Osaka/Kobe, Japan	11.29	Beijing, China	14.55
Beijing, China	11.11	Buenos Aires, Argentina	13.77
Manila, Philippines	11.10	Los Angeles/Long Beach/ Santa Ana (CA)	13.67
Moscow, Russia	10.45	Rio de Janeiro, Brazil	13.41
Istanbul, Turkey	10.06	Jakarta, Indonesia	12.36
New megacities in 2025		Istanbul, Turkey	12.10
		Guangzhou/Guangdong, China	11.84
		Osaka/Kobe, Japan	11.37
		Moscow, Russia	10.53
		Lahore, Pakistan	10.51
		Shenzhen, China	10.20
Source: UN-HABITAT		Chennai, India	10.13

Global Population Is Shifting to Cities

Half a century ago, less than a third of the world's population lived in cities. By 2005, nearly half inhabited urban areas, and in 2030, at least 60 percent of the world's population will be living in cities, reflecting an unprecedented scale of urban growth in the developing world. This will be particularly notable in Africa and Asia, where the urban population will double between 2000 and 2030.

Worldwide Urban and Rural Populations

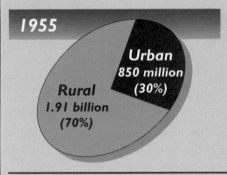

1955

Urban
850 million
(30%)

Rural
1.91 billion
(70%)

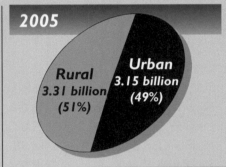

2005

Rural
3.31 billion
(51%)

Urban
3.15 billion
(49%)

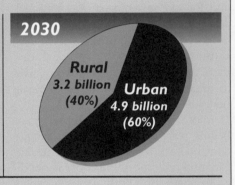

2030

Rural
3.2 billion
(40%)

Urban
4.9 billion
(60%)

Source: U.N. Department of Economic and Social Affairs; U.N. Population Fund

which were located in a flood plain, than in wealthier neighborhoods of the Louisiana port city that were built on higher ground. As global warming raises sea levels, many of the world's largest cities are expected to be increasingly at risk from flooding.

Paradoxically, economic growth also can pose a risk for some cities. Large cities can be attractive targets for terrorist attacks, especially if they are symbols of national prosperity and modernity, such as New York City, site of the Sept. 11, 2001, attack on the World Trade Center. Last November's coordinated Islamic terrorist attacks in Mumbai followed a similar strategy: Landmark properties frequented by foreigners were targeted in order to draw worldwide media coverage, damage India's economy and send a message that nowhere in India was safe. [12]

Today the global economic recession is creating a new problem for city dwellers: Entry-level jobs are disappearing as trade contracts evaporate and factories shut down. Unable to find other jobs, many recent migrants to cities are returning to rural areas that are ill-

prepared to receive them, and laborers who remain in cities have less money to send to families back home. [13]

As national leaders, development experts and city officials debate how to manage urban growth, here are some issues they are considering:

Does urbanization make people better off?

With a billion city dwellers worldwide trapped in slums, why do people keep moving to cities? Demographic experts say that newcomers hope to earn higher incomes and find more opportunities than rural areas can offer.

"Often people are fleeing desperate economic conditions," says David Bloom, a professor of economics and demography at Harvard University's School of Public Health. "And the social attractions of a city — opportunities to meet more people, escape from isolation or in some cases to be anonymous — trump fears about difficult urban conditions. If they have relatives or friends living in cities already, that reduces some of the risk."

When nations attract foreign investment, it creates new jobs. In the 1990s both China and India instituted broad economic reforms designed to encourage foreign investment, paving the way for rapid economic growth. That growth accelerated as information technology advances like the Internet, fiber-optic networks and e-mail made it faster and cheaper to communicate worldwide in real time. [14] As a result, thousands of manufacturing and white-collar jobs were "outsourced" from the United States to India, China and other low-wage countries over the past decade. [15]

These jobs spurred major growth in some cities, especially in areas with educated, English-speaking work forces. The large southern Indian city of Bangalore became a center for information technology — dubbed "India's Silicon Valley." Other cities in India, Singapore and the Philippines now host English-language call centers that manage everything from computer technical support to lost-baggage complaints for airlines. In a twist on this model, the Chinese city of Dalian — which was controlled by Japan from

1895 through World War II and still has many Japanese speakers — has become a major outsourcing center for Japanese companies. [16]

Some observers say an increasingly networked world allows people to compete for global "knowledge work" from anywhere in the world instead of having to emigrate to developed countries. In his best-seller *The World Is Flat*, author and *New York Times* columnist Thomas Friedman cites Asian call centers as an example of this shift, since educated Indians can work at the centers and prosper at home rather than seeking opportunity abroad. While he acknowledges that millions of people in developing countries are poor, sick and disempowered, Friedman argues that things improve when people move from rural to urban areas.

"[E]xcess labor gets trained and educated, it begins working in services and industry; that leads to innovation and better education and universities, freer markets, economic growth and development, better infrastructure, fewer diseases and slower population growth," Friedman writes. "It is that dynamic that is going on in parts of urban India and urban China today, enabling people to compete on a level playing field and attracting investment dollars by the billions." [17]

But others say it's not always so simple. Educated newcomers may be able to find good jobs, but migrants without skills or training often end up working in the "informal economy" — activities that are not taxed, regulated or monitored by the government, such as selling goods on the street or collecting garbage for recycling. These jobs are easy to get but come without minimum wages or labor standards, and few workers can get credit to grow their businesses. Members of ethnic minorities and other underprivileged groups, such as lower castes in India, often are stuck with the dirtiest and most dangerous and difficult tasks. [18]

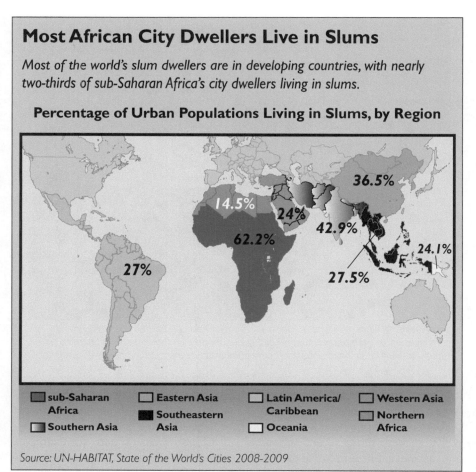

Most African City Dwellers Live in Slums

Most of the world's slum dwellers are in developing countries, with nearly two-thirds of sub-Saharan Africa's city dwellers living in slums.

Percentage of Urban Populations Living in Slums, by Region

36.5%
14.5%
24%
42.9%
62.2%
24.1%
27%
27.5%

- sub-Saharan Africa
- Southern Asia
- Eastern Asia
- Southeastern Asia
- Latin America/ Caribbean
- Oceania
- Western Asia
- Northern Africa

Source: UN-HABITAT, *State of the World's Cities 2008-2009*

And some countries have experienced urban growth without job growth. Through the late 1980s, many Latin American countries tried to grow their economies by producing manufactured goods at home instead of importing them from abroad.

"Years of government protection insulated these industries from outside competition, so they did not feel pressure to become more productive. Then they went under when economies opened up to trade," says Steven Poelhekke, a researcher with DNB, the national bank of the Netherlands. "In Africa, industrialization has never really taken off. And without job creation governments cannot deliver benefits for new urbanites." [19]

Meanwhile, when cities grow too quickly, competition for land, space, light and services increases faster than government can respond. Real estate prices rise, driving poor residents into squatter neighborhoods, where crowding and pollution spread disease. "When cities get too big, the downsides to city life are bigger than the benefits for vulnerable inhabitants," says Poelhekke.

Broadly, however, urbanization has reduced the total number of people in poverty in recent years. According to a 2007 World Bank study, about three-quarters of the world's poor still live in rural areas. Poor people are urbanizing faster than the population as a whole, so some poverty is shifting to cities. Yet, clearly, many of those new urbanites are finding higher incomes — even if they end up living in city slums — because overall poverty rates (urban plus rural) fall as countries urbanize. While the persistence of urban poverty is a serious concern, the authors concluded, if people moved to

AFP/Getty Images/Shafiq Alam

Packed buses in Dhaka take residents in the Bangladeshi capital to their homes in outlying villages on the eve of the Muslim holiday Eid al-Adha — the "Festival of Sacrifice." Rapidly growing cities have trouble keeping up with the transportation needs of residents.

the cities faster, overall poverty rates would decline sooner. [20]

Many development advocates say policy makers must accept urbanization as inevitable and strive to make it more beneficial. "We need to stop seeing migration to cities as a problem," says Priya Deshingkar, a researcher at the Overseas Development Institute in Hyderabad, India. "These people were already vulnerable because they can't make a living in rural areas. Countries need to rethink their development strategies. The world is urbanizing, and we have to make more provisions for people moving to urban areas. They can't depend on agriculture alone."

Should governments limit migration to cities?

Many governments have tried to limit urban problems by discouraging migration to cities or regulating the pace of urban growth. Some countries use household registration policies, while others direct aid and economic development funds to rural areas. Political leaders say limiting migration reduces strains on city systems, slows

the growth of slums and keeps villages from languishing as their most enterprising residents leave.

China's *hukou* system, for example, requires households to register with the government and classifies individuals as rural or urban residents. Children inherit their *hukou* status from their parents. Established in the 1950s, the system was tightly controlled to limit migration from agricultural areas to cities and to monitor criminals, government critics and other suspect citizens and groups. [21]

In the late 1970s China began privatizing farming and opened its economy to international trade, creating a rural labor surplus and greater demand for city workers. The government offered rural workers temporary residence permits in cities and allowed wealthy, educated citizens to buy urban *hukou* designations. Many rural Chinese also moved to cities without changing their registration. According to recent government estimates, at least 120 million migrant workers have moved to Chinese cities since the early 1980s. [22] Today *hukou* rules are en-

forced inconsistently in different Chinese cities, where many rural migrants cannot get access to health care, education, affordable housing or other urban services because they are there illegally.

Chinese officials say they must manage growth so all areas of the country will benefit. In a 2007 report to the 17th Communist Party Congress, President Hu Jintao promised to promote "a path of urbanization with Chinese characteristics" that emphasized "balanced development of large, medium-sized and small cities and towns." [23]

But critics say the *hukou* system has created an urban underclass and should be scrapped. When the municipality of Chongqing (which omits an estimated 4.5 million migrant workers from its official population figures) established November 4 as Migrant Workers' Day in 2007, the *Asia Times* commented, "By not changing the [*hukou*] system and instead giving the migrant workers a special holiday, it's a bit like showing starving people menus instead of feeding them." [24]

India and Vietnam also control migration to urban areas by requiring people to register or show local identity cards to access social services. "They're both trying to promote rural development and keep from overburdening urban areas," says Deshingkar at the Overseas Development Institute. "But it doesn't work. People move despite these regulations. It just makes it harder for them, and if they can access services it's at a price."

Many experts say governments should not try to halt rural-to-city migration because when migrant workers send large shares of their wages home to their families in the country it helps reduce rural poverty and inequality. In Dhaka, Bangladesh, for example, remittances from city workers provide up to 80 percent of rural households' budgets, according to the Coalition for the Urban Poor. [25]

Urban growth also helps rural economies by creating larger markets for agricultural products — including high-value products like meat, chicken and fish that people tend to add to their diets as their incomes rise. Cities can promote economic growth in surrounding areas by creating a demand for local farmers' products. For instance, South Africa's Johannesburg Fresh Produce Market offers vendors stalls, overnight storage space, business-skills training and financing; it also requires market agents to buy at least 10 percent of their produce from small, low-income farms. [26]

However, the rootless lifestyle adopted by so-called circular migrants — those who move back and forth between the city and the country — makes people vulnerable, Deshingkar points out. "There are roughly 100 million circular migrants in India now, and they're completely missed by official statistics because the government only counts permanent migrants," she says. "They can't get any insurance or social services, so they carry all the risk themselves."

Beyond the fact that anti-migration policies usually fail, experts say the biggest factor driving population increase in many fast-growing cities is not new residents moving in but "natural increase" — the rate at which people already living there have children. Natural increase accounts for about 60 percent of urban growth worldwide, while 20 percent comes from domestic and international migration and 20 percent results from reclassification of rural areas as urban. [27]

Family-planning programs helped reduce poverty rates in several developing Asian countries — including South Korea, Taiwan, Thailand, Singapore, Indonesia and Malaysia — where having smaller families increased household savings and reduced national education costs. [28] In contrast, artificial birth control is difficult to obtain in the Philippines, where the population is 80 percent Catholic and the government supports only "natural" family planning.

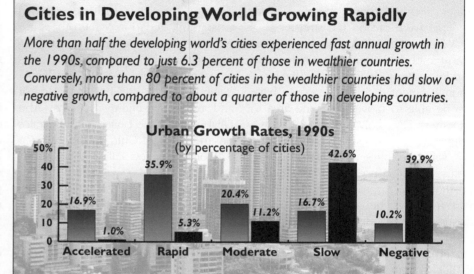

Cities in Developing World Growing Rapidly

More than half the developing world's cities experienced fast annual growth in the 1990s, compared to just 6.3 percent of those in wealthier countries. Conversely, more than 80 percent of cities in the wealthier countries had slow or negative growth, compared to about a quarter of those in developing countries.

Urban Growth Rates, 1990s
(by percentage of cities)

	Accelerated	Rapid	Moderate	Slow	Negative
In developing countries (1,408 cities)	16.9%	35.9%	20.4%	16.7%	10.2%
In developed countries (1,287 cities)	1.0%	5.3%	11.2%	42.6%	39.9%

* Figures may not total 100 due to rounding.

Source: UN-HABITAT

Several professors at the University of the Philippines have calculated that if Filipinos had followed Thailand's example on family planning in the 1970s, the Philippines would have at least 4 million fewer people in poverty and would be exporting rice rather than importing it. Instead, the Philippine government's opposition to family planning "contributed to the country's degeneration into Southeast Asia's basket case," said economist Arsenio Balisacan. [29]

Can we make large cities greener?

Many fast-growing cities are unhealthy places to live because of dirty air, polluted water supplies and sprawling waste dumps. City governments worldwide are increasingly interested in making their cities greener and more sustainable.

Greening cities has many up-front costs but can provide big payoffs. For example, energy-efficient buildings cost less to operate and give cities cachet as centers for advanced technology and design.

Green policies also may help cities achieve broader social goals. When

Enrique Peñalosa was elected mayor of Bogotá, Colombia, in 1998, the city was overrun with traffic and crime. Wealthy residents lived in walled-off neighborhoods, while workers were squeezed into shanties on the city's outskirts. Under Peñalosa's rule, the city built hundreds of new parks and a rapid-transit bus system, limited automobile use, banned sidewalk parking and constructed a 14-mile-long street for bicyclists and pedestrians that runs through some of the city's poorest neighborhoods. The underlying goal of the programs: Make Bogotá more people-friendly for poor residents as well as the rich.

"[A]nything that you do in order to increase pedestrian space constructs equality" said Peñalosa, who now consults with city officials in other developing countries. "It's a powerful symbol showing that citizens who walk are equally important to those who have a car." [30] His administration also invested funds that might otherwise have been spent building highways in social services like schools and libraries. Air pollution decreased as more residents shifted to mass transit. Crime rates also fell, partly because more people were out on the streets. [31]

"Mobility and land use may be the most important issues that a mayor can address, because to unlock the economic potential of cities people have to be able to move from one area to another," says Polytechnic University's Bugliarello. "You also have to take care of water supplies and sanitation, because cities concentrate people and pathologies. Appropriate technologies aren't always the most expensive options, especially if cities get together and form markets for them."

For example, bus rapid transit (BRT) systems, which create networks of dedicated lanes for high-speed buses, are much cheaper than subways but faster than conventional buses that move in city traffic. By 2007 some 40 cities worldwide had developed BRT systems, including Bogotá; Jakarta, Indonesia; and Guayaquil, Ecuador. Many others are planned or under construction. [32]

Some developing countries are planning entire green cities with walkable neighborhoods, efficient mass transit and renewable-energy systems. Abu Dhabi, part of the United Arab Emirates on the Persian Gulf, is designing a $20 billion project called Masdar City, which it bills as the world's first carbon-neutral, zero-waste city. Located on the coast next to Abu Dhabi's airport, Masdar City will be a mixed-use community with about 40,000 residents and 50,000 commuters traveling in to work at high-tech companies. Plans call for the city to be car-free and powered mainly by solar energy. [33]

Abu Dhabi wants to become a global hub for clean technologies, according to Khaled Awad, property development director for the Masdar initiative. "It lets us leverage our energy knowledge [from oil and gas production] and our research and development skills and adapt them to new energy markets," he said.

"If we can do it there, we can do it anywhere," said Matthias Schuler, an engineer with the German climate-engineering firm Transsolar and a member of the international Masdar City design and planning team. [34] He points out that average daytime summer temperatures in Abu Dhabi are well over 100 degrees Fahrenheit, and coastal zones are very humid. "You can't find a harsher climate."

In China, meanwhile, green urban design is gaining support as a way to attract foreign investment and demonstrate environmental awareness. But some showpiece projects are falling short of expectations.

China Aggressively Tackles Air Pollution

"No country in developing Asia takes those challenges more seriously."

China's large cities have some of the world's worst air pollution, thanks to rapid industrial growth, heavy use of coal and growing demand for cars.

The capital, Beijing, lost its 1993 bid to host the 2000 Summer Olympic Games partly because the city was so polluted. A chronic grey haze not only sullied Beijing's international image but also threatened to cause health problems for athletes and impair their performances.

When Beijing was chosen in 2001 to host the 2008 Summer Games, it pledged to put on a "green Olympics," which was widely understood to include clearing the air.

Between 2001 and 2007, however, China's economy grew beyond all predictions, with its gross domestic product expanding by up to 13 percent a year. [1] Beijing's air pollution worsened as new factories, power plants and cars crowded into the city. Winds carried in more pollutants from other burgeoning cities, including nitrogen oxides and sulfur dioxide — which contribute to acid rain and smog — and fine particulates, which can cause or worsen heart and lung problems.

With the Olympic deadline looming, many observers predicted Beijing would not meet its targets even if it relied heavily on authoritarian measures like shutting down factories and limiting auto use. [2] International Olympic Committee President Jacques Rogge said some outdoor endurance sports might have to be postponed if they occurred on high-pollution days — an embarrassing prospect for Chinese leaders. [3]

But China met its promised target, keeping Beijing's daily air pollution index — based on combined measurements of sulfur dioxide, nitrogen dioxide and fine particulates — below 100 during the month the Olympics took place. A 100 index score means air quality will not affect daily activities, compared to a maximum score of 500, when officials warn residents to stay indoors. In fact, during the Olympics in August 2008 Beijing's daily air pollution reached the lowest August measurements since 2000, sometimes even dropping into the 20s. [4]

"No country in Asia has bigger air quality challenges than China, but no country in developing Asia takes those challenges more seriously," says Cornie Huizenga, executive director of the Clean Air Initiative for Asian Cities (CAI-Asia), an international network based in the Philippines and founded by the Asian Development Bank, the World Bank and the U.S. Agency for International Development. "China has taken a whole series of long-term structural measures to address air pollution. The Olympics put a magnifying glass on Beijing and made them focus there, but its programs are much bigger."

For instance, China continuously monitors air quality in more than 100 cities, requires high-polluting provinces and companies to close small, inefficient emission sources and install pollution-control equipment and has new-car emissions standards roughly equivalent to U.S. and Western European laws.

"For the Olympics China took temporary measures on top of those policies, like closing down large facilities and keeping cars off the roads. All of this plus good weather let Beijing deliver what it promised for the Games," says Huizenga.

Now China is further expanding air pollution regulations. During the Olympics, the Ministry of Environment announced that in 2009 it would start monitoring ultra-fine particle and ozone pollution — persistent problems in many developed countries. And Beijing officials plan to increase spending on public transportation.

Local pollution sources, weather patterns and geography influence air pollution, so China's policies for cleaning up Beijing's air might not work in other large cities. Mexico City, for instance, also has tried to reduce its severe air pollution but is hampered by the city's high altitude (7,200 feet). Car engines burn fuel inefficiently at high altitudes, so they pollute more than at sea level. And while automobiles are the biggest emission sources, scientists also found that leaking liquefied petroleum gas (LPG) — which most Mexican households burn for cooking and heating — also contributes to Mexico City's air pollution. [5]

"We need better-harmonized air quality monitoring in developing countries before we can compare them," says Huizenga. "But other cities should be able to make progress on a large scale like Beijing. There's a lot of low-hanging fruit, such as switching to cleaner transportation fuels, getting rid of vehicles with [high-polluting] two-stroke engines, managing dust at construction sites and cutting pollution from coal-fired power plants. But to make them work, you also need effective agencies with enough people and money to carry [out] policies."

[1] Michael Yang, "China's GDP (2003-2007)," forum.china.org.cn, Nov. 10, 2008; "China Revises 2007 GDP Growth Rate to 13%," Jan. 15, 2009, http://english.dbw.cn.

[2] Edward Russell, "Beijing's 'Green Olympics' Test Run Fizzles," *Asia Times*, Aug. 10, 2007; Jim Yardley, "Beijing's Olympic Quest: Turn Smoggy Sky Blue," *The New York Times*, Dec. 29, 2007; David G. Streets, *et al.*, "Air Quality during the 2008 Beijing Olympic Games," *Atmospheric Environment*, vol. 41 (2007).

[3] "IOC President: Beijing Air Pollution Could Cause Events to Be Delayed During 2008 Olympics," The Associated Press, Aug. 7, 2007.

[4] "Summary: AQ in Beijing During the 2008 Summer Olympics," Clean Air Initiative for Asian Cities, www.cleanairnet.org/caiasia/1412/article-72991.html. Weather conditions are important factors in air pollution levels — for example, summer heat and humidity promote the formation of ground-level ozone, a major ingredient of smog — so to put conditions during the Olympics in context, scientists compared them to readings taken in August of previous years.

[5] Tim Weiner, "Terrific News in Mexico City: Air Is Sometimes Breathable," *The New York Times*, Jan. 5, 2001.

Huangbaiyu was supposed to be a sustainable "green village" that would provide new homes for a farming town of more than 1,400 in rural northeast China. But the master plan, produced by a high-profile U.S. green architecture firm, called for 400 densely clustered bungalows without enough yard space for livestock. This meant that villagers would lose their existing income from backyard gardens, sheep flocks and trout ponds. The plan also proposed to use corncobs and stalks to fuel a biogas plant for heat, but villagers needed these crop wastes as winter feed for their goats.

By December 2008 the Chinese builder had constructed 42 houses, but only a few were occupied. The designer blamed the builder for putting up low-quality houses, but others said the plan did not reflect what villagers wanted or needed. [35] Planners "inadvertently designed an ecologically sound plan — from the perspectives of both birds and the green movement — that would devastate the local economy and bankrupt the households whose lives were to be improved," wrote Shannon May, an American graduate student who lived in the old village of Huangbaiyu for two years and wrote her dissertation on the project. [36]

Dongtan, a larger Chinese city designed as a green project with zero-carbon-emission buildings and transit systems, has also been sidetracked. Groundbreaking on the model city of 500,000 on a Manhattan-sized island near Shanghai is more than a year behind schedule. High-rise towers are sprouting up around the site, leading some observers to call the project expensive "greenwashing" — attempting to make lavish development acceptable by tacking on environmentally friendly features.

" 'Zero-emission' city is pure commercial hype," said Dai Xingyi, a professor at Fudan University in Shanghai. "You can't expect some technology to both offer you a luxurious and comfortable life and save energy at the same time. That's just a dream." [37]

Construction is also under way on a new green city southeast of Beijing for 350,000 residents, co-developed by China and Singapore. Tianjin's features include renewable-energy sources, efficient water use and green building standards. Premier Wen Jiabao attended the 2008 groundbreaking. [38]

Although China's green development projects have a mixed record so far, "The government is starting to recognize that it has responsibility for environmental impacts beyond its borders, mainly by promoting renewable energy," says Alastair MacGregor, associate vice president of AECOM, an international design firm with large building projects in China. "Chinese culture is playing catch-up on sustainability."

More than 130 buildings designed to LEED (Leadership in Energy and Environmental Design) standards — which measure energy efficiency and healthy indoor working conditions — are planned or under construction in Beijing, Shanghai, Chongqing, Wuhan and other Chinese cities. [39] Chinese investors see LEED buildings as premium products, not as an everyday model, said MacGregor.

Some Chinese cities are developing their own green standards. About half of worldwide new construction between 2008 through 2015 is projected to occur in China, so even greening a modest share of that development would be significant.

"China could end up being a sustainability leader just by virtue of its size," MacGregor predicted. [40] ∎

BACKGROUND

From Farm to Factory

At the beginning of the 19th century only 3 percent of the world's population lived in cities, and only Beijing had more than a million inhabitants. [41] Then new technologies like the steam engine and railroads began to transform society. As the Industrial Revolution unfolded, people streamed from rural areas to manufacturing centers in Europe and the United States seeking a better income and life. This first great wave of urbanization established cities like London, Paris and New York as centers of global commerce.

It also spawned horrific slums in factory towns and large cities. Tenement houses became a feature of working-class neighborhoods, with lit-

tle access to fresh air or clean drinking water. Often whole neighborhoods shared a single water pump or toilet, and trash was usually thrown into the streets. [42]

German social scientist and a co-founder of communist theory Friedrich Engels graphically described urban workers' living conditions in cities like London and Manchester in 1844: "[T]hey are penned in dozens into single rooms. . . . They are given damp dwellings, cellar dens that are not waterproof from below or garrets that leak from above. . . . They are supplied bad, tattered or rotten clothing, adulterated or indigestible food. . . . Thus are the workers cast out and ignored by the class in power, morally as well as physically and mentally." [43]

Engels and his collaborator Karl Marx later predicted in *The Communist Manifesto* that oppression of the working class would lead to revolution in industrialized countries. Instead, public health movements began to develop in Europe and the United States in mid-century. Seeking to curb repeated cholera and typhoid epidemics, cities began collecting garbage and improving water-supply systems. A new medical specialty, epidemiology (the study of how infections are spread) developed as scientists worked to track and contain illnesses. Cities built green spaces like New York's Central Park to provide fresh air and access to nature. To help residents navigate around town, electric streetcars and subway trains were built in underground tunnels or on elevated tracks above the streets.

Many problems persisted, however. Homes and factories burned coal for heat and power, blanketing many large cities in smoky haze. Horse-drawn vehicles remained in wide use until the early-20th century, so urban streets were choked with animal waste. Wealthy city dwellers, seeking havens from the noise, dirt and crowding of

inner cities, moved out to cleaner suburban neighborhoods.

Despite harsh conditions, people continued to pour into cities. Economic growth in industrialized countries had ripple effects in developing countries. As wealthier countries imported more and more raw materials, commercial "gateway cities" in developing countries grew as well, including Buenos Aires, Rio de Janeiro and Calcutta (now Kolkata). By 1900, nearly 14 percent of the world's population lived in cities. [44]

End of Empires

Worldwide migration from country to city accelerated in the early-20th century as automation spread and fewer people were needed to grow food. But growth was not uniform. Wars devastated some of Europe's major cities while industrial production swelled others. And when colonial empires dissolved after World War II, many people were displaced in newly independent nations.

Much of the fighting during World War I occurred in fields and trenches, so few of Europe's great cities were seriously damaged. By the late 1930s, however, long-range bombers could attack cities hundreds of miles away. Madrid and Barcelona were bombed during the Spanish Civil War, a prelude to intensive air attacks on London, Vienna, Berlin, Tokyo and elsewhere during World War II. In 1945 the United States dropped atomic bombs on the Japanese cities of Hiroshima and Nagasaki, destroying each. For centuries cities had walled themselves off against outside threats, but now they were vulnerable to air attacks from thousands of miles away.

After 1945, even victorious nations like Britain and France were greatly weakened and unable to manage over-

Continued on p. 104

Chronology

1700s-1800s
Industrial Revolution spurs rapid urban growth in Europe and the U.S. Expanding slums trigger reforms and public health laws.

1804
World population reaches 1 billion.

1854
British doctor John Snow discovers the connection between contaminated drinking water and a cholera outbreak in London.

1897
Brazil's first *favela* (shanty town), is established outside Rio de Janeiro.

1900-1960s
Europe and the United States are the most urbanized. Africa and Asia begin gaining independence and struggle to develop healthy economies.

1906
An earthquake and subsequent fire destroy much of San Francisco, killing more than 3,000 people.

1927
World population reaches 2 billion.

1949
Chinese communists defeat nationalists, establishing the People's Republic of China, which aggressively promotes industrial development.

1960
World population hits 3 billion.

1964
Tokyo becomes first Asian city to host the Olympic Games and soon after that displaces New York as the world's largest city.

1970s-1990s
Urbanization accelerates in Asia and Africa. Many U.S. and European cities shrink as residents move to suburbs.

1971
East Pakistan secedes from West Pakistan and becomes the independent nation of Bangladesh; populations in Dhaka and other cities grow rapidly.

1974
World population reaches 4 billion.

1979
China initiates broad economic reforms, opens diplomatic and trade relations with the United States and starts to ease limits on migration to cities.

1985
An earthquake in Mexico City kills some 10,000 people and damages water-supply and transit systems.

1987
World population reaches 5 billion.

1991
India institutes sweeping market reforms to attract foreign investors and spur rapid economic growth.

1999
World population reaches 6 billion.

2000s
Most industrialized countries stabilize at 70-80 percent urban. Cities continue to grow in Asia and Africa.

2000
International community endorses the U.N. Millennium Development Goals designed to end poverty by 2015, including improving the lives of slum dwellers.

2001
Many international companies shift production to China after it joins the World Trade Organization; migration from rural areas accelerates. . . . Terrorists destroy World Trade Center towers in New York City, killing thousands. . . . Taiwan completes Taipei 101, the world's tallest skyscraper (1,671 feet), superseding the Petronas Towers in Kuala Lumpur, Malaysia (1,483 feet).

2005
United Nations condemns Zimbabwe for slum-clearance operations that leave 700,000 people homeless.

2007
The nonprofit group One Laptop Per Child unveils a prototype $100 laptop computer designed for children in developing countries to help close the "digital divide" between cities and rural areas.

2008
More than half of the world's population lives in cities. . . . Beijing hosts Summer Olympic Games. . . . Coordinated terrorist attacks in Mumbai kill nearly 170 people and injure more than 300.

2009
A global recession leaves millions of urban workers jobless, forcing many to return to their home villages.

2030
World's urban population is expected to reach 5 billion, and its slum population could top 2 billion.

2070
About 150 million city dwellers — primarily in India, Bangladesh, China, Vietnam, Thailand, Myanmar and Florida — could be in danger due to climate change, according to a 2008 study.

Continued from p. 102

seas colonies, where independence movements were underway. As European countries withdrew from their holdings in the Middle East, Asia and Africa over the next 25 years, a wave of countries gained independence, including Indonesia, India, Pakistan, the Philippines, Syria, Vietnam and most of colonial Africa. Wealthy countries began providing aid to the new developing countries, especially in Asia and Latin America. But some nations, especially in Africa, received little focused support.

By mid-century most industrialized countries were heavily urbanized, and their populations were no longer growing rapidly. By 1950 three of the world's largest cities — Shanghai, Buenos Aires and Calcutta — were in developing countries. Populations in developing countries continued to rise through the late 1960s even as those nations struggled to industrialize. Many rural residents moved to cities, seeking work and educational opportunities.

In the 1950s and '60s U.S. urban planners heatedly debated competing approaches to city planning. The top-down, centralized philosophy was espoused by Robert Moses, the hard-charging parks commissioner and head of New York City's highway agency from 1934 to 1968. Moses pushed through numerous bridge, highway, park and slum-clearance projects that remade New York but earned him an image as arrogant and uncaring. [45] His most famous critic, writer and activist Jane Jacobs, advocated preserving dense, mixed-use neighborhoods, like New York's Greenwich

Cities Need to Plan for Disasters and Attacks

Concentrated populations and wealth magnify impact.

Flash floods in 1999 caused landslides in the hills around Caracas, Venezuela, that washed away hundreds of hillside shanties and killed an estimated 30,000 people — more than 10 times the number of victims of the Sept. 11, 2001, terrorist attacks in the United States.

Because cities concentrate populations and wealth, natural disasters in urban areas can kill or displace thousands of people and cause massive damage to property and infrastructure. Many cities are located on earthquake faults, flood plains, fire-prone areas and other locations that make them vulnerable. The impacts are magnified when high-density slums and squatter neighborhoods are built in marginal areas. Political instability or terrorism can also cause widespread destruction.

Protecting cities requires both "hard" investments, such as flood-control systems or earthquake-resistant buildings, and "soft" approaches, such as emergency warning systems and special training for police and emergency-response forces. Cities also can improve their forecasting capacity and train officials to assess different types of risk. [1] Although preventive strategies are expensive, time-consuming and often politically controversial, failing to prepare for outside threats can be far more costly and dangerous.

Global climate change is exacerbating flooding and heat waves, which are special concerns for cities because they absorb more heat than surrounding rural areas and have higher average temperatures — a phenomenon known as the urban heat island effect. According to a study by the Organization for Economic Cooperation and Development (OECD), about 40 million people living in coastal areas around the world in 2005 were exposed to so-called 100-year floods — or major floods likely to occur only once every 100 years. By the 2070s, the OECD said, the population at risk from such flooding could rise to 150 million as more people move to cities, and climate

change causes more frequent and ferocious storms and rising sea levels.

Cities with the greatest population exposure in the 2070 forecast include Kolkata and Mumbai in India, Dhaka (Bangladesh), Guangzhou and Shanghai in China, Ho Chi Minh City and Hai Phong in Vietnam, Bangkok (Thailand), Rangoon (Myanmar) and Miami, Florida. Cities in developed countries tend to be better protected, but there are exceptions. For example, London has about the same amount of flooding protection as Shanghai, according to the OECD. [2]

"All cities need to look at their critical infrastructure systems and try to understand where they're exposed to natural hazards," says Jim Hall, leader of urban research at England's Tyndall Centre for Climate Change Research. For example, he says, London's Underground subway system is vulnerable to flooding and overheating. Fast-growing cities planning for climate change, he adds, might want to control growth in flood-prone areas, improve water systems to ensure supply during droughts or build new parks to help cool urban neighborhoods. "Risks now and in the future depend on what we do to protect cities," says Hall.

In some cities, residents can literally see the ocean rising. Coastal erosion has destroyed 47 homes and more than 400 fields in recent years in Cotonou, the capital city of the West African nation of Benin, according to a local nonprofit called Front United Against Coastal Erosion. "The sea was far from us two years ago. But now, here it is. We are scared," said Kofi Ayao, a local fisherman. "If we do not find a solution soon, we may simply drown in our sleep one day." [3]

Social violence can arise from within a city or come as an attack from outside. For example, in 2007 up to 600 people were killed when urban riots erupted in Kenya after a disputed national election. [4]

Urban leaders often justify slum-clearance programs by claiming that poor neighborhoods are breeding grounds for unrest. Others say slums are fertile recruiting grounds for terrorist groups. Slums certainly contain many who feel ill-treated, and extreme conditions may spur them into action. Overall, however, experts say most slum dwellers are too busy trying to eke out a living to riot or join terrorist campaigns.

"Poverty alone isn't a sufficient cause [for unrest]," says John Parachini, director of the Intelligence Policy Center at the RAND Corp., a U.S. think tank. "You need a combination of things — people with a profound sense of grievance, impoverishment and leaders who offer the prospect of change. Often the presence of an enemy nearby, such as an occupying foreign power or a rival tribal group or religious sect, helps galvanize people."

Last November's terrorist attacks in Mumbai, in which 10 gunmen took dozens of Indian and foreign hostages and killed at least 164 people, showed an ironic downside of globalization: Wealth, clout and international ties can make cities terrorist targets.

"Mumbai is India's commercial and entertainment center — India's Wall Street, its Hollywood, its Milan. It is a prosperous symbol of modern India," a RAND analysis noted. Mumbai also was accessible from the sea, offered prominent landmark targets (historic hotels frequented by foreigners and local elites) and had a heavy media presence that guaranteed international coverage. [5]

But serendipity can also make one city a target over another, says Parachini. "Attackers may know one city better or have family links or contacts there. Those local ties matter for small groups planning a one-time attack," he says.

Developing strong core services, such as police forces and public health systems, can be the first step in strengthening

A Bangladeshi boy helps slum residents cross floodwaters in Dhaka. Rising waters caused by global warming pose a significant potential threat to Dhaka and other low-lying cities worldwide.

most cities against terrorism, he says, rather than creating specialized units to handle terrorist strikes.

"Basic governance functions like policing maintain order, build confidence in government and can pick up a lot of information about what's going on in neighborhoods," he says. "They make it harder to do bad things."

[1] George Bugliarello, "The Engineering Challenges of Urban Sustainability," *Journal of Urban Technology*, vol. 15, no. 1 (2008), pp. 64-65.

[2] R. J. Nicholls, *et al.*, "Ranking Port Cities with High Exposure and Vulnerability to Climate Extremes: Exposure Estimates," *Environment Working Papers No. 1*, Organization for Economic Cooperation and Development, Nov. 19, 2008, pp. 7-8, www.olis.oecd.org/olis/2007doc.nsf/LinkTo/NT0000588E/$FILE/JT03255617.PDF.

[3] "Rising Tides Threaten to Engulf Parts of Cotonou," U.N. Integrated Regional Information Network, Sept. 2, 2008.

[4] "Chronology: Kenya in Crisis After Elections," Reuters, Dec. 31, 2007; "The Ten Deadliest World Catastrophes 2007," Insurance Information Institute, www.iii.org.

[5] Angel Rabasa, *et al.*, "The Lessons of Mumbai," *RAND Occasional Paper*, January 2009.

Village, and consulting with residents to build support for development plans. [46] Similar controversies would arise later in developing countries.

By the 1960s car-centered growth characterized many of the world's large cities. "Circle over London, Buenos Aires, Chicago, Sydney, in an airplane," wrote American historian Lewis Mumford in 1961. "The original container has completely disappeared: the sharp division between city and country no longer exists." City officials, Mumford argued, only measured improvements in quantities, such as wider streets and bigger parking lots.

"[T]hey would multiply bridges, highways [and] tunnels, making it ever easier to get in and out of the city but constricting the amount of space available within the city for any other purpose than transportation itself," Mumford charged. [47]

Population Boom

In the 1970s and '80s, as populations in developing countries continued to grow and improved agricultural methods made farmers more productive, people moved to the cities in ever-increasing numbers.

Some national economies boomed, notably the so-called Asian tigers — Hong Kong, Singapore, Taiwan and South Korea — by focusing on manufacturing exports for industrialized markets and improving their education systems to create productive work forces. Indonesia, Malaysia, the Philippines and Thailand — the "tiger cubs" — went through a similar growth phase in the late 1980s and early '90s.

After China and India opened up their economies in the 1980s and '90s, both countries became magnets for foreign investment and created free-trade areas and special economic zones to attract business activity. Cities in those areas expanded, particularly along China's southeast coast where such zones were clustered.

As incomes rose, many Asian cities aspired to global roles: Seoul hosted the 1988 Summer Olympics, and Malaysia built the world's tallest skyscrapers — the Petronas Twin Towers, completed in 1998, only to be superseded by the Taipei 101 building in Taiwan a few years later.

Some Asian countries — including Malaysia, Sri Lanka and Indonesia — implemented programs to improve living standards for the urban poor and helped reduce poverty. However, poverty remained high in Thailand and the Philippines and increased in China and Vietnam. [48]

Cities in South America and Africa also expanded rapidly between 1970 and 2000, although South America was farther ahead. By 1965 Latin America was already 50 percent urbanized and had three cities with populations over 5 million (Buenos Aires, São Paulo and Rio de Janeiro) — a marker sub-Saharan Africa would not achieve for several decades. [49] Urban growth on both continents followed the "primacy" pattern, in which one city is far more populous and economically and politically powerful than all the others in the nation. The presence of so-called primate cities like Lima (Peru), Caracas (Venezuela) or Lagos (Nigeria) can distort development if the dominant city consumes most public investments and grows to a size that is difficult to govern.

Latin America's growth gradually leveled out in the 1980s: Population increases slowed in major urban centers, and more people moved to small and medium-sized cities. [50] On average the region's economy grew more slowly and unevenly than Asia's, often in boom-and-bust cycles. [51] Benefits accrued mostly to small ruling classes who were hostile to new migrants, and income inequality became

deeply entrenched in many Latin American cities.

Africa urbanized quickly after independence in the 1950s and '60s. But from the mid-1970s forward most countries' incomes stagnated or contracted. Such "urbanization without growth" in sub-Saharan Africa created the world's highest rates of urban poverty and income inequality. Corruption and poor management reinforced wealth gaps that dated back to colonial times. Natural disasters, wars and the spread of HIV/AIDS further undercut poverty-reduction efforts in both rural and urban areas. [52]

New Solutions

As the 21st century began, calls for new antipoverty efforts led to an international conference at which 189 nations endorsed the Millennium Development Goals, designed to end poverty by 2015. Experts also focused on bottom-up strategies that gave poor people resources to help themselves.

An influential proponent of the bottom-up approach, Peruvian economist Hernando de Soto, stirred debate in 2000 with his book *The Mystery of Capital: Why Capitalism Triumphs in the West and Fails Everywhere Else.* Capitalist economies did not fail in developing nations because those countries lacked skills or enterprising spirit, de Soto argued. Rather, the poor in those countries had plenty of assets but no legal rights, so they could not prove ownership or use their assets as capital.

"They have houses but not titles; crops but not deeds; businesses but not statutes of incorporation," de Soto wrote. "It is the unavailability of these essential representations that explains why people who have adapted every other Western invention, from the paper clip to the nuclear reactor, have not been able to

produce sufficient capital to make their domestic capitalism work." But, he asserted, urbanization in the developing world had spawned "a huge industrial-commercial revolution" which clearly showed that poor people could contribute to economic development if their countries developed fair and inclusive legal systems. [53]

Not all experts agreed with de Soto, but his argument coincided with growing interest in approaches like microfinance (small-scale loans and credit programs for traditionally neglected customers) that helped poor people build businesses and transition from the "extra-legal" economy into the formal economy. Early microcredit programs in the 1980s and '90s had targeted mainly the rural poor, but donors began expanding into cities around 2000. [54]

The "digital divide" — the gap between rich and poor people's access to information and communications technologies (ICTs) — also began to attract the attention of development experts. During his second term (1997-2001), U.S. President Bill Clinton highlighted the issue as an obstacle to reducing poverty both domestically and at the global level. "To maximize potential, we must turn the digital divide among and within our nations into digital opportunities," Clinton said at the Asia Pacific Economic Cooperation Forum in 2000, urging Asian nations to expand Internet access and train citizens to use computers. [55] The Millennium Development Goals called for making ICTs more widely available in poor countries.

Some ICTs, such as mobile phones, were rapidly adopted in developing countries, which had small or unreliable landline networks. By 2008, industry observers predicted, more than half of the world's population would own a mobile phone, with Africa and the Middle East leading the way. [56]

Internet penetration moved much more slowly. In 2006 some 58 percent of the population in industrial countries used the Internet, compared to 11 percent in developing countries and only 1 percent in the least developed countries. Access to high-speed Internet service was unavailable in many developing regions or was too expensive for most users. [57] Some antipoverty advocates questioned whether ICTs should be a high priority for poor countries, but others said the issue was not whether but when and how to get more of the world's poor wired.

"The more the better, especially broadband," says Polytechnic University's Bugliarello.

While development experts worked to empower the urban poor, building lives in fast-growing cities remained difficult and dangerous in many places. Some governments still pushed approaches like slum clearance, especially when it served other purposes.

Notoriously, in 2005 President Robert Mugabe of Zimbabwe launched a slum-clearance initiative called Operation Murambatsvina, a Shona phrase translated by some as "restore order" and others as "drive out the trash." Thousands of shacks in Zimbabwe's capital, Harare, and other cities across the nation were destroyed, allegedly to crack down on illegal settlements and businesses

"The current chaotic state of affairs, where small-to-medium enterprises operated outside of the regulatory framework and in undesignated and crime-ridden areas, could not be countenanced much longer," said Mugabe. [58]

But critics said Mugabe was using slum clearance as an excuse to intimidate and displace neighborhoods that supported his opponents. In the end, some 700,000 people were left homeless or jobless by the action, which the United Nations later said violated international law. [59] Over the next several years Mugabe's government failed to carry out its pledges to build new houses for the displaced families. [60] ■

Security officers forcibly remove a woman from her home during land confiscations in Changchun, a city of 7.5 million residents in northeast China, so buildings can be demolished to make way for new construction. Some rapidly urbanizing governments use heavy-handed methods — such as land confiscation, eviction or slum clearance — so redevelopment projects can proceed.

CURRENT SITUATION

Economic Shadow

The current global economic recession is casting a dark cloud over worldwide economic development prospects. Capital flows to developing countries have declined sharply, and falling export demand is triggering layoffs and factory shutdowns in countries that produce for Western markets. But experts say even though the overall picture is sobering, many factors will determine how severely the recession affects cities.

In March the World Bank projected that developing countries would face budget shortfalls of $270 billion to $700 billion in 2009 and the world economy would shrink for the first time since World War II. According to the bank, 94 out of 116 developing countries were already experiencing an economic slowdown, and about half of them already had high poverty levels. Urban-based exporters and manufacturers were among the sectors hit hardest by the recession. [61]

These trends, along with an international shortage of investment capital, will make many developing countries increasingly dependent on foreign aid at a time when donor countries are experiencing their own budget crises. As workers shift out of export-oriented sectors in the cities and return to rural areas, poverty may increase, the bank projected.

The recession could mean failure to meet the Millennium Development Goals, especially if donor countries pull back on development aid. The bank urged nations to increase their foreign aid commitments and recommended that national governments:

- Increase government spending where possible to stimulate economies;
- Protect core programs to create social safety nets for the poor;

Slum Redevelopment Plan Stirs Controversy

Conditions for the 60,000 families living in Mumbai's Dharavi neighborhood (top) — one of Asia's largest slums — are typical for a billion slum dwellers around the globe. Slums often lack paved roads, water-distribution systems, sanitation and garbage collection — spawning cholera, diarrhea and other illnesses. Electric power and telephone service are usually poached from available lines. Mumbai's plans to redevelop Dharavi, located on 600 prime acres in the heart of the city, triggered strong protests from residents, who demanded that their needs be considered before the redevelopment proceeds (bottom). The project has stalled recently due to the global economic crisis.

- Invest in infrastructure such as roads, sewage systems and slum upgrades; and
- Help small- and medium-size businesses get financing to create opportunities for growth and employment. [62]

President Barack Obama's economic stimulus package, signed into law on Feb. 17, takes some of these steps and contains at least $51 billion for programs to help U.S. cities. (Other funds are allocated by states and may provide more aid to cities depending on each state's priority list.) Stimulus programs that benefit cities include $2.8 billion for energy conservation and energy efficiency, $8.4 billion for public transportation investments, $8 billion for high-speed rail and intercity passenger rail service, $1.5 billion for emergency shelter grants, $4 billion for job training and $8.8 billion for modernizing schools. [63]

Governments in developing countries with enough capital may follow suit. At the World Economic Forum in Davos, Switzerland, in January, Chinese Premier Wen Jibao announced a 4 trillion yuan stimulus package (equivalent to about 16 percent of China's GDP over two years), including money for housing, railways and infrastructure and environmental protection. " 'The harsh winter will be gone, and spring is around the corner,' " he said, predicting that China's economy would rebound this year. [64]

But according to government figures released just a few days later, more than 20 million rural migrant workers had already lost their jobs in coastal manufacturing areas and moved back to their home towns. [65] In March the World Bank cut its forecast for China's 2009 economic growth from 7.5 percent to 6.5 percent, although it said China was still doing well compared to many other countries. [66]

In India "circular migration" is becoming more prevalent, according to the Overseas Development Institute's Deshingkar. "Employment is becoming more temporary — employers like to hire temporary workers whom they can hire and fire at will, so the proportion of temporary workers and circular migrants is going up," she says. "In some Indian villages 95 percent of migrants are circular. Permanent migration is too expensive and risky — rents are high, [people are] harassed by the police, slums are razed and they're evicted. Keeping one foot in the village is their social insurance."

Meanwhile, international development aid is likely to decline as donor countries cut spending and focus on their own domestic needs. "By rights the financial crisis shouldn't undercut development funding, because the total amounts given now are tiny compared to the national economic bailouts that are under way or being debated in developed countries," says Harvard economist Bloom. "Politically, however, it may be hard to maintain aid budgets."

At the World Economic Forum billionaire philanthropist Bill Gates urged world leaders and organizations to keep up their commitments to foreign aid despite the global financial crisis. "If we lose sight of our long-term priority to expand opportunity for the world's poor and abandon our commitments and partnerships to reduce inequality, we run the risk of emerging from the current economic downturn in a world with even greater disparities in health and education and fewer opportunities for people to improve their lives," said Gates, whose Bill and Melinda Gates Foundation supports efforts to address both rural and urban poverty in developing nations. [67]

In fact, at a summit meeting in London in early April, leaders of the world's 20 largest economies pledged $1.1 trillion in new aid to help developing countries weather the global recession. Most of the money will be channeled through the International Monetary Fund.

"This is the day the world came together to fight against the global recession," said British Prime Minister Gordon Brown. [68]

Reflecting China's stunningly rapid urbanization, Shanghai's dramatic skyline rises beside the Huangpu River. Shanghai is the world's seventh-largest city today but will drop to ninth-place by 2025, as two south Asian megacities, Dhaka and Kolkata, surpass Shanghai in population.

AP Photo/Zhou Junxiang

Slum Solutions

As slums expand in many cities, debate continues over the best way to alleviate poverty. Large-scale slum-clearance operations have long been controversial in both developed and developing countries: Officials typically call the slums eyesores and public health hazards, but often new homes turn out to be unaffordable for the displaced residents. Today development institutions like the World Bank speak of "urban upgrading" — improving services in slums instead of bulldozing them. [69]

This approach focuses on improving basic infrastructure systems like water distribution, sanitation and electric power; cleaning up environmental hazards and building schools and clinics. The strategy is cheaper than massive demolition and construction projects and provides incentives for residents to invest in improving their own homes, advocates say. [70]

To do so, however, slum dwellers need money. Many do not have the basic prerequisites even to open bank accounts, such as fixed addresses and minimum balances, let alone access to credit. Over the past 10 to 15 years, however, banks have come to recognize slum dwellers as potential customers and have begun creating microcredit programs to help them obtain small loans and credit cards that often start with very low limits. A related concept, micro-insurance, offers low-cost protection in case of illness, accidents and property damage.

Now advocates for the urban poor are working to give slum dwellers more financial power. The advocacy group, Shack/Slum Dwellers International (SDI), for example, has created Urban Poor Funds that help attract direct investments from banks, government agencies and international donor groups. [71] In 2007 SDI received a $10 million grant from the Gates foundation to create a Global Finance Facility for Federations of the Urban Poor.

The funds will give SDI leverage in negotiating with governments for land, housing and infrastructure, according to Joel Bolnick, an SDI director in Cape Town, South Africa. If a government agency resists, said Bolnick, SDI can reply, " 'If you can't help us here, we'll take the money and put it on the table for a deal in Zambia instead.' " [72]

And UN-HABITAT is working with lenders to promote more mortgage lending to low-income borrowers in developing countries. "Slum dwellers have access to resources and are resources in themselves. To maximize the value of

Two-thirds of sub-Saharan Africa's city dwellers live in slums, like this one in Lagos, Nigeria, which has open sewers and no clean water, electric power or garbage collection. About 95 percent of today's rapid urbanization is occurring in the developing world, primarily in sub-Saharan Africa and Asia.

Reuters/George Esiri

slums for those who live in them and for a city, slums must be upgraded and improved," UN-HABITAT Executive Director Tibaijuka said in mid-2008. [73]

Nevertheless, some governments still push slum clearance. Beijing demolished hundreds of blocks of old city neighborhoods and culturally significant buildings in its preparations to host the 2008 Olympic Games. Some of these "urban corners" (a negative term for high-density neighborhoods with narrow streets) had also been designated for protection as historic areas. [74] Developers posted messages urging residents to take government resettlement fees and move, saying, "Living in the Front Gate's courtyards is ancient history;

moving to an apartment makes you a good neighbor," and "Cherish the chance; grab the good fortune; say farewell to dangerous housing." [75]

Beijing's actions were not unique. Other cities hosting international "mega-events" have demolished slums. Like Beijing, Seoul, South Korea, and Santo Domingo in the Dominican Republic were already urbanizing and had slum-clearance programs under way, but as their moments in the spotlight grew nearer, eviction operations accelerated, according to a Yale study. Ultimately, the study concluded, the benefits from hosting big events did not trickle down to poor residents and squatter communities who were "systematically removed or concealed from high-profile areas in order to construct the appearance of development." [76]

Now the debate over slum clearance has arrived in Dharavi. Developers are circling the site, which sits on a square mile of prime real estate near Mumbai's downtown and airport. The local government has accepted a

$3 billion redevelopment proposal from Mukesh Mehta, a wealthy architect who made his fortune in Long Island, N.Y., to raze Dharavi's shanties and replace them with high-rise condominiums, shops, parks and offices. Slum dwellers who can prove they have lived in Dharavi since 1995 would receive free 300-square-foot apartments, equivalent to two small rooms, in the new buildings. Other units would be sold at market rates that could reach several thousand dollars per square foot. [77]

Mehta contends his plan will benefit slum residents because they will receive new homes on the same site. "Give me a better solution. Until then you might want to accept this one," he said last summer. [78] But many Dharavi residents say they will not be able to keep small businesses like tanneries, potteries and tailoring shops if they move into modern high-rises, and would rather stay put. (*See "At Issue," p. 111.*)

"I've never been inside a tall building. I prefer a place like this where I can work and live," said Usman Ghani, a potter born and raised in Dharavi who has demonstrated against the redevelopment proposals. He is not optimistic about the future. "The poor and the working class won't be able to stay in Mumbai," he said. "Many years ago, corrupt leaders sold this country to the East India Company. Now they're selling it to multinationals." [79] ∎

OUTLOOK

Going Global

In an urbanizing world, cities will become increasingly important as centers of government, commerce and

Continued on p. 112

At Issue:

Will redevelopment of the Dharavi slum improve residents' lives?

MUKESH MEHTA
CHAIRMAN, MM PROJECT CONSULTANTS

WRITTEN FOR *CQ GLOBAL RESEARCHER*, APRIL 2009

KALPANA SHARMA
AUTHOR, REDISCOVERING DHARAVI:
STORIES FROM ASIA'S LARGEST SLUM

WRITTEN FOR *CQ GLOBAL RESEARCHER*, APRIL 2009

slum rehabilitation is a challenge that has moved beyond the realm of charity or meager governmental budgets. It requires a pragmatic and robust financial model and a holistic approach to achieve sustainability.

Dharavi — the largest slum pocket in Mumbai, India, and one of the largest in the world — houses 57,000 families, businesses and industries on 600 acres. Alarmingly, this accounts for only 4 percent of Mumbai's slums, which house about 7.5 million people, or 55 percent of the city's population.

Mumbai's Slum Rehabilitation Authority (SRA) has undertaken the rehabilitation of all the eligible residents and commercial and industrial enterprises in a sustainable manner through the Dharavi Redevelopment Project (DRP), following an extensive consultative process that included Dharavi's slum dwellers. The quality of life for those residents is expected to dramatically improve, and they could integrate into mainstream Mumbai over a period of time. Each family would receive a 300-square-foot home plus adequate workspace, along with excellent infrastructure, such as water supply and roads. A public-private partnership between the real estate developers and the SRA also would provide amenities for improving health, income, knowledge, the environment and socio-cultural activities. The land encroached by the slum dwellers would be used as equity in the partnership.

The primary focus — besides housing and infrastructure — would be on income generation. Dharavi has a vibrant economy of $600 million per annum, despite an appalling working environment. But the redevelopment project would boost the local gross domestic product to more than $3 billion, with the average family income estimated to increase to at least $3,000 per year from the current average of $1,200. To achieve this, a hierarchy of workspaces will be provided, including community spaces equivalent to 6 percent of the built-up area, plus individual workspaces in specialized commercial and industrial complexes for leather goods, earthenware, food products, recycling and other enterprises.

The greatest failure in slum redevelopment has been to treat it purely as a housing problem. Improving the infrastructure to enable the local economy to grow is absolutely essential for sustainable development. We believe this project will treat Dharavi residents as vital human resources and allow them to act as engines for economic growth. Thus, the DRP will act as a torchbearer for the slums of Mumbai as well as the rest of the developing world.

the controversy over the redevelopment of Dharavi, a slum in India's largest city of Mumbai, centers on the future of the estimated 60,000 families who live and work there.

Dharavi is a slum because its residents do not own the land on which they live. But it is much more than that. The settlement — more than 100 years old — grew up around one of the six fishing villages that coalesced over time to become Bombay, as Mumbai originally was called. People from all parts of India live and work here making terra-cotta pots, leather goods, garments, food items and jewelry and recycling everything from plastic to metal. The annual turnover from this vast spread of informal enterprises, much of it conducted inside people's tiny houses, is an estimated $700 million a year.

The Dharavi Redevelopment Plan — conceived by consultant Mukesh Mehta and being implemented by the Government of Maharashtra state — envisages leveling this energetic and productive part of Mumbai and converting it into a collection of high-rise buildings, where some of the current residents will be given free apartments. The remaining land will be used for high-end commercial and residential buildings.

On paper, the plan looks beautiful. But people in Dharavi are not convinced. They believe the plan has not understood the nature and real value of Dharavi and its residents. It has only considered the value of the land and decided it is too valuable to be wasted on poor people.

Dharavi residents have been left with no choice but to adapt to an unfamiliar lifestyle. If this meant a small adjustment, one could justify it. But the new form of living in a 20-story high-rise will force them to pay more each month, since the maintenance costs of high-rises exceed what residents currently spend on housing. These costs become unbearable when people earn just enough to survive in a big city.

Even worse, this new, imposed lifestyle will kill all the enterprises that flourish today in Dharavi. Currently, people live and work in the same space. In the new housing, this will not be possible.

The alternatives envisaged are spaces appropriate for formal, organized industry. But enterprises in Dharavi are informal and small, working on tiny margins. Such enterprises cannot survive formalization.

The real alternative is to give residents security of tenure and let them redevelop Dharavi. They have ideas. It can happen only if people are valued more than real estate.

Continued from p. 110

culture, but some will be more influential than others. Although it doesn't have a precise definition, the term "global city" is used by city-watchers to describe metropolises like New York and London that have a disproportionate impact on world affairs. Many urban leaders around the world aspire to take their cities to that level.

ern cities like New York, London and Paris but also includes developing-country cities such as Beijing, Shanghai, Bangkok, Mexico City and São Paulo. Many of these cities, the authors noted, are taking a different route to global stature than their predecessors followed — a shorter, often state-led path with less public input than citizens of Western democracies expect to have.

gic plan to make Abu Dhabi a world leader in clean-energy technology. "There is no question of any rollback or slowing down of any of our projects in the renewable-energy sector," said Sultan Ahmed Al Jaber, chief executive officer of the initiative, on March 16. [81] Last year the crown prince of Abu Dhabi created a $15 billion fund for clean-energy investments, which included funds for Masdar City.

Money is the front-burner issue during today's global recession. "Unless a country's overall economic progress is solid, it is very unlikely that a high proportion of city dwellers will see big improvements in their standard of living," says Harvard's Bloom. In the next several years, cities that ride out the global economic slowdown successfully will be best positioned to prosper when world markets recover.

In the longer term, however, creating wealth is not enough, as evidenced by conditions in Abu Dhabi's neighboring emirate, Dubai. Until recently Dubai was a booming city-state with an economy built on real estate, tourism and trade — part of the government's plan to make the city a world-class business and tourism hub. It quickly became a showcase for wealth and rapid urbanization: Dozens of high-rise, luxury apartment buildings and office towers sprouted up seemingly overnight, and man-made islands shaped like palm trees rose from the sea, crowded with multi-million-dollar second homes for jetsetters.

But the global recession has brought development to a halt. The real estate collapse was so sudden that jobless expatriate employees have been fleeing the country, literally abandoning their cars in the Dubai airport parking lot. [82]

Truly global cities are excellent in a variety of ways, says O'Rourke. "To be great, cities have to be places where people want to live and work." They need intellectual and cultural attractions

AP Photo/Kamran Jebreili

In addition to Dubai's glittering, new downtown area filled with towering skyscrapers, the city's manmade, palm-tree-shaped islands of Jumeirah sport hundreds of multi-million-dollar second homes for international jetsetters. Development has skidded to a temporary halt in the Arab city-state, much as it has in some other rapidly urbanizing cities around the globe, due to the global economic downturn.

The 2008 *Global Cities Index* — compiled by *Foreign Policy* magazine, the Chicago Council on Global Affairs and the A. T. Kearney management consulting firm — ranks 60 cities on five broad criteria that measure their international influence, including:
- Business activity,
- Human capital (attracting diverse groups of people and talent),
- Information exchange,
- Cultural attractions and experiences, and
- Political engagement (influence on world policy making and dialogue). [80]

The scorecard is topped by West-

"Rulers in closed or formerly closed societies have the power to decide that their capitol is going to be a world-class city, put up private funds and spell out what the city should look like," says Simon O'Rourke, executive director of the Global Chicago Center at the Chicago Council on Global Affairs. "That's not necessarily a bad path, but it's a different path than the routes that New York or London have taken. New global cities can get things done quickly — if the money is there."

Abu Dhabi's Masdar Initiative, for example, remains on track despite the global recession, directors said this spring. The project is part of a strate-

as well as conventional features like parks and efficient mass transit, he says, and, ultimately, they must give residents at least some role in decisionmaking.

"It will be very interesting to see over the next 20 years which cities can increase their global power without opening up locally to more participation," says O'Rourke. "If people don't have a say in how systems are built, they won't use them."

Finally, great cities need creative leaders who can adapt to changing circumstances. Mumbai's recovery after last November's terrorist attacks showed such resilience. Within a week stores and restaurants were open again in neighborhoods that had been raked by gunfire, and international travelers were returning to the city. [83]

The Taj Mahal Palace & Tower was one of the main attack targets. Afterwards, Ratan Tata, grand-nephew of the Indian industrialist who built the five-star hotel, said, "We can be hurt, but we can't be knocked down." [84] ∎

Notes

[1] Ben Sutherland, "Slum Dwellers 'to top 2 billion,' " BBC News, June 20, 2006, http://news.bbc.co.uk/2/hi/in_depth/5099038.stm.

[2] United Nations Population Fund, *State of World Population 2007: Unleashing the Potential of Urban Growth* (2007), p. 6.

[3] UN-HABITAT, *State of the World's Cities 2008/2009* (2008), p. xi.

[4] UN-HABITAT, *op cit.*, p. 90.

[5] *Ibid.*, p. 92.

[6] *Ibid.*, pp. 90-105.

[7] Anna Tibaijuka, "The Challenge of Urbanisation and the Role of UN-HABITAT," lecture at the Warsaw School of Economics, April 18, 2008, p. 2, www.unhabitat.org/downloads/docs/5683_16536_ed_warsaw_version12_1804.pdf.

[8] For background see Peter Katel, "Ending Poverty," *CQ Researcher*, Sept. 9, 2005, p. 733-760.

[9] For details, see www.endpoverty2015.org. For background, see Peter Behr, "Looming Water Crisis," *CQ Global Researcher*, February 2008, pp. 27-56.

[10] Tobias Just, "Megacities: Boundless Growth?" Deutsche Bank Research, March 12, 2008, pp. 4-5.

[11] Commission on Legal Empowerment of the Poor, *Making the Law Work for Everyone* (2008), pp. 5-9, www.undp.org/legalempowerment/report/Making_the_Law_Work_for_Everyone.pdf.

[12] Angel Rabasa, *et al.*, "The Lessons of Mumbai," *RAND Occasional Paper*, 2009, pp. 1-2, www.rand.org/pubs/occasional_papers/2009/RAND_OP249.pdf.

[13] Wieland Wagner, "As Orders Dry Up, Factory Workers Head Home," *Der Spiegel*, Jan. 8, 2009, www.spiegel.de/international/world/0,1518,600188,00.html; Malcolm Beith, "Reverse Migration Rocks Mexico," *Foreign Policy.com*, February 2009, www.foreignpolicy.com/story/cms.php?story_id=4731; Anthony Faiola, "A Global Retreat As Economies Dry Up," *The Washington Post*, March 5, 2009, www.washingtonpost.com/wp-dyn/content/story/2009/03/04/ST2009030404264.html.

[14] For background, see David Masci, "Emerging India, *CQ Researcher*, April 19, 2002, pp. 329-360; and Peter Katel, "Emerging China," *CQ Researcher*, Nov. 11, 2005, pp. 957-980.

[15] For background, see Mary H. Cooper, "Exporting Jobs," *CQ Researcher*, Feb. 20, 2004, pp. 149-172.

[16] Ji Yongqing, "Dalian Becomes the New Outsourcing Destination," *China Business Feature*, Sept. 17, 2008, www.cbfeature.com/industry_spotlight/news/dalian_becomes_the_new_outsourcing_destination.

[17] Thomas L. Friedman, *The World Is Flat: A Brief History of the Twenty-First Century*, updated edition (2006), pp. 24-28, 463-464.

[18] Priya Deshingkar and Claudia Natali, "Internal Migration," in *World Migration 2008* (2008), p. 183.

[19] Views expressed here are the speaker's own and do not represent those of his employer.

[20] Martin Ravallion, Shaohua Chen and Prem Sangraula, "New Evidence on the Urbanization of Global Poverty," World Bank Policy Research Working Paper 4199, April 2007, http://siteresources.worldbank.org/INTWDR2008/Resources/2795087-1191427986785/RavallionMEtAl_UrbanizationOfGlobalPoverty.pdf.

[21] For background on the *hukou* system, see Congressional-Executive Commission on China, "China's Household Registration System: Sustained Reform Needed to Protect China's Rural Migrants," Oct. 7, 2005, www.cecc.gov/pages/news/hukou.pdf; and Hayden Windrow and Anik Guha, "The Hukou

System, Migrant Workers, and State Power in the People's Republic of China," *Northwestern University Journal of International Human Rights*, spring 2005, pp. 1-18.

[22] Wu Zhong, "How the Hukou System Distorts Reality," *Asia Times*, April 11, 2007, www.atimes.com/atimes/China/ID11Ad01.html; Rong Jiaojiao, "Hukou 'An Obstacle to Market Economy,' " *China Daily*, May 21, 2007, www.chinadaily.com.cn/china/2007-05/21/content_876699.htm.

[23] "Scientific Outlook on Development," "Full text of Hu Jintao's report at 17th Party Congress," section V.5, Oct. 24, 2007, http://news.xinhuanet.com/english/2007-10/24/content_6938749.htm.

[24] Wu Zhong, "Working-Class Heroes Get Their Day," *Asia Times*, Oct. 24, 2007, www.atimes.com/atimes/China_Business/IJ24Cb01.html.

[25] "Internal Migration, Poverty and Development in Asia," *Briefing Paper no. 11*, Overseas Development Council, October 2006, p. 3.

[26] Clare T. Romanik, "An Urban-Rural Focus on Food Markets in Africa," The Urban Institute, Nov. 15, 2007, p. 30, www.urban.org/publications/411604.html.

[27] UN-HABITAT, *op. cit.*, pp. 24-26.

[28] "How Shifts to Smaller Family Sizes Contributed to the Asian Miracle," *Population Action International*, July 2006, www.popact.org/Publications/Fact_Sheets/FS4/Asian_Miracle.pdf.

[29] Edson C. Tandoc, Jr., "Says UP Economist: Lack of Family Planning Worsens Poverty," *Philippine Daily Inquirer*, Nov. 11, 2008, http://newsinfo.inquirer.net/breakingnews/nation/view/20081111-171604/Lack-of-family-planning-worsens-poverty; Blaine Harden, "Birthrates Help Keep Filipinos in Poverty," *The Washington Post*, April 21, 2008, www.washingtonpost.com/wp-dyn/content/story/2008/04/21/ST2008042100778.html.

[30] Kenneth Fletcher, "Colombia Dispatch 11: Former Bogotá Mayor Enrique Peñalosa," Smithsonian.com, Oct. 29, 2008, www.smithsonianmag.com/travel/Colombia-Dispatch-11-Former-Bogota-mayor-Enrique-Penalosa.html.

[31] Charles Montgomery, "Bogota's Urban Happiness Movement," *Globe and Mail*, June 25, 2007, www.theglobeandmail.com/servlet/story/RTGAM.20070622.whappyurbanmain0623/BNStory/lifeMain/home.

[32] Bus Rapid Transit Planning Guide, 3rd edition, Institute for Transportation & Development Policy, June 2007, p. 1, www.itdp.org/documents/Bus%20Rapid%20Transit%20Guide%20%20complete%20guide.pdf.

[33] Project details at www.masdaruae.com/en/home/index.aspx.

[34] Awad and Schuler remarks at Greenbuild 2008 conference, Boston, Mass., Nov. 20, 2008.

[35] "Green Dreams," Frontline/World, www.pbs.org/frontlineworld/fellows/green_dreams/; Danielle Sacks, "Green Guru Gone Wrong: William McDonough," *Fast Company*, Oct. 13, 2008, www.fastcompany.com/magazine/130/the-mortal-messiah.html; Timothy Lesle, "Cradle and All," *California Magazine*, September/October 2008, www.alumni.berkeley.edu/California/200809/lesle.asp.

[36] Shannon May, "Ecological Crisis and Eco-Villages in China," *Counterpunch*, Nov. 21-23, 2008, www.counterpunch.org/may11212008.html.

[37] Rujun Shen, "Eco-city seen as Expensive 'Green-Wash,' " *The Standard* (Hong Kong), June 24, 2008, www.thestandard.com.hk/news_detail.asp?we_cat=9&art_id=67641&sid=19488136&con_type=1&d_str=20080624&fc=8; see also Douglas McGray, "Pop-Up Cities: China Builds a Bright Green Metropolis," *Wired*, April 24, 2007, www.wired.com/wired/archive/15.05/feat_popup.html; Malcolm Moore, "China's Pioneering Eco-City of Dongtan Stalls," *Telegraph*, Oct. 19, 2008, www.telegraph.co.uk/news/worldnews/asia/china/3223969/Chinas-pioneering-eco-city-of-Dongtan-stalls.html; "City of Dreams," *Economist*, March 19, 2009, www.economist.com/world/asia/displaystory.cfm?story_id=13330904.

[38] Details at www.tianjinecocity.gov.sg/.

[39] "LEED Projects and Case Studies Directory," U.S. Green Building Council, www.usgbc.org/LEED/Project/RegisteredProjectList.aspx.

[40] Remarks at Greenbuild 2008 conference, Boston, Mass., Nov. 20, 2008.

[41] Population Reference Bureau, "Urbanization," www.prb.org; Tertius Chandler, *Four Thousand Years of Urban Growth: An Historical Census* (1987).

[42] Lewis Mumford, *The City In History: Its Origins, Its Transformations, and Its Prospects* (1961), pp. 417-418.

[43] Frederick Engels, *The Condition of the Working Class in England* (1854), Chapter 7 ("Results"), online at Marx/Engels Internet Archive, www.marxists.org/archive/marx/works/1845/condition-working-class/ch07.htm.

[44] Population Reference Bureau, *op. cit.*

[45] Robert A. Caro, *The Power Broker: Robert Moses and the Fall of New York* (1975).

[46] Jane Jacobs, *The Death and Life of Great American Cities* (1961).

[47] Mumford, *op. cit.*, pp. 454-455.

[48] Joshua Kurlantzick, "The Big Mango Bounces Back," *World Policy Journal*, spring 2000, www.worldpolicy.org/journal/articles/kurlant.html; UN-HABITAT, *op. cit.*, pp. 74-76.

[49] BBC News, "Interactive Map: Urban Growth," http://news.bbc.co.uk/2/shared/spl/hi/world/06/urbanisation/html/urbanisation.stm.

[50] Licia Valladares and Magda Prates Coelho, "Urban Research in Latin America: Towards a Research Agenda," MOST Discussion Paper Series No. 4 (undated), www.unesco.org/most/valleng.htm#trends.

[51] Jose de Gregorie, "Sustained Growth in Latin America," Economic Policy Papers, Central Bank of Chile, May 2005, www.bcentral.cl/eng/studies/economic-policy-papers/pdf/dpe13eng.pdf.

[52] UN-HABITAT, *op cit.*, pp. 70-74.

[53] Hernando de Soto, *The Mystery of Capital: Why Capitalism Triumphs in the West and Fails Everywhere Else* (2000), excerpted at http://ild.org.pe/en/mystery/english?page=0%2C0.

[54] Deepak Kindo, "Microfinance Services to the Urban Poor," *Microfinance Insights*, March 2007; World Bank, "10 Years of World Bank Support for Microcredit in Bangladesh," Nov. 5, 2007; "Micro Finance Gaining in Popularity," *The Hindu*, Aug. 25, 2008, www.hindu.com/biz/2008/08/25/stories/2008082550121600.htm.

[55] Michael Richardson, "Clinton Warns APEC of 'Digital Divide,' " *International Herald Tribune*, Nov. 16, 2000, www.iht.com/articles/2000/11/16/apec.2.t_2.php.

[56] Abigail Keene-Babcock, "Study Shows Half the World's Population With Mobile Phones by 2008," Dec. 4, 2007, www.nextbillion.net/news/study-shows-half-the-worlds-population-with-mobile-phones-by-200.

[57] "Millennium Development Goals Report 2008," United Nations, p. 48, www.un.org/millenniumgoals/pdf/The%20Millennium%20Development%20Goals%20Report%202008.pdf.

[58] Robyn Dixon, "Zimbabwe Slum Dwellers Are Left With Only Dust," *Los Angeles Times*, June 21, 2005, http://articles.latimes.com/2005/jun/21/world/fg-nohomes21.

[59] Ewen MacAskill, "UN Report Damns Mugabe Slum Clearance as Catastrophic," *Guardian*, July 23, 2005, www.guardian.co.uk/world/2005/jul/23/zimbabwe.ewenmacaskill.

[60] Freedom House, "Freedom in the World 2008: Zimbabwe," www.freedomhouse.org/uploads/press_release/Zimbabwe_FIW_08.pdf.

[61] "Crisis Reveals Growing Finance Gaps for Developing Countries," World Bank, March 8, 2009, http://web.worldbank.org/WBSITE/EXTERNAL/NEWS/0,,contentMDK:22093316~menuPK:34463~pagePK:34370~piPK:34424~theSitePK:4607,00.html.

[62] "Swimming Against the Tide: How Developing Countries Are Coping with the Global Crisis," World Bank, background paper prepared for the G20 finance Ministers meeting, March 13-14, 2009, http://siteresources.worldbank.org/NEWS/Resources/swimmingagainstthetide-march2009.pdf.

[63] "Major Victories for City Priorities in American Recovery and Reinvestment Act," U.S. Conference of Mayors, Feb. 23, 2009, www.usmayors.org/usmayornewspaper/documents/02_23_09/pg1_major_victories.asp.

[64] Carter Dougherty, "Chinese Premier Injects Note of Optimism at Davos," *The New York Times*, Jan. 29, 2009, www.nytimes.com/2009/01/29/business/29econ.html?partner=rss.

[65] Jamil Anderlini and Geoff Dyer, "Downturn Causes 20m Job Losses in China," *Financial Times*, Feb. 2, 2009, www.ft.com/cms/s/0/19c25aea-f0f5-11dd-8790-0000779fd2ac.html.

[66] Joe McDonald, "World Bank Cuts China's 2009 Growth Forecast," The Associated Press, March 18, 2009.

[67] "Bill and Melinda Gates Urge Global Leaders to Maintain Foreign Aid," Bill and Melinda Gates Foundation, Jan. 30, 2009, www.gatesfoundation.org/press-releases/Pages/2009-world

About the Author

Jennifer Weeks is a *CQ Researcher* contributing writer in Watertown, Mass., who specializes in energy and environmental issues. She has written for *The Washington Post*, *The Boston Globe Magazine* and other publications, and has 15 years' experience as a public-policy analyst, lobbyist and congressional staffer. She has an A.B. degree from Williams College and master's degrees from the University of North Carolina and Harvard. Her previous *CQ Global Researcher* examined "Carbon Trading."

-economic-forum-090130.aspx.

[68] Mark Landler and David E. Sanger, "World Leaders Pledge $1.1 Trillion to Tackle Crisis," *The New York Times*, April 4, 2009, www.nytimes.com/2009/04/03/world/europe/03summit.html?_r=1&hp.

[69] "Is Demolition the Way to Go?" World Bank, www.worldbank.org/urban/upgrading/demolition.html.

[70] "What Is Urban Upgrading?" World Bank, www.worldbank.org/urban/upgrading/what.html.

[71] For more information, see "Urban Poor Fund," *Shack/Slum Dwellers International*, www.sdinet.co.za/ritual/urban_poor_fund/.

[72] Neal R. Peirce, "Gates Millions, Slum-Dwellers: Thanksgiving Miracle?" *Houston Chronicle*, Nov. 22, 2007, www.sdinet.co.za/static/pdf/sdi_gates_iupf_neal_peirce.pdf.

[73] "Statement at the African Ministerial Conference on Housing and Urban Development," UN-HABITAT, Abuja, Nigeria, July 28, 2008, www.unhabitat.org/content.asp?cid=5830&catid=14&typeid=8&subMenuId=0.

[74] Michael Meyer, *The Last Days of Old Beijing* (2008), pp. 54-55; Richard Spencer, "History is Erased as Beijing Makes Way for Olympics," *Telegraph* (London), June 19, 2006, www.telegraph.co.uk/news/worldnews/asia/china/1521709/History-is-erased-as-Beijing-makes-way-for-Olympics.html; Michael Sheridan, "Old Beijing Falls to Olympics Bulldozer," *Sunday Times* (London), April 29, 2007, www.timesonline.co.uk/tol/news/world/asia/china/article1719945.ece.

[75] Meyer, *op. cit.*, pp. 45, 52.

[76] Solomon J. Greene, "Staged Cities: Mega-Events, Slum Clearance, and Global Capital," *Yale Human Rights & Development Law Journal*, vol. 6, 2003, http://islandia.law.yale.edu/yhrdlj/PDF/Vol%206/greene.pdf.

[77] Slum Rehabilitation Authority, "Dharavi Development Project," www.sra.gov.in/html pages/Dharavi.htm; Porus P. Cooper, "In India, Slum May Get Housing," *Philadelphia Inquirer*, Sept. 22, 2008.

[78] Mukul Devichand, "Mumbai's Slum Solution?" BBC News, Aug. 14, 2008, http://news.bbc.co.uk/2/hi/south_asia/7558102.stm.

[79] Henry Chu, "Dharavi, India's Largest Slum, Eyed By Mumbai Developers," *Los Angeles Times*, Sept. 8, 2008, www.latimes.com/news/nationworld/world/la-fg-dharavi8-2008sep08,0,1830588.story; see also Dominic Whiting, "Dharavi Dwellers Face Ruin in Development Blitz," Reuters, June 6, 2008, http://in.reuters.com/article/topNews/idINIndia-33958520080608; and

Mark Tutton, "Real Life 'Slumdog' Slum To Be Demolished," CNN.com, Feb. 23, 2009, www.cnn.com/2009/TRAVEL/02/23/dharavi.mumbai.slums/.

[80] Unless otherwise cited, this section is based on "The 2008 Global Cities Index," *Foreign Policy*, November/December 2008, www.foreignpolicy.com/story/cms.php?story_id=4509.

[81] T. Ramavarman, "Masdar To Proceed with $15 Billion Investment Plan," *Khaleej Times Online*, March 16, 2009, www.khaleejtimes.com/biz/inside.asp?xfile=/data/business/2009/March/business_March638.xml&section=business&col=; Stefan Nicola, "Green Oasis Rises From Desert Sands," *Washington Times*, Feb. 2, 2009, www.washingtontimes.com/themes/places/abu-dhabi/; Elisabeth Rosenthal, "Gulf Oil States Seeking a Lead in Clean Energy," *The New York Times*, Jan. 13, 2009, www.nytimes.com/2009/01/13/world/middleeast/13greengulf.html.

[82] David Teather and Richard Wachman, "The Emirate That Used to Spend It Like Beckham," *The Guardian*, Jan. 31, 2009, www.guardian.co.uk/world/2009/jan/31/dubai-global-recession; Robert F. Worth, "Laid-Off Foreigners Flee as Dubai Spirals Down," *The New York Times*, Feb. 12, 2009, www.nytimes.com/2009/02/12/world/middleeast/12dubai.html; Elizabeth Farrelly, "Dubai's Darkening Sky: The Crane Gods are Still," *Brisbane Times*, Feb. 26, 2009, www.brisbanetimes.com.au/news/opinion/dubais-darkening-sky-the-crane-gods-are-still/2009/02/25/1235237781806.html.

[83] Raja Murthy, "Taj Mahal Leads India's Recovery," *Asia Times*, Dec. 3, 2008, www.atimes.com/atimes/South_Asia/JL03Df01.html.

[84] Joe Nocera, "Mumbai Finds Its Resiliency," *The New York Times*, Jan. 4, 2009, http://travel.nytimes.com/2009/01/04/travel/04journeys.html.

Bibliography

Selected Sources

Books

Meyer, Michael, *The Last Days of Old Beijing: Life in the Vanishing Backstreets of a City Transformed*, Walker & Co., 2008.

An English teacher and travel writer traces Beijing's history and describes life in one of its oldest neighborhoods as the city prepared to host the 2008 Olympic Games.

Silver, Christopher, *Planning the Megacity: Jakarta in the Twentieth Century*, Routledge, 2007.

An urban scholar describes how Indonesia's largest city grew from a colonial capital of 150,000 in 1900 into a megacity of 12-13 million in 2000, and concludes that overall the process was well-planned.

2007 State of the World: Our Urban Future, Worldwatch Institute, Norton, 2007.

Published by an environmental think tank, a collection of articles on issues such as sanitation, urban farming and strengthening local economies examines how cities can be healthier and greener.

Articles

"The 2008 Global Cities Index," *Foreign Policy*, November/December 2008, www.foreignpolicy.com/story/cms.php?story_id=4509.

Foreign Policy magazine, the Chicago Council on World Affairs and the A.T. Kearney management consulting firm rank the world's most "global" cities in both industrialized and developing countries, based on economic activity, human capital, information exchange, cultural experience and political engagement.

"Mexico City Bikers Preach Pedal Power in Megacity," The Associated Press, Dec. 28, 2008.

Bicycle activists are campaigning for respect in a city with more than 6 million cars, taxis and buses.

Albright, Madeleine, and Hernando De Soto, "Out From the Underground," *Time*, July 16, 2007.

A former U.S. Secretary of State and a prominent Peruvian economist contend that giving poor people basic legal rights can help them move from squatter communities and the shadow economy to more secure lives.

Bloom, David E., and Tarun Khanna, "The Urban Revolution," *Finance & Development*, September 2007, pp. 9-14.

Rapid urbanization is inevitable and could be beneficial if leaders plan for it and develop innovative ways to make cities livable.

Chamberlain, Gethin, "The Beating Heart of Mumbai," *The Observer*, Dec. 21, 2008, www.guardian.co.uk/world/2008/dec/21/dharavi-india-slums-slumdog-millionaire-poverty.

Eight boys growing up in Dharavi, Asia's largest slum, talk about life in their neighborhood.

Osnos, Evan, "Letter From China: The Promised Land," *The New Yorker*, Feb. 9, 2009, www.newyorker.com/reporting/2009/02/09/090209fa_fact_osnos.

Traders from at least 19 countries have set up shop in the Chinese coastal city of Guangzhou to make money in the export-import business.

Packer, George, "The Megacity," *The New Yorker*, Nov. 13, 2006, www.newyorker.com/archive/2006/11/13/061113fa_fact_packer.

Lagos, Nigeria, offers a grim picture of urban life.

Schwartz, Michael, "For Russia's Migrants, Economic Despair Douses Flickers of Hope," *The New York Times*, Feb. 9, 2009, www.nytimes.com/2009/02/10/world/europe/10migrants.html?n=Top/Reference/Times%20Topics/People/P/Putin,%20Vladimir%20V.

Russia has an estimated 10 million migrant workers, mainly from former Soviet republics in Central Asia — some living in shanty towns.

Reports and Studies

"Ranking of the World's Cities Most Exposed to Coastal Flooding Today and in the Future," Organization for Economic Cooperation and Development, 2007, www.rms.com/Publications/OECD_Cities_Coastal_Flooding.pdf.

As a result of urbanization and global climate change, up to 150 million people in major cities around the world could be threatened by flooding by 2070.

"State of World Population 2007," U.N. Population Fund, 2007, www.unfpa.org/upload/lib_pub_file/695_filename_sowp2007_eng.pdf.

A U.N. agency outlines the challenges and opportunities presented by urbanization and calls on policy makers to help cities improve residents' lives.

"State of the World's Cities 2008/2009: Harmonious Cities," UN-HABITAT, 2008.

The biennial report from the U.N. Human Settlements Programme surveys urban growth patterns and social, economic and environmental conditions in cities worldwide.

The Next Step:

China's Air Quality

"Beijing's Bad Air Days," *Bangkok Post* (Thailand), Sept. 1, 2007.

Beijing's air pollution isn't any worse that is has been over the past two decades, but it has become more political and integral to China's international image.

Kono, Hiroko, "Cooperation Needed on China's Dirty Air," *Yomiuri Shimbun* (Japan), April 24, 2008.

Despite local efforts to curb Chinese air pollution, many experts insist that international cooperation is necessary to fully address the problem.

Oster, Shai, "China Disputes Criticism of Its Air-Pollution Data," *The Wall Street Journal*, July 11, 2008, p. A7.

Officials dispute charges that China's air pollution has worsened due to loose regulations and data manipulation.

City Migration

"Migration From Rural to Urban Areas Reversed," *Turkish Daily News*, Dec. 27, 2008.

The global financial crisis has caused many Turkish families to return to their rural roots to survive the economic hard times.

"Small Cities to Drive Urban Growth," *The Times of India*, June 29, 2007.

While megacities continue to dominate the urban boom, many smaller cities are growing at much higher rates, according to the U.N. Population Fund.

Kashif, Alia, "Urbanisation Increasing at 58 Percent," *Business Recorder* (Pakistan), Sept. 30, 2007.

Increased migration to Islamabad, Pakistan, has raised the cost of accommodations, inflated food prices and increased transportation costs.

Obaid, Thoraya Ahmed, and Jean-Michel Severino Le Monde, "Urban Growth Is a Chance for Change," *The Guardian* (England), July 6, 2007.

Poverty may go hand-in-hand with increased migration into cities, but it is not the direct cause of it.

Greening

"Activists Demand More Green Spaces," *Jakarta Post* (Indonesia), Dec. 15, 2008.

The Indonesian Forum for the Environment is imploring the Jakarta city administration to implement more consistent and sustainable pro-environment policies.

"The Green Giant," *South China Morning Post*, July 23, 2007.

Megacities and their efficient use of resources are the best way for China to contain damage to the environment without hindering growth.

"Lagos — From 'Ghetto' to Garden City?" *This Day* (Nigeria), Sept. 11, 2008.

The city government of Lagos has embarked upon an aggressive campaign to plant 1 million trees in the region within a decade.

Rogers, Monica Kass, "It's Not Easy Building Green," *Chicago Tribune*, Jan. 31, 2008, p. C1.

A lack of education about energy conservation and bureaucratic obstacles hinder Chicago's progress in becoming a greener city.

Natural Disasters

"China Heightens Grid-Building Standard to Guard Against Disasters," *Asia Pulse* (Australia), July 3, 2008.

China's State Council has agreed to improve the disaster-prevention capacity of the country's power grid.

"World Bank Says Asian Cities at Risk," *Bangkok Post* (Thailand), July 19, 2008.

CITING *CQ GLOBAL RESEARCHER*

Sample formats for citing these reports in a bibliography include the ones listed below. Preferred styles and formats vary, so please check with your instructor or professor.

<u>MLA STYLE</u>

Flamini, Roland. "Nuclear Proliferation." <u>CQ Global Researcher</u> 1 Apr. 2007: 1-24.

<u>APA STYLE</u>

Flamini, R. (2007, April 1). Nuclear proliferation. *CQ Global Researcher, 1*, 1-24.

<u>CHICAGO STYLE</u>

Flamini, Roland. "Nuclear Proliferation." *CQ Global Researcher,* April 1, 2007, 1-24.

DAVID DODMAN
Researcher, International Institute for Environment and Development, England

Cities aren't to blame for climate change.

"Blaming cities for climate change is far too simplistic. There are a lot of economies of scale associated with energy use in cities. If you're an urban dweller, particularly in an affluent country like Canada or the U.K., you're likely to be more efficient in your use of heating fuel and in your use of energy for transportation."

Toronto Star, March 2009

BABATUNDE FASHOLA
State Governor
Lagos, Nigeria

Megacities create many challenges.

"Because of human activities there will be conflict and there will be the issue of security, everybody fighting for control, and these are some of the challenges that come with the status of a megacity. It is really a status that creates certain challenges that the government must respond to."

This Day (Nigeria), November 2007

JONATHAN WOETZEL
Director, McKinsey & Company, Shanghai

Migration to China could cause problems.

"The fact that 40 to 50 per cent of [Chinese] cities [by 2025] could be made up of migrant workers is a real

wake-up call. Smaller cities in particular are going to face a growing challenge if they are to provide equal access to social services."

Irish Times, March 2008

THE WORLD BANK

Singapore does it right.

"Improving institutions and infrastructure and intervening at the same time is a tall order for any government, but Singapore shows how it can be done. . . . Multi-year plans were produced, implemented and updated. For a city-state in a poor region, it is also not an exaggeration to assert that effective urbanization was responsible for delivering growth rates that averaged 8 per cent a year throughout 1970s and 1980s."

World Development Report 2009

THORAYA AHMED OBAID
Executive Director, U.N. Population Fund

Informal work has value.

"Many of tomorrow's city dwellers will be poor, swelling the ranks of the billion who already live in slums, but however bad their predicament, experience shows that newcomers do not leave the city once they have moved. . . . They are also remarkably productive. Economists agree that informal work makes a vital contribution to the urban economy and is a key growth factor in developing countries."

The Guardian (England), July

2007

ZHU TONG
Environmental Scientist, Peking University

Different air standards cause confusion.

"Different countries vary in their air quality standards, and the WHO does not have a binding set of standards. China's national standards are not as high as those in developed countries, which has led to disagreements, confusion or even misunderstandings."

South China Morning Post, July 2008

SUDIRMAN NASIR
Lecturer, University of Hasanuddin, Indonesia

Opportunities lead to migration.

"The lack of job and economic opportunities in rural areas justifies migration to the cities as a survival strategy. It is a rational choice made by villagers because cities generally have more jobs to offer. It's impossible to reduce urbanization through

ENERGY NATIONALISM

BY PETER BEHR

Excerpted from the CQ Global Researcher. Peter Behr. (July 2007). "Energy Nationalism." *CQ Global Researcher*, 151-180.

Energy Nationalism

BY PETER BEHR

THE ISSUES

Westerners saw the Soviet Union's 1991 collapse as a defining triumph of democracy, but Russian President Vladimir Putin has called it "the greatest geopolitical catastrophe of the century." [1] Today, to the growing unease of leaders in Washington and Europe, Putin is bent on erasing the wounds of what some Russian leaders call the "16 lost years" since the breakup and reclaiming Russia's position as a superpower. His weapon: the country's considerable energy resources.

With $500 million pouring into its coffers daily from oil and gas exports, Moscow is raising its voice — and using its elbows — in international business negotiations. During the winter of 2005-06, Russia temporarily cut off natural gas deliveries to Ukraine and Western Europe over a pricing dispute. [2] Putin also jailed Russian oil tycoon Mikhail Khodorkovsky after he challenged government energy plans and political control. And to the dismay of Washington, Moscow is considering energy investments in increasingly bellicose Iran and enticing former Soviet states Turkmenistan and Kazakhstan to channel new Caspian Sea natural gas production through Russia's existing and planned pipelines — supplies that will be vital to Europe.

"The truth is that Russia, having first scared its neighbors into [joining] NATO by its bullying behavior, is currently outmaneuvering a divided and indecisive West on almost every front, and especially on energy," said *The Economist*, the respected British newsweekly. [3]

The Caspian Sea oil town of Neft Dashlari ("Oil Rocks") produces more than half of Azerbaijan's crude oil. Built in 1947 on a chain of artificial islands, the facility contains 124 miles of streets, schools, libraries and eight-story apartments housing some 5,000 oil workers. Energy companies are targeting the Caspian Sea and other areas in the search for non-Persian Gulf oil sources.

Getty Images/Reza

Oil and politics have always made a volatile blend — particularly in the Middle East. But Russia's recent in-your-face actions represent a new strain of energy nationalism being practiced by Russia and a handful of emerging petrostates in Africa, Central Asia and Latin America that are nationalizing or taking greater control over their oil resources. Moreover, the leaders of some petrostates are imposing new political agendas on their oil sectors, notably Putin and Venezuela's combative socialist president Hugo Chávez.

"Everywhere there is a return to oil nationalism," says Jean-Marie Chevalier, director of the energy geopolitics center at Paris-Dauphine University. [4]

In the three decades since the world's first great oil shock in 1973, oil prices have periodically climbed and crashed as shortages were followed by surpluses. But this time around, the high prices are likely to stay high, many experts warn. To be sure, the war in Iraq and a looming confrontation over Iran's nuclear program are feeding the high prices. And escalating global markets, led by booming China and India, also intensify demand.

But rising energy nationalism is also triggering anxiety in global oil markets. A dramatic shift has occurred in world oil supplies since 30 years ago, when roughly three-quarters of the world's oil production was managed by private multinational oil companies — the so-called Seven Sisters — and the rest belonged to a handful of state-owned oil companies. "Today, that is about reversed," Former CIA Director John M. Deutch succinctly told the House Foreign Affairs Committee. [5]

As of 2005, 12 of the world's top 20 petroleum companies were state-owned or state-controlled, according to *Petroleum Intelligence Weekly (PIW)*. [6] (*See chart, p. 166.*) "There has been a very significant change in the balance of power between international oil companies, and it's clear today that it is the national companies that have the upper hand," said Olivier Appert, president of the French Oil Institute. [7]

"One of the favorites of headline writers is 'Big Oil,' " says Daniel Yergin, author of *The Prize: The Epic Quest for Oil, Money & Power*. "But it's the wrong Big Oil. 'Big Oil' today means the national oil companies."

The nationalization of foreign oil com-

'Hot Spots' to Supply Most of World's Energy

To reduce dependence on the unstable Persian Gulf, an oil-hungry world is turning to sources in Central Asia, Africa and Russia. But most of these "emerging" producers have either nationalized their oil industries or are considered vulnerable to terrorists or dissidents. By 2010, according to the U.S. Energy Information Agency, 58 percent of global daily oil production will be at risk because it originates or passes through one of the world's oil "hot spots," including Saudi Arabia, Russia, Iraq, Nigeria, the Caspian region, Venezuela and the straits of Hormuz and Malacca.

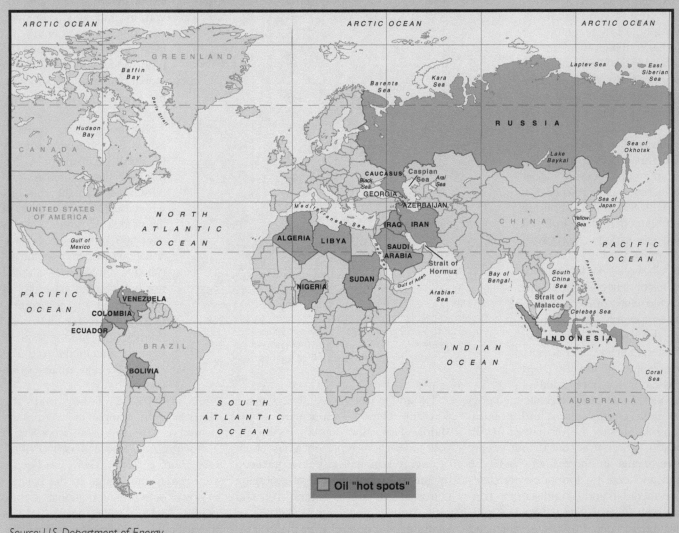

Oil "hot spots"

Source: U.S. Department of Energy

pany interests in Venezuela and Bolivia in the past two years is the hard edge of this new chapter in oil politics, echoing the same raging denunciations of Western governments and oil companies that accompanied Iran and Libya's nationalizations of foreign oil interests in the 1950s and '60s. [8] "The nationaliza-

tion of Venezuela's oil is now for real," said Chávez at a ceremony in May marking the takeover of the country's last foreign-run oil fields. "Down with the U.S. empire!" he shouted as newly purchased Russian jet fighters roared overhead. [9]

Oil-production arrangements vary widely among the dozen leading na-

tional oil companies. In Nigeria and Brazil, the government invites foreign companies to develop their oil regions, while Kuwait keeps them out. Ecuadorian President Rafael Correa, a Chávez ally who took office in January, has demanded a higher share of revenues from foreign oil companies but needs outside help to expand refining facilities. [10] Russia is forcing Shell and BP to give up majority positions in oil and gas joint ventures but hasn't thrown them out. And neither have Chávez and Correa.

Kazakhstan, after becoming independent in 1991, combined existing state firms into KazMunaiGaz — a new company that it intends to take public — while maintaining government influence through a parent company. The China National Offshore Oil Corp. is publicly traded but state-controlled.

But whatever model a petrostate adopts, *PIW* says the trend is largely the same: Major oil companies are finding their interests "increasingly subordinated to the nationalistic political agendas of key reserve-holding host countries." [11]

The new oil nationalism has been fed by energy prices at or near peak levels — when adjusted for inflation — reached after the 1970s oil shocks. (*See "Background," p. 162, and chart, p. 156.*) [12]

Rising energy prices also have produced a vast shift in wealth — over $970 billion in 2006 — from consuming nations to producing countries, a $670 billion jump in four years, and most has gone to a handful of countries, according to the Federal Reserve Bank of New York. [13]

Some industry experts say new sources of oil coming online — often from politically unstable hot spots in Africa and Central Asia — could mean lower consumer prices if Russia and the Central Asian petrostates remain independent of the Organization of Petroleum Exporting Countries (OPEC), which seeks to set international oil prices. On the other hand, the dramatic changes occurring in the industry could boost prices and — eventually — lead to declining supplies if state-run companies reduce exploration investments or botch operations, as some have done.

The International Energy Agency estimates that at least $2.2 trillion will need to be invested in the global oil

Pipeline Politics Play Pivotal Role

New and proposed oil and gas pipelines from fields in Russia, the Caspian region and Africa will likely play crucial roles in meeting the world's future energy needs. But global politics will influence when, where and whether the pipelines will be built. For instance, China covets oil and gas from eastern Siberia, but Russia's leaders have delayed building a proposed pipeline into Daqing, China. They want the pipeline to go to Russia's Pacific coast, to serve competing customers in Asia and the United States.

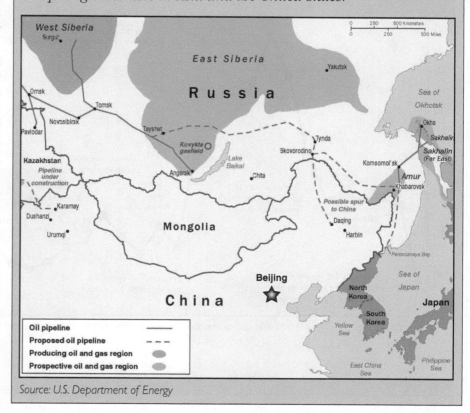

Source: U.S. Department of Energy

sector over the next 30 years to meet rising demand for oil, but oil nationalism "is slowing or even discouraging this needed investment," according to the James A. Baker III Institute for Public Policy at Rice University. [14]

Consolidation of the world's oil supplies into government hands also raises questions about whether the new oil producers will break the historic "curse of oil" pattern, in which petrostate leaders used oil profits to line their pockets and buy arms rather than lift indigent populations out of poverty. Still others worry that intensified

competition for energy between nations will sow new conflicts around the globe.

In addition to oil shortages and high prices, the International Energy Agency says Earth is facing "twin energy-related threats" — inadequate and insecure supplies of affordable oil and, paradoxically, environmental harm caused by excessive oil consumption. [15]

High prices and dangerous climate-changing energy emissions are fostering conservation-oriented responses similar to those prompted

World Oil Prices Respond to Events

Oil prices reached an all-time high of $78 a barrel in 1981, two years after the U.S.-Iran hostage crisis began. Prices dropped for the next 17 years as new non-OPEC (Organization of Petroleum Exporting Countries) supplies came online and demand declined. After bottoming out at $15.50 a barrel in 1998, prices have risen, largely due to increased demand from India and China, Middle East conflicts and the growing state control of oil operations around the world.*

World Crude Oil Prices, 1973-2006
(in $U.S. per barrel, adjusted for inflation)

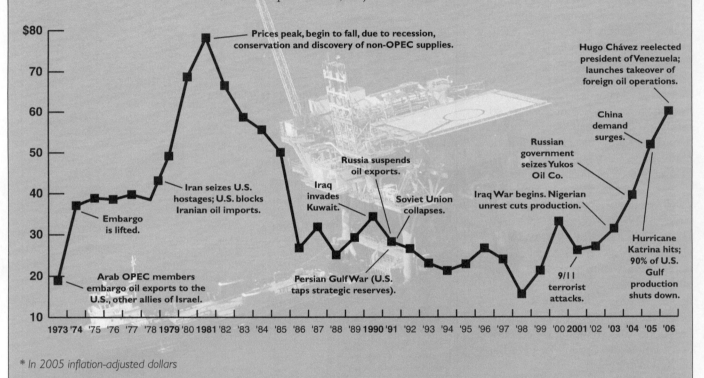

Prices peak, begin to fall, due to recession, conservation and discovery of non-OPEC supplies.

Hugo Chávez reelected president of Venezuela; launches takeover of foreign oil operations.

China demand surges.

Russian government seizes Yukos Oil Co.

Russia suspends oil exports.

Iran seizes U.S. hostages; U.S. blocks Iranian oil imports.

Iraq invades Kuwait.

Soviet Union collapses.

Iraq War begins. Nigerian unrest cuts production.

Embargo is lifted.

Arab OPEC members embargo oil exports to the U.S., other allies of Israel.

Persian Gulf War (U.S. taps strategic reserves).

9/11 terrorist attacks.

Hurricane Katrina hits; 90% of U.S. Gulf production shuts down.

* In 2005 inflation-adjusted dollars

Source: Energy Information Administration

by skyrocketing oil prices in the 1970s, including the use of smaller cars and investments in energy-efficient manufacturing, construction and appliances. [16]

But short supplies also can trigger intense competition between consuming nations, and experts are closely watching the political fallout as major powers vie for control over oil and gas resources. The construction of new pipelines to carry oil and gas from Central Asia to Asia and Europe has already sparked disputes among Russia, China and the United States, and more could follow. [17]

China's worldwide search for oil is causing particular concern because its aggressive attempts to secure important new reserves in countries such as Sudan and Myanmar (formerly Burma) have ignored human rights abuses in those countries that the international community is trying to halt, critics say. [18]

Trends in oil discoveries and price moves have long defied accurate forecasting. An escalation of Persian Gulf conflicts, a terrorist attack on Saudi Arabian oil facilities or congested sea channels could shoot oil prices past $100 a barrel. [19]

For the moment, the world is consuming oil faster than it is finding new supplies, and the historic trend of gradual increases in the world's hydrocarbon reserves has shifted to one of "stagnation and modest decline." Global oil reserves were down by nearly 1 percent in 2006, according to the *PIW's* latest reserves survey. [20]

As increased oil nationalism and global conditions trigger tight supplies, high prices, nervous markets and potential conflict, here are some questions being asked by the oil industry, its investors and critics:

Will emerging petrostates undermine OPEC's control over oil prices, benefiting consumers?

The first Arab oil embargo, in 1973, established oil as a pivotal political and economic lever. Since then, the OPEC cartel has sought to keep world oil prices high enough to maximize producers' returns without tipping global economies into recession.

It is widely assumed that OPEC's continued control over prices depends on whether emerging African, Caspian and Latin American producers reject OPEC membership and create excess global supply.

Of course, a widespread economic recession or financial crisis could slash oil demand, generating a surplus and a collapse in oil prices. In the past, OPEC has responded by cutting production to shore up prices, with mixed results.

So far, Russia has rejected OPEC requests to limit production. Neither Russia nor its Caspian neighbors are strong candidates to join OPEC, says Robert E. Ebel, senior adviser at the Center for Strategic and International Studies (CSIS). "Why would they want to join? Why would they want other people telling them what they can produce and export? They can derive all the benefits [of OPEC's pricing strategies] without being a member."

But experts disagree over whether Russia might support creation of an OPEC-style cartel for natural gas — of which it has the world's largest supply. In November 2006, a confidential NATO economic study warned Russia may be seeking to assemble a gas cartel with Algeria, Qatar, Libya, the countries of Central Asia and perhaps Iran. [21]

But Dmitry Peskov, deputy Kremlin spokesman, has denied the suggestion. "Our main thesis is interdependence of producers and consumers. Only a madman could think that Russia would start to blackmail Europe using gas, because we depend to the same extent on European customers." [22]

A banner at a natural gas plant in Tarija, Bolivia, proclaims: "Nationalized: Property of the Bolivians," after President Evo Morales nationalized foreign oil and gas operations in May 2006. "The looting by the foreign companies has ended," he declared.

AFP/Getty Images/Presidencia

Whatever Russia does, the supply-demand balance is running tight for oil and gas, even with new petrostate supplies coming online, and new conflicts in oil-rich "hot spots" would only worsen conditions. "Many of the world's major oil-producing regions are also locations of geopolitical tension," said Daniel S. Sullivan, assistant secretary of State for economic, energy and business affairs. "Instability in producing countries is the biggest challenge we face, and it adds a significant premium to world oil prices." [23]

When supplies are tight, consumers lose. And tight supplies could persist since government-controlled energy operations may not develop new reserves or build pipelines as aggressively as the international oil companies. Instead, national oil companies tend to use more of their profits to fund social improvements and provide cheap, subsidized energy for citizens. [24]

To make matters worse, demand for more energy — led by booming China and India — is accelerating. Assuming their growth bubbles don't burst, experts predict China's energy use would have to grow by 150 percent by 2020 and India's to double to maintain current economic expansion. [25]

However, China's continued growth is not a certainty, according to a study from the Stanley Foundation, in Muscatine, Iowa. [26] "China faces immense problems, including pollution, disease, poverty, inequality, corruption, abuses of power, an aging population and a shrinking labor force," contend authors Michael Schiffer, a foundation program officer, and Gary Schmitt, director of the American Enterprise Institute's advanced-strategic studies program. "China's leaders today are, thus, holding a tiger by the tail. They have built the legitimacy of their continued rule largely on meeting the rising expectations of a billion-plus people, but to meet those expectations they eventually have to release the reins of economic and political power they are clutching so tightly." [27]

Some experts hope China and India — which are eyeing Persian Gulf oil — could eventually add their considerable consumer weight to efforts by others to restrain OPEC's pricing strategies. "Much of the recent discussion in Washington about the growing oil demand of China — and to a lesser extent India — has focused on the threats posed to the U.S. economy and foreign policy, but that often obscures the fact that the oil interests of China, India and the United States are also broadly aligned," writes Xuecheng Liu, a senior fellow at China's Institute of International Studies. [28]

Will nationalizing oil wealth help the poor?

In May 2006, newly elected President Evo Morales ordered troops to occupy Bolivia's oil and gas fields and gave foreign companies 180 days to renegotiate their energy leases or leave the country. "The looting by the foreign companies has ended," he declared. [29]

Morales was elected partly on a populist platform to take over energy resources in Bolivia, which has Latin America's second-largest gas reserves after Venezuela. "We are the owners of this noble land," he said during the campaign, "and it is not possible that [natural resources] be in the hands of the transnationals." [30]

Echoing former Mexican President Lazaro Cardenas, who nationalized 17 foreign oil companies in 1938, leaders of Bolivia, Ecuador and Venezuela have called their energy reserves a critical tool for helping poor, indigenous populations.

Since the oil age began more than a century ago, governments in the developing world — on both the right and left — have promised their people a fair share of the wealth created by geological forces. But few leaders have followed through. Instead, "black gold" has spawned corruption, economic hardship, vast class differences and civil war.

"Look what oil is doing to us, to the oil-exporting countries," said OPEC

Have the World's Oil Supplies Peaked?

After 50 years, the debate continues

It's called the "peak oil" theory, and ever since American geologist M. King Hubbert developed it in 1956, oil experts have been divided into two camps — those who believe Earth's oil supplies have peaked, and those who don't.

If proponents of the theory are correct — that the world has used up half of the planet's oil stocks and the remaining supplies will face rapid depletion — the future promises even higher prices and more energy shocks. But critics of the theory say the high point in oil production is still 20 or 30 years away, that oil production is not likely to decline precipitously thereafter and that political events and energy prices — not hydrocarbon shortages — will dictate the industry's course until near the mid-century mark.

According to industry estimates, world oil reserves increased by 24 billion barrels during 2006 to 1.3 trillion barrels — a gain of about 2 percent over 2005. [1] Reserve estimates are periodically recalculated based on new geological and engineering data and new discoveries. But the 2006 increase cannot be documented because two of the countries reporting the greatest increases were Iran and Saudi Arabia, and their governments don't let outsiders check their figures.

"No one knows the amount of oil really contained in reservoirs," says Leonardo Maugeri, an economist and oil industry analyst with the Italian oil and gas company ENI. Such knowledge evolves over time after new wells are drilled and more sophisticated technology is developed.

"In fact," he adds, "countries such as Saudi Arabia or Iraq (which together hold about 35 percent of the world's proven reserves of oil) produce petroleum only from a few old fields,

although they have discovered, but not developed, more than 50 new fields each." [2]

The peak oil argument begins with the controversial 1956 prediction by Hubbert that oil production from the lower U.S. 48 states would top out in 1968. The actual peak occurred two to four years later, depending on which measure of oil production is used. As a result of Hubbert's controversial prediction, "He found himself being harassed and vilified," says one of Hubbert's champions, Chris Skrebowski, editor of the monthly magazine *Petroleum Review*, published by the Energy Institute in London. [3]

But Peter M. Jackson, a director of the international research firm Cambridge Energy Research Associates (CERA), argues Hubbert erred in not considering how new drilling technologies could increase output from older fields or how energy prices affect exploration and production. [4]

He is even more critical of Hubbert's present-day disciples who say an oil field peaks when half of its available oil has been extracted. Their model is illustrated with a simple, smoothly rising and falling bell-shape curve.

Jackson says Hubbert's curve ignores the typical expansion of oil field dimensions as more exploration and development occurs. Oil production from the lower 48 states since 1970 has been 66 percent higher and 15 billion barrels greater that Hubbert predicted, Jackson writes, citing U.S. Geological Survey findings and his company's oil field analysis.

When admittedly high-priced, "unconventional" sources such as shale and tar sands or Arctic fields are counted, the world's total supply of oil is 4.8 trillion barrels, Jackson stated. That is enough, at current growth rates, to delay a peak until 2030 or later, and even then, the peak will not be followed by a sharp decline, he said.

BP chief economist Peter Davies complains that Hubbert's theory also ignores the impact of increased conservation and the switching to alterative fuels that occurs as oil prices rise, which tend to extend oil supplies. Since 1980, for instance, the world's economic output has doubled while oil consumption has only increased by a third, he noted in a June 14, 2006, speech in London. "Year by year, a combination of exploration investment and the application of technology is ensuring that every unit of oil and gas that is produced is replaced by new proved reserves," he said. [5]

Jackson likened peak oil advocates to sidewalk doomsayers who predict the end of the world. "Peakists continue to criticize those who disagree, but their projections of the date of the peak continue to come and go," he said in his CERA report. "One of the most recent peak oil dates was supposed to have occurred just after the U.S. Thanksgiving Day 2005, and we still wait for the evidence."

Skrebowski replied furiously that Jackson and the anti-peak oil crowd were either Polyannas or paid shills for an oil industry that must persuade investors that untapped oil abounds. [6]

But, when one gets beyond the name calling, the two sides appear less far apart. Skrebowski says Jackson's 4.8 trillion barrels may be technically available "but is only of interest if it can be discovered, mobilised and marketed within a reasonable time period. "This," he says, "is the entire debate: Can all the unfound and unproven resources be exploited quickly enough to more than offset the peaking and decline of the known and proven reserves?"

A leading peak oil advocate, Dallas energy financier Matthew

R. Simmons, argues that Saudi Arabia's reserves are being greatly overestimated. [7] But he also says more than half the world's conventional oil and a larger share of its unconventional oil remain to be extracted. "What the world is running out of is cheap oil — the $20 oil we built our civilization around," he writes. [8]

That sounds close to the views of CERA chairman Daniel Yergin. However, he asks, will economics and government decisions in a politicized oil world permit enough new exploration and production to keep pipelines full?

Although energy companies will be prospecting in more difficult environments, he says, "the major obstacle to the development of new supplies is not geology but what happens above ground: namely, international affairs, politics, decision-making by governments and energy investment and new technological development." [9]

[1] "World Proved Reserves of Oil and Natural Gas," Energy Information Administration. Jan. 9, 2007; www.eia.doe.gov/emeu/international/reserves.html.
[2] "The Cheap Oil Era is Far from Over," *Alexander's Oil and Gas Connections*, June 2, 2004; www.gasandoil.com/goc/features/fex42299.htm.
[3] Chris Skrebowski, "Open letter to Peter Jackson of CERA," *Energy Bulletin*, Dec. 21, 2006; www.energybulletin.net/23977.html.
[4] Peter M. Jackson, "Why the 'Peak Oil' Theory Falls Down," Cambridge Energy Research Associates, Inc., Nov. 10, 2006; http://cera.ecnext.com/coms2/summary_0236-821_ITM.
[5] Peter Davies, "BP Statistical Review of World Energy 2005," presentation, London, June 14, 2006, p. 9.
[6] Skrebowski, *op. cit.*
[7] "Twilight in the Desert," *The Oil Drum*, June 13, 2005; www.theoildrum.com/classic/2005/06/twilight-in-desert.html.
[8] Randy Udall and Matthew R. Simmons, "CERA's Rosy Oil Forecast — Pabulum to the People," *ASPO-USA's Peak Oil Review/Energy Bulletin*, Aug. 21, 2006.
[9] Daniel Yergin, "Ensuring Energy Security," *Foreign Affairs*, March/April 2006, p. 75 www.foreignaffairs.org/20060301faessay85206/daniel-yergin/ensuring-energy-security.html.

founder Juan Pablo Pérez Alfonzo, a Venezuelan, nearly 30 years ago. [31] "It is the excrement of the devil."

Oil bonanzas often leave developing economies worse off — a phenomenon economists call the "resource curse." [32] PEMEX, Mexico's state-run oil company, pays an estimated 60 percent of oil earnings to fund government programs. But Mexico has overborrowed to keep production going and has more than $30 billion in pension liabilities, leaving it with a huge longstanding debt and too little money for maintaining old oil fields or finding new ones. [33] And Mexico's biggest field is in decline, raising fears that a chronic slippage

in oil revenues could trigger a budget disaster. [34]

Similarly, OPEC members had an average gain of 1.3 percent in per capita gross domestic product (GDP) between 1965 and 1980, while the rest of the world saw GDP grow 2.2 percent annually. [35]

Sudden oil windfalls have also triggered what economists call the "Dutch disease" — skyrocketing currency values that depress local manufacturers' exports and trigger huge jumps in imports. The economic paradox got its nickname from a drastic decline in economic growth in the Netherlands after natural gas was discovered there in the 1960s. [36]

Oil's easy money also often ends up filling government officials' Swiss bank accounts rather than benefiting public health or education. Some of the most egregious excesses are in Africa. Since oil was discovered in Nigeria's Niger Delta in 1956, for example, the country's infamous kleptocracy has used oil billions to enrich elites, leaving delta residents trapped in pollution and poverty. "Everything looked possible — but everything went wrong," *National Geographic's* Tom O'Neill reports. [37]

Now the situation "has gone from bad to worse to disastrous," said Senan Murray of BBC News. [38] The Movement for the Emancipation of the Niger Delta (MEND) has stepped up

Saudi Arabia and Russia Have Biggest Reserves

Saudi Arabia and Canada lead the world in oil reserves, with nearly 450 billion barrels — more than half as much as the next 10 nations combined. Russia has the most natural gas reserves with 1.68 quadrillion cubic feet — almost three-quarters more than Iran.

Oil Reserves*			Natural Gas Reserves*		
Rank	Country	Barrels (in billions)	Rank	Country	Cubic ft. (in trillions)
1.	Saudi Arabia	262.3	1.	Russia	1,680.0
2.	Canada	179.2	2.	Iran	974.0
3.	Iran	136.3	3.	Qatar	910.5
4.	Iraq	115.0	4.	Saudi Arabia	240.0
5.	Kuwait	101.5	5.	United Arab Emirates	214.4
6.	United Arab Emirates	97.8	6.	United States	204.4
7.	Venezuela	80.0	7.	Nigeria	181.9
8.	Russia	60.0	8.	Algeria	161.7
9.	Libya	41.5	9.	Venezuela	152.4
10.	Nigeria	36.2	10.	Iraq	112.0
11.	Kazakhstan	30.0	11.	Kazakhstan (tie)	100.0
12.	United States	21.8	11.	Turkmenistan (tie)	100.0

* As of Jan. 1, 2007

Source: "World Proved Reserves of Oil and Natural Gas, Most Recent Estimates," *Energy Information Administration,* Jan. 9, 2007

attacks on foreign oil facilities and the police who protect them, including an oil rig 40 miles offshore. In May, six Chevron employees were kidnapped and released after a month, but other kidnappings followed. [39] The oil companies — in conjunction with the Nigerian government — have pledged to support rural education, environmental cleanup and other social programs, but armed rebels in the delta say improvements aren't being implemented fast enough. [40]

In Venezuela, Chávez has kept his promises to channel petrodollars to health care, roads and housing. The percentage of Venezuelans living in poverty has shrunk from 42.8 percent to 30.4 percent under Chávez, according to government statistics. Researchers at Catholic University, near Caracas, es-

timate that about 45 percent of the population lives in poverty, less than in 1999. [41]

Chávez also uses oil money to promote his anti-capitalism ideology by investing in social programs in other Latin American countries. But he hasn't made a dent in Venezuela's chronic corruption, according to Transparency International. The Berlin-based nonprofit puts Venezuela in the bottom quarter of its 2002 and 2006 rankings. [42]

At the same time, the Washington-based advocacy group Freedom House says Chávez has presided over the "deterioration of the country's democratic institutions," replacing the Supreme Court, filling civilian government posts with military personnel, blacklisting political opponents from government positions and shutting

down a leading opposition television station. [43]

Russia, China, Mexico and Iran also provide cheap, subsidized energy to their populations, in a tradeoff that carries a stiff economic price. The policy has backfired in Iran, where the government imposed gasoline rationing in June 2007, triggering violent protests that led to more than a dozen gas stations being set on fire. [44] Iran's subsidized gasoline prices are among the lowest in the world, so Iranian motor fuel consumption has been climbing fast. But the government was forced to ration gasoline because it has not used its oil profits to build enough refinery capacity, and gasoline imports have not kept up with demand.

Oil wealth has generated violence and even civil war in many developing countries. For instance, factions from northern and southern Sudan, where oil was discovered in 1978, fought a civil war in the 1980s over the nation's oil revenue. Although a peace accord was signed in 2005, the largely Arab and Islamist ruling party in the north has dragged its feet on sharing the oil wealth with the largely black, Christian southerners.

Meanwhile, some analysts say oil has played a key role in the international community's failure to stop the rape, murder and wholesale destruction of villages in western Sudan's Darfur region, where the Coalition for Darfur says as of 2005 Sudanese militia reportedly had killed 140,000 villagers, 250,000 have perished from disease, famine or exposure and 2 million more are homeless. The Sudanese government disputes the figures. [45]

Until recently, U.N. Security Council efforts to sanction Sudan have been hampered by China, which buys two-thirds of Sudan's oil and has invested more than $8 billion in its oil sector. [46] "Business is business," said Deputy Foreign Minister Zhou Wenzhong in 2004. "We try to separate politics from business." [47]

But this year, after critics threatened to make Darfur an issue during China's preparations to host the 2008 Summer Olympic Games, China shifted course. It now supports a combined U.N.-African Union peacekeeping force in Sudan, which Sudan agreed to accept in June. However, skeptics doubt the agreement will be fully carried out. [48]

In an effort to buffer the negative impact of oil wealth on developing countries, industrialized nations have launched the Extractive Industries Transparency Initiative, announced by then British Prime Minister Tony Blair in October 2002. By requiring oil, gas and other "extractive" companies to report what they pay foreign governments for their natural resources, the initiative aims to expose corruption and foster accurate reporting of oil revenues and spending.

"Knowing what companies pay and what governments receive is a critical first step" to creating accountability in the handling of oil wealth, says the initiative's statement of purpose. [49] Members include industrialized countries as well as the World Bank, major oil companies and about 20 oil-producing developing nations.

However, transparency efforts are still hampered by national oil companies that keep their energy books closed and ignore international accountability guidelines.

Nevertheless, BP chief economist Peter Davies is optimistic about the initiative. "There is still a broad tendency toward transparency," he says. "There are forces that counteract this from time to time, [but] the forces for progress are there."

Will the growing competition for energy trigger new international conflicts?

The Cold War that dominated the last half of the 20th century was about ideology. As a new century begins, a widely shared concern is that energy will become a new arena for super-power or regional confrontations.

A forest of oil derricks lines the Caspian shore just outside of Azerbaijan's capital Baku. The oil-rich Caucasus republic is expected to be a significant source of the world's oil in the future, some of it delivered via new pipelines.

AFP/Getty Images/Mladen Antonov

Conflicts over oil historically have centered in the Middle East. Now, because of the new petrostates, other hot spots claim attention in Central Asia, Africa and Latin America. The risks are magnified by the recent escalation of energy prices, which have made oil and natural gas resources an even bigger prize for rulers seeking to take or keep power.

New York Times columnist Thomas L. Friedman recently described a perverse relationship between oil prices and democracy: The higher oil prices go, the more democracy suffers and authoritarianism grows in the countries with oil. "Not only will some of the worst regimes in the world have extra cash for longer than ever to do the worst things," Friedman wrote, "but decent, democratic countries — India and Japan, for instance — will be forced to kowtow or turn a blind eye to the behavior of petro-authoritarians, such as Iran or Sudan, because of their heavy dependence on them for oil. That cannot be good for global stability." [50]

Japan and China see themselves competing for access to natural gas

reserves in eastern Russia. Poland fears that Russia's construction of a new "North Stream" natural gas pipeline to Germany, now under way, will enable Russia to cut gas deliveries to Poland if tensions between those two countries erupt. [51] (A large portion of Russia's lucrative gas sales to Germany now transit through Poland, but that route could be bypassed by the North Stream project, Polish leaders fear.)

In Latin America, Bolivia's seizure of majority control over its natural gas industry in 2006 was a direct challenge to Brazil, which needs Bolivia's gas and whose state energy company Petrobras is a major gas producer in Bolivia. [52]

Some experts especially worry about the possibility of conflicts over energy between the United States and China, which is on a path to challenge U.S. economic and military leadership within two decades unless its hyper-growth spins out of control. Maureen S. Crandall, a professor of economics at the Industrial College of the Armed Forces, says that while China badly wants to import oil and natural gas from eastern Russia, it is not clear that pipelines will be built to deliver those resources. So China is looking hard at Caspian gas production and at the prospects for a pipeline through Iran to bring gas to seaports for export in liquefied form aboard tankers. [53]

That puts China in opposition to the Bush administration's top-priority campaign to isolate Iran to prevent it from developing nuclear weapons — a goal Iran denies it is seeking. The Iran issue headed America's agenda for the U.S.-China Senior Dialogue between top diplomats from both nations in June 2007, while China pushed for assurances the United States was not boosting its support for China's rival, Taiwan. [54]

The two nations are not consciously pointed toward conflict, says the National Intelligence Council's 2020 Project report — the most recent public forecast by the CIA's research arm. "[T]he growing dependence on global financial and trade networks increasingly will act as a deterrent to conflict among the great powers — the U.S., Europe, China, India, Japan and Russia," says the report. [55]

But, the report adds, inadvertent conflicts could erupt as a result of growing oil nationalism, the lack of effective international conflict-resolution processes or raw emotions exploding over key issues. For instance, a naval arms race could develop between China, intent on protecting vital seaborne oil shipments, and the United States, determined to maintain strategic leverage in Asian waters. While China's interest "lies with a peaceful and stable regional and international order," write Schiffer and Schmitt, China's ambitions or internal political conflicts could take it in a different direction. [56] Prudence favors maintaining a credible U.S. military posture in Asia, they argue, but if U.S. actions are seen as a bid for supremacy or a check on China's rightful regional role, "it might fuel further resentments and

incite precisely the reaction we don't seek, a redoubling of countervailing military, economic and diplomatic strategies."

"The United States and China are not seeking to make war on one another," agrees Michael Klare, a political science professor at Hampshire College. "But they are inadvertently contributing to the risk of conflict in Africa and Central Asia by using arms transfers as an instrument of influence."

China, for instance, has sent troops to Sudan to protect its energy investment there, he points out, and the U.S. military maintains a presence in Central Asia. In the same vein, former Chinese deputy chief of staff Gen. Xiong Guangkai told an international conference on energy security last December that "the strategic race for the world's energy may result in regional tension and even trigger a military clash." [57]

The recent deterioration of U.S.-Russian relations is a case study of what should not be allowed to happen between the United States and China, say some experts. The dialogue has grown raw, escalated by Russia's sharp swing toward an aggressive nationalism. But the division also has been fostered by arrogant and shortsighted U.S. moves over the past 15 years that treated Russia as a defeated world power and dictated terms to them instead of seeking a working relationship, says Blair Ruble, director of the Kennan Institute in Washington.

"It has been a bipartisan failure," adds Ruble's colleague, program associate F. Joseph Dresen. After the Soviet Union's collapse, the United States "had tons of leverage" but "we needed more influence. It starts with diplomacy."

A win-win relationship with China that minimizes potential for conflict "will take far more sophistication than U.S. policymakers from either political party have previously shown," Schiffer and Schmitt conclude. [58] ∎

BACKGROUND

OPEC Is Born

In 1960, representatives of Iran, Iraq, Kuwait, Saudi Arabia and Venezuela met in Baghdad to form a cartel designed to stabilize world oil markets. Today the 12-member Organization of Petroleum Exporting Countries — now based in Vienna, Austria — also includes Qatar, Indonesia, Libya, the United Arab Emirates, Algeria, Nigeria and Angola. Ecuador and Gabon joined in the '70s but dropped out in the '90s. [59]

Despite the cartel's promise of stability, oil markets have been chaotic since the 1970s, characterized by four distinct periods.

Two oil shocks hit world energy markets in the 1970s. Resentful of U.S. efforts to suppress oil prices and angered by U.S. support for Israel in the 1973 Yom Kippur War, several Arab OPEC members on Oct. 17, 1973, imposed an oil embargo on the United States and other countries aiding Israel, followed by a production cut. [60] The world suddenly faced a crude-oil shortage of 4 million barrels a day, 7 percent below demand. Prices shot up from $3 a barrel to $12. [61] Long lines formed at gasoline pumps in the United States and some European countries.

To limit the impact on American consumers, President Richard M. Nixon imposed price controls on the U.S. economy, and President Gerald Ford created the U.S. Strategic Petroleum Reserve, which today holds more than 688 million barrels of crude oil in underground caverns. [62]

The embargo ended five months later — in March 1974 — after Arab-Israeli tensions eased. Egyptian President Anwar el-Sadat, intent on moving toward a peace agreement, argued

Continued on p. 164

Chronology

1951-1979

Oil surpluses keep crude prices low; U.S. restricts oil production to maintain prices.

1951
Soviet Union builds first deep-sea oil platform.

1956
Geologist M. King Hubbert's "peak oil" theory contends half of U.S. oil stocks would be depleted by the 1960s, and the remaining supplies face rapid depletion.

1960
Iran, Iraq, Kuwait, Saudi Arabia and Venezuela form the Organization of Petroleum Exporting Countries (OPEC) to stabilize world oil prices.

1970s
OPEC gains control of global oil pricing; Arab countries begin using oil as a political weapon.

1972
Oil production from Lower 48 states peaks; limits on U.S. production are lifted.

1973
Major Arab oil producers impose embargo on oil exports to United States and several allies in retaliation for their support of Israel in Yon Kippur War; oil prices quadruple.

1979
Shah flees Iran; Iranian students seize hostages at U.S. Embassy, triggering more price shocks.

1980s

Oil from non-OPEC sources breaks the cartel's market hold, helping to create an oil glut.

1980
Iraq attacks Iran, triggering an eight-year war.

1981
Global oil prices drop after a severe recession.

1983
Production from the North Sea and Alaska's North Slope swells global oil supply.

1985
Saudis boost output; prices plummet.

1988
Iran-Iraq War ends.

1990s

Breakup of Soviet Union raises hope for development of Caspian Sea oil and gas; oil production increases in Africa; global warming emerges as environmental issue.

1990
Iraq invades Kuwait.

1991
U.S.-led coalition drives Iraq from Kuwait; Soviet Union collapses.

1993
Crude prices drop to $15 a barrel.

1996
Giant Sakhalin oil project announced in Russian Far East.

1997
Violence, protests disrupt Nigerian and Colombian production; Caspian pipeline consortium formed to deliver Caspian Sea oil to Black Sea ports; Kyoto global warming protocol drafted.

1999
Oil production flattens; prices rise.

2000s

Terrorist attacks in U.S. lead to new Iraq war; China becomes fastest-growing oil importer; oil prices climb.

Sept. 11, 2001
Arab terrorists attack World Trade Center, Pentagon; oil prices surge.

2002
Oil workers strike in Venezuela.

2003
Iraq War begins; attacks close some oil platforms in Nigeria. . . . Major Iraq pipeline is sabotaged; violence escalates.

2004
Oil production in Russia, former Soviet states continues to recover, surpassing 1991 Soviet Union totals.

2005
China's oil demand soars. . . . Hurricane Katrina strikes the U.S. Gulf Coast, shutting down nearly 90 percent of oil and gas production in federal waters.

2006
Venezuelan President Hugo Chávez reelected, launches takeover of foreign-run oil operations. . . . Bolivian President Evo Morales announces the nationalization of all remaining natural gas reserves in the country. . . . Baku-Tblisi-Ceyhan pipeline opens, bypassing the Bosporus Strait.

2007
In a tariff dispute with Belarus, Russia's state-owned Transneft oil company shuts down a pipeline supplying oil to several European countries. . . . Dissidents attack three major pipelines in Nigeria's Niger Delta. . . . On May 1, Chávez takes control of the last remaining privately run oil operations in Venezuela.

Continued from p. 162

successfully that the "oil weapon had served its purpose." [63]

But memories of the embargo continued to drive a search for new energy policies. On April 18, 1977, shortly after being inaugurated, President Jimmy Carter warned about America's overdependence on foreign oil supplies, calling the energy crisis "the moral equivalent of war." With the exception of preventing war, Carter said, "this is the greatest challenge our country will face during our lifetimes." [64]

Then in early 1979, after a year of paralyzing strikes and demonstrations by supporters of militant Iranian Shia Muslim cleric Ayatollah Ruhollah Khomeini, Iran's Shah Mohammad Reza Pahlavi fled Tehran, opening the door to the founding of an Islamic republic.

As the impact of the Iranian Revolution on world oil prices began to be felt, Carter in July 1979 unveiled a comprehensive energy plan to help America combat its overdependence on unstable Middle Eastern oil, promoting conservation, alternative fuels and higher taxes on gasoline and gas-guzzling cars. [65]

Four months later, on Nov. 4, Islamist zealots and students took over the U.S. Embassy in Tehran, holding 52 hostages for 444 days — until Ronald Reagan replaced Carter. [66] During the crisis, oil prices nearly doubled. [67] World oil markets got even tighter in 1980, when Iran's oil production nearly dried up after Iraq invaded — beginning an eight-year-long conflict. Panic purchases by governments, companies and consumers made the shortage worse, and,

World Crude Supplies Remain Vulnerable

Oil 'hot spots' are most at risk

On Feb. 24, 2006, a small band of al Qaeda gunmen attacked Saudi Arabia's giant oil processing facility at Abqaiq — the first such attack since terrorist leader Osama bin Laden publicly targeted Saudi oil installations in a 2004 audio message.

Although the Saudis repulsed the assault, the incident was a wake-up call as to what terrorists' intentions were concerning oil supplies, warned Simon Henderson, director of the Gulf and Energy Policy Program at the Washington Institute for Near East Policy. "Saudi oil production remains extremely vulnerable to sabotage," he wrote shortly after the attack, and the kingdom's estimated 12,000 miles of pipelines are also "at particular risk." A Saudi police raid on a terrorist hideout the previous year had reportedly uncovered copies of maps and plans of the new Shaybah oil field, he pointed out. [1]

Had the terrorists succeeded in destroying the sulfur-clearing towers at Abqaiq — through which about two-thirds of Saudi crude passes — it would have driven the price of crude to more than $100 a barrel for months, perhaps even up to bin Laden's goal of $200 a barrel, according to R. James Woolsey, a former CIA director. [2]

World leaders have been warning since the onset of the Industrial Age that the key to energy security lies in diversification of supplies. When Winston Churchill — then the First Lord of the Admiralty — shifted the Royal Navy from coal to oil on the eve of the First World War, he presciently warned, "Safety and certainty in oil lie in variety and variety alone." [3]

The conflicts and crises that have periodically disrupted Middle East oil supplies — from the oil shocks of the 1970s to Saddam Hussein's invasion of Kuwait in 1990 — have repeatedly reinforced the wisdom of Churchill's advice: find more sources of oil outside the Persian Gulf.

Today, the world is once again seeking to diversify its energy supplies, turning to sources in Central Asia, Africa and Russia. But while the emergence of these rising petrostates has increased the diversity of energy supplies, it has not increased energy security. Many of those new producers appear along with Saudi Arabia on the U.S. Energy Information Administration (EIA) list of various "hot spots" in world oil markets.

Saudi Arabia tops the list, but it is followed by other "emerging" oil producing states: Russia, Iran, Iraq, Nigeria, the Caspian region, Sudan, Venezuela and seven other countries where energy facilities are considered at risk from saboteurs or unstable domestic policies. [4] The EIA projects that by 2010 at least 50 million barrels of oil per day — 58 percent of worldwide daily production — will be in jeopardy because it originates or passes through oil hot spots.

"The security of the energy infrastructure is becoming progressively in doubt," says Massachusetts Institute of Technology Professor John Deutch, also a former CIA director. "Oil facilities, pipelines [and] control systems for the energy distribution systems are all very much more vulnerable to terrorist attack and national disaster." [5]

The choke points for seaborne oil — and, increasingly, natural gas — create some of the worst risks. According to Daniel Yergin, author of *The Prize: The Epic Quest for Oil, Money & Power*, those ocean chokepoints include the:

- Strait of Hormuz, at the entrance to the Persian Gulf;
- Suez Canal, which connects the Red Sea and the Mediterranean;
- Bab el Mandeb Strait at the Red Sea's entrance;
- Bosporus Strait, a major transit channel for Russian and Caspian oil; and
- Strait of Malacca between Malaysia and Indonesia, a conduit for 80 percent of the oil used by Japan and South Korea and about half of China's oil. [6]

The Malacca strait is only 15 miles wide at its narrowest point, and if terrorists or pirates scuttled a ship at that choke point it could disrupt supplies for a long time, Yergin warns.

"It may take only one asymmetric or conventional attack on a Ghawar [Saudi oil field] or tankers in the Strait of Hormuz to throw the market into a spiral, warns Anthony H. Cordesman, a scholar at the Center for Strategic and International Studies in Washington. [7]

"Assuring the security of global energy markets will require coordination on both an international and a national basis among companies and governments, including energy, environmental, military, law enforcement and intelligence agencies," Yergin writes. "But in the United States, as in other countries, the lines of responsibility — and the sources of funding — for protecting critical infrastructures, such as energy, are far from clear."

Countries are trying a wide range of policies and practices to increase security of energy production and delivery, experts say. Colombia has military units — trained and partly supplied by the United States — tasked with combating rebel attacks on oil pipelines. The natural gas networks of Qatar and the United Arab Emirates are being connected to shipping terminals in Oman that lie outside the vulnerable Strait of Hormuz. [8] China is expanding its naval forces in order to protect oil shipments through Asian sea lanes where piracy is a threat.

But Gal Luft, executive director of the Institute for the Analysis of Global Security in Washington, says security efforts have been hampered by uncertainty over whether private companies or governments should pay for the additional security.

"NATO is looking into defining the roles of industry and government," Luft says. "Each wants the other to do more. In places where you can introduce technology or more manpower economically, you do it. But on the ground not a lot is happening."

Building in redundancy and the availability of alternative sources are also popular strategies for assuring energy deliv-

Separatist rebels show their firepower in Nigeria's oil-rich Niger Delta in February 2006. Insurgents have kidnapped foreign oil workers and sabotaged oil facilities to protest the slow pace of economic development in the delta.

eries, says Mariano Gurfinkel, associate head of the Center for Energy Economics at the University of Texas. "Since it is very hard to avoid all incidents on all elements of the energy infrastructure, efforts are made to minimize the consequences."

[1] Simon Henderson, "Al-Qaeda Attack on Abqaiq: The Vulnerability of Saudi Oil," Washington Institute for Near East Policy, www.washingtoninstitute.org/templateC05.php?CID=2446.

[2] R. James Woolsey, "Global implications of Rising Oil Dependence and Global Warming," testimony before the House Select Committee on Energy Independence and Global Warming, April 18, 2007, p. 2.

[3] Daniel Yergin, "Ensuring Energy Security," Foreign Affairs, March/April 2006, p. 69.

[4] "World Energy Hotspots," Energy Information Administration, Sept. 2005, www.eia.doe.gov/emeu/cabs/World_Energy_Hotspots/Full.html.

[5] John M. Deutch, testimony before the House Foreign Affairs Committee, March 22, 2007.

[6] Yergin, op. cit., p. 79.

[7] Anthony H. Cordesman, "Global Oil Security," Center for Strategic and International Studies, Nov. 13, 2006, p. 14.

[8] Energy Information Administration, "Oman" country analysis, April 2007, www.eia.doe.gov/emeu/cabs/Oman/NaturalGas.html.

once again, motorists in industrialized countries queued up at gas stations.

The 1970s price shocks triggered a determined campaign to reduce energy dependence. Congress in 1975 directed U.S. auto manufacturers to double the efficiency of their cars within a decade, and businesses made serious efforts to shrink energy use. [68]

But the pendulum would soon be reversed. A sharp recession stunted energy demand, the search for oil outside the Persian Gulf intensified and

the balance between supply and demand was set to shift again.

Oil Glut

Discoveries and exploitation of vast oil and gas reserves in the North Sea, Mexico and Alaska's North Slope in the early 1980s led to a tide of new production, tipping events in consumers' favor.

North Sea development, called "one of the greatest investment pro-

jects in the world," required intrepid drilling crews, path-breaking technology and platforms able to withstand crushing waves and 130-mile-per-hour winds. [69] By the early 1980s, daily North Sea production had reached 3.5 million barrels — more than Kuwait and Libya combined — and a new 800-mile pipeline to the port of Valdez from Alaska's landlocked North Slope was supplying up to 2 million barrels of oil a day to the Lower 48 states — a quarter of U.S. production. [70] In 1985, non-OPEC production had increased

Majority of Oil Companies Are State-Owned

Thirteen of the world's 25-largest oil companies are entirely owned or controlled by national governments, including all the companies in the Middle East; three other oil firms are partially state-owned. In 1973, by comparison, roughly three-quarters of the world's oil production was managed by the privately owned "Seven Sisters" — the seven major Western oil companies.*

World's Largest Oil Companies

Rank (2005)	Company	Country of origin	Percentage of firm owned by state
1	Saudi Aramco	Saudi Arabia	100
2	Exxon Mobil	United States	0
3	NIOC	Iran	100
4	PDVSA	Venezuela	100
5	BP	United Kingdom	0
6	Royal Dutch Shell	United Kingdom/Netherlands	0
7	PetroChina	China	90
8	Chevron	United States	0
8	Total	France	0
10	Pemex	Mexico	100
11	ConocoPhillips	United States	0
12	Sonatrach	Algeria	100
13	KPC	Kuwait	100
14	Petrobras	Brazil	32
15	Gazprom	Russia	50.002
16	Lukoil	Russia	0
17	Adnoc	United Arab Emirates	100
18	Eni	Italy	0
19	Petronas	Malaysia	100
20	NNPC	Nigeria	100
21	Repsol YPF	Spain	0
22	Libya NOC	Libya	100
23	INOC	Iraq	100
24	EGPC	Egypt	100
24	QP	Qatar	100

* The Seven Sisters were: Exxon, Mobil, Chevron, Texaco, Gulf, Shell, British Petroleum

Source: Petroleum Intelligence Weekly

by 10 million barrels a day over 1974 levels, more than double the cartel's daily output. [71]

Moreover, by 1983 energy conservation was working. Americans were consuming less gasoline than in 1973, even with more cars on the road, and the U.S. economy had become 25 percent more energy efficient. Conservation efforts in Europe and Japan also were cutting consumption. [72] The two trends sent energy prices into a nosedive. By 1985 crude was below $10 a barrel ($20 in inflation-adjusted, 2006 prices), prompting Saudi Arabia to abandon efforts to control cartel production and boost its own output. Analysts have since interpreted the Saudis' decision as a strategic move to hamper the ability of Iran and Iraq to continue their war, raging just across the Saudi border. Others say the move hastened the demise of communism — by draining the Soviet Union's treasury at a time when it was facing rising internal pressures and fighting a war in Afghanistan.

But the lower oil prices also knocked the wind out of the conservation movement. The push to continue raising vehicle performance stalled in Congress, and gas-slurping minivans and SUVs became wildly popular. [73]

New Petrostates

Oil prices spiked briefly in 1991 after Saddam Hussein invaded Kuwait, and the U.S.-led coalition counterattacked, knocking out 3 percent of world oil output. After Iraq's defeat, the oil industry focused on the rising petrostates in Africa and Central Asia and on the collapse — and stunning recovery — of Russia's oil production. [74]

The Caspian Sea — about the size of California — holds one of the world's oldest-known concentrations of petroleum. The Caspian has long triggered fears that its oil, known since Alexander the Great's day, would become a conflict flashpoint. "It will be sad to see how the magnet of oil draws great armies to the Caucasus," wrote journalist Louis Fischer in 1926. [75]

The Caspian is bordered by Russia on the northwest, Kazakhstan on the north and east, Turkmenistan to the east, Iran to the south and Azerbaijan in the west. (*See map, p. 154.*) The rise of the independent former Soviet satellites triggered extravagant hopes that the Caspian could become "the Middle

East of the next millennium." The State Department fanned the hyperbole, estimating Caspian oil reserves at 200 billion barrels, or 10 percent of the world's total potential reserves. [76]

Then developers began hitting dry holes, and war and separatist violence spread through the region. Caspian countries disagreed over how to divide the Caspian's energy reserves and whether the Caspian is, in fact, a "sea" or a "lake" — a definition that could affect the ultimate distribution. "The dreams have faded as the hard realities of energy development and politics have set in," says economist Crandall at the Industrial College of the Armed Forces, who predicts Caspian reserves will top out at 33-48 billion barrels, or 3 percent of the world's total. [77]

But even with the lower estimates, the Caspian reserves still are larger than Alaska's North Slope, big enough to attract not only Russia and Iran but also Europe and China. By 2010, the Energy Information Agency projects the Caspian region will be producing 2.9-3.8 million barrels a day — more than Venezuela. [78]

Dreams for a birth of democracy in the region also have faded. Most of the region's governments have become more authoritarian and corrupt since the demise of the Soviet Union, says Martha Brill Olcott, a senior associate at the Carnegie Endowment for International Peace. [79] Indeed, says Crandall, most Central Asian states are "one-bullet regimes" that would fall into chaos if current leaders were deposed. [80]

In Africa, the discovery of oil in Algeria in 1955 — and later in the Niger Delta and Libya — seemed like gifts from the gods for the planet's poorest continent. The riches lured flocks of petroleum companies.

As exploration expanded, Africa's proven reserves more than doubled from 1980 to 2005, to 114.3 billion barrels, far ahead of overall reserve gains worldwide. In 2004, Nigeria ranked eighth among the world's

biggest oil exporters, followed by No. 10 Algeria and 12th-place Libya. [81] Angola soon joined Africa's oil club: In the past 10 years, Angola's estimated oil reserves have nearly tripled and its crude oil production doubled. [82]

Motorists in London queue up for petrol in 1973. The world's first oil shock was caused by an Arab oil embargo, which established oil as a pivotal political and economic lever.

Oil also was found in Sudan, where production has been climbing since completion in 1999 of an oil pipeline for exports, despite years of civil war. In 2006, estimates of proven reserves topped 5 billion barrels, a 10-fold increase over the year before. [83]

Africa also has abundant natural gas. Nigeria has the continent's largest reserves and the world's seventh-biggest, while Algeria's reserves rank eighth. [84] Both are on a par with Saudi Arabia and the United States. Algeria in 1964 became the first nation to ship liquified natural gas (LNG) aboard tankers. But Nigeria, convulsed by tribal wars and coups,

has been unable to capitalize on its gas deposits until recently. It still "flares," or burns away, 40 percent of the natural gas produced with its oil, although Nigeria is beginning to expand LNG production. [85]

The New Nationalism

China's staggering expansion and modernization have overtaken its energy resources. Twenty years ago, China was the largest oil exporter in East Asia. Now it is the world's second-largest oil purchaser, accounting for nearly one-third of the global increase in oil demand, note David Zweig and Bi Jianhai of the Hong Kong University of Science. [86] Similarly, India's oil consumption doubled between 1990 and 2004, and other industrializing Asian nations nearly matched that pace. [87]

Fortunately for the world's consumers, the explosive growth of China's oil demand was matched by a remarkable recovery in Russia's oil output. The fall of the Soviet Union and a financial credit crisis had devastated Russia's oil industry. Starved for capital and leadership, it was producing only 6 million barrels a day in 1995. But oil output had rebounded to average 9.4 million barrels a day this year, making Russia currently the world's largest oil producer, ahead of Saudi Arabia, which has trimmed its output.

deputy director of the Carnegie Moscow Center. [88]

Today, says Leonid Grigoriev, president of the Institute for Energy and Finance in Moscow, "We see ourselves as a great power." [89]

That power has frightened Russia's neighbors, especially after the Putin government took control of major petroleum reserves and energy pipelines, forcing Western energy companies to surrender equity positions in the country's largest new gas fields. Putin "has a very traditional Soviet view of the na-

BP economist Davies. "That debate is continuing."

Like Putin, Venezuela's Chávez is an architect of oil's rising nationalism. Following in the footsteps of Argentinean strongman Juan Perón and Cuba's Fidel Castro, Chávez is using Venezuela's oil and gas reserves — the Western Hemisphere's largest — to promote his socialist "Bolivarian Revolution." [90] While Chávez delights in confronting U.S. policy goals in Latin America, he also finds willing listeners in the Middle East and Asia. Having survived a coup attempt and an oil-workers' strike that stunted output in the winter of 2002-03, the former rebel paratrooper is firmly in control.

At home, Chávez has steered energy export earnings toward the three-quarters of the population that comprise Venezuela's poor. Their plight worsened in the 1980s and '90s despite market reforms recommended by globalization advocates at the International Monetary Fund (IMF) and World Bank. Chávez rejects free-market, capitalist economic approaches and vows to establish a socialist, classless society. [91] Social spending by Petróleos de Venezuela S.A. (PDVSA), the state-run oil and natural gas company, has increased 10-fold since 1997.

Abroad, Chávez seeks a coalition of allies who will help him parry opposition from the United States and pursue his agenda. He has offered low-priced oil to Latin America. (He also has donated heating oil to the poor in the United States.) PDVSA has forced major oil companies to give up majority holdings in Venezuela's oil fields and has signed oil deals with China, Iran, Vietnam, Brazil and Belarus. [92]

But Venezuela now spends more on social programs than on maintaining and expanding its oil production capacity, according to the Baker Institute. The current production rate of 2.4 million barrels a day is down from

Pollution and poverty abound in the swampy Niger Delta region of Nigeria, where international oil companies are drilling for the country's rich oil resources. Shanties reflect the slow rate of development, which has sparked violent protests in recent years.

Corbis Images/Ed Kashi

With the world's largest production and reserves of natural gas, it is poised to be Europe's prime supplier while developing its immense Far East gas reserves for eventual use by China, the rest of Asia and North America.

Russia's energy wealth also has transformed its self-image and ambitions, as it pulls away from the West. "In the late 19th century, Russia's success was said to rest on its army and its navy; today, its success rests on its oil and gas," writes Dmitri Trenin,

ture of power," says the Kennan Institute's Ruble. "He views oil and gas as strategic playing cards to reassert Russia in the world scene."

Now Europe anxiously faces growing dependence on Russia for its energy. "Russia is a natural, reliable and stable supplier" for Europe, insists Grigoriev.

"They see things strictly through the eyes of Russia: What is in their national interest?" responds CSIS's Ebel.

"The issue of security of supply is critical for European consumers," says

3.1 million barrels when Chávez took office in 1999. [93]

"He is good at giving oil away, but he's not good at producing oil," says Chávez opponent Luis Giusti, who headed PDVSA in the 1990s.

While Chávez is a thorn in the side to the U.S. government and international oil companies, his overtures to China and Iran and his willingness to slow future development in favor of higher returns today represent a new reality in the world's energy story. [94] ∎

CURRENT SITUATION

Majors Shut Out

Government-owned or controlled petroleum companies today control a majority of the world's hydrocarbon reserves and production. By 2005, nationalized oil companies had taken over 77 percent of the world's 1.1 trillion barrels of oil reserves. And, while Western oil companies have absorbed their share of the short-term windfall created by recent higher prices, their long-term future does not look particularly rosy. Major oil firms now control only 10 percent of global petroleum reserves. [95]

"International majors have been relegated to second-tier status," concluded the Baker Institute. In the 1970s and '80s, Western companies were invited to explore the new fields in the North Sea, Alaska and the Gulf of Mexico, but today key future resources in Russia and Central Asia are government-controlled.

"The bulk of the resources remain in a number of key countries, which are dominated by states, and we have

to be dependent on governments and state companies to deliver the capacity," says BP's Davies.

"Access really is a consideration," adds Cambridge Energy Research Associates chairman Yergin. "Where can you go to invest money, apply technology and develop resources and bring them to market? Terms get very tough. The decision-making slows down, if you can get there at all."

China's strategy for feeding its oil appetite is a major source of concern, says former CIA Director Deutch. Its

In the shadow of Istanbul's historic Blue Mosque (left), the Hagia Sophia Museum (center) and Topkapi Palace (right), an oil tanker enters the Bosporus Strait. The 21-mile-long waterway is the sole route for Caspian oil shipped through pipelines to the Black Sea, where it is then loaded onto tankers for the trip through the strait to the Mediterranean. Turkey fears increased tanker traffic could bring an environmental catastrophe to the already busy Bosporus, so it has encouraged development of an overland pipeline that would bypass the strait.

oil companies scour the world seeking access to oil and gas resources, effectively reducing supplies on the world market.

"China — and now India — are making extensive efforts in Africa and elsewhere in the world to lock up oil supplies," says Deutch. These state-to-state deals typically are not based solely on market terms but include sweeteners such as political incentives,

military assistance, economic aid or trade concessions, he explains.

International oil companies, while banking record profits, are facing higher taxes or demands to surrender parts of their stakes in projects. For instance, say Western analysts, Putin's government has shown its knuckles to Royal Dutch Shell and Exxon Mobil in disputes over control of two huge projects on Sakhalin Island, off Russia's Pacific coast. Shell had to give up controlling interest in the Sakhalin-2 pipeline project to Russia's

natural gas monopoly Gazprom after suffering cost overruns. Russia wants to determine where the gas goes, says its oil minister. [96]

Emboldened by rising oil prices, Russia and nations in South America and West Africa that once relied on Western oil companies are now "increasingly calling the shots," said *The Wall Street Journal*. [97]

AP Photo/Vladimir Rodionov

President Vladimir Putin addresses executives of Russia's Rosneft oil company in September 2005 after visiting the Tuapse oil terminal, pictured behind him. Putin is attempting to reclaim Russia's position as a superpower by harnessing its considerable energy resources.

Producer Windfalls

Oil prices have more than tripled since 2002, sparking an unprecedented transfer of wealth from consuming to producing nations. The amount energy-importing nations must spend for oil has leaped from $300 billion in 2002 to nearly $1 trillion in 2006 (roughly the gross domestic product of Spain or South Korea). [98] The higher prices, of course, affect not only oil production but also the current value of oil in the ground. The IMF reports the value of energy exporters' oil reserves increased by more than $40 trillion between 1999 and 2005. Thus if prices stay at current

levels, it would translate into an enormous increase in future wealth for the exporting nations, concentrated in the Middle East, Russia, Central Asia and Africa. [99]

Higher oil prices leave all consumers with less to spend and save, but the impact is harshest in poor countries with no oil. Without oil or other high-value exports to offset increased energy costs, poor countries go deeper in debt.

"Debt is the central inhibitor of economic development," says former CIA Director R. James Woolsey Jr. "Importing expensive oil is helping bind hundreds of millions of the world's poor more firmly into poverty." [100]

The flow of petrodollars also is profoundly affecting the United States

— the world's richest nation but also its largest oil consumer. When the U.S. buys oil from abroad, dollars pile up in the exporting nations' coffers. The oil-dollar outflow has added enormously to the U.S. "current account" deficit, or the dollar difference between U.S. imports and exports and international financial transactions. The United States is the only major nation that pays for its oil imports by borrowing heavily from the rest of the world. [101]

"The U.S. now borrows from its creditors — such as China and Saudi Arabia — over $300 billion per year, approaching a billion dollars a day of national IOU-writing, to import oil," according to Woolsey. [102]

A consequence of the increasing role of national oil companies is that most U.S. dollars paid for oil go into accounts controlled by foreign governments, according to the Federal Reserve Bank of New York. A crucial question is what those governments will do with their petrodollars, bank experts said. [103]

The outward tide of U.S. petrodollars has been matched by purchases of U.S. securities and properties by exporting countries, providing crucial support for U.S. stock and bond markets, the Federal Reserve report notes. In a dramatic example, China recently purchased $3 billion in stock of the Blackstone Group, a prominent U.S. equity firm that buys and turns around distressed companies.

"Officials in Beijing have $1.2 trillion of reserves they want to invest more profitably than in U.S. Treasuries. They lack the expertise to do it themselves and don't want to pay money managers millions in fees," said financial columnist William Pesek. [104]

For its part, Blackstone will get the increased access to China's surging economy that it covets. [105]

But such purchases are full of complexities, the Federal Reserve report

Continued on p. 172

At Issue:

Has a new cold war begun over oil that could lead to conflict?

MICHAEL T. KLARE
FIVE COLLEGE PROFESSOR OF PEACE AND WORLD SECURITY STUDIES (AMHERST, HAMPSHIRE, MOUNT HOLYOKE AND SMITH COLLEGES AND THE UNIVERSITY OF MASSACHUSETTS, AMHERST)

WRITTEN FOR *CQ GLOBAL RESEARCHER*, JUNE 2007

*t*wo simultaneous developments are likely to intensify future conflicts over oil. On one hand, increasing competition for a finite resource will become more intense in the years ahead. With China and India leading the growth in demand, competition is going to soar, and supply isn't likely to expand nearly as fast as demand. In addition, oil supplies increasingly will be located in areas of tension and inherent friction — the Middle East, Africa, Central Asia and other unstable places.

During the Cold War, the superpowers competed for influence by providing arms for various proxies in Africa, the Middle East and Asia. We are seeing the same thing happening in the oil cold war.

The United States, Russia and China, in their pursuit of oil allies, are again providing arms to proxies and suppliers, which is intensifying the risk of internal conflicts. It is an exceedingly dangerous development. The United States and China are not seeking to make war on one another. But they are inadvertently contributing to the risk of conflict in Africa and Central Asia by using arms transfers as an instrument of influence.

Ultimately, the only solution will be to reduce our craving for imported oil. That is easier said than done. It is a craving, and cravings lead to irrational behavior. For China, its close embrace of the Sudanese government — including supplying arms — is bringing the Chinese terrible criticism. The United States, for its part, engages in equally irrational behavior in creating close ties with — for example — the leaders of Kazakhstan and Azerbaijan — alienating the pro-democracy movements in those countries.

Ultimately, the most dangerous piece in all of this is the U.S.-China competition for energy. We have a cold war today, but it could become a hot war, although not through a deliberate act over oil. But we are engaged in competitive arms competition in Africa and Asia, and this could lead to inadvertent local conflicts and an accidental clash between the United States and China, much the way World War I began.

Neither side would choose such a conflict, but it would arise from a clash of proxies, eventually involving U.S. and Chinese advisers and troops.

Such an outcome may not be highly probable, but it is an exceedingly dangerous possibility.

AMY MYERS JAFFE
WALLACE S. WILSON FELLOW IN ENERGY STUDIES, JAMES A. BAKER III INSTITUTE FOR PUBLIC POLICY; ASSOCIATE DIRECTOR, RICE UNIVERSITY ENERGY PROGRAM

WRITTEN FOR *CQ GLOBAL RESEARCHER*, JUNE 2007

*c*ompetition over energy may contribute to fundamental global conflicts, but the conflicts would have happened with or without the energy situation. North Korea is not an oil issue. Kosovo was not about oil. The Iran confrontation is not an oil issue.

The Persian Gulf remains a special case. Saudi Arabia controls most of the world's excess oil production. In the case of the Iraq War in 1991 — which followed Iraq's invasion of oil-rich Kuwait — the United States was not going to let Saddam Hussein control 40-50 percent of the world's oil reserves. Saudi oil facilities have been targeted by terrorists, and Iran has threatened in the past to use military force to interfere with oil shipments through the Strait of Hormuz.

A future confrontation with Iran would greatly increase the risk to essential oil exports through the Persian Gulf. On the other hand, Iran is critically dependent upon revenues from its own oil sales and must import gasoline from foreign refiners to meet its population's requirements.

The overriding concern, however, is that the sudden loss of the Saudi oil network would paralyze the global economy. The United States — and the rest of the world — has a concrete interest in preventing that. But most conflicts facing the United States today, like North Korea or Afghanistan, are not going to change whether the price of oil is $50 a barrel or $70 a barrel.

Nor is Central Asia likely to become a flash point. We have been watching a revival of the so-called Great Game competition over Caspian oil for a decade. Why? Because the leaders in those countries have chosen to delay, trying to get the best economic and geopolitical deals they can. The Russians play the Japanese off the Chinese. The Chinese are trying to take care of their needs. Their motivations vary from country to country, but it is a dynamic that is very unlikely to lead to conflict. We are not likely to go to war with Russia over a pipeline in Kazakhstan.

The United States and China, as the world's largest oil importers, are economic partners by virtue of their trade and, consequently, potential political rivals. But both share a common interest in reasonable oil prices.

If the United States and China ever go to war over Taiwan, oil will not be the trigger.

Continued from p. 170

notes. The Blackstone stock purchase was made by China's new state-owned investment fund, and other oil exporters have set up similar "sovereign wealth funds" to make direct investments in the United States and other oil-buying countries, writes columnist Sebastian Mallaby. "Chunks of corporate America could be bought by

Nigerians work on a French gas-drilling installation in Nigeria's Niger Delta, aided by Chinese contractors. The Chinese are competing with other international oil companies for the delta's rich oil reserves.

Beijing's government — or, for that matter, by the Kremlin." The economic and political fallout could be seismic, he adds. [106]

If events make oil exporters less willing to put dollars back into the United States, U.S. interest rates could increase to keep the foreign investment coming. Otherwise, U.S. consumers would have to cut their spending to reduce the outflow of dollars. A big shift of petrodollars away from the United States would pull a vital prop out from under stock markets. [107]

The flow of untraceable petrodollars also affects world security. Because so much oil revenue goes into

the Middle East and from there into untraceable channels, some of it is being used to finance terrorist organizations opposed to the United States. "Thus . . . when we pay for Middle Eastern oil today, this long war in which we are engaged becomes the only war the U.S. has ever fought in which we pay for both sides," Woolsey says. [108]

Pipeline Politics

Over the next quarter-century, the world will rely on new oil and gas fields in Russia, Central Asia and Africa for a critical part of its energy needs. But uncertainty over when, where and if new pipelines will be built to access those new fields is heightening political tensions among the central players in the global competition for energy.

China covets oil and gas from eastern Siberia, but Russia's leaders have delayed building a pipeline into China, unwilling to hinge such a costly project on a single customer. Instead, Rus-

sia wants to channel those resources to its Pacific coast, where they can be shipped to competing customers not only in China but also in Japan, the rest of Asia and the United States.

Europe depends on Russian natural gas delivered over a Soviet-era pipeline network, which must be expanded to handle future growth. But Russia itself needs more gas and thus wants to build new pipelines into Central Asia to transport gas from Caspian fields — at market prices — to Europe.

Many Russian pipelines carrying Caspian oil terminate at the Black Sea. But Turkey opposes plans to expand that route because it fears a catastrophic oil spill from tanker traffic through the Bosporus Strait.

The most direct export route for Caspian oil is southward, by pipeline through Iran — a project China would welcome. But the United States opposes the route because it seeks to block Tehran's suspected nuclear-weapons development. [109]

Pipeline infighting is further reflected in BP's controversial $4 billion, 1,100-mile BTC pipeline from Baku, Azerbaijan, past Tbilisi, Georgia, and on to the Mediterranean port of Ceyhan, Turkey. The world's second-longest pipeline threads through mountains and volcanic regions and had to withstand unrest in Georgia, environmental opposition and sabotage threats. Its completion in 2005 fulfilled a hardball strategy by the United States to keep the pipeline out of Russian territory and block a shorter, cheaper route through Iran. [110] Leaders in Moscow and Tehran were infuriated.

"We really put all our cards on the table on that one," said Ebel, of CSIS.

But Russia has high cards to play, too, in the current pipeline tug-of-war over the undeveloped natural gas riches on the Caspian's eastern coast. The only gas pipeline through this region now leads north from Turkmenistan, through Kazakhstan into Russia, and Moscow controls it. [111]

The United States is pushing for a new pipeline across the Caspian seabed to carry Turkmenistan's gas westward to Baku.

From there it could travel into Turkey and connect with a new pipeline that the European Union wants to see built into Austria, (the "Nabucco" project), thus completing a pathway for Caspian gas to Europe without setting foot on Russian soil.

"We would love to see the Trans-Caspian Gas Pipeline put in place," Deputy Assistant Secretary of State for European and Eurasian Affairs Matthew Bryza said in January. [112]

Putin has other ideas. He is pressing Turkmenistan and Kazakhstan to support a Russian-built pipeline around the north end of the Caspian into Russia to move the gas into Europe over old and new Russian lines. Russia insists that no pipeline may cross the Caspian Sea unless all five adjoining nations agree — and Moscow is ready with its veto.

Speaking in Turkey recently, Bryza took a shot at Gazprom's natural gas pipeline monopoly, saying Russia uses its pipelines to intimidate governments in Europe that depend on them. "Europeans are finally waking up to the reality, I'm sorry to say, that Gazprom isn't always the most reliable partner for them. The more gas that moves from Central Asia and Azerbaijan to Europe via Turkey, the better." [113]

Russian officials contend the United States is still trying to throw its weight around, telling Moscow what to do.

In May, Moscow claimed the advantage after Putin and President Gurbanguly Berdymukhammedov of Turkmenistan agreed to the Russian plan for moving Caspian gas, but the United States says the door is still open for its favored route. [114] "These two pipelines are different," Bryza said in June, speaking of the Russian plan and trans-Caspian pipeline.

Meanwhile, Turkmenistan continues to talk with China about an eastward pipeline connection for its gas.

Uncertainty over the construction of new oil and gas pipelines is heightening political tensions among the central players in the global competition for energy. New pipelines will be increasingly vital in moving energy resources from new fields in Russia, Central Asia and Africa.

Getty Images/Stanton R. Winter

Currently, however, most of the pipeline-route disputes remain on paper. Soaring steel prices continue to inflate the costs of the billion-dollar pipeline networks. "The watchword today is delay," says Yergin of Cambridge Energy Research Associates, "not only because of political issues, but also because construction costs are going through the roof." ∎

OUTLOOK

Curbing Demand

The race for Earth's remaining energy resources increasingly is splitting the world into two camps: countries that sell oil and natural gas and those that buy them.

The buyers — led by the United States, Europe, China and India and Japan — have a clear imperative, according to the International Energy Agency (IEA) and other energy experts: Start curbing demand. [115]

Energy security is the primary reason. Two-thirds of the growth in oil supplies over the next quarter-century will likely come from the Middle East, Russia, the Caspian region, Africa and Venezuela — areas beset by conflict or political instability. [116]

Climate change is also driving the need to curb demand. Total world economic output is projected to more than double by 2030, accelerating the discharge of greenhouse gases into the atmosphere. Eighty percent of the growth will come from China, India, Brazil and other developing countries. [117]

To avert potentially catastrophic climate disasters before the end of this century, both industrial and developing countries must agree on strategies for conserving energy and reducing greenhouse gases without halting economic growth, says the Intergovernmental Panel on Climate Change. [118]

Both energy insecurity and climate threats demand greater international cooperation than in the past decade, experts say. The United States has, until now, mainly sought to deal with its energy challenges by producing more oil and gas outside its borders, said a 2004 report for the Baker Institute for Public Policy.

After the Sept. 11, 2001, terrorist attacks, influential members of the Bush administration saw regime change in Iraq as a way to shake OPEC's hold on oil production, the Baker Institute authors wrote. Instead of taking responsibility for reducing energy consumption, however, the U.S. addressed the challenge "by attempting to control the Middle East." [119] But the strategy "has fallen flat on its face," the authors have asserted.

No matter how the Iraq War ends, the authors continue, the United States must move more decisively to reduce its energy demands if it wants credibility in seeking cooperation from China. China is quickly catching up to the United States in energy production and greenhouse-gas emissions, according to a recent report by U.S. climate experts Jeffrey Logan, Joanna Lewis and Michael B. Cummings. [120]

Gas stations in Tehran were torched and looted on June 26, 2007, after the Iranian government announced plans to begin fuel rationing. The state-controlled National Iranian Oil Co. has subsidized consumer fuel prices, sparking increased demand for oil.

AFP/Getty Images/Behrouz Mehri

China has been building on average one new electric power plant a week for the past few years, and its automobile sales are booming (though they're small by U.S. standards). [121] But by the end of the decade, China will have 90 times more motor vehicles than it had in 1990, and by 2030 — or sooner — there may be more cars in China than in the United States. [122]

This year China announced new climate goals, including a 10 percent reduction in carbon-dioxide emissions over five years. "[W]e have to take responsibility for lowering greenhouse emissions," said Zhang Zhang Guobao, vice chairman of the energy-policy-setting National Development and Reform Commission. [123]

But China has adopted a "wait-and-see" attitude toward international climate-change agreements, unwilling to make binding commitments until it is clear what the United States and the developed world will do, according to Logan, Lewis and Cummings. The United States must lead by example, they said.

"Thinking about how to alter our energy-consumption patterns to bring down the price of oil is no longer simply a hobby for high-minded environmentalists or some personal virtue," says *Times* columnist Friedman. "It is a national-security imperative." [124]

"It must be recognized," says Yergin of Cambridge Energy Research Associates, "that energy security does not stand by itself but is lodged in the larger relations among nations and how they interact with one another." [125] ■

Notes

[1] The Associated Press, "Putin: Soviet Collapse a 'Genuine Tragedy,'" MSNBC, April 25, 2005, www.msnbc.msn.com/id/7632057.

[2] "Russia Cuts Ukraine Gas Supplies," BBC News, Jan. 1, 2006; http://news.bbc.co.uk/1/hi/world/europe/4572712.stm.

[3] "Russia and the West; No Divide, No Rule," *The Economist*, May 17, 2007, p. 12.

[4] "Oil Nationalism Troubling Multinationals," *Iran Daily*, Oct. 23, 2006, p. 11, http://irandaily.ir/1385/2691/pdf/i11.pdf.

[5] John M. Deutch, testimony before the House Foreign Affairs Committee, March 22, 2007.

[6] "PIW Ranks the World's Top Oil Companies," *Energy Intelligence*, www.energyintel.com/DocumentDetail.asp?document_id=137158.

[7] *Iran Daily, op. cit.*

[8] Peter Katel, "Change in Latin America," *CQ Researcher*, July 21, 2006, pp. 601-624.

[9] Natalie Obiko Pearson, "Chávez takes over Venezuela's last private oil fields," The Associated Press Worldstream, May 2, 2007.

[10] Alexandra Valencia, "Ecuador says started review of oil contracts," Reuters, June 6, 2007; www.reuters.com/article/companyNewsAndPR/idUSN0645081020070607.

[11] *Energy Intelligence, op. cit.*

[12] U.S. motorists were paying over $1.42 a gallon for regular gasoline in March 1981. Adjusted at 2006 price levels to account for inflation, that cost would be $3.22 a gallon; www.eia.doe.gov/emeu/steo/pub/fsheets/petroleumprices.xls.

[13] Matthew Higgins, Thomas Klitgaard and Robert Lerman, "Recycling Petrodollars: Current Issues in Economics and Finance," Federal Reserve Bank of New York, December 2006, p. 1; www.newyorkfed.org/research/current_issues/ci12-9.pdf.

[14] "The Changing Role of National Oil Companies in International Energy Markets," James A. Baker III Institute for Public Policy, April 2007; http://bakerinstitute.org/Pubs/BI_Pol%20Rep_35.pdf, page 2; see all reports www.rice.edu/energy/publications/nocs.html.

[15] "World Energy Outlook 2006," International Energy Agency, p. 1; www.worldenergyoutlook.org/summaries2006/English.pdf.

[16] For background, see Colin Woodard, "Curbing Climate Change," *CQ Global Re-*

searcher, February 2007, pp. 27-50, and the following *CQ Researchers*: Barbara Mantel, "Energy Efficiency," May 19, 2006, pp. 433-456; Marcia Clemmitt, "Climate Change," Jan. 27, 2006, pp. 73-96; Mary H. Cooper, "Energy Policy," May 25, 2001, pp. 441-464; Mary H. Cooper, "Global Warming Treaty," Jan. 26, 2001, pp. 41-64; Mary H. Cooper, "Global Warming Update," Nov. 1, 1996, pp. 961-984.

[17] Maureen S. Crandall, *Energy, Economics and Politics in the Caspian Region: Dreams and Realities* (2006), pp. 23, 46.

[18] Amy Myers Jaffe and Matthew E. Chen, James A. Baker III Institute for Public Policy, testimony before the U.S.-China Economic and Security Review Commission, hearing on China's Role in the World, Aug. 4, 2006; www.uscc.gov/hearings/2006hearings/written_testimonies/06_08_3_4wrts/06_08_3_4_jaffe_amy_statement.php.

[19] R. James Woolsey, "Global implications of Rising Oil Dependence and Global Warming," testimony before the House Select Committee on Energy Independence and Global Warming, April 18, 2007, p. 2.

[20] "PIW Survey: Oil Reserves Are Not Rising," *Petroleum Intelligence Weekly*, April 16, 2007; www.energyintel.com/DocumentDetail.asp?document_id=199949. See also "Performance Profiles of Major Energy Producers 2005," Energy Information Agency, pp. 20-21; www.eia.doe.gov/emeu/perfpro/020605.pdf.

[21] Michael Connolly, "Fragmented Market Would Hamper Russian-Iranian 'Gas OPEC'," *Wall Street Journal Online*, Feb. 2, 2007.

[22] Daniel Dombey, Neil Buckley, Carola Hoyos, "NATO fears Russian plans for 'gas OPEC'," *Financial Times*, Nov. 13, 2006.

[23] Daniel S. Sullivan addressed the Energy Council's Federal Energy & Environmental Matters Conference, March 9, 2007.

[24] James A. Baker III Institute for Public Policy Report, *op. cit.*; also Baker Institute Report, "Introductions and Summary Conclusions," pp. 7-19; www.rice.edu/energy/publications/docs/NOCs/Presentations/Hou-Jaffe-KeyFindings.pdf.

[25] "Mapping the Global Future: Report of the National Intelligence Council's 2020 Project," National Intelligence Council, December 2004; www.dni.gov/nic/NIC_globaltrend2020.html.

[26] Michael Schiffer and Gary Schmitt, "Keeping Tabs on China's Rise," The Stanley Foundation, May 2007, p. 1; www.stanleyfoundation.org/publications/other/SchifferSchimitt07.pdf.

[27] *Ibid.*, p. 9.

[28] Xuecheng Liu, "China's Energy Security and Its Grand Strategy," The Stanley Foundation, September 2006, p. 13; www.stanleyfoundation.org/publications/pab/pab06chinasenergy.pdf.

[29] Quoted in Paulo Prada, "Bolivian Nationalizes the Oil and Gas Sector," *The New York Times*, May 2, 2006, p. A9.

[30] Quoted in Juan Forero, "Presidential Vote Could Alter Bolivia, and Strain Ties With U.S.," *The New York Times*, Dec. 18, 2005, p. A13.

[31] Alfonzo quoted by Stanford University's Terry Lynn Karl, Senior Fellow at the Institute for International Studies, Stanford University, in "The Oil Trap," Transparency International, September 2003; ww1.transparency.org/newsletters/2003.3/tiq-Sept2003.pdf.

[32] Richard M. Auty, *Sustaining Development in Mineral Economies: The Resource Curse Thesis* (Routledge), 1993. Summarized in Richard M. Auty, "The 'Resource Curse' in Developing Countries Can Be Avoided," United Nations University, Helsinki; www.wider.unu.edu/research/pr9899d2/pr9899d2s.htm.

[33] "Country Analysis Briefs: Mexico," Energy Information Administration, January 2007; www.eia.doe.gov/emeu/cabs/Mexico/Oil.html; and "Major Non-OPEC Countries' Oil Revenues," www.eia.doe.gov/cabs/opecnon.html.

[34] Robert Collier, "Mexico's Oil Bonanza Starts to Dry Up," *San Francisco Chronicle*; www.sfgate.com/cgi-bin/article.cgi?file=/c/a/2006/06/30/MNGAAJN9JG1.DTL.

[35] Karl, Transparency International, *op. cit.*, p. 1.

[36] See "The 'Dutch Disease': Theory and Evidence," *Poverty and Growth Blog*, The World Bank, http://pgpblog.worldbank.org/the_dutch_disease_theory_and_evidence.

[37] Tom O'Neill, "Hope and Betrayal in the Niger Delta," *National Geographic*, February 2007, p. 97.

[38] Senan Murray, "Tackling Nigeria's Violent Oil Swamps," BBC News, May 30, 2007; http://news.bbc.co.uk/2/hi/africa/6698433.stm.

[39] Karl Maier, "Nigeria Militants Release Six Chevron Oil Workers," Bloomberg, June 2, 2007; www.bloomberg.com/apps/news?pid=20601087&sid=aXT6yOlwMVGY&refer=home.

[40] Daniel Balint Kurti, "New Militia is a Potent Force," *The Christian Science Monitor*, March 7, 2007; www.csmonitor.com/2006/0307/p04s01-woaf.html.

[41] Bernd Debusmann, "In Venezuela, obstacles to 21st Century socialism," Reuters, June 20, 2007.

[42] Transparency International, Corruption Perceptions Index, 2006; www.transparency.org/policy_research/surveys_indices/cpi/2006.

[43] Freedom House, "Countries at the Crossroads 2006; Country Report: Venezuela," www.freedomhouse.org/template.cfm?page=140&edition=7&ccrpage=31&ccrcountry=141.

[44] "Iran fuel rations spark anger, pump stations burn," Reuters, June 27, 2007, www.reuters.com/article/worldNews/idUSDAH72595420070627.

[45] "New Analysis Claims Darfur Deaths Near 400,000," Coalition for Darfur, April 25, 2005, http://coalitionfordarfur.blogspot.com/2005/04/new-analysis-claims-darfur-deaths-near.html.

[46] Jaffe, *op. cit.*

[47] David Zweig and Bi Jianhai, "China's Global Hunt for Energy," *Foreign Affairs*, Sept./Oct. 2005, p. 32.

[48] Scott McDonald, "China Welcomes Darfur Agreement," The Associated Press, June 14, 2007; www.boston.com/news/world/asia/articles/2007/06/14/china_welcomes_darfur_agreement/.

[49] "Fact Sheet," Extractive Industries Transparency Initiative, 2007; www.eitransparency.org/section/abouteiti.

[50] Thomas L. Friedman, "The First Law of Petropolitics," *Foreign Policy*, May/June 2006, p. 4; www.foreignpolicy.com/story/cms.php?story_id=3426.

[51] Ariel Cohen, the Heritage Foundation, "The North Eureopean Gas Pipeline Threatens Europe's Energy Security," Oct. 26, 2006; www.heritage.org/Research/Europe/bg1980.cfm.

[52] Alexandre Rocha, "Burned by Bolivia, Brazil Goes to Africa and Middle East Looking for Gas," *Brazzil Magazine* (online), June 20, 2007; www.brazzilmag.com/content/view/8368/1/.

[53] Crandall, *op. cit.*, p. 143.

[54] Foster Klug, "U.S. Presses China on Iran in Latest Talks," The Associated Press, June 20, 2007.

[55] "Mapping the Global Future," *op. cit.*

[56] Schiffer and Schmitt, *op. cit.*, p 14.

[57] Evan Osnos, "U.S., China vie for oil, allies on new Silk Road," *Chicago Tribune*, Dec. 19, 2006, p. 4.

[58] Schiffer and Schmitt, *op. cit.*, p. 15.

[59] "About Us," Organization of Petroleum Exporting Countries, www.opec.org/aboutus/history/history.htm.

[60] Until 1972 production limits set by the Texas Railroad Commission effectively set a ceiling on oil prices in the United States and

the rest of the world. But U.S. output peaked then, opening the way for OPEC's moves to control oil markets; http://tonto.eia.doe.gov/dnav/pet/hist/mcrfpus1m.htm.

[61] For background, see Mary H. Cooper, "OPEC: Ten Years After the Arab Oil Boycott," *Editorial Research Reports*, Sept. 23, 1983; available in *CQ Researcher Plus Archive*, www.cqpress.com.

[62] "U.S. Strategic Petroleum Reserve," Fact Sheet, U.S. Department of Energy, May 30, 2007; www.fossil.energy.gov/programs/reserves.

[63] Daniel Yergin, *The Prize: The Epic Quest for Oil, Money & Power* (1991), p. 631.

[64] "Carter Energy Program," *CQ Historic Documents Series Online Edition*. Originally published in *Historic Documents of 1977*, CQ Press (1978), CQ Electronic Library; http://library.cqpress.com/historicdocuments/hsdc77-0000106610.

[65] *Ibid.*

[66] "Iranian Hostage Crisis, 1980 Special Report," *Congress and the Nation, 1977-1980* (Vol. 5); CQ Press; available at CQ Congress Collection, CQ Electronic Library, http://library.cqpress.com/congress/catn77-0010173673.

[67] "Real Gasoline Prices," Energy Information Administration; www.eia.doe.gov/emeu/steo/pub/fsheets/real_prices.html.

[68] For background, see R. Thompson, "Quest for Energy Independence," *Editorial Research Reports*, Dec. 23, 1983, available in *CQ Researcher Plus Archive*, CQ Electronic Library, http://library.cqpress.com.

[69] Yergin, *op. cit.*, p. 669.

[70] *Ibid.*, p. 666.

[71] "Annual Energy Review 2005, World Crude Oil Production, 1960-2005," Energy Information Administration; www.eia.doe.gov/emeu/aer/pdf/pages/sec11_11.pdf.

[72] Yergin, *op. cit.*, p. 718.

[73] Mary H. Cooper, "SUV Debate," *CQ Researcher*, May 16, 2003, pp. 449-472.

[74] For background, see Kenneth Jost, "Russia and the Former Soviet Republics," *CQ Researcher*, June 17, 2005; pp. 541-564.

[75] Louis Fischer, *Oil Imperialism* (1926), cited by Robert E. Ebel, Center for Strategic and International Studies, July 25, 2006.

[76] Bruce W. Nelan, "The Rush for Caspian Oil," *Time*, May 4, 1998, p. 40.

[77] Crandall, *op. cit.*, p. 1.

[78] "Caspian Sea," Energy Information Administration, 2007; www.eia.doe.gov/emeu/cabs/Caspian/Full.html.

[79] Martha Brill Olcott, "Will Central Asia Have Another 'Second Chance'?" speech, Carnegie Endowment for International Peace, Sept. 15, 2005.

[80] Crandall, *op. cit.*, p. 3.

[81] "Top World Oil Producers, Exporters, Consumers, and Importers 2004," Information Please Database, 2007; www.infoplease.com/ipa/A0922041.html.

[82] "BP Statistical Review 2006," British Petroleum, p. 8; www.bp.com/sectiongenericarticle.do?categoryId=9017903&contentId=7033469.

[83] "Sudan," Energy Information Administration, April 2007; www.eia.doe.gov/emeu/cabs/Sudan/Background.html.

[84] "Libya — Natural Gas," Energy Information Administration, March 2006; www.eia.doe.gov/emeu/cabs/Libya/NaturalGas.html.

[85] "Nigeria/Natural Gas," Energy Information Administration, April 2007, www.eia.doe.gov/emeu/cabs/Nigeria/NaturalGas.html.

[86] Zweig and Jianhai, *op. cit.*, p. 25.

[87] "International Energy Outlook, 2007," Energy Information Administration, p. 83; www.eia.doe.gov/oiaf/ieo/pdf/ieorefcase.pdf.

[88] Dmitri Trenin, Deputy Director, Carnegie Moscow Center, "Russia Leaves the West," *Foreign Affairs*, July/August 2006.

[89] Leonid Grigoriev, speaking at the Kennan Institute, Feb. 5, 2007; www.wilsoncenter.org/index.cfm?topic_id=1424&fuseaction=topics.event_summary&event_id=215229.

[90] *Oil and Gas Journal*, quoted in www.eia.doe.gov/emeu/cabs/Venezuela/Oil.html. Con-

ventional reserves do not include the extensive Canadian tar sands or Venezuela's extra-heavy oil and bitumen deposits.

[91] Michael Shifter, "In search of Hugo Chávez," *Foreign Affairs*, May/June 2006, p. 47. For background, see Peter Katel, "Change in Latin America," *CQ Researcher*, July 21, 2006, pp. 601-624.

[92] Baker Institute for Public Policy, *op. cit.*, p. 6.

[93] "Venezuela," Energy Information Administration, September 2006; www.eia.doe.gov/emeu/cabs/Venezuela/Oil.html.

[94] Baker Institute, *op. cit.*, p. 5.

[95] *Ibid.*, p. 1.

[96] Gregory L. White and Jeffrey Ball, "Huge Sakhalin Project Is Mostly on Track, As Shell Feels Pinch," *The Wall Street Journal*, May 7, 2007, p. 1.

[97] *Ibid.*, p. 1. Also see Amy Myers Jaffe, James A. Baker III Institute for Public Policy, "Russia: Back to the Future?" testimony before the Senate Committee on Foreign Relations, June 29, 2006, p. 1.

[98] Higgins, Klitgaard and Lerman, *op. cit.*, p. 1.

[99] "World Economic Outlook, April 2006," Chapter 2, p. 24, International Monetary Fund; www.imf.org/external/pubs/ft/weo/2006/01/pdf/c2.pdf.

[100] Woolsey, *op. cit.*, p. 3.

[101] Higgins, Klitgaard and Lerman, *op. cit.*, p. 6.

[102] Woolsey, *op. cit.*, p. 3.

[103] Higgins, Klitgaard and Lerman, *op. cit.*, pp. 3-4.

[104] William Pesek, "Blackstone + China = Bubble," Bloomberg, May 23, 2007; www.bloomberg.com/apps/news?pid=20601039&sid=aU7bs9CJazGI&refer=columnist_pesek.

[105] Ransdell Pierson and Tamora Vidaillet, "China flexes FX muscle with $3 bln Blackstone deal," Reuters, May 21, 2007.

[106] Sebastian Mallaby, "The Next Globalization Backlash," *The Washington Post*, June 25, 2007, p. A19.

[107] Higgins, Klitgaard and Lerman, *op. cit.*, p. 6.

[108] Woolsey, *op. cit.*, p. 4.

[109] The United States has its own huge pipeline project on the table, a plan to transport natural gas from Alaska's North Slope into the U.S. Midwest, which would reduce some of the future need for natural gas imports by LNG tankers from Russia and the Middle East.

[110] Robert E. Ebel, "Russian Energy Policy," Center for Strategic and International Studies, testimony before the U.S. Senate Com-

About the Author

Peter Behr recently retired from *The Washington Post*, where he was the principal reporter on energy issues and served as business editor from 1987-1992. A former Nieman Fellow at Harvard University, Behr worked at the Woodrow Wilson Center for Scholars and is working on a book about the history of the U.S. electric power grid.

mittee on Foreign Relations, June 21, 2005; Crandall, *op. cit.*, p. 23.

[111] "Central Asia," Energy Information Administration, September 2005; www.eia.doe.gov/emeu/cabs/Centasia/NaturalGas.html.

[112] "Washington Pushes for Trans-Caspian Pipeline," *New Europe*, Jan. 15, 2007; www.neurope.eu/view_news.php?id=69019.

[113] Press statement, State Department, Consulate General-Istanbul, Remarks by Matthew Bryza, deputy assistant secretary of State for European and Eurasian affairs, May 11, 2007; http://istanbul.usconsulate.gov/bryza_speech_051107.html.

[114] "Turkmenistan open oil, gas to Russia," UPI, June 13, 2007.

[115] "World Energy Outlook 2006," International Energy Agency, p. 3; www.worldenergyoutlook.org/summaries2006/English.pdf.

[116] "International Energy Outlook 2007," Energy Information Administration, p. 187; www.eia.doe.gov/oiaf/ieo/pdf/ieopol.pdf.

[117] "Fighting Climate Change Through Energy Efficiency," United Nations Environment Program, May 30, 2006; www.unep.org/Documents.Multilingual/Default.asp?DocumentID=477&ArticleID=5276&l=en.

[118] "Working Group III Report," Intergovernmental Panel on Climate Change, May 2007; www.mnp.nl/ipcc/pages_media/AR4-chapters.html.

[119] Joe Barnes, Amy Myers Jaffe, Edward L. Morse, "The Energy Dimension in Russian Global Strategy," James A. Baker III Institute for Public Policy," 2004, p. 5; www.rice.edu/energy/publications/docs/PEC_BarnesJaffeMorse_10_2004.pdf.

[120] Jeffrey Logan, Joanna Lewis and Michael B. Cummings, "For China, the Shift to Climate-Friendly Energy Depends on International Collaboration," *Boston Review*, January/February 2007; www.pewclimate.org/press_room/discussions/jlbostonreview.cfm.

[121] Logan, Lewis and Cummings, *op. cit.*

[122] Global Insight Forecast, "Outlook Still Buoyant for Chinese Auto Market," March 2007; www.globalinsight.com/SDA/SDADetail9307.htm.

[123] Catherine Brahic, "China to promise cuts in greenhouse gases," NewScientist.com news services, Feb. 14, 2007, http://environment.newscientist.com/article/dn11184.

[124] Friedman, *op. cit.*, p. 10.

[125] Yergin, "Ensuring Energy Security," *op cit*, p. 69.

FOR MORE INFORMATION

American Enterprise Institute, 1150 17th St., N.W., Washington, DC 20036; (202) 862-5800; www.aei.org. Public-policy research group studying economic and social issues.

American Petroleum Institute, 1220 L St., N.W., Washington, DC 20005-4070; (202) 682-8000; www.api.org. Industry group representing oil and gas producers.

James A. Baker III Institute, 6100 Main St., Rice University, Baker Hall, Suite 120, Houston, TX 77005; (713) 348-4683; http://bakerinstitute.org. Academic research group specializing in energy.

Cambridge Energy Research Associates, 55 Cambridge Parkway, Cambridge, MA 02142; (617) 866-5000; www.cera.com. Renowned energy consultancy to international energy firms, financial institutions, foreign governments and technology providers.

Center for Strategic and International Studies, 800 K St., N.W., Washington, DC 20006; (202) 887-0200; www.csis.org. Public-policy research group specializing in defense, security and energy issues.

Council on Foreign Relations, 1779 Massachusetts Ave., N.W., Washington, DC 20036; (202) 518-3400; www.cfr.org. Think tank focusing on international issues; publishes *Foreign Affairs*.

Energy Future Coalition, 1800 Massachusetts Ave., N.W., Washington, DC 20036; (202) 463-1947; www.energyfuturecoalition.org. A bipartisan advocacy group for energy conservation and alternative fuels.

Energy Information Administration, 1000 Independence Ave., S.W., Washington, DC 20585; (202) 586-8800; www.eia.doe.gov. The primary source of federal data and analysis on energy.

Extractive Industries Transparency Initiative, Ruseløkkveien 26, 0251 Oslo, Norway; +47 22 24 2110; www.eitransparency.org. Advocates responsible energy use and public disclosure of energy-based revenues and expenditures on behalf of more than 20 nations.

Human Rights Watch, 350 Fifth Ave., 34th floor, New York, NY 10118-3299; (212) 290-4700; www.hrw.org. Advocates for human rights.

International Energy Agency, 9 rue de la Fédération, 75739 Paris Cedex 15, France; 33 1 40 57 65 00/01; www.iea.org. The principal international forum for global energy data and analysis.

Kennan Institute, Woodrow Wilson International Center for Scholars, Ronald Reagan Building and International Trade Center, One Woodrow Wilson Plaza, 1300 Pennsylvania Ave., N.W., Washington, DC 20004-3027; (202) 691-4000; www.wilsoncenter.org. Think tank specializing in social, political and economic developments in Russia and the former Soviet states.

Organization of the Petroleum Exporting Countries, Obere Donaustrasse 93, A-1020 Vienna, Austria; +43-1-21112-279; www.opec.org. Coordinates and unifies petroleum policies among its 12 oil-exporting member nations.

Transparency International, Alt-Moabit 96, 10559 Berlin, Germany; 49-30-34-38 20-0; www.transparency.org. Advocacy group that campaigns against corruption worldwide.

World Bank, 1818 H St., N.W., Washington, DC 20433; (202) 473-1000; www.worldbank.org. Provides financial and technical assistance to developing countries.

Bibliography

Selected Sources

Books

Crandall, Maureen S., *Energy, Economics, and Politics in the Caspian Region: Dreams and Realities*, Praeger Security International, 2006.
An economics professor at the National Defense University argues that the Caspian region's oil development will accelerate global and regional military, ethnic and religious conflict.

Klare, Michael, *Resource Wars: The New Landscape of Global Conflict*, Henry Holt, 2001.
A political science professor describes how the demand for scarce resources among growing populations has led to wars over the past century.

Yergin, Daniel, *The Prize: The Epic Quest for Oil, Money & Power*, Simon & Schuster, 1991.
In a Pulitzer Prize-winning work, the chairman of Cambridge Energy Research Associates chronicles the political and economic history of the oil industry.

Articles

"PIW Ranks the World's Top Oil Companies," *Energy Intelligence*, www.energyintel.com.
Petroleum Intelligence Weekly, a leading industry publication, ranks Saudi Aramco of Saudi Arabia and Exxon Mobil of the United States as the world's top two oil companies.

O'Neill, Tom, "Curse of the Black Gold," *National Geographic*, February 2007, p. 88.
The writer examines the politics and corruption of multinational petroleum companies that critics claim have created poverty and violence in the wake of Nigeria's oil boom.

Schiffer, Michael and Gary Schmitt, "Keeping Tabs on China's Rise," The Stanley Foundation, May 2007, www.stanleyfoundation.org.
Two foreign policy experts encourage the West to continue diplomatic relations with the Beijing government amid China's rise as a global superpower.

Shifter, Michael, "In Search of Hugo Chávez," *Foreign Affairs*, May-June 2006, p. 45.
According to a vice president of the Inter-American Dialogue, the profits from nationalization of Venezuela's oil have yielded only modest gains for the country's poor.

Trenin, Dmitri, "Russia Leaves the West," *Foreign Affairs*, July-Aug. 2006, p. 87.
Russia's vast energy resources make it a potential threat to the United States and other Western nations, according to the deputy director of the Carnegie Moscow Center.

Udall, Randy, and Matthew R. Simmons, "CERA's Rosy Oil Forecast — Pabulum to the People," *ASPO-USA's Peak Oil Review/Energy Bulletin*, Aug. 21, 2006, www.energybulletin.net.
Two energy experts refute a recent optimistic oil study by Cambridge Energy Research Associates, contending that in actuality oil will be in shorter supply and more expensive by 2015.

Yergin, Daniel, "Ensuring Energy Security," *Foreign Affairs*, March-April 2006, p. 69.
The chairman of Cambridge Energy Research Associates explores new tactics for safeguarding the world's energy supplies and alleviating energy-related conflicts.

Zweig, David, and Bi Jianhai, "China's Global Hunt for Energy," *Foreign Affairs*, Sept.-Oct. 2005, p. 25.
Two foreign policy professors at Hong Kong University argue that China must find new energy sources if it wants to maintain rapid economic growth.

Reports

"Challenge and Opportunity, Charting a New Energy Future," Energy Future Coalition, 2002, www.energyfuturecoalition.org.
A bipartisan energy research group advocates alternative energy strategies to reduce dependence on foreign oil.

"The Changing Role of National Oil Companies in International Markets," James A. Baker III Institute for Public Policy, Rice University, May 1, 2007, www.rice.edu.
Energy researchers provide case studies analyzing the problems of private petroleum companies amid the rise of oil nationalism.

Ebel, Robert E., "Russian Energy Policy," testimony before Senate Foreign Relations Committee, June 21, 2005.
A senior energy adviser at the Center for Strategic and International Studies stresses the United States' need for a diplomatic energy-policy dialogue with Russia.

Jaffe, Amy Myers, "Russia: Back to the Future?" testimony before Senate Foreign Relations Committee, June 29, 2006.
A noted energy analyst reviews Russia's increasingly nationalistic energy policies.

Woolsey, R. James, "Geopolitical Implications of Rising Oil Dependence and Global Warming," testimony before Select Committee on Energy Independence and Global Warming, April 18, 2007.
A former CIA director offers solutions for curbing the United States' dependence on oil and natural gas.

The Next Step:

Additional Articles from Current Periodicals

Energy Dependency

Diehl, Jackson, "The New Threat to Europe," *The Washington Post*, Dec. 25, 2007, p. A29.
Russia's temporary cutoff of natural gas to Western Europe demonstrated how much the West depends on imported energy.

Neikirk, William, "Cheap Alternatives Key to Oil Plan," *The Chicago Tribune*, Jan. 24, 2007, p. C6.
In his 2007 State of the Union address, President Bush called for a 20 percent drop in U.S. gasoline consumption over the next 10 years.

Tollefson, Jeff, "Senate Passes Compromise Bill," *CQ Weekly*, June 25, 2007, p. 1920.
The U.S. Senate passed a comprehensive energy bill that would beef up automotive fuel-efficiency standards and create subsidies for biofuels and other petroleum alternatives.

White, Aoife, "EU: Days of Secure, Cheap Energy Are Over," The Associated Press, Jan. 10, 2007.
The European Union is investing in more diverse and renewable sources of fuel.

Energy Diplomacy

"No Divide, No Rule; Russia and the West," *The Economist*, May 19, 2007, p. 12.
A troubling new trans-Caspian pipeline deal is a symbol of the West's inability to cope with Russia.

The Associated Press, "Russian Minister: Gas Exporting Countries Plan to Form Gas-Pricing Group," April 9, 2007.
A high-level Russian official says the largest natural-gas exporters plan to establish a cooperative agreement on gas pricing similar to the Organization of Petroleum Exporting Countries (OPEC), but skeptics doubted the deal with ever be finalized.

Felsted, Andrea, and David Victor, "Gas and Oil do not Mix in the Chaotic World of Energy Policy," *The Financial Times*, May 9, 2006, p. 17.
Global energy troubles are due to the West's inability to reduce demand.

Nationalization

"Chávez-Style Oil Nationalism is Endangering World Economic Growth," *Newsweek International*, May 21, 2007.
Political interference, lack of reinvestment and instability in nationalized oil extraction raise the cost of fuel for the rest of the world.

"Yukos Buy Makes Rosneft Russia's Top Oil Producer," *Turkish Daily News* (Newswire), May 5, 2007.
The state-owned Russian oil giant Rosneft becomes the country's largest oil firm by buying a part of the defunct Yukos oil company amid accusations the Kremlin illegally imprisoned the company's owner and seized its assets.

Prada, Paulo, "Bolivian Nationalizes the Oil and Gas Sector," *The New York Times*, May 2, 2006, p. A9.
President Evo Morales nationalized Bolivia's oil and gas operations, giving foreign firms 180 days to renegotiate their contracts.

The Resource Curse

"IMF Hails Botswana's Growth," *The Reporter* (Botswana), Aug. 14, 2006.
Botswana has implemented anti-corruption policies that have kept it from contracting the "Dutch Disease."

Agence France-Presse, "Progress Made in Fight Against Oil Corruption: Conference," Oct. 17, 2006.
Several resource-rich states have made great strides in improving transparency and fighting corruption.

Clayton, Jonathan, "Chaos and Kidnap: How the Swamp Boys Hold a Government to Ransom," *The Times* (London), June 8, 2007, p. 35.
Rebels in the Niger Delta have hobbled oil production and brought the government to the negotiating table through guerrilla warfare and kidnappings.

Sershen, Daniel, "Turkmenistan's Natural Gas: Mixed Blessing," *The Christian Science Monitor*, May 15, 2007, p. 7.
Though Turkmenistan possesses some of the world's largest natural gas reserves, the wealth of resources hasn't translated into improved living standards for citizens.

CITING CQ GLOBAL RESEARCHER

Sample formats for citing these reports in a bibliography include the ones listed below. Preferred styles and formats vary, so please check with your instructor or professor.

MLA STYLE
Flamini, Roland. "Nuclear Proliferation." CQ Global Researcher 1 Apr. 2007: 1-24.

APA STYLE
Flamini, R. (2007, April 1). Nuclear proliferation. *CQ Global Researcher*, 1, 1-24.

CHICAGO STYLE
Flamini, Roland. "Nuclear Proliferation." *CQ Global Researcher*, April 1, 2007, 1-24.

Voices From Abroad:

ABDALLA SALEM EL-BADRI
Secretary General, OPEC

Oil is important to all

"Any talk of energy security must take into account both supply and demand perspectives. The role of oil is equally important to the economic growth and prosperity of consuming-importing countries, as well as to the development and social progress of producing-exporting countries."

Speech, Second Asian Ministerial Energy Roundtable, May 2007

NGOZI OKONJO-IWEALA
Minister of Finance, Nigeria

Niger Delta problems present an opportunity

"The government is determined to address the genuine problems of the Niger Delta people but will not allow gangsterism to prevail. . . . There will be no going back on the present reforms no matter what political configuration is in place, because Nigeria can not afford to miss this opportunity."

This Day (Nigeria), April 2006

EDITORIAL
Gazeta (Russia)

Russia: 'Oil is its everything'

"In 2009 oil prices will fall. [Putin] understands perfectly well that it is the falling of oil prices and not at all the elections that could return the demand for liberal reforms. This is the kind of country Russia is — oil is its everything."

February 2006

RAFAEL RAMÍREZ
Minister of Energy and Petroleum, Venezuela

Oil stability requires social stability

"There cannot be stability in the international oil market if there is no stability within the oil producing countries, which in turn presupposes political and social stability, justice and a truly national and fair distribution of the oil rent."

Speech during Third OPEC International Seminar, September 2006

XU WEIZHONG
African Studies Director China Institute of Contemporary International Relations

It's not just about oil

"[Western media] believed that China became interested in Africa only because of oil. But . . . Africa has always been a focus of China's foreign policy over the past half-century. . . . China has broad cooperation with African countries, including both energy-rich countries and resource-lacking ones. Western media's accusation against China [regarding Darfur] was not objective."

Xinhua news agency (China), October 2006

LUIZ INACIO LULA DA SILVA
President, Brazil

Brazil has rights too

"Bolivia's nationalization of its gas reserves was a necessary adjustment for a suffering people seeking a greater measure of control over their own resources. However, the fact that Bolivia has rights does not deny the fact that Brazil has rights in the matter as well."

AP Worldstream, May 2006

DMITRY PESKOV
Deputy Presidential Spokesman, Russia

Russia depends on Europe, too

"I think the authors of such an idea [gas OPEC] simply fail to understand our thesis about energy security. Our main thesis is interdependence of producers and consumers. Only a madman could think that Russia would start to blackmail Europe using gas, because we depend to the same extent on European customers."

Financial Times, November 2006

CARLOS LOPES
Political Analyst, Brazil

Lula is weak

"Presidential meetings don't resolve technical questions. They're symbolic, and the symbolism was bad from Brazil's viewpoint. Bolivia's sovereignty defends Bolivia, not Brazil. Brazil's role is to defend its interests. . . . Its attitude was very weak, but that's Lula."

AP Worldstream, May 2006

UZEIR JAFAROV
Military Expert, Azerbaijan

Iran threatens U.S. oil interests

"Even if Azerbaijan gives no consent to using its territory by U.S. troops, it should be not ruled out that Tehran, being in a desperate situation, would strike objects of U.S. economic interests in Azerbaijan: works in the Caspian Sea, the Baku-Tbilisi-Ceyhan pipeline."

United Press International, April 2006

Arcadio Esquivel/La Prensa, Panama

RACE FOR THE ARCTIC

BY BRIAN BEARY

Excerpted from the CQ Global Researcher. Brian Beary. (August 2008). "Race for the Arctic." *CQ Global Researcher*, 213-242.

Race for the Arctic

BY BRIAN BEARY

THE ISSUES

Along with several other nations, Russia claims a vast swath of the oil-rich Arctic. But last summer the Russians got fed up with the glacial pace of international efforts to settle the claims. In a swashbuckling move that outraged other Arctic players, Russia sent a pair of submersible vessels more than two miles under the Arctic ice cap to plant a titanium Russian flag in the seabed.

"This isn't the 15th century," fumed Canadian Foreign Minister Peter MacKay. "You can't go around the world and just plant flags and say, 'We're claiming this territory.' " [1]

But while MacKay scoffed at Moscow's antics, Canada, albeit more discreetly, also has been asserting its sovereignty in the Arctic — as are Norway, Denmark and the United States — prompted by high energy prices and the melting ice cap.

In recent decades the Arctic's climate has changed more dramatically than other parts of the world. Alaska, for instance, has warmed by 4.9 degrees Fahrenheit since 1950, compared to a 1.8-degree increase since 1908 in the rest of the United States. [2] Average Arctic air temperatures were 10.4 degrees higher in November 2007 than during the same period in the 1980s and '90s. More Arctic sea ice melted in 2007 than in any other year on record, with summer ice levels 20 percent lower than the previous record, set in 2005. [3]

"Many scientists who track Arctic change recognized that an abrupt decline in sea ice was possible, but nearly all were surprised that a dramatic sea-ice decline could occur so fast," according to James E. Overland, an oceanographer at the National Oceanic and Atmospheric Administration's (NOAA) Pacific Marine Environmental Laboratory in Seattle. [4]

Unlike Antarctica, which is a continent covered by mile-high glaciers, the Arctic is mostly an ocean covered with sea ice that has declined in minimum thickness from about 12 feet in the 1980s to eight feet today. *

But melting ice has been a boon in Greenland, a huge, glacier-covered Danish island located almost entirely within the Arctic Circle, making ex-

A young Nenets woman harnesses her reindeer in Siberia, Russia. Indigenous groups worry that the race for Arctic riches will affect their traditional way of life and deprive them of their fair share of the resources.

© B&C Alexander/Arcticphoto.com

traction of the rich resources beneath the ice simpler. "Climate change has a positive impact on Greenland," says Foreign Affairs Minister Aleqa Hammond. "But we are aware of severe impacts both globally and locally."

As the ice melts, Hammond says global warming's "winners and losers" are becoming obvious, such as the polar bear. Last May, the U.S. Department of Interior listed the iconic Arctic predator as a threatened species because it relies on sea ice as both a home and a feeding area. [5] For indigenous Arctic peoples, global warming has its advantages and disadvantages. On the one hand, the shrinking ice cap makes access to oil, gas and minerals easier, and warmer weather allows more agriculture. But the loss of sea ice also disrupts the habitats of seals and other marine mammals, threatening the livelihood of indigenous hunters.

In the end, however, environmentalists say the Earth itself could be the biggest loser, as a vicious cycle plays itself out: Melting ice triggers more oil and gas drilling, causing more global warming when the carbon-based fuels are burned. [6]

The most sought-after Arctic area is the huge Lomonosov Ridge — an underwater mountain range as big as California, Indiana and Texas combined that straddles the North Pole.

---

* Arctic sea ice is melting much faster than Antarctic ice because the South Pole is protected somewhat by a hole in the region's ozone layer, Overland notes, which has caused winds to increase, keeping the warmer temperatures out.

Continued on p. 217

Huge Area at Stake in Race for Arctic Resources

Eight nations have territory within the Arctic Circle, a vast region that encompasses the Arctic Ocean, the North Pole, 24 time zones, 5 million people, 30 ethnic groups and three transcontinental shipping routes. A recent U.S. Geological Survey report estimated the area could contain 22 percent of the world's undiscovered oil and gas deposits. Of the eight Arctic countries, five with borders on the mostly ice-covered Arctic Ocean — Russia, Canada, the United States, Norway and Denmark (which owns Greenland) — are scrambling to extend their offshore boundaries beyond the traditional 200-mile limit in order to claim potential offshore resources.

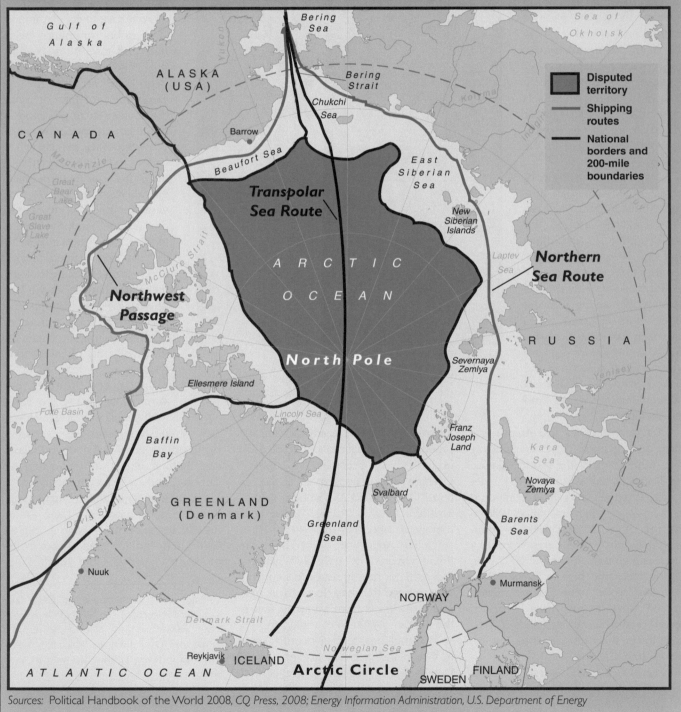

Sources: Political Handbook of the World 2008, CQ Press, 2008; Energy Information Administration, U.S. Department of Energy

Continued from p. 215

When Russia planted its flag in the middle of the Lomonosov, it angered the Danes and Canadians, who also claim the area. In total, Russia claims sovereignty over half of the Arctic Ocean. [7] Norway and the United States — the other two countries with Arctic Ocean coastlines — claim more southerly Arctic waters. *

Under the United Nations Convention on the Law of the Sea (UNCLOS), countries can claim an area 350 nautical miles or more from their shores if they can prove the adjacent seabed is an extension of their continental shelves. [8] But the United States cannot file such a claim because the Senate has refused to ratify UNCLOS. Were such a claim filed and accepted, the United States would gain almost as much territory as it did when it purchased Alaska from Russia in 1867 for $7.2 million.

Many senators regret the Senate's refusal, including Sen. Lisa Murkowski, R-Alaska. "The Arctic is one of the last spaces on Earth whose borders are not set," she says. "The U.S. needs to be a player, not an outsider. We have an opportunity that is unparalleled around the world."

Meanwhile, the Arctic's indigenous communities fear being sidelined once again in the rush to develop the region's resources (*See sidebar, p. 220.*) "This is Inuit territory," says Aqqaluk Lynge, president of the Inuit Circumpolar Council (ICC) in Greenland, which represents the Arctic's 150,000 Inuit, once known as Eskimos. "While we are very loyal to our respective governments, they must assist us in helping build Inuit unity and help the Inuit use the resources in a sustainable manner."

With oil prices now topping $127 a barrel, gasoline at nearly $4 a gallon and global oil consumption rising,

---

* The Danes are involved because Greenland, which is located mostly within the Arctic Circle, is an independent Danish province.

Arctic Region at a Glance

Country	Population (in millions, 2006, unless otherwise indicated)	Area (in square miles)	Net petroleum exports/imports (-), 2007 (thousand barrels per day)
Canada	32.8**	3,855,081	1,026
Russia	142.4	6,592,800	7,018
Norway	4.7	149,282	2,321
Greenland (Danish)	0.06*	840,000	-4
United States	304.6**	3,732,396	-12,210
Iceland	0.3	39,768	-19
Finland	5.2	130,119	-224
Sweden	9.1	173,731	-357

* 2005 estimate

** 2007 estimate

Sources: Political Handbook of the World 2008, CQ Press, 2008; Energy Information Administration, U.S. Department of Energy

---

a rush to find new supplies in the Arctic has begun. [9] Unlike in Antarctica, where mining is banned until 2041, no international moratorium on Arctic drilling exists. Arctic states already extract "black gold" in large quantities and are stepping up their operations. (*See table above.*)

Some 90 billion barrels of undiscovered oil and 1,670 trillion cubic feet of natural gas could lie onshore and offshore within the Arctic Circle region, according to a July report from the U.S. Geological Survey (USGS). That represents 30 percent of the world's undiscovered gas reserves and 13 percent of the oil reserves. [10] The United States produces 1.85 billion barrels of oil each year and 19.3 trillion cubic feet of natural gas. [11]

Russia is already the biggest player in the Arctic, with 75 percent of the known Arctic oil and 90 percent of the gas — and validation of its territorial claims would only enhance its energy-kingpin status. [12] The Norwegians, who operate offshore fields in the North Sea, are moving operations north as old wells dry up. Petroleum,

which represented 31 percent of Norway's revenues in 2007 and 48 percent of its exports, is key to Norway's wealth. [13] Greenland has quintupled the number of exploitation licenses it grants and expanded the area earmarked for oil and gas exploration from 2,657 sq. miles to nearly 39,000 sq. miles. [14] Meanwhile, both the United States and Canada are opening up their sections of the Beaufort Sea to drilling. (*See map, p. 216.*)

The European Union (EU) — which recently predicted the scramble for resources will intensify in the Arctic — will probably end up in the Arctic "loser" category, because it has no territory in the region. Although Denmark is a member, its Arctic province Greenland left the EU in 1985, and Norway stood on the threshold of EU membership twice, but referenda in 1972 and 1994 narrowly failed. [15]

Despite the attention being paid to the Lomonosov Ridge, USGS geologist Don Gautier, says it "is not an interesting place from the petroleum point of view" because the vast majority of Arctic oil and gas lies elsewhere. "Offshore Alaska is the most obvious place to look for oil, while the area with the most gas is the West Siberian Basin in Russia." Although the Lomonosov contains sedimentary rock — a critical component for petroleum reserves to be present — there is no evidence of a previous tectonic event that would have sealed reserves under the seabed, Gautier explains.

And, even if oil and gas are found, countries will drill closer to their coastlines first, Gautier says, because they have undisputed sovereignty over these areas and because it's easier to operate there than in the Lomonosov, which is hundreds of miles offshore.

The shrinking sea ice also is beginning to affect global shipping. In summer 2007 the legendary Northwest Passage through northern Canada, which connects the Atlantic and Pacific oceans, was ice-free for the first time in recorded history, raising the promise of new, shorter trans-Arctic shipping routes. By 2030 the Arctic may be entirely ice-free during the summer months, NOAA's Overland predicts.

Savings on shipping costs could be enormous if the Arctic routes were to become usable for longer periods. A ship sailing from New York to Tokyo, for instance, could shave 2,600 miles off its journey by taking the Northwest Passage instead of a conventional route through the Panama Canal. Vessels traveling from London to Tokyo could reduce their journey by some 5,000 miles by taking the Northern Sea Route — also called the Northeast Passage — through Russian waters instead of the Suez Canal. (*See map, p. 223.*) [16]

And even partly ice-covered passages are navigable now with new "double acting" ships, which sail through ice-free waters using their V-shaped bow, and then turn around when they hit icy waters and navigate with their U-shaped stern, eliminating the need for an accompanying icebreaker.

Satellite images of the top of the Earth show how the polar ice cap has shrunk dramatically over the past 27 years. Scientists say more Arctic ice melted in 2007 than in any other year on record, with summer ice levels 20 percent lower than the previous record — set in 2005.

National Snow and Ice Data Center

Initially, with expanded offshore energy exploitation and warmer waters attracting more warm-water fish, most of the increased trans-Arctic sea travel would probably be petroleum-laden tankers or commercial fishing vessels. The jury is still out, however, on whether trans-Arctic shipping of consumer goods will be commercially viable anytime soon — because savings in distance would be offset by other expenses, such as building more ice-resistant ships.

In any case, Russia — which already has 18 icebreakers that escort cargo ships along the Northern Sea Route — seems best positioned to exploit new opportunities. [17] A rise in commercial traffic along North America's Northwest Passage seems less likely, because the route is more difficult

to navigate, and the United States and Canada disagree about its legal status. The United States says the passage is an international strait open to all; Canada says it is Canadian waters.

As the race for Arctic treasure intensifies, here are some of the key questions political leaders are grappling with:

Is an Arctic treaty needed?

"This is an absolute necessity," says Rob Huebert, a politics professor and Arctic expert at the University of Calgary. "We have no multilateral system of governance for the Arctic. The U.S. is not a party to UNCLOS, and the treaty does not deal with certain things — for example the rights of indigenous peoples."

Huebert insists a new treaty should be concluded between all eight states with territory within the Arctic Circle — Canada, Denmark, Russia, the United States, Norway, Iceland, Sweden and Finland — rather than just the five whose coastlines border the Arctic Ocean (Canada, Denmark, Russia, the United States and Norway), who were invited to a ministerial meeting in May in Greenland to discuss territorial issues. A treaty should also contain a conflict-resolution mechanism and provisions on navigation, fishing and tourism, he says.

"Right now, everyone is going it alone," he continues. "We need a champion like Malta was in the 1960s for UNCLOS. It would be nice to see Canada assume this role." (*See p. 227.*)

Craig Stewart, director of the World Wildlife Fund's (WWF) Ottawa Bureau, adds, "We need a framework convention on resource extraction that sets environmental and oil and gas recovery standards. This should enshrine the 'integrated management planning concept' under which the impact of all activities — fishing, shipping, oil drilling — are assessed, all stakeholders are involved and resource extraction is banned in certain areas."

He says Canada's 1996 Oceans Act would be a good model for such a convention.

Britain's Diana Wallis, vice president of the European Parliament, says an international Arctic treaty would be "appropriate," especially now. "The timing is good, with 2007-2008 being International Polar Year."

So far, she says, at least "people are honoring the legal frameworks" of the UNCLOS territorial-claims procedure. "If it works, OK. But in the long term we should think about drafting an Arctic Charter and strengthening the political dimension of the Arctic Council."

Set up in 1996, the Norway-based council includes representatives from all eight Arctic states plus indigenous community organizations. Yet so far, it has not been much of a player: A U.S. secretary of State has never attended one of the council's biannual ministerial-level meetings. [18] And when asked to outline the council's vision for development of the Arctic, a spokesman said he was "not in position to speak to political issues on behalf of its eight member states."

Environmental groups are favorably disposed to such a treaty and to beefing up the council's role. "The council has not developed policies yet — it just releases studies, although this is useful, too," says Chris Krenz, Arctic Project Manager in Juneau, Alaska, for the Oceana environmental group. "UNCLOS may be useful for deciding territorial rights, but there needs to be something else just for the Arctic."

But Norway's ambassador to the United States, Wegger Strommen, says an Arctic treaty is superfluous. "UNCLOS has worked well," he says. "If disputes arise from overlapping claims, it is best to resolve them through bilateral negotiations. Once we have the technical data, I am confident they will be resolved."

Norway's Foreign Minister Jonas Gahr Store agrees. "We do not exclude future new regulation in particular fields," he said, "but only if real needs have been identified with pre-

© B&C Alexander/Arcticphoto.com (both)

Moving the "Gold"

The huge oil facility at Prudhoe Bay, on Alaska's Beaufort Sea coast, sits next to the largest oil field in the United States (top). With gasoline prices at nearly $4 a gallon and global oil consumption rising, Arctic countries are racing to develop the region's "black gold" — both onshore and offshore. Canada is building a 750-mile natural gas pipeline to transport natural gas from its Northwest Territory fields to southern markets. And in Norway, drilling rigs are being moved northward toward the Arctic as its North Sea wells dry up (bottom).

cision. The actual challenges related to the legal regime in the Arctic may have more to do with a lack of implementation of existing rules than a lack of rules." [19] As evidence, a Norwegian embassy official responsible for fisheries policy cites agreements Norway concluded with other countries in 2005 setting fishing quotas for blue whiting and herring. [20]

The Norwegian view was endorsed in May by Canada, Denmark, the United States and Russia when ministers met in Ilulissat, Greenland, to discuss this issue. A post-meeting joint declaration said UNCLOS provides a "solid foundation for responsible management by the five coastal states and other users of this ocean." It concluded: "We therefore see no need to develop a new, comprehensive international legal regime to govern the Arctic Ocean." [21]

Given such opposition from the key governments, an Arctic treaty looks politically unfeasible.

Oran Young, a professor of environmental science and international governance at the University of California, Santa Barbara, dismissed the idea of an Arctic treaty as "utopian — both politically and legally. Legally binding agreements are hard to negotiate, often lacking in substance, and commonly slow to enter into force. They are clumsy instruments that are apt to cause trouble in highly dynamic settings." [22]

And although Young supported enhancing the Arctic Council's role somewhat, he also recommended "keeping its decision-making authority and organizational capacity to a minimum." [23]

Gunnar Sander, Arctic adviser at the Copenhagen-based European Environment Agency (EEA), a network of EU environmental agencies, suggests something slightly more modest than

Arctic's Indigenous People Fight for Their Resources

Big nations are rushing to exploit offshore gas, oil.

Exploitation of the Arctic's resources is ancient history to its indigenous peoples, says Aqqaluk Lynge, president of the Inuit Circumpolar Council (ICC), which represents 150,000 Inuit. When Arctic-area ministers met recently in breathtakingly picturesque Ilulissat, Greenland, to discuss sharing the region's natural resources, he reminded them the debate stretches back to the 1600s, "when the first foreign whaling ship came to hunt our big whales and decimate our stocks, from which they have never recovered." [1]

Today Arctic governments have their sights set on offshore oil and gas and new shipping and fishing opportunities. And indigenous communities want to know whether they will be winners or losers in the new race for Arctic resources.

"All this is nothing new for us," says Gun-Britt Retter, a member of the indigenous Sami community's parliament in Norway.* "We used to have to pay taxes to four different kings — Sweden, Norway, Denmark and Russia. Today we have governments making nice speeches about indigenous peoples having the right to their culture. But they do not give us the basis for that culture — our land. We have no right to veto drilling operations and no right to revenues from oil and gas extraction on our land."

Indeed, throughout modern times indigenous Arctic peoples have fought to preserve their language, culture and way of life as neighboring colonial powers encroached on their turf. In 1953, for instance, Canada pushed the Inuit in Nunavik, Quebec, into the High Arctic in an effort to assert Canada's sovereignty over the region. The same year Eskimos were forced out of their homes in northwestern Greenland to make room for a Danish-backed U.S. Air Force base in Thule.

Native peoples have scored some successes, however. In 1971, under the Alaska Native Claims Settlement Act, the U.S. government gave the Inuit $962.5 million and 44 million acres

of land in Alaska after complaints that oil developers were robbing locals' land and destroying the environment. [2] In 1999 Canada created the Inuit-dominated Territory of Nunavut by splitting the Northwest Territories in two. Greenland, which is 90 percent Inuit, gained home rule from Denmark in 1979, after nearly three centuries of domination. A referendum this November would give the Inuit in Greenland further autonomy and divide up future oil and gas revenue between the Greenlandic and Danish governments.

Nenets reindeer herders meet with gas company officials in Yamal, Siberia, to discuss how oil development is affecting their community.

But Edward Itta, the mayor of Alaska's North Slope Borough, is critical of how the U.S. government is conducting its current policy review of the Arctic.

"We have not been formally involved in the review," says Itta, an Inuit. "We have lived here for 10,000 years, yet the bureaucrats in Washington think they know it all."

Besides fighting for a seat at the table, indigenous peoples are trying to forge a strong and unified stance among them-

* The Sami were formerly known as Laplanders.

selves. Inuit from Russia, Alaska, Greenland and Canada will meet in Nunavik in November to devise a common position, which is vital to resisting the Arctic powers' "divide-and-conquer approach," says Lynge. [3]

In addition to the Inuit, the 70,000-strong Sami — residing in Norway, Sweden, Finland and Russia — are the second-largest Arctic indigenous group. Other smaller groups include the Aleut, who live on the Pacific Aleutian islands between Russia and Alaska, the Athabaskans and Gwich'in from Alaska and Canada plus 41 indigenous groups who live in Arctic Russia.

The rights of indigenous peoples vary widely. For example, the Sami have their own parliaments in Norway, Sweden and Finland, but not in Russia. "Our people face their biggest challenge in Russia," says Retter. 'The government there draws up maps for pipelines and mining development, ignoring the people who live there. Reindeer herding is a huge part of our culture, but because of this new infrastructure, the reindeer, which move homes between summer and winter, become blocked."

Now global warming is forcing indigenous communities to adapt quickly. According to Kenneth Hoegh, Greenland's agriculture adviser in Qaqortog, climate change has been a mixed blessing in southwest Greenland. The reduction in drift ice has hurt Inuit hunters because the ice calms the sea, enabling the hunters to shoot seals.

"If this continues, they will need to find other livelihoods — maybe fishing, eco-tourism or ethnic tourism," he says.

On the other hand, warmer seas have attracted more cod, which fetch good prices. Fishermen have had to invest in new gear and boats, however, since they previously fished mainly for shrimp, which are eaten by the cod. Global warming also has given a boost to farming, allowing more grazing and hay, silage and vegetable cultivation.

Meanwhile, industrial pollution from faraway regions threatens indigenous peoples' health. Inuit mothers' breast milk has become dangerous because the polar bears, seals, walruses, fish and whales they eat are contaminated by heavy metals,

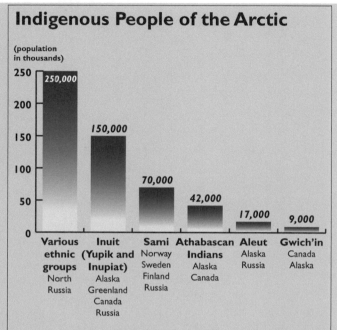

Indigenous People of the Arctic

(population in thousands)

Various ethnic groups	Inuit (Yupik and Inupiat)	Sami	Athabascan Indians	Aleut	Gwich'in
250,000	150,000	70,000	42,000	17,000	9,000
North Russia	Alaska Greenland Canada Russia	Norway Sweden Finland Russia	Alaska Canada	Alaska Russia	Canada Alaska

Sources: Inuit: www.inuitcircumpolar.com; Sami: www.galdu.org; Aleut: www.apiai.com; Athabaskan: www.arcticathabaskancouncil.com; Gwich'in: www.gwichin.org; Russian indigenous: www.raipon.org.

PCBs and other industrial compounds found in seawater and stored in the animals' fat. [4]

While much is known already about the environmental challenges, more research is needed for policymakers to make the right decisions, according to Mayor Itta. "Our ocean is getting more acidic," he says. "We need more baseline data to understand the impact on the entire food chain — from the krill to the bowhead whale."

[1] Aqqaluk Lynge, president, Inuit Circumpolar Council (ICC) Greenland, address to the Ministerial Summit of Arctic Oceans, "Issues relating to the local inhabitants and indigenous communities," May 28, 2008, Ilulissat, Greenland, www.inuit.org/index.asp?lang=eng&num=3 (Inuit version)

[2] Encyclopaedia Britannica Online, www.britannica.com.

[3] Lynge, op. cit.

[4] Colin Woodard, "Oceans in Crisis," CQ Global Researcher, October 2007, pp. 237-264.

an Arctic treaty: a regional protocol for the Arctic Ocean within the UNCLOS framework. Twelve regional-seas conventions already have been adopted under the UNCLOS framework, including a 1992 treaty for protecting the North-East Atlantic. [24] Yet, with the five key Arctic governments clearly favoring bilateral and sector-specific pacts, even this modest suggestion looks to be a long shot.

Should Arctic oil and gas reserves be exploited?

While environmentalists are more enthusiastic than governments about an Arctic treaty, the situation is reversed on the question of oil and gas drilling: Governments are much more enthusiastic than environmentalists on the issue.

International mineral and oil companies "are flocking to Greenland," says Foreign Affairs Minister Hammond. "I

would, too, if I were in their position." Greenland, with a population 56,000, would get a massive windfall if major oil or gas deposits are found, notes Hammond, who is also minister for finance.

Claudia A. McMurray, the U.S. State Department's assistant secretary for oceans, environment and science, supports exploration if it is done "in a sustainable manner." Guidelines are needed on how to conduct operations and

Life is treacherous these days for Arctic polar bears — like this one at Cape Churchill, Canada — as Arctic ice melts away due to global warming. The U.S. Department of Interior in May listed the iconic predator as a threatened species because it relies on sea ice as both a home and a feeding area. The listing could block or delay development of Alaskan oil reserves.

© B&C Alexander/Arcticphoto.com

contingency plans for spills, she says, but "the U.S already has rules for this."

Environmentalists are more fearful. "The rush to exploit Arctic resources can only perpetuate the vicious cycle of human-induced climate change," said Greenpeace International spokesman Mike Townsley. [25] Extracting oil and gas will only lead to more fossil fuels being burned, which will trigger further global warming and more melting of the ice caps, he explained.

But Danish Ambassador to the United States Friis Arne Petersen says, "We cannot give up on oil and gas. We want to drill offshore for oil in western Greenland and gas in eastern Greenland."

Oceana's Krenz worries about the impact exploration will have on the surrounding environment, noting that Alaskan beaches still have not recovered from the 1989 *Exxon Valdez* oil spill. [26] "If you dig a hole, you can still see oil seeping its toxic compounds into the ocean," he says. "This has been very detrimental to pink salmon. Herring stocks have never really recovered either."

Oil spills are especially lethal for seabirds and seals, because the oil covers their feathers and fur, making it harder for them to escape from predators, causing seals to drown by sticking to their flippers and causing

hypothermia in seal pups and birds. "Placing wells, pipelines and vessels in the remote Arctic creates a substantial risk of a catastrophic oil spill, and there is no proven method to clean up an oil spill in the icy conditions often found in the Arctic," Oceana warns. [27] And the noise caused by drilling could drive whales and other marine life away from feeding areas, it added.

For these reasons, the World Wildlife Fund is calling for a moratorium on Arctic development. "The U.S. Minerals Management Service (MMS) estimates that there is a 20 to 52 percent risk of an oil spill in the Chukchi or Beaufort seas," says WWF's Stewart. "The British-based energy company BP did a test in 2000 that concluded you could not clean up a spill if there was 30 percent ice coverage. It is highly irresponsible to proceed without a recovery system."

The European Environmental Agency's Sander says drilling in the High Arctic would be "very risky." The large quantities of ice, 24-hour darkness in winter and extremely low temperatures make conditions for operating a facility treacherous, he says.

EU parliament member Wallis is more circumspect: "In a sense, it is fair for countries to exploit the resources. But you must proceed with caution, getting the best science, not taking risks with the environment and respecting what the local populations want."

According to a recent Arctic Council report, "knowledge about effects on the environment and human health of oil and gas activities is limited." [28] To date there have been no large oil spills in the Arctic Ocean from oil and gas activities (the *Exxon Valdez* ran aground in the Pacific), it noted, and seismic exploration has left "no long-lasting effects on fish stocks or marine ecosystems." [29] Most animals revert to normal behavior when the noise ceases, the report said, except for bowhead whales in the Beaufort Sea, which had been observed changing swimming direction in response to noise sources up to 20 miles away.

In addition, the report continued, the local communities experienced increased employment during the construction phase, but many workers were brought in from the outside because the locals did not have the necessary training. And when oil and gas activity ceases, old sites need to be safely removed and the surrounding environment cleaned up — something the industry does not always bother to do. [30]

Perhaps for these reasons, the Inuit conference's Lynge says tough questions must be answered before energy exploitation gets the green light. "How many new jobs will be created?" he asks. "How many of them will go to Greenlanders? Who will be getting the high-paid jobs?"

In Alaska, some say the issue is complicated by the polar bears, which the U.S. government recently listed as threatened. "If the Department of Interior decides that exploration will further threaten the polar bears' habitat, it will be difficult to grant development leases," says the State Department's McMurray.

"This finding, given the inflexibility of the 1973 Endangered Species Act, makes it easier to legally challenge oil and gas lease sales in the Beaufort and Chukchi seas," says Professor Jonathan Adler, an environmental law expert at Case Western Reserve University in Cleveland. "I think the matter will be resolved in the courts."

Will the melting ice caps revolutionize international shipping routes?

The receding ice caps are likely to affect Arctic shipping, but how quickly things will change and what routes will be affected is unknown.

Three basic routes cross the Arctic Ocean: the Northwest Passage in North America, the Northern Sea Route through Russia's northern wa-

Arctic Melting Opens Up Shorter Shipping Routes

The melting of the Arctic ice cap could create more efficient shipping routes from Europe to the Far East. For example, the distance from Hamburg, Germany, to Yokohama, Japan, via traditional shipping routes — around the Cape of Good Hope or through the Panama or Suez canals — could be as long as 14,500 nautical miles. By contrast, the distance is half that amount via the Northern Sea Route (see below), and even less for the Transpolar Route and Northwest Passage.

Nautical Miles from Hamburg, Germany, to Yokohama, Japan

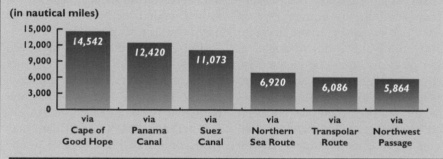

(in nautical miles)

via Cape of Good Hope	14,542
via Panama Canal	12,420
via Suez Canal	11,073
via Northern Sea Route	6,920
via Transpolar Route	6,086
via Northwest Passage	5,864

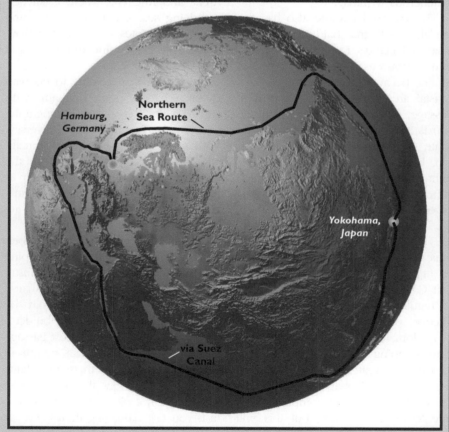

Source: Ray Chartier Jr., "Arctic Sea Ice Recent Trends and Causes; Impact on Arctic Operations," U.S. Naval Ice Center

ters and the Transpolar Route over the North Pole. (*See map, p. 223.*)

"The greatest potential saving is probably over the Northern Sea Route" because it's the most ice-free, according to Mead Treadwell, chairman of the U.S. Arctic Research Commission.

A 2006 study funded by the Alaska-based Institute of the North concluded that shipping containers from the Aleutian Islands in western Alaska to Iceland using the Northern Sea Route would cost $354 to $526 per container compared to the current cost of $1,500 per container from Japan to Europe using the southern route. [31] Other costs also must be factored in, such as having to build and operate new terminals, the study stresses. Until now, the northern route has not been used for shipping goods between continents because its icy waters make it treacherous to navigate.

The Transpolar Route also has potential, says Ragnar Baldursson, an official at Iceland's Department of Natural Resources and Environmental Affairs, because it avoids treacherous coastal areas, such as the Northwest Passage, where ice is often swept towards the straits. "The Transpolar route is easier for political reasons, too, because no one disputes ships' right to sail over the North Pole, which is not the case for the Northwest Passage," he says. In an ongoing dispute, the United States claims the Northwest Passage is an international strait through which all ships can pass, while Canada argues it is Canadian waters and that Canada can impose environmental regulations and demand to be notified of passing ships.

Diplomatic squabbles aside, Canadian shipping executive Thomas Paterson says greater commercial use of the Northwest Passage is not viable. "It would need to be completely ice-free to be economical, but this will not happen," says Paterson, vice president of Montreal-based Fednav Ltd. "Ships coming up Greenland's coast

[to reach the passage] will encounter icebergs broken off from Greenland's glaciers and will have to slow down to avoid them, so the journey could end up taking more time despite the shorter distance."

In addition, icebreaking ships are more expensive to build — $100 million compared to $40 million for a standard ship, he points out. Businesses shipping goods across continents will not take on these extra risks and costs, he says, although oil companies shipping Arctic oil in and around the Northwest Passage may be willing to pay.

The greatest potential for improved Arctic shipping lies with regional, not trans-continental, shipping, says Lawson W. Brigham, director of the Arctic Research Commission's Alaska Office and a former U.S. Coast Guardsman who commanded icebreakers in the Arctic and Antarctic. There is already a thriving seasonal trade in shipping minerals, he notes, including zinc from Alaska's Red Dog mine and nickel and copper from Siberia.

"This trade has nothing to do with climate change," says Brigham, who earned a Ph.D. in polar oceanography from England's Cambridge University. With seven nuclear-powered icebreakers to escort ships, Russia has used the Northern Sea Route for 60 years to transport fuel, food, minerals and machinery during the summer months, according to a Russian official.

The University of Calgary's Huebert predicts that most of the new shipping will support the oil and gas industry rather than regular cargo, a view shared by Brigham. "The distances may be shorter, but there will still be ice," Brigham says. "The Northwest Passage was open for less than a month in 2007, and even then there may have been small ice floes that the satellite images did not pick up. If you go too fast, you risk damaging things."

In 1994 Brigham captained the U.S icebreaker *Polar Sea*, which, in a joint expedition with its Canadian counter-

part the *Louis S. St. Laurent*, became the first surface ships to cross the central Arctic Ocean. They approached the polar sea via the Bering Strait and came out at Svalbard. "It was a slow crossing, about 40 days," he says. "Certain commodities — toys, fresh fruit, clothing, cars — are very time-sensitive and you may not be able to afford to arrive a few days late." Brigham says "there are strong differences of opinion about the use of trans-Arctic navigation on a routine basis" and that solid economic analyses of its viability have yet to be performed.

Ice itself isn't the only obstacle to trans-Arctic shipping, notes Winn Dayton, a director of transportation policy at the U.S. State Department's Bureau of Economic and Business Affairs. Special ice-resistant hulls are expensive, and Arctic ports lack road and rail connections, Dayton says. Extracting materials in such remote areas is costly and dangerous, since it is far from Coast Guard crews who could launch search-and-rescue operations, he adds. [32]

"Conditions are extremely tough up there, with the storms and rough seas," says Petter Meier, fisheries counselor at the Norwegian Embassy in Washington. "In ice-covered waters, the ice can actually screw a ship right down into the sea."

Arctic tourism — which is experiencing something of a boom, with hundreds of cruises crisscrossing Arctic waters last summer — is a potential growth sector. Yet lack of infrastructure in the remote regions of Canada and Greenland limits the potential, according to Capt. Ted Thompson, senior vice president of the Cruise Lines International Association.

"Tourists will have nothing to do if you dump them at the ports there," he says. "Even if the Northwest Passage is open for a week and a half this year, it will not be long enough to make it through all the way. We

Continued on p. 227

Chronology

35,000–10,000 B.C.
First settlers arrive in the Arctic.

11,000 B.C.
Ice cap thaws, and land bridge linking Asia and North America floods.

900-1910s
Explorers and colonial powers arrive in the Arctic.

986 A.D.
Erik the Red establishes a small Norse colony in Greenland.

1609
Dutch jurist Hugo Grotius publishes *Mare Liberum*, establishing free-seas principle.

1845
An expedition led by Britain's Sir John Franklin in search of Canada's Northwest Passage disappears, never to be heard from again.

1867
Russia sells Alaska to the United States for $7.2 million.

1909
U.S. explorer Robert Peary claims to be first to reach North Pole.

1920s-1960s
Governments begin exploiting the Arctic's natural resources and questioning Grotius' free-seas principle.

1920
Oil exploration begins in Canada's Northwest Territories.

1922
Oil exploration begins on Alaska's North Slope.

1945
President Harry S Truman claims all resources on the U.S. continental shelf.

1958
U.S. government issues first lease for oil exploration on the North Slope.

1967
Malta's ambassador to the U.N. gives a groundbreaking speech calling for an international treaty on the sea.

1968
First surface expedition, led by American Ralph Plaisted, reaches the North Pole, using snowmobiles.

1970s-1990s
Global governments regulate the seas. Economic development of the Arctic continues to grow.

1973
U.N. conference convenes to write a global oceans treaty.

1977
Trans-Alaska pipeline begins pumping oil 800 miles from northern Alaska to the ice-free southern seaport of Valdez. . . . Inuit establish the Inuit Circumpolar Council to represent their interests. . . . Soviet nuclear-powered icebreaker *Arktika* is first ship to reach the Pole.

1981
First exploratory well is drilled on Alaska's outer continental shelf.

1982
U.N. adopts the Convention on the Law of the Sea (UNCLOS) and opens it for ratification. Eventually 156 nations, but not the United States, will ratify it.

1984
Snohvit gas field is discovered in Norway's Barents Sea.

1987
Soviet leader Mikhail Gorbachev proposes transforming the Arctic into a "zone of peace."

1994
UNCLOS enters into force.

2000s
Melting ice caused by global warming triggers a new rush for Arctic land and natural resources.

2001
Russia becomes the first Arctic nation to claim sovereignty over the North Pole, using a process set up under UNCLOS.

2007
The Northwest Passage is ice-free for the first time on record. . . . Russia plants its flag on the North Pole seabed, galvanizing other Arctic nations into asserting their own sovereignty. . . . U.S. Senate Foreign Relations Committee approves UNCLOS, bringing it a step closer to ratification.

2008
U.S. authorizes oil and gas exploration in the Chukchi Sea. . . . U.S. lists the polar bear as a threatened species due to loss of its sea-ice habitat. . . . Russia, Canada, Denmark, the United States and Norway agree to use UNCLOS to resolve Arctic territorial claims. . . . U.S. and Canada begin collaborating on Arctic seabed mapping to pursue continental shelf claims.

Territorial Disputes Roil the Arctic

Many nations eying gas and oil reserves.

On a tiny, uninhabited island in the remotest reaches of the Arctic, Danish naval officers stake their nation's claim to the island once a year by planting bottles of Old Danish bitter in the snow. Troops from Canada also claim the barren patch of land, but with Canadian Club whiskey, drinking the bitters to remove all traces of a Nordic colony. The Danes return the favor when they come back, downing the Canadian whiskey and replacing it with more Danish spirits.

This good-humored sparring continued for years until recently, when the dispute between Canada and Denmark over Hans Island — about 300 acres wedged between Canada's Ellesmere Island and Greenland — suddenly was no longer a joke.

At an Arctic foreign ministers meeting in Ilulissat, Greenland, in May, Greenlandic Prime Minister Hans Enoksen declared, "We traditionally have already named the island The Kidney-Shaped Island. . . . should anyone have any claims prior, they would have named it already before we did." Canada's natural resources minister, Gary Lunn, retorted tersely, "I'm not going to comment on that," adding, "we're here to affirm Canada's sovereignty by our strong presence in the North." [1]

While the row may seem as silly as a "Monty Python" sketch, no one is giving a square inch — keenly aware of the precedent it might set for other territorial claims and, ultimately, control of the Arctic's vast resources.

Southwest of Hans Island, the United States and Canada have been entangled in a similar standoff since 1969 over the status of the Northwest Passage, the stretch of water in northern Canada that connects the Atlantic and Pacific oceans. The United States insists the fabled passage is an international strait open to all vessels. [2] Canada insists it is part of Canada's territorial waters.

"If Canada is right," says Professor Rob Huebert, an Arctic sovereignty expert at the University of Calgary, "Canadian authorities will decide who can or cannot pass and what safety and environmental rules they must follow."

Until now, the dispute has been largely academic, because the passage historically has been frozen and unnavigable for most of the year. But global warming made the passage ice-free last summer for the first time in observational record and has catapulted the dispute into the political realm.

"There is no room to move. We have agreed to disagree," says Claudia McMurray, the State Department's assistant secretary for oceans, environment and science. "It is not a major issue right now, but it will be if the ice melts forever and the passage becomes a shipping lane." Neither side has brought a case to the International Court of Justice (ICJ) yet, but this could well happen given the increasing inflexibility Arctic nations are showing in territorial disputes.

A 1949 ICJ ruling on the use of the Corfu Channel in the Mediterranean Sea found that the channel was an international strait only if the passage was used for international navigation. So far the Canadians have argued that is not the case for the Northwest Passage, because so few ships use it. But that may soon change.

Canada and the United States also are at odds over how to draw their border in the Beaufort Sea. The Canadians say it should be a continuation of the land border between Alaska and Canada, while the United States says it should be a line equidistant from both countries' coastlines. At stake are rights to the rich, underwater oil and gas deposits.

Some 10,000 miles to the east, Russia and Norway are contesting the sovereignty of a 60,000-square-mile area in the Barents Sea, with no resolution in sight. Oslo and Moscow have agreed to allow their military, commercial and fishing vessels to use the waters, but the oil or gas reserves remain untouched.

Meanwhile, Iceland contests Norway's sovereignty — gained under a 1920 treaty — over the Svalbard archipelago. "It is not obvious that Norway owns Svalbard, yet it has taken unilateral control of fishing rights there," complains an Icelandic diplomat, who asked that his name not be used. In another Arctic dispute, Norway, Denmark and Iceland have overlapping claims on a section under the Norwegian Sea called the Banana Hole. [3]

But the granddaddy of Arctic territorial disputes is now before the U.N. Continental Shelf Commission. Russia, Canada and Denmark each have their eyes set on the enormous, underwater Lomonosov Ridge, which straddles the North Pole. The ridge is as big as California, Indiana and Texas combined and is thought to contain rich mineral and possibly oil and gas deposits, although the U.S. Geological Survey recently concluded that most Arctic oil and gas are located elsewhere. [4]

For now, diplomacy remains the favored channel for resolving the disputes, but as the Arctic's geopolitical significance grows, that could change. As Denmark's ambassador to the United States, Friis Arne Petersen, notes, "In the 1930s and 1940s when Denmark and Norway contested a part of Greenland in the northeast, we went to the ICJ to get our sovereignty confirmed. If we cannot agree on Hans Island, we could go to the ICJ again."

---

[1] Randy Boswell, "Hans Island ours first: Greenland; Premier rejects Canada's claim to disputed Arctic territory," Canwest News Service, May 29, 2008, www.canada.com/ottawacitizen/news/story.html?id=582509c7-fe1a-46f9-887d335a1b100e72.

[2] "Documents on the law of the sea — historical perspective," United Nations Web site, www.un.org/Depts/los/convention_agreements/convention_historical_perspective.htm#Historical%20Perspective.

[3] "Continental Shelf Submission of Norway in respect of areas in the Arctic Ocean, the Barents Sea and the Norwegian Sea," Government of Norway, 2006, www.un.org/Depts/los/clcs_new/submissions_files/submission_nor.htm.

[4] "Circum-Arctic Resource Appraisal," U.S. Geological Survey, July 2008, http://energy.usgs.gov/arctic.

Continued from p. 224

will see expeditionary cruises but not big-scale tourist cruises." [33]

Bill Sheffield, director of the Port of Anchorage and a former Alaska governor, notes that only "120 ships crossed the Northwest Passage last year, whereas 5,000 passed through the Panama Canal. I do not think the Northwest Passage will be commercially viable in my age. But if it does happen, Anchorage will be very important because it is a deepwater port." [34] ■

BACKGROUND

Exploration and Migration

The first human settlers arrived in the Arctic — from the Greek word *arktos*, meaning "bear" — some 30,000 years ago. [35] At the time a land bridge across the Bering Strait linked the North American and Eurasian continents, but around 11,000 B.C. the climate warmed and melting ice flooded the land bridge. [36] The warmer temperatures enabled forests to grow and led to better hunting and fishing possibilities.

By the Middle Ages (900-1400 A.D.) Europeans looking for fur and foodstuffs made trading and raiding forays into the northern parts of what is now Norway, Sweden, Finland and Russia. During the modern era, Russia emerged as the dominant Arctic power, expanding its borders, subjugating indigenous peoples, exploiting the region's fishing, forestry, mineral and energy resources and establishing a strong military presence. They colonized Alaska after Tsar Peter the Great ordered the exploration of Russia's Pacific coast; a Russian expedition landed in Alaska in 1741. [37] The United States purchased Alaska from a finan-

cially strapped Russia in 1867 for $7.2 million, although Alaska did not become a U.S. state until 1959. [38]

Greenland was settled more than 4,000 years ago by North American Inuit. In 986 A.D. the Vikings settled the huge island during a warm spike, led by Erik the Red, but they left when it got colder in the early 15th century. In 1721, a Danish-Norwegian priest, Hans Egede, settled in Godthab (today's capital, Nuuk), marking the beginning of Danish sovereignty over the island. [39] Norway occupied and tried to claim part of Greenland in 1931, but the International Court of Justice ruled in 1933 that the whole island belonged to Denmark.

During World War II, Nazi Germany's occupation of Denmark blocked Danish contact with Greenland. The United States, recognizing its geopolitical importance, started to trade more with Greenland and in 1946 even offered $100 million for the island, which the Danes rejected. [40] In 1951 the United States opened an Air Force base at Thule in northwest Greenland. [41] Meanwhile, Greenlanders began to seek more autonomy, winning home rule in 1979.

Between the 16th and 20th centuries, explorers mapped out the Arctic sea routes. Dutchman Willem Barents discovered the island of Novaya Zemlya off the north Russian coast in 1594 and the Svalbard archipelago north of Norway in 1596. An Austro-Hungarian expedition led by Karl Weyprecht discovered the Franz Josef Land archipelago north of Novaya Zemlya in 1873. In 1845 Britain's Sir John Franklin led the most famous and tragic expedition, searching vainly for a navigable path through Canada's Northwest Passage. Franklin and his men were never seen again, but expeditions launched to find him greatly expanded knowledge of the area's geography. Remains of the expedition, including bones of crew members, were later discovered, indicat-

ing they perished from a combination of bad weather, disease and starvation. [42] In 1906 Norwegian explorer Roald Amundsen became the first to successfully traverse the passage.

The race to reach the North Pole began in the late 1800s, with American explorer Robert E. Peary declaring he had won in 1909 — although his claim is widely disputed today. The first surface expedition definitely to reach the Pole was led by American Ralph Plaisted in 1968, using snowmobiles. The Soviet nuclear-powered icebreaker *Arktika* in 1977 became the first surface vessel to reach the Pole. [43]

Global Treaties

Historically, the seas were regulated by the "cannon-shot-rule," which held that countries controlled the seas up to three miles from their coast — or the range of a 17th-century cannon. In 1609, Dutch philosopher and jurist Hugo Grotius' influential treatise, *Mare Liberum*, established the right to freely navigate the seas for trade purposes. [44]

In the mid-20th century, however, countries began pushing to expand their maritime territory to exploit offshore natural resources. In 1945 President Harry S Truman extended U.S. jurisdiction over all oil, gas and minerals on the "continental shelf" — a term not clearly defined at the time. [45] In 1970, in an effort to prevent its waters from becoming polluted, Canada asserted its right to regulate navigation for 100 nautical miles from its shores. [46] Iceland then extended its maritime boundary to 200 miles, provoking three bloodless "cod wars" with Britain, which dispatched warships to protect its trawlers against the Icelandic coast guard. [47]

Such tensions underscored the need for a global treaty to regulate the seas. In 1967, a passionate address

A family of indigenous Sami — formerly called Laplanders — prepares a meal during the reindeer migration in northern Norway.

© B&C Alexander/Arcticphoto.com

to the General Assembly by Malta's ambassador to the United Nations, Arvid Pardo, helped launch a 15-year negotiation process that led to the signing of UNCLOS in 1982. [48] Another groundbreaking moment came five years later, when Soviet leader Mikhail Gorbachev declared the Arctic should be transformed into "a zone of peace." At the time, the Soviet and U.S. navies were conducting Cold War maneuvers in the region as a display of their military preparedness. [49]

UNCLOS gave coastal countries the right to exploit all marine resources in their "exclusive economic zone," or the area up to nautical 200 miles from shore. The provision especially benefited coastal states without a big continental shelf, guaranteeing them a minimum of 200 miles of control. States could extend that limit even farther if their continental shelf extended beyond 200 miles. [50]

In 2001 Russia became the first Arctic coastal state to request such an extension, claiming four separate areas of the Arctic Ocean. The U.N. Continental Shelf Commission has asked for more data. [51] In 2006 Norway, whose sovereignty over the Svalbard archipelago in the Barents Sea has proved extremely useful in the claims process, submitted a claim for parts of the Norwegian Sea, Barents Sea and Arctic Ocean. [52]

Under a 1995 UNCLOS agreement, global commercial fishing is also being regulated, with countries obliged to set total allowable catches for certain species. [53]

In 1982 President Ronald Reagan objected to UNCLOS provisions on deep seabed mining in international waters, arguing they went against U.S. economic and security interests. The provisions empowered a new international body to license such activities, including the right to collect and distribute royalties. [54] While President Bill Clinton secured an agreement in 1994 aimed at allaying such concerns, the Senate continued to balk at ratification, fearing it undermined U.S. sovereignty.

Last October, the Senate Committee on Foreign Relations approved UNCLOS, but it still must pass the full Senate by a two-thirds majority. Sen. Richard G. Lugar, R-Ind., who supports passage, noted that "unlike some treaties, such as the Kyoto Agreement and the Comprehensive Test Ban Treaty — where U.S. non-participation renders the treaties virtually ineffective — the Law of the Sea will continue to form the basis of maritime law regardless of whether the U.S. is a party." [55] Lugar was referring to the fact that the United States applies the Law of the Sea in practice, even though it hasn't ratified the treaty.

Another UNCLOS backer in the Senate, Foreign Relations Committee Chairman Joseph R. Biden, D-Del., argues the treaty gives the United States the opportunity to extend its control up to 600 miles off the Alaska coast. "The oil and gas industry is unanimous in support of the convention," said Biden. [56]

Environmental Threats

The reduction in sea ice has devastated marine mammals, causing mass walrus deaths along the Chukotka coast in northeastern Russia, notes Oceana's Krenz. Fish populations have been affected too. Shrimp and crab, which prefer the cold, have become rarer while stocks of cod, salmon, mackerel and pollock have increased. [57] For unknown reasons, fish stocks are moving from the coastal areas to the open sea, causing problems for the seabirds that feed on them, which don't venture that far from shore; sea parrots, for instance, are declining significantly. [58]

Meanwhile, Greenland's stunning Ilulissat glacier, a UNESCO World Heritage Site, moves toward the sea at seven feet per hour as it melts — three times its pace in 2002. [59] When the

Arctic Life Is Changing

Traditional Arctic culture and ways of life are disappearing as temperatures rise and oil and gas exploration intensifies. Inuit hunters in Northwest Greenland today rarely build igloos like this one built during a long 1986 hunting trip because the warmer atmospheric temperatures make construction difficult (top). Reindeer herders like this Tundra Nenets man in Western Siberia say oil and gas pipelines make it difficult to move the herds. An Inuit man in Igloolik, Nunavut, Canada (right), wears a traditional caribou-skin outfit.

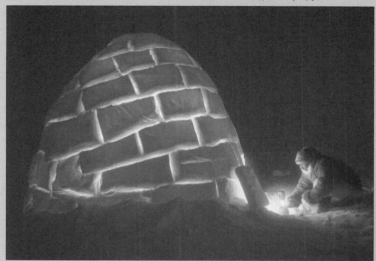

© B&C Alexander/Arcticphoto.com (all)

glacial ice reaches the sea and breaks off into the ocean it raises sea levels, threatening coastal populations. (Melting sea ice, on the other hand, does not raise sea levels because the ice is already floating on the sea.)

The Arctic's permafrost — or frozen soil — is melting as well, buckling highways, bursting pipelines and weakening the foundations of buildings. In Vorkuta, Russia, for example, resident Lyubov I. Denisova complained that "everything is falling apart. The ceiling has warped,

the walls cracked, the window frames splintered." [60] Melting permafrost also releases methane into the atmosphere, further accelerating global warming.

The Arctic environment also has been polluted by industrial emissions transported by air and sea, often from installations thousands of miles away, such as coal-fired power plants. "The Arctic is a sink for a lot of contaminants like mercury, pesticides and PCBs," says Sander at the European Environment Agency. "Arctic species have no ca-

pacity to resist these chemicals. They get stored in fat deposits and enter the food chain." A new study from the University of Northern British Columbia found that mercury levels in the Arctic remain stubbornly high, with coal-fired power plants the main culprit. [61]

Opportunity Knocks

Commercial oil activity in the Arctic began in 1920 in Canada's

Canada and Russia Dominate Arctic Production

The Arctic contains 22 percent of the world's undiscovered oil and gas deposits (top). Russia produces more oil and gas per day than any other Arctic country and has more gas reserves than all the other Arctic nations combined. Canada holds the most known oil reserves. Finland, Sweden, Iceland and Greenland produce almost no oil or gas and have no known reserves, although a recent study estimated that Greenland, which is owned by Denmark, is likely to have large, undiscovered deposits.

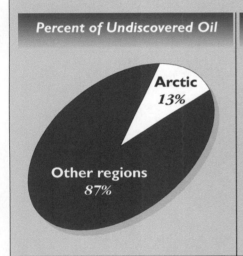

Percent of Undiscovered Oil

Arctic 13%

Other regions 87%

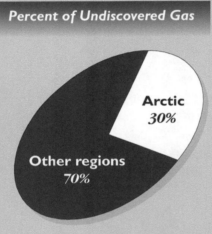

Percent of Undiscovered Gas

Arctic 30%

Other regions 70%

Arctic Energy Production and Known Reserves

Country	Daily oil production, 2007 (thousand barrels)	Proved oil reserves, 2007 (billion barrels)	Annual natural gas production, 2006 (billion cubic feet)	Proved natural gas reserves, 2006 (billion cubic feet)
Canada	3,355.79	179.21	6,548	56,577
Russia	9,875.77	60.00	23,167	1,680,000
Norway	2,565.27	7.85	3,196	84,260
U.S.	8,487.40	21.76	18,531	204,385
Greenland	0	0	0	0
Iceland	0	0	0	0
Finland	8.95	0	0	0
Sweden	2.35	0	0	0

Source: Energy Information Administration, U.S. Department of Energy; U.S. Geological Survey 2008

Northwest Territories, with ventures in Russia and Alaska following soon after.[62] The U.S. government issued its first oil exploration lease on Alaska's North Slope in 1958, bringing in big producers like ConocoPhillips.[63] Given the difficulty of Arctic Ocean shipping, a land-based pipeline was constructed in the 1970s to transport oil from northern Alaska to the ice-free port of Valdez in the south.

Oil companies began drilling off the Alaskan, Canadian, Norwegian and Russian coasts in the 1980s. Interest in Greenland has intensified in recent years, especially after a 2007 U.S. Geological Survey assessment of reserves ranked an area of northeastern Greenland as 19th among the world's 500 biggest oil and gas regions.[64]

Offshore oil and gas drilling in the Arctic was initially concentrated along shorelines due to the harsh climate, but companies are gradually venturing northward. In August 2007 the Snohvit natural gas field went online in the Barents Sea, 90 miles from the Norwegian coast.[65] Snohvit operates some 300 meters below sea level, sending extracted gas via pipeline to a processing plant on Melkoya Island, which liquefies it for export to Europe and the United States.[66] Petroleum has made Norway one of the globe's richest nations, with total revenues from Snohvit alone expected to top $39 billion, or $8,200 per citizen, over its estimated 25-year lifecycle.[67]

Russia, with help from two foreign energy firms, Norway's huge Statoil-Hydro and France's Total, has begun developing the Shtokman gas field in the Barents Sea, 373 miles offshore, although it is not expected to produce for several years.

Russian Foreign Minister Sergey Lavrov has noted how "global warming not only creates additional problems for us but opens up new possibilities as well."[68] As for environmental concerns about resource extraction, Lavrov said he had helped set up a public-private partnership, Emercom, to monitor and respond quickly to risks arising from "oil and gas production, nuclear energy, the transportation and processing of hydrocarbons and other raw materials."[69]

Meanwhile, the development of the double acting ship has made it easier for ships to navigate the Arctic's icy

Offshore Drilling Poses Special Challenges

Access and cleanup are more difficult in the Arctic.

Exploring for oil and gas under the sea usually begins with geological mapping and a search for two key factors: sedimentary rock at least 1.8 miles thick and evidence of an ancient "tectonic event." The thick rock is a precondition for oil and gas to be present, geologists say. And a shift in one or more of the Earth's tectonic plates typically would have sealed the oil into a confined space. [1]

In the early days, oil explorers simply drilled holes to find oil. Today images of the seabed are created by seismic surveys, in which explosions are triggered that send shockwaves into the Earth, which then are reflected back in radio waves that provide a picture of the ocean floor. Next, well data is gathered by boring into the ground to obtain core samples.

Before a well can be drilled, an area must be at least temporarily ice-free. If oil is discovered, facilities and pipelines must be constructed. Rigs may be installed either above the sea surface or on the seabed. Surface rigs can be either fixed or floating units, with the latter providing necessary flexibility to cope with icy Arctic conditions. The Snohvit gas rig in Norway's Barents Sea, which is ice-free, stands entirely on the seabed, with no surface installations. Snohvit is "over-trawlable" and does not interfere with trawl nets and other fishing equipment. Pipes along the sea floor transport the gas from the wells to the shore 90 miles away for processing.

Drillers avoid areas where the sea is permanently ice-covered, because access to oil and gas is more complicated and spillages more difficult to clean up. Extreme cold and the need to work waters. Pat Broe, a Denver businessman, has spent $50 million modernizing a derelict Hudson Bay port that he bought for $7 from Canada in 1997 and hopes will figure prominently in a coming boom in Arctic shipping. Broe has estimated that the port in Churchill, Manitoba, could make $100 million a year serving as a terminal for ships from Murmansk, a major Russian port. [70] Churchill could also service the increasingly popular Arctic tourist cruises, some carrying more than 1,000 passengers.

during the winter months, in 24-hour darkness, also deter Arctic petroleum exploration.

Once extracted, oil is transported by pipeline or tanker to a refinery or storage depot; gas is converted into liquefied natural gas (LNG) and shipped. Once a well is exhausted, the rig must be removed in an environmentally safe manner. Depending on the country, exploration may need to be accompanied by environmental-impact assessments and public consultations with neighboring communities.

Spillages can be caused by oil-well explosions, collisions of oil-laden ships or leaking pipelines. Clean-ups pose particular challenges because ice makes it hard to reach the spills and more difficult to detect spills that are trapped under the ice. On the other hand, if contained by the ice, the oil can be easier to clean up because it is less emulsified than when mixed with the water.

Sometimes a spill is cleaned up by setting it on fire. But the resulting thick, black smoke plume releases toxic chemicals into the atmosphere and may not be feasible if there is a community nearby. An experimental, controlled spill and so-called *in situ* burning are planned for the Barents Sea in May 2009, organized by Norway in collaboration with Statoil, Chevron and ConocoPhillips.

---

[1] Arctic Monitoring and Assessment Programme, "Arctic Oil and Gas 2007," Oslo, Norway, 2007, www.amap.no/oga; Don Gautier, geologist at United States Geological Survey; Amy Merten, co-director, Coastal Response Research Center, National Oceanic and Atmospheric Administration.

The development of huge factory fishing ships that can stay at sea for months has led to severe depletions of fish stocks. [71] Governments responded by setting catch quotas and limiting fishing rights of foreign vessels within their 200-mile boundaries. Norway and Russia now have agreements allowing some non-Arctic nations like Poland, Spain, France, Germany and the U.K. to fish in the Barents Sea. [72] Norway, which exports $6 billion worth of fish a year, also has been clamping down on Russian vessels that poach in Norwegian waters. [73] Meanwhile, the shrinking sea ice is encouraging vessels to move further north.

"It is happening in the Barents Sea — not yet in Canadian and U.S. waters, but the potential is there," says Oceana's Krenz.

Arctic coastal states traditionally have maintained a strong military presence in the region. In the past year, however, Canada has beefed up its military profile. The government announced in October 2007 that "as part of asserting sovereignty in the Arctic . . . new Arctic patrol ships [costing $3.1 billion] and expanded aerial surveillance will guard Canada's Far North and the Northwest Passage." [74]

Canadian forces also have stepped up patrols in the world's most northerly settlement — the community of Alert on Ellesmere Island — to "look for evidence of incursions into the area by Inuit from Greenland to hunt polar bears." [75] It also launched a space satellite, *Polar Epsilon*, to provide land and sea surveillance for Canadian forces beginning this summer. [76] ∎

Arctic Wildlife Abounds

A variety of mammals are able to survive the Arctic's harsh climate, including walruses in Spitsbergen, Norway (top), reindeer in northern Norway being herded by a Sami woman (right) and bull musk-oxen in Canada's Northwest Territories (bottom).

CURRENT SITUATION

Ilulissat Fallout

The joint declaration adopted at the May 27-29 Ilulissat ministerial meeting asserts the primacy of UNCLOS for resolving territorial claims. [77] Danish Foreign Minister Per Stig Moller proclaimed "hopefully we have eradicated all the myths about a 'race for the North Pole.' The legal framework is in place, and the five states have now declared that they will abide by it." [78]

But Huebert at the University of Calgary insists "not everyone is getting along like they pretend. In reality, there is a race to the North Pole."

To begin with, the meeting ruffled feathers by its exclusivity. "This is a very strange way of discussing what is a pan-Arctic issue or indeed an international issue," protested EU Parliament Vice President Wallis. "Why have not Finland, Sweden and Iceland been invited, countries which are also full Arctic Council member states?" [79]

Indeed, an Icelandic diplomat says his government was "not amused" at being left out. "We agree that territorial claims can be resolved by bilateral agreements, but in Ilulissat they also talked about shipping, Inuit rights and security. We should have been invited."

The State Department's McMurray, who attended the meeting, says the most concrete thing to emerge was a green light for Norway to draft a proposal to improve search and rescue services. "This will cover airplane and shipping accidents," she says. "Greenland presently has no capability to cope with the numbers of tourists going there, most of whom are Americans."

New Energy Leases

More than 400 oil and gas fields have been discovered north of the Arctic Circle, and that figure is set to rise. [80] In February 2008 the U.S. government's Minerals Management Service approved the extraction of oil and gas from a portion of the Chukchi Sea off Alaska's northern coast. It plans to open four other sections of the Chukchi and Beaufort seas between now and 2012. [81] The World Wildlife Fund says the leases should not have been awarded, because the impact of exploration on polar bears and indigenous communities has not been determined. [82]

Sen. John Kerry, D-Mass., agrees and has introduced a bill to ban exploration until the assessment is made. [83] But President Bush is calling for more off-shore oil drilling to help bring down high oil prices. "Congress should permit exploration in currently restricted areas of northern Alaska, which could produce roughly the equivalent of two decades of imported oil from Saudi Arabia," Bush recently said. [84]

Although Bush was talking about drilling inland, the USGS's recent conclusion that Arctic Alaska is the region's most oil-rich area will undoubtedly increase pressure to drill — especially offshore, where most of the Arctic's undiscovered oil and gas is thought to be found. The Arctic's 90 billion barrels of undiscovered oil compares to U.S. reserves of 22 billion barrels and annual production of 1.6 billion barrels. [85]

This June Canada awarded a $1.2 billion lease to Britain's BP to develop oil and gas in the Beaufort Sea. [86] The World Wildlife Fund's Stewart criticizes the move and notes that the Canadian government has no consistent energy-exploration policy because the responsible departments disagree over whether drilling should go ahead. One indication of that internal disarray: Despite numerous efforts, the Canadian Embassy in Washington was unable to provide a single Canadian official willing to discuss the topic for this article, because, according to an embassy official, no single agency is in charge of Arctic policy.

Meanwhile, Russia is stepping up its activity. In the Shtokman gas operation in the Barents Sea, it is using the expertise of foreign companies — StatoilHydro of Norway and Total of France — to produce the gas, but they must sell it all to Russia's state-owned energy giant Gazprom. [87] A new oil terminal at Varandey, 14 miles offshore in the Barents Sea, became operational in June 2008. It will load oil onto ships for transport to Europe and America.

"The infrastructure we have been able to establish helps develop new fields in Timan-Pechora oil and gas province," noted Vagit Alekperov, president of Russia's Lukoil, which spearheaded the project. [88]

In other developments, Greenland awarded numerous exploration licenses this year to U.S., Canadian, British, Danish and Swedish companies, and in July StatoilHydro began mapping the seabed of northeast Greenland. [89] Iceland plans to grant licenses within the next year to develop undersea resources on the Jan Mayen Ridge, off its northeastern coast.

Revamping Policies

The EU is paying more attention to the Arctic than ever before, with the Parliament planning to pass a resolution in September providing direction to the European Commission on its Arctic policy paper, due out in the autumn. EU Parliament Vice President Wallis feels the EU, with no Arctic territory, could play the role of an honest broker in future talks.

The Bush administration also is due to unveil its new Arctic policy soon, but none too soon for Alaska's Sen. Murkowski. "We have not accepted the responsibility of being an Arctic nation yet. I want a policy that does not simply say, 'We value the Arctic' or 'The Arctic is a lovely place,' but provides specifics, such as how many icebreakers we will acquire." The United States currently has only three: One is laid up in Seattle for repairs, another was designed mainly for scientific expeditions and a third, a more heavy-duty design, is in use. [90] Consequently, the United States contracts with foreign icebreakers to meet its needs. Meanwhile, Russia has 18, Finland and Sweden each have seven and Canada has six. Apart from helping other ships navigate icy seas, icebreakers can be used to support search and rescue and oil-spill cleanup operations as well as to gather seabed data to evaluate extended continental shelf claims.

The Arctic Council is scheduled to publish an assessment of the long-term potential for Arctic shipping. Inuit leader Lynge believes a moratorium on increased commercial shipping should be imposed until a stricter regime can ensure that only "Arctic-proof" ships enter Arctic waters.

Meanwhile, Arctic governments continue mapping the Arctic seabed in pursuit of continental shelf claims. But there probably won't be global scientific consensus on where those geological borders lie because finding the shelf can be tricky.

"Think of a continent as a big rock sitting in a bathtub, and imagine that a chunk of it rises out of the water," wrote *Wired* reporter Geoffrey Gagnon. "The question for scientists is, where does the rock end and the acrylic tub begin? It sounds simple enough, but imagine now that your tub is also made of rock and that smaller rocks are piled up all over the place." [91]

In Nunavut, Canada, researchers 375 miles north of Grise Fjord are trying to determine whether their shelf extends as far as the Eurasian side of the North Pole, where Russian geologists are also gathering data. [92]

"The need to assert our sovereignty and take action to protect our territorial integrity in the Arctic has never been more important," Canadian Natural Resources Minister Gary Lunn has said. [93]

Equally assertive, Danish Ambassador Petersen says, "We already have a lot of geological data. We believe Russia's claim to the Lomonosov Ridge to be unfounded." The Danes have until 2014 to make their claim, Canada until 2013 and Norway and Russia until 2009.

The United States is at another disadvantage because it does not own any islands near the North Pole. Calculating the continental shelf limit can begin at any of a country's islands.

It will be several years before the U.N. commission assessing the claims passes judgment. And that probably will not be the final word on the matter. In its submission to the commission, Norway said the final boundaries will have to be determined through bilateral agreements with its Arctic neighbors. [95]

point out that they have shared the Arctic's resources for thousands of years without resorting to conflict with one another.

The polar bears' threatened-species designation could throw a monkey wrench into the oil developers' plans, because U.S. law bars government agencies from taking any action that could further endanger a listed species. Conservation groups can argue before the courts that drilling poses a threat — both directly from spillages and indirectly through more fossil fuels being consumed, triggering more global warming and more loss of sea ice. [96]

In June conservationists scored another success when President Bush signed a congressional resolution aimed at preventing a mad dash to exploit Arctic fish stocks. The resolution's sponsor, Republican Sen. Ted Stevens of Alaska said, "with less summer ice in the Arctic, our northern waters will be open for exploitation from pirate fishing fleets. But the passage of the resolution will help protect our marine resources." [97]

The measure calls on the United States to consult with other Arctic nations for an agreement on managing fish stocks. [98] Oceana has called the move "the first significant step the U.S. government has taken to protect the Arctic Ocean," adding, "hopefully this starts a trend towards conservation and away from the 'too much, too fast and too soon' pace we've seen so far." [99] ∎

A fur-clad Inuit hunter in northwest Greenland scans the sea ice for polar bears. His world is rapidly changing as global warming melts the ice and introduces increased tourism, energy development and transcontinental shipping.

© B&C Alexander/Arcticphoto.com

The United States has just begun gathering data for a claim. The Coast Guard's largest icebreaker, *Healy*, is conducting a joint collaboration in the Arctic Ocean this summer with Canada's *Louis S. St. Laurent*. [94] "We are far behind other countries," Assistant Secretary McMurray admits. But the United States cannot submit a claim unless and until the Senate ratifies UNCLOS. Most senators support ratification, but the Democratic leadership has not put it to a vote yet, fearing it will not pass by the necessary two-thirds majority.

McMurray agrees: "It is not going to be the U.N. that sorts out overlapping claims. The countries will have to agree among themselves." Petersen says "if we fail to agree bilaterally we can still go to the International Court of Justice."

Lynge believes indigenous communities are the key to avoiding an ugly dispute. "They must look for partnership with us — otherwise they will simply fight among themselves for decades. The Inuit can be the glue that stops this from disintegrating into a territorial fight." Indigenous community representatives are quick to

OUTLOOK

Strategic Importance

Most scientists believe the Arctic will continue to warm faster than the rest of the planet. "It seems nearly impossible for summer Arctic sea ice to

Continued on p. 236

At Issue:

Should the U.S. Senate ratify the U.N. Convention on the Law of the Sea?

SEN. LISA MURKOWSKI, R-ALASKA
*MEMBER, SENATE FOREIGN RELATIONS
AND ENERGY AND NATURAL RESOURCES
COMMITTEES*

WRITTEN FOR *CQ GLOBAL RESEARCHER*, AUGUST 2008

LAWRENCE A. KOGAN
*PRESIDENT AND CEO, INSTITUTE FOR
TRADE, STANDARDS AND SUSTAINABLE
DEVELOPMENT*

WRITTEN FOR *CQ GLOBAL RESEARCHER*, AUGUST 2008

recent actions by Russia, Canada and other northern-tier nations to strengthen or establish claims in the Arctic Ocean underscore why it's so critical for the U.S. Senate to ratify the U.N. Convention on the Law of the Sea (UNCLOS). Otherwise, we may be watching from the sidelines as other nations divvy up the significant energy resources contained in the Arctic seabed.

It is believed that the Arctic may hold 22 percent of the world's undiscovered oil and gas — a number that could rise considerably as additional survey work is completed in the region. An expedition by the Coast Guard cutter *Healy* last winter showed that the United States could lay claim to an area the size of California as part of our extended continental shelf.

The problem? The United States has no legal claim to most of this area — and the oil or gas it contains — unless we become a party to the UNCLOS. If we don't claim it, others certainly will.

Russia already has claimed almost half of the Arctic, including parts of what we believe to be Alaska's extended continental shelf, while Canada is looking to establish military bases in the north.

For those who think Russia's claims, or those of other nations, will not be recognized, think again. On April 21 of this year, Australia's claim to 2.5 million square kilometers of extended continental shelf — an area three times the size of Texas — was recognized. It is only a matter of time before other claims are accepted as well.

It would be naïve to believe we could reach multiple bilateral agreements with nations once their claims in the Arctic, and to its oil and natural gas, are internationally recognized. What is their incentive? What would the United States need to give up in return?

When we talk about sovereignty, those who say the United States already enjoys the benefits of the Law of the Sea Treaty ignore that we do so only by the grace of other nations. They ignore that our military commanders believe UNCLOS is vital to ensuring the passage of U.S. naval vessels through international waters. They ignore that if we cede the Arctic to Russia and other Arctic nations, we could very well be importing oil that should belong to us in the first place.

In a time of rising energy costs and demands, that does not seem like a sound policy for the United States to follow.

since 2007, a growing body of evidence has revealed that UNCLOS, if ratified by the United States, would ensure much more than what the U.S. Navy recognizes as America's absolute right to freedom of navigation.

Granted, UNCLOS codifies this customary international-law principle. But UNCLOS parties, especially European governments, also increasingly embrace recent U.N. concepts of environment-centric sustainable development, which ultimately are based on environmental and health regulation that reflect hypothetically possible rather than empirically probable harms. These nostrums already have been used to reshape international environmental law, which will now be implemented and enforced through the treaty's dynamic environmental provisions, putting new conditions on the exercise of that right to freedom of navigation.

Indeed, there are 45-plus environmental articles, annexes, appendices, protocols and regulations within UNCLOS that can expand and evolve over time to reflect the most current international environmental law. Ironically, these same provisions have captured the imagination of creative, transatlantic policy makers, who, aided by sensationalist media, have triggered public anxieties about the potential environmental and health hazards posed by observable (but likely cyclical) global warming and melting ice.

UNCLOS ratification raises several important questions. For instance, what connection, if any, exists between UNCLOS, global warming, carbon dioxide emissions and other types of air and water "pollution" generated both within U.S. jurisdiction and beyond? What legislative, regulatory and judicial obligations would UNCLOS place on the U.S. government to prevent such pollution from materializing in the first place? What economic, technological and legal burdens would U.S. businesses and individual Americans consequently face as U.S. laws are made more stringent and costly due to UNCLOS ratification? Why is UNCLOS ratification necessary to drill for oil and gas in Alaska's inland and offshore sites, since most of the untapped reserves are reportedly within 200 miles of the coast — where the United States can drill without U.N. permission? And finally, why do administration officials, congressional representatives and environmental activists oppose holding open, public and transparent Senate and House hearings to investigate the potential impact of UNCLOS' "green" provisions on the U.S. economy, national security and sovereignty?

If UNCLOS isn't the green Trojan Horse that opponents say it is, why not prove it?

Continued from p. 234

return to the climatological extent that existed prior to 1980," according to NOAA's Overland, who predicts that within 12 years the Arctic may be entirely ice-free in summer. [100]

Meanwhile, offshore oil and gas development will expand, especially within Arctic nations' exclusive economic zones, where reserves are easier to access and where most oil and gas is thought to be located. There will be more onshore development as well. A 750-mile natural gas pipeline connecting gas fields in Canada's Northwest Territories with markets to the south is being planned, although its development has slowed recently due to land-ownership disputes with indigenous communities. [101] A rival pipeline starting in Prudhoe Bay in neighboring Alaska is also being touted. [102]

Russia is expanding operations in western Siberia — where most of the Arctic's gas is thought to lie — and in the Timan-Pechora Basin. But concerns about global warming could stall drilling operations, especially in Canada and the United States, where environmental groups are likely to mount strong legal challenges.

Rich oil profits should flow into the region, although foreign energy firms could do well, too, since four of the five Arctic coastal states have no restrictions on foreign ownership of oil companies. [103] Cash-strapped governments may demand a bigger slice of the pie. Currently, their revenue share ranges from 46-65 percent in Greenland to 90-100 percent in Russia. [104] The people of Greenland will vote in November on a revenue-sharing agreement signed in June with Denmark. Exploration could creep higher into the Arctic if the sea ice continues to recede and the seas become more accessible — but only if companies are convinced there is enough oil and gas to make it worth their while.

Shipping may be the more viable option for transporting fuels, given how costly and complicated it is to construct pipelines and how vulnerable they are to being ruptured by melting permafrost. [105] Thus, Arctic shipping routes will become busier, and the necessary support infrastructure must be developed. An immediate boom in transcontinental shipping of other non-fuel cargo looks less likely, because ice-related delays or the cost of extra fuel needed to cut through ice-covered seas will cancel out the cost-savings from the shorter travel distances. Oceana's Krenz says an increase in shipping also could further hasten the melting of the ice cap, because ships' carbon emissions darken the ice, increasing its absorption of heat.

Exporting freshwater — either from icebergs or, more likely, from existing lakes, of which Canada has many — is another potential profit source, says the Arctic Research Commission's Brigham. "No one is doing it yet, but with freshwater increasingly scarce and expensive, it may become commercially interesting to ship it to countries with shortages." [106]

"The amount of ice that comes into the ocean [from melting glaciers] could provide the water supply for any of the largest cities in the world for an entire year," according to Robert Corell, director of the Global Change Program at the H. John Heinz III Center for Science, Economics and the Environment, in Washington, D.C. [107]

The European Environment Agency's Sander predicts Arctic waters will become a center for genetic resources, a growing industry involving the harnessing of plant or animal substances for use in medicines. And, as job opportunities grow in the oil and gas, agriculture, fishing, shipping and tourism industries, so, too, could immigration, which may heighten tensions between indigenous communities and newcomers.

The burgeoning economic activity may motivate countries to step up their military presence, as Canada and Russia are doing. Russian Gen. Vladimir Shamanov has announced plans to deploy more naval vessels, adding, "We are also planning to increase the operational radius of the Northern Fleet's submarines."

A prominent military hawk, Shamanov insists Russia has the capability to defend its claim to half of the Arctic Ocean. [108]

On the governance side, the European Union will seek ways to assert itself, despite owning no Arctic territory. A recent EU policy paper noted "an increasing need to address the growing debate over territorial claims" and new trade routes, which challenge Europe's trade and resource interests in the region and "may put pressure on its relations with key partners." [109]

The Arctic Council may also have a chance to play a more prominent role. "The council could focus on policy development in specific areas like combating mercury pollution," says Sander. Icelandic official Baldursson says the council "could be used as a venue for information exchange and preliminary negotiations." Danish Ambassador Petersen says the council should address not just environmental but also economic activities.

In the inevitable pursuit of the Arctic's resources, Gun-Britt Retter, a member of the Sami indigenous people's parliament in Norway, fears governments will pay scant regard to the environmental impact of it all. "We have lived off the land for 10,000 years without extracting all the resources. We think long-term, not about filling our budgets. Governments only think four years ahead until the next election." ■

Notes

[1] See James Graff, "Fight for the Top of the World," *Time*, Oct. 1, 2007, pp. 28-36, www.time.com/time/world/article/0,8599,1663445,00.html.

[2] Presentation by Thomas Armstrong, Senior Advisor for Global Change Programs, U.S. Geological Survey, at the Danish Embassy to the U.S., May 7, 2008.

[3] James Overland, et al., "The Arctic and Antarctic: Two Faces of Climate Change," EOS, American Geophysical Union, May 6 2008, www.noaanews.noaa.gov/stories2008/images/5-6-08_Overland.pdf.

[4] Ibid.

[5] Documentation concerning the U.S. Interior Department's May 14, 2008, decision to list the polar bear as a threatened species is at www.doi.gov/issues/polar_bears.html.

[6] See Scott. G. Borgerson, "Arctic Meltdown," Foreign Affairs, March/April 2008, www.foreignaffairs.org/20080301faessay87206/scott-g-borgerson/arctic-meltdown.html.

[7] Adrian Blomfield, "Russia plans military build-up in the Arctic," The Daily Telegraph, June 12, 2008, www.telegraph.co.uk/news/worldnews/europe/russia/2111507/Russia-plans-Arctic-military-build-up.html. For a detailed map of the various territorial claims filed by the five Arctic states, go to http://news.bbc.co.uk/go/em/fr/-/2/hi/staging_site/in_depth/the_green_room/7543837.stm.

[8] See "Decision regarding the date of commencement of the ten-year period for making submissions to the Commission on the Limits of the Continental Shelf set out in article 4 of Annex II to the United Nations Convention on the Law of the Sea," May 14-18, 2001, www.un.org/Depts/los/convention_agreements/convention_historical_perspective.htm.

[9] Market price on July 24, 2008.

[10] "Circum-Arctic Resource Appraisal," U.S. Geological Survey, July 2008, http://energy.usgs.gov/arctic.

[11] United States Energy Information Administration, http://tonto.eia.doe.gov/dnav/ng/hist/n9070us2A.htm.

[12] "Arctic Oil and Gas 2007," Arctic Monitoring and Assessment Programme (AMAP), 2007, www.amap.no/oga, p. ix.

[13] For more information, see "Facts 2008 — The Norwegian Petroleum Sector," Norwegian Petroleum Directorate, www.npd.no/English/Produkter+og+tjenester/Publikasjoner/Faktaheftet/Faktaheftet+2008/fakta2008.htm.

[14] "Swedish oil company joins hunt for oil and gas in Greenland," press release, Greenland Home Rule Web site, May 23, 2008, http://uk.nanoq.gl/Emner/News/News_from_Parliament/2008/05/2008_may_Swedish_oil_company.aspx.

[15] "Climate change and international security," paper from the High Representative and the European Commission to the European Council, March 14, 2008, p. 5, www.consilium.europa.eu/ueDocs/cms_Data/docs/pressData/en/reports/99387.pdf.

[16] Graff, op. cit.

[17] The 18 figure is provided by U.S. Coast Guard, Department of Homeland Security. Presentation by Niels Bjorn Mortensen, Head of Marine, BIMCO (Baltic and International Maritime Council), at a conference on Arctic Transportation, U.S. Maritime Administration, June 5, 2008.

[18] Interview with Friis Arne Petersen, Danish Ambassador to the United States.

[19] Remarks at a seminar, "Arctic Governance in a global world: is it time for an Arctic Charter?" European Parliament, Alliance of Liberals and Democrats for Europe, May 7, 2008. For seminar presentations, see www.alde.eu/index.php?id=42&L=0&L=ht&tx_ttnews[tt_news]=9348&cHash=76d92ab815.

[20] Norwegian Ministry of Fisheries and Coastal Affairs, press release, Dec. 16, 2005, www.regjeringen.no/se/dep/fkd/Preassaguovdda/Preassadieahusat/2005/Broad-agreement-on-fisheries-between-Norway-and-the-EU.html?id=419750.

[21] "The Ilulissat Declaration," Governments of Denmark, United States, Canada, Russia and Norway, May 28, 2008, www.um.dk/NR/rdonlyres/BE00B850-D278-4489-A6BE-6AE230415546/0/ArcticOceanConference.pdf.

[22] "Arctic Governance in a global world: is it time for an Arctic Charter?" op. cit.

[23] Ibid.

[24] Ibid.

[25] Quoted in Graff, op. cit.

[26] NOAA Fisheries, National Marine Service Regional Alaska Office, www.fakr.noaa.gov/oil/default.htm.

[27] See Web page of Oceana, www.protecthearctic.org.

[28] AMAP, op. cit., p. xii.

[29] Ibid., p. 25.

[30] Ibid., p. vii.

[31] "Arctic shuttle container link from Alaska US to Europe," Aker Arctic Technology Inc, March 2006, p. 28, www.institutenorth.org/servlet/content/studies.html.

[32] Ibid.

[33] Ibid.

[34] Ibid.

[35] Jeff Hechts, "Ancient site hints at first North American settlers," New Scientist, January 2004, www.newscientist.com/article.ns?id=dn4526.

[36] Britannica Online Encyclopaedia, www.britannica.com.

[37] "Alaska's Heritage," Alaska History and Cultural Studies, www.akhistorycourse.org/articles/article.php?artID=155.

[38] Ibid.

[39] Rasmus Ole Ramussen, "Factsheet: Denmark — Greenland," Royal Danish Ministry of Foreign Affairs, January 2004.

[40] John J. Miller, "Let's Buy Greenland!" National Review Online, May 7, 2007, www.nationalreview.com/nr_comment/nr_comment050701b.shtml.

[41] See U.S. Air Force Web site, www.thule.af.mil.

[42] See Derek Hayes, Historical Atlas of the Arctic (2003).

[43] Britannica Online, op. cit., p. 79.

[44] See "Documents on the law of the sea: historical perspective," U.N. Web site, www.un.org/Depts/los/convention_agreements/convention_historical_perspective.htm#Historicalpercent20Perspective.

[45] Ibid.

[46] Ibid.

[47] For background, see Colin Woodard, "Oceans in Crisis," CQ Global Researcher, October 2007, pp. 237-264.

[48] Speech of Malta's Ambassador to the U.N., Arvid Prado, Nov. 1, 1967, www.un.org/Depts/los/convention_agreements/texts/pardo_ga1967.pdf. Also see "Documents on Law of the Sea," op. cit.

[49] Philip Taubman, "Soviet Proposes Arctic Peace Zone," The New York Times, Oct. 2, 1987, http://query.nytimes.com/gst/fullpage.html?res=9B0DE0DC173CF931A35753C1A961948260.

[50] "Documents on Law of the Sea," op. cit.

[51] "Russia's 2001 submission to the Commission on the Limits of the Continental Shelf," www.un.org/Depts/los/clcs_new/submissions_files/submission_rus.htm.

[52] "Continental Shelf Submission of Norway in respect of areas in the Arctic Ocean, the Barents Sea and the Norwegian Sea, 2006," Government of Norway, www.un.org/Depts/los/clcs_new/submissions_files/submission_nor.htm.

[53] "Documents on Law of the Sea — historical perspective," op. cit.

[54] See www.oceanlaw.org/downloads/references/reagan/PresidentalStmt-Jan82.pdf.

[55] Senate Committee on Foreign Relations, hearing on Convention on the Law of the Sea, Oct. 4, 2008. For full testimonies go to http://foreign.senate.gov/hearings/2007/hrg071

004a.html.

56 Biden statement, Oct. 31, 2007. See http://biden.senate.gov/press/press_releases/release/?id=15d1b23d-4d04-4e3b-8727-932dd1352bd2.

57 Woodard, *op. cit.*

58 Petter Meier, Fisheries Counselor, Embassy of Norway to the United States, Washington, D.C.

59 See Colin Woodard, "In Greenland, an Interfaith Rally for Climate Change," *The Christian Science Monitor,* Sept. 12, 2007, Ilulissat, Greenland, www.csmonitor.com/2007/0912/p06s01-woeu.html.

60 Steven Lee Myers, Andrew C. Revkin, Simon Romero and Clifford Krauss, "Old Ways of Life are Fading as the Arctic Thaws," *The New York Times,* Oct. 20, 2005, www.nytimes.com/2005/10/20/science/earth/20arctic.ready.html.

61 Bob Weber, "Toxic chemical levels in Arctic food animals dropping: study," The Canadian Press, July 14, 2008, http://cnews.canoe.ca/CNEWS/Canada/2008/07/14/6155656-cp.html.

62 AMAP, *op. cit.*

63 *Ibid.*

64 "New Oil and Gas Assessment of Northeastern Greenland," U.S. Geological Survey, press release, Aug. 28, 2007, www.usgs.gov/newsroom/article.asp?ID=1750.

65 Graff, *op. cit.*

66 For more information on the Snohvit operation, see www.statoil.com/STATOILCOM/snohvit/svg02699.nsf?OpenDatabase&lang=en.

67 Revenue estimates from Statoil Web site, www.statoil.com/STATOILCOM/snohvit/svg02699.nsf?OpenDatabase&lang=en.

68 Remarks, Russian Minister of Foreign Affairs Sergey Lavrov at Conference of Five Arctic Coastal States, Ilulissat, Greenland, May 28, 2008, www.mid.ru/brp_4.nsf/0/A7DABB275A1E95CFC325745800497B84.

69 *Ibid.*

70 Clifford Krauss, Steven Lee Myers, Andrew C. Revkin and Simon Romero, "As Polar Ice Turns to Water, Dreams of Treasure Abound," *The New York Times,* Oct. 20, 2005, www.nytimes.com/2005/10/10/science/10arctic.html?pagewanted=1&_r=1.

71 Woodard, *CQ Global Researcher, op. cit.*

72 Meier, *op. cit.*

73 *Ibid.*

74 See "Ottawa buying up to 8 Arctic patrol ships," CBC News, www.cbc.ca/canada/story/2007/07/09/arctic-cda.html. Also see "Strong Leadership. A Better Canada — Speech from the Throne," Government of Canada, Oct. 16, 2007, www.sft-ddt.gc.ca/eng/media.asp?id=1364.

75 "Canadian Forces Patrol to Confirm Arctic Sovereignty," National Defence and the Canadian Forces, March 22, 2007, www.dnd.ca/site/Newsroom/view_news_e.asp?id=2224.

76 "Polar Epsilon to assert Canada's arctic sovereignty," National Defence and the Canadian Forces, press release, Jan. 10, 2008, www.forces.gc.ca/site/newsroom/view_news_e.asp?id=2547.

77 "The Ilulissat Declaration," *op. cit.*

78 "Conference in Ilulissat, Greenland: Landmark political declaration on the future of the Arctic," Ministry of Foreign Affairs of Denmark, edited June 6, 2008, www.missionfn-newyork.um.dk/en/menu/statements/CONFERENCEINILULISSATGREENLAND.htm.

79 Diana Wallis, European Parliament Vice President, statement on Arctic Five meeting in Greenland, May 28, 2008, http://dianawallismep.org.uk/news/000590/diana_wallis_responds_to_meeting_of_arctic_five.html.

80 "Circum-Arctic Resource Appraisal," *op. cit.*

81 For map of planned oil and gas exploration areas in Chukchi and Beaufort seas, U.S. Minerals Management Services Scoping Report, Environmental Impact Assessment, March 2008, p. 17, www.mms.gov/alaska/cproject/ArcticMultiSale/scoping_rpt.pdf.

82 "Native and Conservation Groups Voice Opposition to Lease Sale 193 in the Chukchi Sea," World Wildlife Fund, press release, Feb. 2, 2008, www.worldwildlife.org/who/media/press/2008/WWFPresitem5921.html.

83 Sen. John Kerry, press release, Jan. 30, 2008, http://kerry.senate.gov/cfm/record.cfm?id=291475.

84 "President Bush Discusses Outer Continental Shelf Exploration," White House, July 14, 2008, www.whitehouse.gov/news/releases/2008/07/20080714-4.html.

85 Circum-Arctic Resource Appraisal, *op. cit.*

86 David Ebner, "BP signals start of Arctic oil rush," *The Globe and Mail* (Canada) June 7, 2008, www.uofaweb.ualberta.ca/govrel/news.cfm?story=79420.

87 Guy Chazan, "Oil Sees End of Sweet Deals," *The Wall Street Journal,* July 14, 2008, http://online.wsj.com/public/search/page/3_0466.html?KEYWORDS=Shtokman&mod=DNH_S.

88 "LUKoil starts oil exports through Varandey terminal," *New Europe,* June 16, 2008, www.neurope.eu/articles/87870.php.

89 "Swedish oil company joins hunt for oil and gas in Greenland," Greenland Home Rule Web site, May 23, 2008, http://uk.nanoq.gl/Emner/News/News_from_Parliament/2008/05/2008_may_Swedish_oil_company.aspx. Also see "Geological investigations offshore North East Greenland in the summer of 2008," Greenland Home Rule Web site, May 2, 2008, http://uk.nanoq.gl/Emner/News/News_from_Parliament/2008/05/2008_apr_geological_investigation_offshore.aspx.

90 Testimony by Admiral Thad Allen, U.S. Coast Guard Commandant, hearing, House Subcommittee on Coast Guard and Maritime Transportation, July 16, 2008, http://transportation.house.gov/News/PRArticle.aspx?NewsID=681.

About the Author

Brian Beary — a freelance journalist based in Washington, D.C. — specializes in EU-U.S. affairs and is the U.S. correspondent for *Europolitics,* the EU-affairs daily newspaper. Originally from Dublin, Ireland, he worked in the European Parliament for Irish MEP Pat "The Cope" Gallagher in 2000 and at the EU Commission's Eurobarometer unit on public opinion analysis. A fluent French speaker, he appears regularly as a guest international-relations expert on television and radio programs. Apart from his work for Congressional Quarterly, Beary also writes for the *European Parliament Magazine* and the *Irish Examiner* daily newspaper. His last report for *CQ Global Researcher* was "Separatist Movements."

[91] Geoffrey Gagnon, "The Last Great Land-grab," *Wired*, February 2008, www.wired.com/science/planetearth/magazine/16-02/mf_continentalshelf.

[92] Randy Boswell, "Scientist warns over Arctic quest," Canwest News Service, June 2, 2008, www.canada.com/vancouversun/news/story.html?id=5131cf5a-d7fc-47f1-b487-39c4d0c972bc.

[93] "Minister Lunn Visits Canadian Scientists in Far North: Research Supports Canada's Claim to Arctic Sovereignty," Natural Resources Canada, press release, April 17, 2008, www.nrcan-rncan.gc.ca/media/newcom/2008/200824-eng.php.

[94] Kathy Eagen, public affairs officer, U.S. State Department, June 13, 2008.

[95] Government of Norway, 2006, *op. cit.*

[96] See Kenneth P. Green, "Is the Polar Bear Endangered, or Just Conveniently Charismatic?" American Enterprise Institute, May 2008, www.aei.org/publications/filter.all,pubID.27918/pub_detail.asp.

[97] Sen. Ted Stevens, press release, June 4, 2008, www.stevens.senate.gov/public/index.cfm?FuseAction=NewsRoom.PressReleases&ContentRecord_id=5538fa34-d757-3e73-f6cd-ec6c3628762c&Region_id=&Issue_id=.

[98] S.J. Res. 17, "A joint resolution directing the United States to initiate international discussions and take necessary steps with other nations to negotiate an agreement for managing migratory and transboundary fish stocks in the Arctic Ocean," P.L. 110-243.

[99] Oceana, press release, June 3, 2008, www.oceana.org/north-america/media-center/press-releases/press_release/0/788/.

[100] Overland, *et al.*, *op. cit.*

[101] See www.mackenziegasproject.com/theProject/index.html.

[102] Ed Struzik, "Pipeline or Pipe Dream?" Canwest News Service, July 18, 2008, www.canada.com/topics/news/national/story.html?id=45752856-72d2-4cbd-b988-d374b73a03e9.

[103] Rachel Halpern, " 'Above-Ground' issues and Arctic Oil and Gas Development," International Trade Administration, U.S. Department of Commerce, presentation at National Defense University and Forces Transformation and Resources Seminar, May 14, 2008, pp. 135-145, www.ita.doc.gov/td/energy/arctic%20paper.pdf.

[104] *Ibid.*

[105] For background, see Trans-Alaska Pipeline Web site, at www.alyeska-pipe.com/Pipeline-

FOR MORE INFORMATION

Arctic Council, Polarmiljøsenteret, NO-9296 Tromsø, Norway; +47-77-75-01-40; www.arctic-council.org. High-level forum for cooperation between Arctic states and their indigenous communities.

Embassy of Denmark to the United States, 3200 Whitehaven St., N.W., Washington, DC 20008; (202) 234-4300; www.ambwashington.um.dk. Denmark owns the self-governing Arctic territory of Greenland, making the Danes a major player in the race for Arctic resources.

Embassy of Russia to the United States, 2650 Wisconsin Ave., N.W., Washington, DC 20007; (202) 598-5700; www.russianembassy.org. The largest Arctic nation, which has claimed sovereignty over the North Pole.

International Maritime Organization, 4, Albert Embankment, London, SE1 7SR, United Kingdom; +44 (0)20-7735-7611; www.imo.org. U.N. agency responsible for improving maritime safety and preventing pollution from ships.

Inuit Circumpolar Council, Dronning Ingridsvej 1, P.O. Box 204, 3900 Nuuk, Greenland; +11 299-3-23632; www.inuit.org. Greenlandic branch of the organization that represents 150,000 Inuits in Russia, Canada, Alaska and Greenland.

National Snow and Ice Data Center, 449 UCB University of Colorado, Boulder, CO 80309-0449; (303) 492-6199; www.nsidc.org. Studies snow, ice, glacier, frozen ground and climate interactions around the world; provides regular updates on extent of Arctic sea ice.

Norwegian Petroleum Directorate, P.O. Box 600, 4003 Stavanger, Norway; +47 51-87-60-00; www.npd.no. Norwegian government agency responsible for managing the country's abundant petroleum resources.

Oceana, 1350 Connecticut Ave., 5th floor, N.W., Washington, DC 20036; (202) 833-3900; www.oceana.org. Environmental advocacy group with offices in the United States, Chile, Spain and Belgium; dedicated to protecting and restoring the world's oceans.

U.S. Arctic Research Commission, Alaska Office, 420 L St., Suite 315, Anchorage, AK 99501; (907) 271-4577; www.arctic.gov. Government agency set up in 1984 to promote research and provide policy guidance on Arctic matters.

WWF Canada, 245 Eglinton Ave. East, Suite 410, Toronto, ON M4P 3J1, Canada; (416) 489-8800; www.wwf.ca. A leading Canadian conservation organization and a member of the World Wildlife Fund's global network.

facts/Permafrost.html.

[106] See Lawson W. Brigham, "Thinking about the Arctic's Future: Scenarios for 2040," *The Futurist*, Sept-Oct 2007, www.wfs.org/SeptOct07percent20files/FuturecontSO07.htm.

[107] Woodard, *The Christian Science Monitor, op. cit.*

[108] Blomfield, *op. cit.*

[109] "Climate change and international security," paper from the High Representative and the European Commission to the European Council, March 14, 2008, p. 8, www.consilium.europa.eu/ueDocs/cms_Data/docs/pressData/en/reports/99387.pdf.

Bibliography

Selected Sources

Books

Brandt, Anthony, ed., *North Pole, A Narrative History*, National Geographic Society, 2005.

Drawing on extensive Society archives, an adventure expert chronicles the race to the North Pole using memoirs, letters, ships' logs and diaries of the great Arctic explorers.

Hayes, Derek, *Historical Atlas of the Arctic*, University of Washington Press, 2003.

An award-winning author and book designer uses nearly 200 historical maps to illustrate all the significant Arctic explorations from the 16th century well into the 20th.

Vaughan, Richard, *The Arctic: A History*, Phoenix Mill, 1994.

A former history professor describes man's struggle to survive in the Arctic from the Stone Age until modern times, including an examination of the impact of exploration on the lives of indigenous peoples.

Articles

Borgerson, Scott G., "Arctic Meltdown," *Foreign Affairs*, March/April 2008, www.foreignaffairs.org/20080301faessay87206/scott-g-borgerson/arctic-meltdown.html.

A fellow at the Council on Foreign Relations discusses how the melting Arctic ice cap is opening up access to natural resources and shipping shortcuts.

Brigham, Lawson W., "Thinking about the Arctic's Future: Scenarios for 2040," *The Futurist Magazine*, September-October 2007, www.wfs.org/Sept-Oct07%20files/FuturecontSO07.htm.

The deputy director of the U.S. Arctic Research Commission — a Ph.D. in polar oceanography and former icebreaker commander — describes how the Arctic might look in 30 years if global warming continues.

Gagnon, Geoffrey, "The Last Great Landgrab," *Wired*, February 2008, www.wired.com/science/planetearth/magazine/16-02/mf_continentalshelf.

A science writer charts the ongoing efforts by Arctic nations to map the sea floor in an effort to bolster claims to expand their continental shelves.

Graff, James, "Fight for the Top of the World," *Time*, Oct. 1, 2007, www.time.com/time/world/article/0,8599,1663445,00.html.

Arctic nations are racing to assert their sovereignty over large swaths of unclaimed Arctic territory, including the North Pole.

Green, Kenneth P., "Is the Polar Bear Endangered, or Just Conveniently Charismatic?" American Enterprise Institute, May 2008, www.aei.org/publications/filter.all,pubID.27918/pub_detail.asp.

A scholar at a conservative think tank examines whether scientific evidence justifies the designation of the polar bear as an endangered species — a step the U.S. government has taken since the article was published.

Myers, Steven Lee, Andrew C. Revkin, Simon Romero and Clifford Krauss, "Old Ways of Life are Fading as the Arctic Thaws," *The New York Times*, Oct. 20, 2005, www.nytimes.com/2005/10/20/science/earth/20arctic.ready.html.

This installment of a series examining the impact of climate change on Arctic communities focuses on how melting permafrost threatens buildings, highways and pipelines.

Overland, James, *et al.*, "The Arctic and Antarctic: Two Faces of Climate Change," *EOS*, National Oceanic and Atmospheric Administration, May 6, 2008, www.noaanews.noaa.gov/stories2008/images/5-6-08_Overland.pdf.

Scientists explain how the polar ice caps are reacting in different ways to climate change.

Reports and Studies

"Arctic Climate Impact Assessment, Impacts of a Warming Arctic," Cambridge University Press, 2004.

A joint collaboration among more than 300 scientists — commissioned by the Arctic Council and the International Arctic Science Committee — evaluates the global impact of climate change.

"Arctic Oil and Gas 2007," Arctic Monitoring and Assessment Programme, 2008, www.amap.no/oga.

A report mandated by the Arctic Council describes past, present and future oil and gas exploration projects around the Arctic.

"Circum-Arctic Resource Appraisal: Estimates of Undiscovered Oil and Gas North of the Arctic Circle," U.S. Geological Survey, July 2008, http://energy.usgs.gov/arctic.

The first publicly available resource assessment of the area north of the Arctic Circle estimates it contains 22 percent of the world's undiscovered oil and gas.

"Climate change and international security," Policy Paper, EU High Representative and European Commission, March 14, 2008, www.consilium.europa.eu/ueDocs/cms_Data/docs/pressData/en/reports/99387.pdf.

The European Union examines international security issues relating to resource-exploitation opportunities in the Arctic.

The Next Step:

Additional Articles from Current Periodicals

Arctic Treaty

"Closed-Door Arctic Deal Denounced as 'Carve-Up,'" *The Guardian* (England), May 28, 2008.

Environmentalists have denounced a deal signed by five Arctic nations seeking to resolve opposing claims for the region, saying it is nothing more than a carve-up.

"Drawing Lines in Melting Ice," *The Economist*, Aug. 18, 2007.

Countries that want to make a territorial claim under the U.N. Convention on the Law of the Sea must do so within a decade of ratifying it.

Arsana, Andi, "The Constitution of the Oceans," *Jakarta Post* (Indonesia), Oct. 29, 2007.

More geoscientists need to examine the technical aspects of the U.N. Convention on the Law of the Sea in order to make it more effective.

Bellinger, John B., "Treaty on Ice," *The New York Times*, June 23, 2008, p. A21.

While increased cooperation is needed for search and rescue, there is already an extensive framework governing the Arctic region, according to ministers who met in Greenland.

Indigenous Groups

"Trooping the Tribal Colours," *The Economist*, June 7, 2008.

An Inuit representative was dismayed at being given only 10 minutes to address the five-nation ministerial meeting in Greenland in May.

Allagui, Slim, "Stop Stealing Our Land, Inuits Say," Agence France-Presse, June 16, 2008.

The Arctic region's indigenous Inuits are demanding that nations bordering the area respect their land and their way of life.

Dugan, Emily, "Climate Change Plea From Tribe of Herders Who Face Extinction," *The Independent* (England), May 10, 2008.

Global warming in the Arctic has put the Sami, one of the largest remaining indigenous communities in Europe, at risk.

Resource Battle

"Gas and Glory Fuel Race for the Pole," *Moscow Times*, July 27, 2007.

Competition for Arctic resources is heating up as climate change and new technologies make previously unfathomable exploration now possible.

"Race Is on for Arctic Resources," *Canberra Times* (Australia), Sept. 10, 2007.

The Arctic contains much of the world's undiscovered oil and gas, and the race for those resources is speeding up.

Dyer, Gwynne, "Race to the Pole is Just a Rush for Fool's Black Gold," *South China Morning Post*, May 27, 2008.

It is unlikely that the Arctic Ocean holds much oil and gas, given that it accounts for only 3 percent of the Earth's surface.

Weir, Fred, "As Ice Caps Melt, Russia Stakes Its Claim to Oil-Rich Arctic," *The Christian Science Monitor*, Aug. 3, 2007, p. 25.

Russian explorers have planted a titanium copy of their country's flag more than two miles below the Arctic surface to symbolize Moscow's claim to the territory's resources.

Shipping Routes

"Shipping Industry Fails to Warm to Northwest Passage," Chinadaily.com.cn, Oct. 4, 2007.

Shipping companies and commerce experts say using the Northwest Passage through Canada's Arctic archipelago is still too dangerous, despite the melting of polar sea ice caused by global warming.

McKie, Robin, "Arctic Thaw Opens Fabled Trade Route," *The Observer* (England), Sept. 16, 2007.

The clearing of Arctic shipping passages could fuel further animosity between countries competing over resources.

Miller, Hugo, "Ships Intrude on Arctic's Warming Waters," *Los Angeles Times*, March 10, 2008, p. C4.

A Moscow-based company has ordered five reinforced cargo vessels to plow through the waters north of Siberia now that new sea routes have opened up.

CITING *CQ GLOBAL RESEARCHER*

Sample formats for citing these reports in a bibliography include the ones listed below. Preferred styles and formats vary, so please check with your instructor or professor.

MLA STYLE

Flamini, Roland. "Nuclear Proliferation." CQ Global Researcher 1 Apr. 2007: 1-24.

APA STYLE

Flamini, R. (2007, April 1). Nuclear proliferation. *CQ Global Researcher*, 1, 1-24.

CHICAGO STYLE

Flamini, Roland. "Nuclear Proliferation." *CQ Global Researcher*, April 1, 2007, 1-24.

Voices From Abroad:

VLADIMIR CHAMANOV
General, Russia
Wars are won beforehand

"After the reaction of a certain number of heads of state to Russia's territorial claims to the continental plateau of the Arctic, the training division has immediately set out (training) plans for troops that could be engaged in Arctic combat missions."

Krasnaya Zvezda (Russia), June 2008

FRANK-WALTER STEINMEIER
Foreign Minister Germany

Arctic is not law-free

"Everybody should respect international law. The North Pole is not a law-free zone; there are international accords which must be respected by all nations who have interests here. If everybody sticks to the rules, there will be no conflict."

Agence France-Presse, August 2007

GARY LUNN
Minister of Natural Resources, Canada

The Arctic will bring prosperity to Canada

"The need to assert our sovereignty and take action to protect our territorial integrity in the Arctic has never been more important. . . . Our commitment to this initiative, as well as other investments in the North, is ultimately about turning potential into prosperity for this remarkable region and for our country as a whole."

Marketwire (Canada), April 2008

MINIK ROSING
Geologist, Greenland

Oil will curse Greenland

"As soon as we find oil, that will end independence. Everyone thinks that oil will buy us independence, but how would we absorb all of this wealth? As everyone gets more desperate for that commodity, you don't want to be a very small, very independent country, very far from anywhere else. Independence based on oil is probably not a good idea."

The Independent (England), September 2007

PER STIG MOELLER
Foreign Minister Denmark

All parties must act responsibly

"I am sure we will be able to identify ways ahead for future development in and around the Arctic Ocean which will be peaceful, secure and to the benefit of all our countries. We need to send a common political signal to both our own populations and the rest of the world that the five coastal states will address the opportunities and changes in a responsible manner."

Turkish Daily News, May 2008

AQQALUK LYNGE
Vice President, Inuit Circumpolar Conference

Enough is enough

"We no longer want to accept the isolation and harsh treatment that has been inflicted upon us in the past. We paid the price of the sovereignty of these governments who steal our land, our resources. Enough is enough; we don't want to be displaced by force, as was the case in Thule, and we demand to be treated humanely."

Agence France-Presse, June 2008

SERGEI LAVROV
Foreign Minister, Russia

Expedition is proof

"The goal of the expedition is not to reserve Russia's rights but to prove that our shelf reaches the North Pole. The Arctic region is rich in natural resources, but we must find a reliable method of their development. This expedition is very important for the solution of this complicated task."

The Christian Science Monitor, August 2007

ALEQA HAMMOND
Minister for Finance and Foreign Affairs, Greenland

The North Pole belongs to Greenland

"The Russians came and planted their flag up there on the North Pole, but everyone knows it's Greenlandic. The last land before you reach the Pole is Greenlandic land."

Business Line (India), October 2007

MIKE TOWNSLEY
Spokesman, Greenpeace International

Nothing more than a carve-up

"It's clear what's going on. They are going to use the Law of the Sea to carve up the raw materials, but they are ignoring the law of common sense — these are the same fossil fuels driving climate change in the first place. The closed-door nature of this is doubly troubling. It's clear they know what they're trying to do is unacceptable."

The Guardian (England), May 2008

10

BUYING GREEN

BY JENNIFER WEEKS

Excerpted from the CQ Researcher. Jennifer Weeks. (February 29, 2008). "Buying Green." *CQ Researcher*, 193-216.

Buying Green

BY JENNIFER WEEKS

THE ISSUES

During Lent, many Christians commemorate the time that Jesus spent fasting and praying in the desert, according to the Bible, before taking up his ministry. Most churchgoers mark Lent by giving up alcohol, red meat or other luxuries. But this year two prominent British bishops called on the faithful to sacrifice something else: carbon emissions. Through steps such as insulating hot-water heaters, sealing drafts in their houses and changing to energy-efficient light bulbs, the church leaders urged observers to reduce their carbon footprints — the greenhouse gases (GHGs) emitted from human activities that contribute to global climate change. "We all have a pivotal role to play in tackling the stark reality of climate change," said Richard Chartres, Bishop of London. "Together we have a responsibility to God, to future generations and to our own well-being on this earth to take action." [1]

Although they may not cast the issue in religious terms, Americans are increasingly willing to take personal action to protect the environment. And while conservation has long been associated with sacrifices, such as driving smaller cars and turning down the heat, today some advocates argue that a comfortable lifestyle can be eco-friendly. The key, they say, is "buying green" — choosing products designed to reduce pollution, waste and other harmful impacts.

Activists have long recognized that consumer spending, which accounts for about two-thirds of the $14 trillion

Actor Brad Pitt is spearheading the construction of 150 "affordable and sustainable" homes in hurricane-battered New Orleans. Activists say the key to protecting the environment is "buying green" — choosing products designed to reduce pollution and waste. Consumer spending accounts for about two-thirds of the $14 trillion U.S. gross domestic product, making eco-consumerism a potentially powerful influence on policy and the economy.

AP Photo/Alex Brandon

U.S. gross domestic product, can be a powerful influence on national policy. Consumer campaigns often stigmatize a product to highlight suppliers' unacceptable behavior. For example, civil rights activists in the 1950s and '60s boycotted segregated buses in Montgomery, Ala., and held sit-ins at lunch counters that refused to serve African-Americans. Both strategies drew national attention to segregation in the South and built support for new civil rights laws.

Consumers can also reward positive behavior with their dollars. In the 1970s, the garment workers' union urged Americans to "Look for the union label" that identified clothing made in the United States instead of choosing products from low-wage foreign sources. Today eco-conscious shoppers are buying organically grown food, fuel-efficient cars

and shares in socially responsible investment funds that target companies with strong environmental records.

According to the annual Green Brands Survey, U.S. consumers will spend about $500 billion on environmentally friendly products and services in 2008, double last year's amount. [2] A typical American family spends roughly $50,000 each year on food, clothing, shelter, transportation, health care, entertainment and other items. [3] (*See graph, p. 197.*) And consumers frequently use buying power to communicate their opinions: Boycotting or "buycotting" (deliberately choosing) products for political or ethical reasons are among the most common ways in which Americans express political views. [4] (*See graph, p. 196.*)

"The consumer movement has quietly become part of the fabric of American society," says Caroline Heldman, an assistant professor of politics at Occidental College in Los Angeles and author of a forthcoming book on consumer activism. "Environmental concerns are the most important motives that drive people to engage in consumer activism, and with concern about global warming so high, the public is primed to act if environmental groups can find tangible things for people to do."

However, not all green products deliver on their promises. Since it first issued guidelines for environmental marketing in 1990, the Federal Trade Commission (FTC) has acted against 37 companies for misleading consumers with green claims. [5] A recent survey by TerraChoice, an environmental marketing firm, suggests that "greenwashing" — making misleading environmental claims about a company or

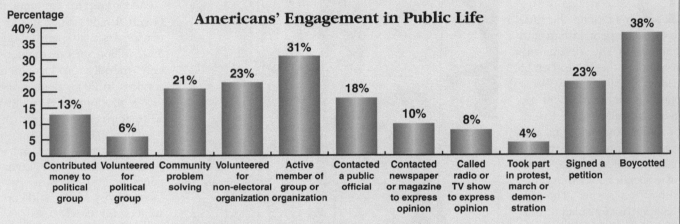

Boycotting Leads Civic Engagement Activities

Nearly 40 percent of Americans participated in some form of boycotting in 2002, and nearly a quarter signed a petition or volunteered for a non-electoral organization.

Americans' Engagement in Public Life

Percentage

- Contributed money to political group: 13%
- Volunteered for political group: 6%
- Community problem solving: 21%
- Volunteered for non-electoral organization: 23%
- Active member of group or organization: 31%
- Contacted a public official: 18%
- Contacted newspaper or magazine to express opinion: 10%
- Called radio or TV show to express opinion: 8%
- Took part in protest, march or demonstration: 4%
- Signed a petition: 23%
- Boycotted: 38%

Source: Scott Keeter, "Politics and the 'DotNet' Generation," Pew Research Center, May 2006, based on National Civic Engagement Survey 2002

product — is becoming more pervasive as companies bring new green products to market. In a review of 1,108 consumer products that made environmental claims, TerraChoice found that all but one provided some form of false or misleading information. (*See sidebar, p. 204.*)

"Green labeling today is where auto-safety information was in the 1950s. Standards and certification programs are still emerging," says TerraChoice Vice President Scot Case. "This is unexplored territory, so marketers may be stretching the truth unintentionally. We think that the sudden interest in green just caught a lot of people off guard, and marketers were busy slapping buzzwords on packaging. But FTC's guidelines are clearly 15 to 20 years out of date."

Many issues are spurring interest in green products. In 2007 the Intergovernmental Panel on Climate Change, an international scientific association created to advise national governments, called global warming unequivocal and concluded with at least 90 percent certainty that human activities since 1750 had warmed the planet. [6] Repeated

warnings about climate change are prompting many companies and individuals to shrink their carbon footprints. New products like renewable energy certificates and carbon offsets, which allow buyers to pay for green actions that happen elsewhere, make this task easier. (*See glossary, p. 198.*) But critics say that these commodities are feel-good gestures and do not always promote new, clean technologies.

Recent cases of contaminated food and toxic ingredients in common household products like pet food and toothpaste also are spurring consumers to seek out green alternatives. [7] Green consumption is a logical response to environmental threats, but Andrew Szasz, a sociologist at the University of California, Santa Cruz, believes that it could actually threaten environmental progress if consumers see it as a substitute for political action.

"A lot of people get environmentally conscious enough to get worried. Then they go buy everything green that they can afford and move on to something else," says Szasz, who calls the trend an example of "inverted quarantine" — citizens protecting themselves

from danger by building barriers instead of organizing to reduce the threat. "Pressure from social movements to take toxic substances out of our water and air will create more progress than individual consumer actions," he argues.

Eco-consumption mirrors a similar trend in the business sector. Many U.S. companies are working to green their operations, both to appeal to the fast-growing market and because leaders are finding that environmental strategies can help cut costs and make their operations more efficient. [8] Many large corporations that have clashed with environmentalists in the past, such as DuPont, Monsanto and Waste Management, Inc., now highlight their commitments to environmental stewardship and sustainability. [9]

In a notable sign of corporate greening, the U.S. Climate Action Partnership (a coalition including Alcoa, General Electric, Shell and Xerox) called in early 2007 for prompt mandatory limits to slow and reverse the growth of GHG emissions. Many large companies have opposed mandatory GHG limits in the past, arguing that putting a price on carbon emissions would drive up energy costs. [10] However, U.S.-CAP members contended that addressing climate change "will create more economic opportunities than risks for the U.S. economy." [11]

Corporate greening appears to be widespread but hard to measure because there is no authoritative definition of a green business. A recent report by Greener World Media found that green businesses are making progress toward some milestones, such as disclosing their carbon emissions and investing in new clean technologies. It also judged, however, that corporate America is treading water or falling behind on other targets, such as using more renewable energy and emitting fewer GHGs per unit of economic activity. "Green business has shifted from a movement to a market. But there is much, much more to do," the authors asserted. [12]

As environmentalists, business executives and consumers ponder what buying green can accomplish, here are some issues they are considering:

Do carbon offsets slow climate change?

Curbing climate change is difficult because greenhouse gases, especially carbon dioxide (CO_2), are produced from many routine activities like powering appliances and driving cars. Every year the average American generates roughly 10 to 20 metric tons of CO_2 through day-to-day activities, mainly through home energy use and transportation. [13]

Consumers can shrink their carbon footprints through steps such as adding insulation to their houses, buying more energy-efficient appliances and using public transit for some trips instead of cars. But if people want to do more, or have carbon-intensive lifestyles because they own large homes or travel frequently, they can buy carbon offsets from brokers, who use the money to fund projects elsewhere that reduce GHG emissions. Pollution offsets date back to the mid-1970s, when the Environmental Protection Agency (EPA)

allowed industries to build new emission sources in regions with serious air pollution if they made larger reductions at existing sources nearby. This policy was written into the Clean Air Act in 1977 and later expanded to let companies earn and trade emission-reduction credits if they cut emissions below thresholds required by law.

"Offsets have an important role to play as we try to shrink our carbon footprint," says Mike Burnett, executive director of the Climate Trust, an Oregon nonprofit created to implement a 1997 state law that requires new power plants to offset some CO_2 emis-

sions. The trust invests money from power plants, as well as businesses and individuals, in energy efficiency, renewable energy and other low-carbon projects to offset clients' emissions. "Oregon has pledged to reduce its GHG emissions 75 percent below 1990 levels by 2050. Investing in high-quality offsets can help us address climate change at the lowest overall cost, which will leave more money for other priorities," says Burnett.

The Climate Trust uses strict criteria to screen potential investments. Emission reductions must be rigorously quantified, and sponsors have to show that

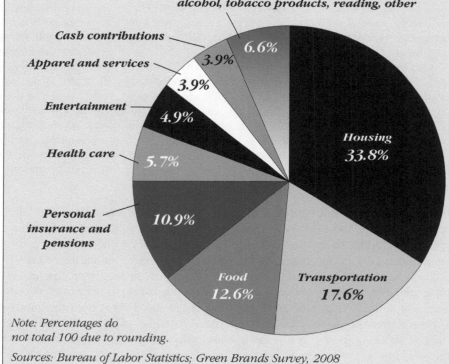

Average Household Spends Nearly $50,000

The average household spent $48,398 in 2006, including more than one-third on housing and 18 percent on transportation. According to a recent survey, U.S. consumers will spend about $500 billion on environmentally friendly products and services in 2008, double last year's amount.

Average Household Expenditures, 2006

- Education, personal care products and services, alcohol, tobacco products, reading, other — 6.6%
- Cash contributions — 3.9%
- Apparel and services — 3.9%
- Entertainment — 4.9%
- Health care — 5.7%
- Personal insurance and pensions — 10.9%
- Housing — 33.8%
- Food — 12.6%
- Transportation — 17.6%

Note: Percentages do not total 100 due to rounding.

Sources: Bureau of Labor Statistics; Green Brands Survey, 2008

A 21st-Century Carbon Glossary

The pollutant plays a key role in today's environmental efforts.

Carbon footprint — The sum of all greenhouse gas (GHG) emissions caused during a specified time period by a person's activities, a company's operations or the production, use and disposal of a product.

Carbon neutral — Operating in a way that does not produce any net addition of GHGs to the atmosphere. For both businesses and individuals, becoming carbon neutral typically involves two steps: reducing GHG emissions that they generate directly, through steps such as conserving energy; and buying carbon offsets that equal whatever direct GHG emissions they cannot eliminate.

Carbon offset — An activity that reduces GHG emissions, such as planting trees to take up atmospheric carbon dioxide or producing energy from carbon-free fuels like wind and solar energy. Buying carbon offsets is a way of contracting out GHG emission reductions, typically because the offset project can reduce emissions more cheaply than the buyer can.

Carbon trading — Buying and selling GHG emission allowances (government permits to release a specific quantity of pollution) or emission-reduction credits, which may be issued by government under mandatory regulations or created by companies and individuals through voluntary trading schemes.

Greenhouse gases (GHGs) — Heat-trapping gases that absorb solar energy in the atmosphere and warm earth's surface. Six major GHGs are controlled under the Kyoto Protocol, but since carbon dioxide (CO_2) is the most abundant and causes the most warming, companies and governments convert their total emissions into CO_2 equivalents.

Renewable energy certificates (RECs) — Certificates that represent the environmental attributes of electricity produced from renewable sources and can be sold separately from the electricity itself. Investors can buy RECs to support green energy whether or not they are located close to the source. Some companies may market themselves as "powered by green energy," even though they use electricity from coal- or gas-fired power plants, because they buy RECs to equal their total electric power usage (thus helping to put that amount of carbon-free energy into the electric power grid).

---

offset projects would not happen without funding from the trust — a concept called "additionality" to indicate that resulting GHG reductions must be additional to business as usual. For example, although installing underground systems at landfills to capture methane (a potent greenhouse gas produced when waste decomposes) is a popular type of offset, the trust would not invest in a methane-capture project if regulations already required the landfill operator to control methane emissions.

Not all providers are as strict. A 2006 study commissioned by Clean Air-Cool Planet (CACP), a New England nonprofit group, found that the market for voluntary carbon offsets was largely unregulated and had no broadly accepted standards for defining or measuring offsets. Prices to offset a ton of carbon varied widely, as did the types of offsets available and the amount of information companies provided to customers. [14]

"There clearly are good offsets and not-so-good ones on the market, so the problem for buyers is finding the good ones," says CACP Chief Executive Officer Adam Markham. "If they don't buy good ones, they're not making a difference, and they're wasting their money."

A popular strategy that has raised questions is paying to plant trees. Growing plants absorb CO_2 from the atmosphere to make plant tissue, and trees also offer many other benefits, such as stabilizing soils and providing habitat for animals and birds. Movie stars Brad Pitt and Jake Gyllenhall, along with Home Depot, Delta Airlines and other corporations, have funded tree-planting projects from suburban Atlanta to Bhutan.

But trees don't always help the environment. Planting non-native species can soak up local water supplies and replace other valuable ecosystems such as prairie grassland. Moreover, calculating how much carbon various types of forests take up is an inexact science. And since trees eventually release carbon when they die and decompose (or are logged or burned down), they cycle carbon quickly and only remove it from the atmosphere for a matter of decades.

In contrast, today's oil, coal and natural gas supplies represent much more permanent carbon reserves that formed when carbon-based plant materials were compressed in ancient, underground fossil beds. Burning these fossil fuels permanently releases carbon stores that have been sequestered for thousands of years and will not be recreated in the foreseeable future. [15]

"Forest offsets tend to be more risky because we know less about how much carbon they displace than we do for energy projects, and they're less likely to be permanent," says Markham. Instead, he prefers energy projects because it's easier to quantify the emissions that they displace and demonstrate additionality. "Wind power and methane-capture projects tend to be pretty high-quality investments," Markham says.

But nothing is guaranteed. After the for-profit broker TerraPass provided offsets to help green the 2007 Academy Awards ceremony, an investigation by *Business Week* magazine found that six projects that generated TerraPass offsets would have taken place in any case. One, a methane-capture system installed by Waste Management, Inc. at an Arkansas landfill, was initiated in response to pressure from state regulators. TerraPass's investment was "just icing on the cake" for another project, a county official in North Carolina told *Business Week*. [16]

"There are a lot of new entrants into the market, so some offerings probably aren't as robust as others, and it's causing some confusion," says Burnett. "If this sector doesn't become more standardized within the next five years, government will have to step in. We don't necessarily need a single federal scheme, but it would be very useful to have a federally sanctioned panel of experts who could review offset products."

Beyond the characteristics of specific projects, some critics argue that carbon offsets don't reduce climate change because they let people keep doing high-carbon activities, which the offsets counterbalance at best. Worse, offsets may serve as cover for carbon-intensive activities. For example, a recent report from the Transnational Institute in Amsterdam, the Netherlands, points out that British Airways offers passengers an option to buy carbon offsets for their flights but is also pushing to expand British airports and short-haul flights, which will increase the company's total GHG emissions. [17]

"Offsets may be tarnished by revelations of practices that aren't credible. That would be a problem, because these tools can be quite useful if they're applied effectively," says Thomas Tietenberg, a professor of economics at Colby College in Waterville, Maine. "The consumer offset market is facing an important moment in terms of its credibility. It needs to get some agreement about what the standards are."

Should government require green purchases?

Government officials often want to boost demand for green products, even if they cost somewhat more, be-

cause these goods reduce pollution, conserve energy or keep waste out of landfills. One option is to mandate the use of green goods and services. But critics argue that government interference distorts markets and that setting environmental performance standards may deliver inferior products.

Renewable energy is perhaps the most widely mandated green commodity. As of January 2008, 26 states and the District of Columbia had adopt-

ed renewable portfolio standards (RPSs) requiring electricity suppliers to generate certain fractions of their power from renewable fuels like wind, solar energy and biomass. [18] Advocates would like to see a national renewable-energy requirement, but so far Congress has failed to enact one.

Most recently, in 2007 the House passed an energy bill that included a 15 percent RPS requirement by 2020, with utilities allowed to meet up to 4 percent of their

---

Some Buyers Are Greener Than Others

A 2006 study by the Natural Marketing Institute classified adult U.S. consumers into five categories based on their attitudes toward ethical consumption.

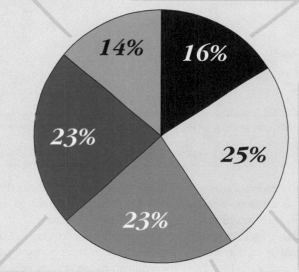

Unconcerneds — Do not consider social or environmental values in buying decisions.

LOHAS (Lifestyles of health and sustainability) — Make purchases based on belief systems and values, including environmental protection and social responsibility.

Naturalites — Are interested in natural and healthy products, but their choices are driven more strongly by personal and family health concerns than by broader environmental views.

Conventionals — May recycle or give money to environmental groups, but do not shop based on a cohesive set of values; sometimes buy green products, especially items that offer economic savings.

Drifters — May believe in protecting the environment, but often think that measuring the impact of their consumer choices is too hard or don't know how to do it.

Note: Percentages add to more than 100 due to rounding.

Source: LOHAS Forum, "Understanding the LOHAS Consumer: The Rise of Ethical Consumerism," www.lohas.com

---

GHG Emissions Still Rising

Greenhouse gas emissions have risen by 15 percent since 1990, reaching over 7 billion metric tons in 2006.

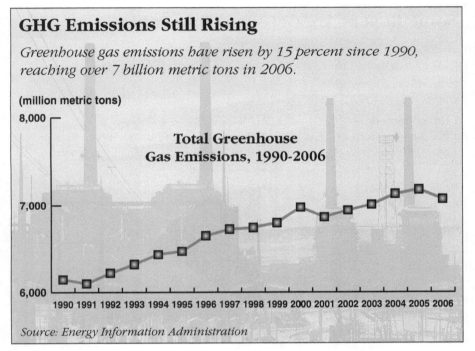

(million metric tons)

Total Greenhouse Gas Emissions, 1990-2006

Source: Energy Information Administration

targets through energy conservation. Supporters argued that the measure would reduce air pollutants and greenhouse gas emissions from fossil fuel combustion and spur the growth of a domestic renewable-energy industry. But the provision was dropped after critics charged that it would raise electricity prices and penalize regions with fewer renewable resources. (*See "At Issue," p. 209.*)

"The market should be allowed to work things out. We don't support having the government impose a mandate that says, "Thou shalt do this," says Keith McCoy, vice president for energy and resources policy at the National Association of Manufacturers. "Utilities and regulators in RPS states are looking at the right fuel mixes for their regions, but we need to take into account what's possible in different parts of the country."

RPS advocates want a national standard to push states that have been less aggressive in developing renewable energy. A national RPS "is absolutely achievable," said Rep. Tom Udall, D-N.M., a sponsor of the measure, during House debate. "[B]ut the full potential for renewable electricity

will be left unrealized without the adoption of a federal program to enhance the efforts of these states." [19]

Governments can also ensure that products are at least somewhat green by establishing content or performance requirements. Measures such as building codes and energy-efficiency standards for appliances are one way to remedy a common problem: Many buyers don't know much about products, so it's hard to choose the best even if they want to. "If you're walking around a house looking at it, you have no idea what kind of insulation is in the walls or how efficient the heating system is, but building codes set some basic thresholds for performance," says Colby College economist Tietenberg.

Forcing manufacturers to comply with new standards may spur technical advances, but it can also challenge businesses to meet the new goals. When new energy-efficiency standards for top-loading washing machines went into effect in 2007, *Consumer Reports* gave low performance ratings to the first models that it tested. The Competitive Enterprise Institute (CEI), a

think tank that opposes excessive regulation, accused the Energy Department of ruining a once-dependable home appliance. "Send your underwear to the undersecretary," CEI urged dissatisfied consumers. [20]

"If these technologies really are that good, we shouldn't need laws to force them down people's throats," says CEI General Counsel Sam Kazman. "We don't think that promoting energy efficiency is an appropriate role for government, but if that's the goal, the way to do it is with an energy tax, which would reduce energy use and create incentives to develop energy-saving technologies. One big attraction of regulations is that the public doesn't see them as tax increases — people perceive them as relatively cost-free."

Today, however, those energy-efficient washers look better. "What a difference a year makes," *Consumer Reports* commented in February 2008. The best high-efficiency top-loading washers were performing better, testers found, and *CR* pointed out that high-efficiency models could end up costing the same or less than standard machines over their lifetime when energy savings were factored in. [21]

Posing the issue as a choice between a free market or regulations is misleading, says Bill Prindle, deputy director of the American Council for an Energy-Efficient Economy (ACEEE). "The real issue is what the rules should be for market players. When you set boundaries and targets, manufacturers come up with very ingenious solutions that give customers great value," Prindle contends. He also notes that manufacturers and conservation advocates have negotiated some two dozen energy-efficiency standards since 2005 that subsequently were enacted into law. "These are largely consensus-based agreements. They wouldn't have passed otherwise," Prindle argues.

Another way to promote green technologies is through voluntary labeling programs that identify environmentally preferable products. The Energy Star program, administered by EPA and the Department of Energy, was launched in 1992 in response to a Clean Air Act provision directing EPA to find non-regulatory strategies for reducing air pollution. Energy Star defines superior energy efficiency standards for more than 50 types of residential, commercial and industrial equipment, including consumer electronics, heating and cooling systems and lighting. [22] EPA estimates that over 2 billion products with Energy Star labels were sold in 2006, saving 170 billion kilowatt-hours of electricity, or enough to power more than 15 million average American households for a year. [23]

Another program, Leadership in Energy and Environmental Design (LEED), was developed by the U.S. Green Building Council to identify highly energy-efficient buildings with extremely healthy indoor environments. [24] More than 800 buildings in the U.S. and worldwide have received LEED certification by scoring points on a fixed scale for features like energy and water conservation and indoor air quality. Many large corporations and universities have built LEED buildings to demonstrate environmental commitments.

Labeling programs complement requirements to use green products, says Tietenberg. "Mandates make sense as a floor, but you don't want to stop there. Labels like LEED provide something that performs above the minimum," he says. "They let buyers know that they are getting a certain value for their investment and communicate that fact to other people."

CEI's Kazman argues that green labeling programs can also be problematic. "Consumers don't get the full story if labels omit repair issues and the risk that very new technologies will have problems," he says. "And once items earn stars, there's a risk that the next step will be to mandate them. But we'd rather have government give advice and make recommendations than impose mandates."

Growing numbers of eco-conscious shoppers are buying organically grown food today. Demand for organic and natural groceries has made Whole Foods the nation's largest natural food market chain.

Is buying green better for the environment than buying less?

Most observers agree that today's green consumption boom signals the mainstreaming of environmental values. In the 1980s eco-friendly products like soy milk and recycled paper were of uneven quality and were viewed as niche goods for a small subset of dedicated customers. Today megastores like Wal-Mart and Target offer green cleaning supplies, organic food and energy-efficient light bulbs.

"We've seen green waves before, but today there's better understanding of environmental issues, higher quality products and more consumer understanding," says Case at TerraChoice Environmental Marketing. "This issue has penetrated the heads of the average consumer and business executive."

With more consumers buying more earth-friendly products, some advocates say that environmental protection no longer has to mean scaling back affluent lifestyles. Instead, they assert, we can shop our way to sustainability. "We all need to be presented with better product choices that enable us to maintain the way of life to which we're accustomed without overtaxing the planet's ability to sustain it," said entrepreneur and satellite radio host Josh Dorfman, the self-styled "Lazy Environmentalist," in a 2007 interview. "As a nation, we don't really want to deal with [global warming]. We have neither the political leadership nor the political will, which is why I think that for now the environmental solutions presented have to be both effective and painless." [25]

This is a new perspective for the environmental movement, which has long argued that rampant economic growth and high consumption are root causes of environmental harm. Not all environmentalists agree that so-called checkbook environmentalism can save the planet. For one thing, critics argue, the green product boom has had little impact so far on U.S. greenhouse gas emissions. Since 1990 the emissions intensity of the U.S. economy (the amount of GHG emissions produced for every dollar of economic activity) has declined, but total GHG emissions have increased nearly every year due to overall economic growth. (*See graph, p. 200.*)

"True, as companies and countries get richer they can afford more efficient machinery that makes better use of fossil fuel, like the hybrid Honda Civic I drive," writes bestselling author Bill McKibben. "But if your appliances have gotten more efficient, there are also far more of them: The furnace is better than it used to be, but the average size of the house it heats has doubled since 1950. The 60-inch TV? The always-on cable modem? No need for you to do the math — the electric company does it for you, every month." [26]

Complicating the issue, many green living guides fail to distinguish between actions that have a major impact, like insulating your house, and those with smaller effects such as buying a natural-fiber shower curtain or dog leash. "People tend not to sort through which choices are important and which are insignificant. They view most actions as equally important," says Warren Leon, coauthor of *The Consumer's Guide to Effective Environmental Choices* and director of the Massachusetts Renewable Energy Trust. [27] (*See sidebar, p. 206.*) "I worry about products that are sold as green, often by promoters who sincerely believe in them, but that either don't work well or don't have a serious impact. Mediocre or trivial green products will turn consumers off in the long run," he warns.

"Greenwashing" further undercuts the impact of buying green by marketing products with vague claims like "All Natural," "Earth Smart" and other labels that are too general to document whether goods will help the environment or not. Some consumers analyze these slogans critically, but many are likely to take them at face value. According to a 2006 study by the Natural Marketing Institute, LOHAS (Lifestyles of Health and Sustainability) buyers, who make purchases based on belief systems and values, including environmental protection and social responsibility, account for only about 16 percent of U.S. consumers. (*See graph, p. 199.*)

Although LOHAS consumers are a relatively small segment of the market, green business experts say that they have significant influence. "LOHAS consumers push the envelope. They're always testing the boundaries, and they make decisions for the sake of

the mission," says Ted Ning, who directs an annual business gathering in Colorado called the LOHAS Forum. "Once their items become mainstream, they move on to the next issue. For example, instead of just buying organic food or locally grown food, now they're choosing food based on its carbon footprint."

LOHAS buyers also size up companies critically, says Ning. "They expect a lot of in-depth information to show whether products are authentic. Blogs and Web sites give people lots of ways to communicate, so if companies don't make that data available, there's an assumption that they have something to hide. And LOHAS consumers are evangelists, so they're proud to share their information. If you get on their wrong side, they'll bad-mouth you to death."

Businesses are keenly interested in LOHAS consumers, who represent an estimated $209 billion market for goods including organic food, personal and home care products, clean energy technologies, alternative transportation and ecotourism. [28] But it's not clear that this group's preferences can steer the entire U.S. economy toward sustainability.

"Consumers are most interested in high-quality, affordable products. That's still a larger driver than other environmental considerations, although green aspects often are tie-breakers," says TerraChoice's Case. Recent polls show that while Americans are increasingly willing to make lifestyle changes to protect the environment, they prefer easy actions like recycling over more demanding steps like reducing their carbon footprints. [29]

"Green labeling and marketing are market-based instruments that can be adopted quickly as our environmental knowledge grows, but in the long term they'll be seen as transitional steps," says Case. "Ultimately, we'll address these issues with other mechanisms like cap-and-trade systems and taxes." ∎

BACKGROUND

Conservation Focus

Before the United States was a century old, early conservationists began to warn about threats to precious lands and resources. In his 1854 classic *Walden*, Henry David Thoreau decried loggers and railroads that encroached on his forest retreat. In 1876 naturalist John Muir wrote that California's forests, which he called "God's first temples," were "being burned and cut down and wasted like a field of unprotected grain, and once destroyed can never be wholly restored even by centuries of persistent and painstaking cultivation." [30]

Congress began putting lands under federal protection with the creation of Yellowstone National Park in 1872. It also established scientific agencies to manage natural resources, including the U.S. Fisheries Commission (later the Fish & Wildlife Service) in 1871 and the U.S. Geological Survey and Division of Forestry (later the Forest Service) in 1879.

But politicians mainly sought to develop and use resources, not to protect them in their natural states. To settle the West, Congress passed laws like the 1872 Mining Law, which allowed prospectors to buy mining rights on public lands for $5 per acre, and the 1878 Timber and Stone Act, which made land that was "unfit for farming" available for $2.50 per acre for timber and stone resources. These statutes often allowed speculators and large corporations to exploit public resources at far less than fair market value. [31]

Environmental advocates formed many important conservation groups before 1900, including the Appalachian Mountain Club, American Forests and the Sierra Club. Their members, mainly affluent outdoorsmen, focused on preserving land for hunting, fishing and

Continued on p. 204

Chronology

1960-1980
Environmentalists use lobbying, litigation and citizen action to curb pollution. . . . Congress imposes new regulations on businesses.

1967
Congress enacts Clean Air Act.

1969
Congress passes National Environmental Policy Act, requiring environmental-impact studies for federal projects with potentially significant effects on the environment.

1970
Millions of Americans celebrate Earth Day on April 22. . . . Congress establishes Environmental Protection Agency and expands Clean Air Act.

1973
Endangered Species Act enacted.

1974
Safe Drinking Water Act enacted.

1978
Homeowners in New York's Love Canal neighborhood force federal government to pay for evacuating them from houses built atop toxic-waste dump.

1979
Three Mile Island nuclear power plant in Pennsylvania partially melts down, stalling the growth of nuclear energy.

1980
Superfund law assigns liability and fund cleanup at hazardous-waste sites. . . . Ronald Reagan is elected president on platform calling for reducing government's role.

1980-2000
Global climate change emerges as major environmental issue.

1987
Twenty-four nations initially sign Montreal Protocol, pledging to phase out chemicals that deplete Earth's ozone layer; dozens more sign in subsequent years.

1989
Exxon Valdez runs aground in Alaska, contaminating more than 5,000 kilometers of pristine coast with oil and killing thousands of animals and birds.

1990
Congress creates market-based allowance trading system to reduce emissions that cause acid rain. . . . Federal Trade Commission (FTC) brings first enforcement case against deceptive green marketing, challenging claims for "pesticide free" produce sold by Vons supermarkets.

1992
Delegates to the Earth Summit in Rio de Janeiro, Brazil, adopt first international pledge to cut greenhouse gas (GHG) emissions. . . . FTC issues marketing guides for green products and services.

1997
International conference approves Kyoto Protocol requiring GHG reductions but lets wealthy nations meet some of their obligations with offset projects in developing countries; U.S. signs but fails to ratify pact.

1998
U.S. Green Building Council launches Leadership in Energy and Environmental Design (LEED) program for rating energy-efficient, healthy buildings.

2000
British Petroleum re-brands itself BP and pledges to go "Beyond Petroleum" by investing in clean energy.

———— • ————

2001-2007
As environmental concern grows, more companies offer eco-friendly products. Skeptics warn of "greenwashing."

2000
Department of Agriculture issues final rule for certifying organic food.

2001
President George W. Bush rejects mandatory controls on GHG emissions. . . . Following the Sept. 11 terrorist attacks, Bush urges Americans to shop to help fend off economic recession.

2005
General Electric launches "Ecomagination" advertising campaign to demonstrate its environmental commitment. . . . Kyoto Protocol enters into force, including credits for carbon offset projects in developing countries. . . . European Union members begin trading carbon credits.

2006
Democrats recapture control of Congress, increasing support for policies to boost renewable energy and curb greenhouse gas emissions.

2007
FTC initiates review of green marketing guidelines and environmental products, including carbon offsets. . . . Toyota Prius hybrids surpass top-selling sport-utility vehicles.

The Six Sins of 'Greenwashing'

Misleading environmental claims are common.

A perfectly green product may not exist, but some certainly are much greener than others, according to a recent study by Pennsylvania-based TerraChoice Environmental Marketing. [1] It examined 1,018 consumer products that made a total of 1,753 environmental claims and found that every product but one offered false or misleading information. The firm identified six broad categories of misleading environmental claims, or "greenwashing":

- **The hidden trade-off:** Marketing a product as eco-friendly based on a single green attribute like recycled content, without addressing other issues such as where its materials come from or how much energy is required to produce it.

- **No proof:** Making environmental claims without providing information backing them up at the point of purchase or on the manufacturer's Web site.

- **Vagueness:** Touting products based on claims that are too vague to have any real meaning, such as "Non-Toxic," "All Natural" or "Earth-Friendly."

- **Irrelevance:** Offering a claim that is true but not important or helpful to consumers. For example, some products are labeled "CFC-Free," but ozone-destroying chlorofluorocarbons (CFCs) have been outlawed in the U.S. for several decades.

- **Lesser of two evils:** Selling a product with an environmental label even though it belongs to a class of goods that is generally bad for consumers' health or the environment, such as organic cigarettes.

- **Fibbing:** Providing false information or claiming a certification, such as USDA Organic, that the product has not actually earned.

Greenwashing matters for several reasons, the study contends. First, consumers will waste money and may conclude that environmentally friendly products do not work. Second, greenwashing takes business away from legitimate green products. This makes it harder for honest manufacturers to compete and slows the rate at which high-quality products penetrate the market.

Indeed, greenwashing was a factor in the demise of an early wave of green consumerism in the 1980s, says TerraChoice Vice President Scot Case, but more scrutiny this time may deter cheaters. "We'll know if things are improving when we repeat the study in a few months," says Case. "We're hopeful that attention from the media and the Federal Trade Commission [FTC] will help."

Consumers who want to ensure that they are getting green products have several options. First, they can look for seals of approval from organizations such as EcoLogo and Green Seal, both of which certify green products based on multiple criteria. [2] These eco-labeling programs are standardized under a set of principles developed by the International Organization for Standards.

Consumers also can check product labels and manufacturers' Web sites for information that supports green marketing claims. "Companies should be very careful not to claim that things are green, only that they are greener," says Case. "They shouldn't suggest that just because they've addressed one issue, it's a green product." The FTC has published guidance to help consumers sort through green advertising claims. [3]

Although greenwashing may be pervasive today, TerraChoice argues that green marketing can be a positive force. "[G]reen marketers and consumers are learning about the pitfalls of greenwashing together," the report states. "This is a shared problem and opportunity. When green marketing overcomes these challenges, consumers will be better able to trust green claims, and genuinely environmentally preferable products will penetrate their markets more rapidly and deeply. This will be great for consumers, great for business and great for the planet." [4]

---

[1] TerraChoice Environmental Marketing, "The 'Six Sins of Greenwashing," November 2007, www.terrachoice.com.

[2] For more information see www.ecologo.org and www.greenseal.org.

[3] U.S. Federal Trade Commission, "Sorting Out 'Green' Advertising Claims," www.ftc.gov/bcp/edu/pubs/consumer/general/gen02.pdf.

[4] TerraChoice, op. cit., p. 8.

Continued from p. 202

expeditions. One notable exception, the Massachusetts Audubon Society, was founded in 1896 by two Boston society women who opposed killing exotic birds to provide feathers for fashionable ladies' hats. Within a year the group persuaded the state legislature to ban commerce in wild bird feathers. Its work later spurred Congress to pass national legislation and support a treaty protecting migratory birds. [32]

Although early groups won some notable victories, most conservation work was mandated by the federal government. Congress and Presidents Theodore Roosevelt (1901-1909) and William Howard Taft (1909-1913) set aside many important tracts of land as parks and monuments. Congress created the National Park Service in 1916 to manage these new preserves. But national policy also spurred harmful development, such as federally funded

irrigation projects to help settlers farm in dry Western states. With government agencies urging them on, farmers plowed up the Great Plains, destroying their natural grass cover and helping to create the Dust Bowl when drought struck in the 1930s.

During the long tenure of President Franklin D. Roosevelt, (1933-1945), several important conservation programs were launched even as Western dam building accelerated. The Civilian

Conservation Corps (CCC), also known as "Roosevelt's Tree Army," hired more than 3 million unemployed Americans to build fire towers, plant trees and improve parks across the nation. More than 8 million people worked for the Works Progress Administration on projects including roads, bridges and park lodges.

These were top-down programs, writes anthropologist Michael Johnson: "The federal government defined the problems, defined the solutions and then 'fixed' the problems by employing lots of people . . . the average citizen had an almost blind trust in the federal definition of problems and solutions." Moreover, while the CCC and other initiatives treated symptoms such as soil erosion, they failed to address human actions like plowing and over-grazing that caused the problems. [33]

Environmental Awakening

As the economy grew rapidly after World War II, human impacts on the environment became obvious. Pollutants from power plants, factories and passenger cars mixed in the atmosphere to create toxic smog. Offshore oil-drilling platforms appeared along California's scenic coastline. And Rachel Carson's 1962 book *Silent Spring* warned that widespread use of pesticides threatened ecosystems and human health.

Alarmed environmentalists began fighting back. In 1955 they rallied against a hydropower dam that would have flooded part of Dinosaur National Monument in Utah. A decade later, a coalition led by the Sierra Club helped to block a dam that would have inundated the Grand Canyon. Conservationists won a big victory in 1965 when a federal court allowed them to sue against a proposed electric power plant on Storm King Mountain in New York's Hudson Valley. [34] Courts previously had decided such siting issues on narrow

technical grounds but in this case held that groups not directly involved in development projects could intervene to protect scenic resources. Litigation quickly became an important tool for environmental advocates.

Congress passed several key environmental laws in the 1960s, including the Wilderness Act (1964), which created a process for protecting land permanently from development, and the National Environmental Policy Act (1969), which subjected major federal actions such as building dams to environmental-impact studies. But new disasters spurred calls for further action. In 1969 an offshore oil well near Santa Barbara, Calif., ruptured and spilled oil along 30 miles of coastline. Five months later Ohio's Cuyahoga River caught fire when flammable chemicals on its surface ignited.

On April 22, 1970, the first Earth Day, more than 20 million Americans attended rallies and teach-ins designed to force environmental issues onto the national agenda. Activists followed up with lobbying and lawsuits. In response Congress passed a flurry of new laws, including an expanded Clean Air Act (1970), the Endangered Species Act (1973), the Safe Drinking Water Act (1974), the Resources Conservation and Recovery Act (1976) and the Clean Water Act (1977).

"Citizens across the country became aware of what was happening to their physical surroundings," writes journalist Philip Shabecoff. "Equally important, they also acquired a faith — not always requited — that in the American democracy change was possible, that they could act as individuals and communities to obtain relief from the environmental dangers with which they were threatened." [35] Slogans like "Reduce, reuse, recycle" and "Think globally, act locally" underlined the importance of personal action.

While national groups pressured Congress and the new EPA, grassroots activists attacked local problems. In 1978 residents of the Love Canal neighbor-

hood in upstate New York, led by housewife Lois Gibbs, forced the federal government to pay for moving them out of homes that had been built on top of an industrial-waste site. Groups with names like the Abalone Alliance sprang up to oppose new nuclear power plants, blocking some and delaying others.

Businesses and free-market advocates pushed back. President Ronald Reagan was elected in 1980 on a platform that called for reforming regulation and ensuring that benefits from environmental controls justified their costs. [36] During the campaign Reagan argued that air pollution had been "substantially controlled" in the United States and that laws like the Clean Air Act were forcing factories to shut down. [37] When he installed anti-regulation appointees like Interior Secretary James G. Watt and EPA Administrator Anne Gorsuch, many environmentalists worried that their recent victories would be reversed.

Working With Markets

As the Reagan administration learned, most Americans did not support a broad rollback of environmental laws. Public backlash against proposals such as selling off millions of acres of public lands drove Watt and Gorsuch from office. But environmentalists still faced a Republican administration and Senate majority that opposed new controls.

In response some groups began working with the private sector and developing market-based policies. Proponents of this environmental "third wave" contended that if regulations were more cost-effective and flexible, industries could be persuaded to cut pollution instead of having to be forced.

Their most visible success was promoting tradable permits to cut pollution. EPA had started experimenting in the 1970s with programs that allowed companies to earn and trade credits for reducing air pollutants such as carbon monoxide and particulates. Business

Guidelines for Eco-minded Consumers

Here's how to have the most impact.

For many consumers, the biggest challenge of buying green is not finding earth-friendly goods but figuring out which choices have the biggest environmental impact. Green buying choices can be complicated, and green products often cost more than conventional alternatives.

Moreover, as journalist Samuel Fromartz observes in his history of the organic food business, few shoppers buy everything from premium suppliers like Whole Foods. Instead, regardless of income level, they buy organic in categories that matter to them, such as milk for their children, and choose other items of lower concern from conventional or discount stores. [1]

To help eco-minded consumers focus on purchases with the biggest environmental impact, *The Consumer's Guide to Effective Environmental Choices* identifies the biggest environmental problems related to household consumption: air and water pollution, global warming and habitat alteration. Then, by quantifying environmental impacts and linking these impacts to consumer products and services, authors Michael Brower and Warren Leon identify three household activity areas that account for most of these impacts: food, household operations and transportation.

To address these issues, Brower and Leon urge consumers to take steps such as driving fuel-efficient, low-polluting cars, eating less meat and making their homes energy-efficient. [2]

"A green purchase can have at least three results," says Leon. "First, it can favor a lower-impact product over conventional options. Second, it may allow you to consume fewer resources over the lifetime of the product. That's why energy choices are important — not only does energy use have significant environmental impacts, but you will use less energy every time you turn that appliance on."

As another example, consider a gardener who spends several hundred dollars on outdoor furniture. If she chooses items made from sustainably harvested wood, she may preserve several trees in a threatened forest. But if she uses the

Installing compact fluorescent light bulbs is one of several tips for responsible consumption recommended by the Center for a New American Dream.

Getty Images/Steve Wisbauer

same money to buy a backyard composting bin, she can divert hundreds of pounds of food waste from landfills (which produce greenhouse gases as wastes break down and can leak and contaminate groundwater) during the years that she uses the bin.

Third, Leon argues, some green purchases can favor new environmentally friendly technologies or industries with big growth potential. "By joining the early adapters who reinforce demand for a new product, you can help create a perception that it's a success," he says. However, it is important to note that some products will never become market phenomena because they have small niche markets. Only a small fraction of the Americans who drink wine will buy organic wine, but nearly everyone has to clean a bathroom at some point, so green cleaning supplies have a bigger prospective market.

The nonprofit Center for a New American Dream, which advocates for responsible consumption, offers a similar list of personal steps to "Turn the Tide":

- Drive less.
- Eat less feedlot beef.
- Eat eco-friendly seafood.
- Remove your address from bulk mailing lists.
- Install compact fluorescent light bulbs.
- Use less energy for home heating and cooling.
- Eliminate lawn pesticides.
- Reduce home water usage.
- Inspire your friends.

"None of Turn the Tide's nine actions involve drastic changes in your life, yet each packs an environmental punch," says the center. "In fact, every thousand participants prevent the emission of 4 million pounds of climate-warming carbon dioxide every year." [3]

---

[1] Samuel Fromartz, *Organic, Inc.: Natural Foods and How They Grew* (2006), pp. 248-53.

[2] Michael Brower and Warren Leon, *The Consumer's Guide to Effective Environmental Choices* (1999), pp. 43-85.

[3] Center for a New American Dream, "Turn the Tide," www.newdream.org/cnad/user/turn_the_tide.php.

leaders preferred this approach because instead of mandating specific control technologies, it let them decide how and where to make reductions. For example, instead of installing pollution controls a company might make its operations more efficient or switch to cleaner methods or products.

When Congress amended the Clean Air Act in 1990, some environmentalists supported a cap-and-trade system to reduce sulfur dioxide (SO_2)

and nitrogen oxide emissions that caused acid rain. This approach set an overall cap on emissions and issued a fixed number of tradable emission permits to sources. Factories emitting less pollution than their allotments could sell extra permits to other sources — giving polluters an economic incentive to clean up, advocates asserted. [38]

The SO_2 trading program went into effect in 1995 and expanded in 2000. Many supporters praised it for cutting SO_2 releases sharply at a lower cost than industry had predicted. [39] However, acid rain remained a problem in areas located downwind from major pollution sources, such as the Adirondack Mountains, and emissions trading did not prove to be a panacea for other U.S. air pollution problems. [40]

During the 1990s some economic experts began to argue that going green made sound business sense. By reducing pollution, the theory held, companies would make their operations more efficient, which meant that they would use less energy and waste fewer raw materials. "Innovation to comply with environmental regulation often improves product performance or quality," business professors Michael Porter and Claas van der Linde asserted in 1995. [41]

As one step, some companies forged relationships with large environmental groups. [42] McDonald's worked with Environmental Defense to design a paperboard alternative to its polystyrene "clamshell" hamburger package, and the Rainforest Alliance helped Chiquita Brands develop social and environmental standards for its banana farms in Latin America. [43] However, critics argued that by accepting corporate donations and putting business executives on their boards of directors, environmentalists risked becoming too sympathetic to private interests. [44]

Some smaller groups stuck to more aggressive tactics. San Francisco's Rainforest Action Network carried out scrappy direct-action campaigns that persuaded Burger King to stop using beef raised on former rainforest lands and Home Depot to sell only sustainably produced wood. The Earth Island Institute used negative publicity and a consumer boycott to make tuna companies adopt fishing practices that avoided killing dolphins in tuna nets. And major groups continue to vilify companies like oil giant Exxon/Mobil, whose opposition to action on global warming and support for oil drilling in the Arctic National Wildlife Refuge made it a prime environmental target. [45]

The 1997 Kyoto Protocol applied offsets to climate change in a provision called the Clean Development Mechanism (CDM), under which developed countries could meet part of their commitments by paying for projects that reduced GHG emissions in developing countries. This process was designed to reduce costs by letting industrialized nations cut GHG emissions in locations where environmental upgrades were cheaper. (GHGs dissipate widely throughout the atmosphere, so eliminating a ton of CO_2 emissions has the same impact on climate change wherever it occurs.)

Shopping for Change

National environmental policy became more contentious after George W. Bush was elected president in 2000 with strong support from energy- and resource-intensive industries. Many administration appointees pushed to loosen environmental regulations, and President Bush reversed a campaign pledge to limit greenhouse gas emissions that caused global warming, arguing that doing so would hurt the economy. [46]

Stymied at the federal level, environmentalists looked for other ways to leverage public support for green policies. Many advocacy groups deepened ties with businesses to influence corporate policies and earn political support from the private sector. They also urged members to target their buying power toward green goals. "People got tired of the gloom and doom approach. They wanted to hear about solutions," explained Bud Ris, executive director of the Union of Concerned Scientists from 1984 through 2003. [47]

Even as scientific consensus increased that human actions were causing global climate change, President George W. Bush opposed calls for mandatory controls on U.S. GHG emissions. Instead, in 2002 Bush pledged to reduce U.S. GHG emissions per dollar of economic activity by 18 percent by 2012. "This will set America on a path to slow the growth of our greenhouse gas emissions and, as science justifies, to stop and then reverse the growth of emissions," Bush said. However, many analysts noted, even if the American economy became 18 percent less carbon-intensive, its total GHG emissions would increase during that time as a result of normal economic growth.

Many corporations joined voluntary initiatives, however, like EPA's Climate Leaders program or the privately funded Pew Center on Global Climate Policy, both to show stockholders that they were paying attention to the environment and to discuss what kind of climate change policies would be most workable for businesses. [48] These partnerships required companies to measure their GHG emissions and develop strategies for reducing them. Companies also began exploring options like renewable energy certificates (RECs) and carbon offsets to reduce their carbon footprints.

Some companies turned growing concerns about pollution and climate change to their advantage with products that were both high-quality and green. Toyota's gas-electric hybrid Prius hatchback, which promised drivers 60 miles per gallon in city driving, debuted with limited sales in U.S. markets in 2000. By 2005 the Prius had become a symbol of green chic, and Toyota was selling

100,000 per year. And after the U.S. Department of Agriculture finalized standards for certifying organic food in 2000, Whole Foods rode growing demand for organic and natural groceries to become the largest natural food market chain in the nation. ∎

CURRENT SITUATION

Keeping Standards High

As consumer interest in green products rises, regulators and environmentalists are taking a critical look at definitions and marketplace practices. Strong standards are needed, observers say, to prevent a new wave of greenwashing and help consumers avoid wasting money.

"The nature of marketing is to puff up products. That's why we have labeling laws, and there are struggles over who regulates what," says University of California sociologist Szasz. "The first struggle is over how regulated a product like organic food will be and who will do it. Then once the rules are written, debate over practices like greenwashing takes place within those boundaries."

In late 2007 the Federal Trade Commission (FTC) announced plans to review its green marketing guidelines and new green products such as carbon offsets. [49] FTC's guidelines offer advice for manufacturers on a variety of green products, but the commission may issue specific guidance on carbon offsets and RECs.

"We want to learn more about what these products are, how they work and how much activity is going on in the marketplace. We're also exploring how marketers are substantiating their claims and what consumers need to know about these products that we can provide," says FTC attorney Hampton Newsome.

The FTC is not an environmental agency, so it will not set specifications for individual products. Rather, it considers questions such as whether labels provide enough clear information for consumers to make judgments. For example, according to agency guidelines, a bottle labeled "50% more recycled content" would be ambiguous because the comparison could refer to a competing brand or to a prior version of the product. A label reading "50% more recycled content than our previous package" would be clearer. [50]

"Marketers have to substantiate express or implied marketing claims with competent and reliable evidence," says Newsome. "How consumers understand the claim is key, because that determines their purchasing decisions, not what the seller intended." Under the Federal Trade Commission Act, which outlaws unfair and deceptive trade practices, companies that make false or misleading claims could face penalties including injunctions or forfeiture of profits.

Many organizations are working to help standardize carbon offsets and define high-quality versions. There are a number of issues to consider, says Colby College's Tietenberg. "Quantification is important. The fact that something reduces greenhouse gases is useful, but you need to quantify how much it reduces them," he says. "You need to ensure that the initial reductions prevail through the life of the offset — for example, if you plant trees and the forest burns down, you don't get the offset. And you need a tracking system to keep people from selling the same offsets to multiple buyers."

Advocates also want to make green certification programs more rigorous. Some have criticized the LEED rating system for green buildings, saying that its checklists are simplistic and give too much weight to small steps, like installing bicycle racks, and not enough to bigger ones, such as renovating a historic building instead of razing it. [51] But the green building movement remains strong: By 2010, trade publications estimate that about 10 percent of commercial construction starts will be green projects (not all of which may seek LEED ratings). [52]

Watchdogs also see room for improvement in the Energy Star program. In 2007 EPA's inspector general reported that the agency was not doing enough to confirm that Energy Star products (which are tested by manufacturers, not EPA) performed at the promised level, or to prevent unqualified products from being labeled as Energy Star models. [53] The Government Accountability Office also criticized relying on manufacturers to test products and urged EPA and the Energy Department to look more closely at issues such as how many products are purchased because of Energy Star ratings. [54]

More Mandates?

Congressional supporters of a national renewable electricity portfolio standard have pledged to bring RPS legislation up again this year. Countering the argument that this policy would penalize some states, a study by the American Council for an Energy-Efficient Economy (ACEEE) projects that electricity prices would be lower across the U.S. in 2020 and 2025 under a standard like that passed by the House in 2007 (combining renewable electricity and conservation) than without an RPS. A more aggressive standard that met 15 percent of electricity demand with renewable fuels and 15 percent through conservation would push prices even lower, ACEEE found. [55]

"Including energy efficiency brings down wholesale prices," explains ACEEE Deputy Director Prindle. Efficiency and renewables also complement each other, he says, because conservation projects can be put in place more quickly while new renewable energy projects are sited and built.

Continued on p. 210

At Issue:

Does the United States need a national renewable electricity portfolio standard?

GOV. BILL RITTER, JR., D-COLO.

FROM TESTIMONY BEFORE HOUSE SELECT COMMITTEE ON ENERGY INDEPENDENCE AND GLOBAL WARMING, SEPT. 20, 2007

*i*t has been our experience that [a renewable electricity portfolio standard] creates new jobs, spurs economic development and increases the tax base all while saving consumers and businesses money and protecting our environment. In 2004, following three years of failed legislative efforts, the people of Colorado placed the nation's first citizen-initiated renewable portfolio standard (RPS), Amendment 37, on the ballot. While the effort was opposed by virtually all Colorado utilities, including the state's largest utility — Xcel Energy — the effort passed by a wide margin. The Colorado RPS established a goal of 10 percent renewable resources by 2015 for Xcel Energy (along with the other Colorado Public Utilities Commission-regulated utility, Aquila).

In 2004, 10 percent was an ambitious goal: a little over 1 percent of Xcel's electricity was generated from renewable sources at that time. Today, it is the country's leading provider of wind energy. Xcel will meet the 10-percent-by-2015 goal at the end of 2007 — nearly eight years ahead of schedule.

Xcel has done what all successful businesses do — it adapted. While Xcel originally viewed the RPS as a burden, it soon recognized it as an opportunity, and the utility is now a great example of the successes that will come from our New Energy Economy. . . .

Renewable energy development of the future is not limited to wind. In Colorado, we are fortunate to have a broad mix of renewable resources, including wind on our Eastern Plains, solar in the San Luis Valley and southwest part of the state and geothermal all along our Western Slope. . . .

The committee has asked how a national renewable electricity standard will impact technologies in Colorado. Developments in wind technology have led the industry to be cost competitive with fossil fuel generation, but we need similar developments in both solar electric as well as concentrated solar technology. With the appropriate leadership from the federal government, these resources have the opportunity to join wind as a primary source of renewable power. . . .

As we saw with the RPS in Colorado — we encouraged the market through the RPS, and the market has responded. Investment, research and development are following the establishment of the RPS. A federal RPS provides more markets for renewable energy, prosperity for Americans in the heartland and a more responsible energy future for our nation.

CHRIS M. HOBSON
SENIOR VICE PRESIDENT, RESEARCH AND ENVIRONMENTAL AFFAIRS, SOUTHERN COMPANY

FROM TESTIMONY BEFORE HOUSE SELECT COMMITTEE ON ENERGY INDEPENDENCE AND GLOBAL WARMING, SEPT. 20, 2007

*S*outhern Company opposes a national renewable-energy mandate. We believe that mandates are an inefficient and potentially counterproductive means of increasing the production of cost-effective, reliable electric power from renewable sources. We prefer to seek cost-effective additions to our generation portfolio based on technological maturity, technical performance, reliability and economic cost. . . .

Our estimates show that a 15 percent federal renewable-energy mandate would far exceed the available renewable resources in the Southeastern region. To replace 15 percent of the nation's retail energy by 2020 would require approximately 80,000 wind turbines of 2 megawatt capacity each, or 2,200 square miles of land — an area larger than Delaware — for solar photovoltaic arrays, or 87,000 square miles of switch grass fields — an area the size of Minnesota. To replace 15 percent of just Southern Company's retail energy by 2020 would require approximately 6,900 wind turbines of 2 megawatt capacity each, or 200 square miles of land for solar photovoltaics, or 6,000 square miles of switchgrass fields — an area the size of Connecticut. . . .

Because the renewable resources that would be required to comply with a 15 percent mandate are not available in the Southeast, Southern Company would be required to comply largely by making alternative compliance payments to the federal government. . . . Because of the limited availability of renewable resources in our region and the fact that most of what is available will likely be more expensive than the 3 cents/kilowatt-hour price cap, the majority of the $19 billion cost to our customers will simply be payments to the federal government. Thus a nationwide [renewable portfolio standard] mandate could cost electricity consumers in the Southeast billions of dollars in higher electricity prices, with no guarantee that additional renewable generation will actually be developed. . . .

Not every technology will be well-suited to every region of the country. We do believe that the use of renewable energy to produce electricity can be increased, and we intend to play a key role in the research and development needed to reach such an objective. This is best reached by the enhancement of current strategies to provide incentives for the R&D as well as the use of renewable energy as compared to the adoption of a federal mandate for a single standard across the country.

Continued from p. 208

But opponents are likely to fight any new RPS proposals in 2008. Energy producers in the Southeast maintain that a national RPS will penalize their region, and the White House threatened to veto the 2007 energy bill over its RPS requirement. "A federal RPS that is unfair in its application, is overly prescriptive in its definition by excluding many low-carbon technologies and does not allow states to opt out would hurt consumers and undercut state decisions," National Economic Council Chair Allan Hubbard wrote to congressional leaders in late 2007. [56]

First Congress may have to revisit another controversial green mandate — the Renewable Fuels Standard (RFS), enacted in 2005 and expanded in 2007, which promotes bio-based transportation fuels like ethanol and biodiesel. [57] The original RFS, which was adopted to reduce U.S. dependence on imported oil and cut pollution from transportation, required refiners to use 5.4 billion gallons of renewable fuels (mostly blended with conventional gasoline) in 2008, rising to 7.5 billion gallons by 2012. The new law mandates 9 billion gallons in 2008, increasing to 36 billion gallons by 2022.

Most biofuel sold in the United States is ethanol made from corn, although researchers are starting to make ethanol from cellulosic sources (crop wastes and woody plants), which have a higher energy content and require fewer resources to produce. For the moment, however, much support for the RFS comes from farm-state lawmakers and agribusinesses invested in corn ethanol.

Many observers believe that the RFS is poorly designed and is producing unintended consequences. Two recent studies suggest that the push to expand biofuel crops may trigger such widespread land clearing that it increases climate change (by destroying forests that take up carbon) instead of reducing it. [58]

Many grocery stores have begun selling reusable shopping bags as an alternative to environmentally unfriendly plastic bags.

And by driving up demand for corn, which also is used in animal feed and processed foods, critics say the mandate is increasing food prices. [59]

"The RFS has a narrow focus on a particular technology, and it doesn't strike a balance between demand and supply," says Prindle. "Also, it will take a lot of new capacity to meet the targets, including inputs like water and electricity as well as grain. You have to develop a massive new infrastructure across the middle of the country [where most corn is grown]." Colby College's Tietenberg seconds this perspective. "Mandates should be performance-based instead of requiring a specific input," he says. "You want to make sure the standard is clear but that there are flexible options for meeting it."

Green Is Red-Hot

Amid these debates, green marketing is spreading across much of the nation's economy. Today green labeling is most commonly found on office products, building materials, cleaning products and electronics. "In the 20 years between the last green bubble and this one, the only people who expressed strong interest in green products were large institutional purchasers like government agencies, colleges and hospitals, so green labeling had a very business-centric focus," explains TerraChoice's Case.

But now the message is penetrating into new sectors. Transportation, for example, accounts for about 27 percent of U.S. GHG emissions and is a major contributor to regional air pollution. A decade ago gas-guzzling sport utility vehicles (SUVs) and light trucks dominated the U.S. auto market, but in 2007 sales of gas-electric hybrid Toyota Prius hatchbacks surpassed the Ford Explorer, long the top-selling SUV. [60]

Now, with gas prices high and new fuel-efficiency standards signed into law, U.S. automakers are terminating some SUV lines, converting others to smaller "crossovers" and putting more money into alternative vehicles. In 2007 General Motors unveiled a concept model of the Chevrolet Volt, a plug-in electric car that uses a small gasoline engine as a generator to charge its batteries. GM is still designing the Volt but hopes to have it on the market by late 2010. [61] Other companies, including Toyota and Ford, are developing plug-in hybrids that can be recharged at standard 120-volt outlets.

Home and personal care products are also becoming increasingly green, in response to consumer alarm over recent reports describing toxic, hazardous and untested ingredients in common consumer goods. For example, laboratory testing carried out for the Campaign for Safe Cosmetics in 2007 found detectable levels of lead, a neurotoxic chemical, in many brand-name lipsticks. [62] Another recent study found increased levels of phthalates (chemical softeners that have been linked to reproductive problems and are banned from personal care products in the European Union) in infants who were treated with baby lotions, powders and shampoos. [63] Toys, food and beverage containers, upholstery, and other goods have also been found to contain compounds known or suspected to be hazardous to human health.

But this area is a major greenwashing zone, with marketers often relying on slogans that have no standard meaning. "While splashy terms and phrases such as 'earth-friendly,' 'organic,' 'nontoxic,' and 'no harmful fragrances' can occasionally be helpful, the ugly truth is in the ingredients list," the environmental magazine *Grist* advises in its green living guide. [64] ∎

OUTLOOK

Focus on Carbon

Whichever party wins the White House in 2008, it appears likely that the United States will adopt binding GHG limits sometime after a new president takes office in 2009. All of the front-runners for president, including Democrats Hillary Rodham Clinton and Barack Obama and Republican John McCain, support cap-and-trade legislation that would sharply reduce U.S. emissions by 2050 — a timetable that many scientists believe is needed to avert catastrophic global warming. [65]

Many companies can read the writing on the wall and are working to turn the issue to their advantage. "The business community sees tremendous opportunity in green products. It's a chance for companies to push technology and come up with innovative solutions," says the National Association of Manufacturers' McCoy. "We're on the cusp of some fascinating discoveries that could help solve our energy needs. We need to consider what research and development incentives government can offer to facilitate that, with manufacturers and the business community involved." Even Exxon/Mobil, long one of the strongest foes of binding GHG limits, has started to discuss what national controls should look like. [66]

With mandatory GHG limits in place, will green consumerism still have a role to play? Many observers see buying green as an important piece of the larger solution. "The reality is that we consume products every day. This is not going to change any time soon," says Dorfman, the "Lazy Environmentalist." "So we have to find more environmentally conscious ways to consume if we want to maintain our quality of lives and not see them degraded by climate change. . . . However, the solutions have to fit our lifestyles or the great majority of us won't even consider them." [67]

But green consumption and business/environment partnerships are not substitutes for political action, says Colby College's Tietenberg. "You need policies to level the playing field. If some firms are out there doing more and it costs more, they may have trouble competing and lose market share," he argues. "But if government sets rules that create a level playing field, business will take the ball and run with it. Many businesses are asking for national standards now."

Consumers who want to make a serious impact with their purchases need to learn which steps make the most difference. "People have a limited understanding of their carbon footprints," says Climate Trust Director Burnett. "They don't necessarily know what kind of fuel generates their electricity, or how significant airplane flights are." And comparing products' full life-cycle impacts can get complicated. For example, today many consumers are debating whether it is preferable to buy organically grown food that is shipped over long distances to market (generating GHG emissions in the process) or locally grown food that has been raised using less earth-friendly methods. [68]

"Wisdom is a curse — once you learn about these issues, you can't overlook them," says LOHAS Forum Director Ning. "But as we confront more problems like environmental toxins that stem from manufacturing processes, people are becoming more aware of design impacts. They're trying to understand more about how products work, and producers are trying to learn more about sustainability. Now that these ideas are becoming part of school curriculums, and people are talking about them more, our consciousness is only going to grow." ∎

Notes

[1] Tearfund, "Senior Bishops Call For Carbon Fast This Lent," Feb. 5, 2008, www.tearfund.org.

[2] Penn, Schoen, & Berland Associates, "Consumers Will Double Spending on Green," Sept. 27, 2007.

[3] U.S. Bureau of Labor Statistics, Consumer Expenditure Survey, 2000-2006, www.bls.gov/cex/2006/standard/multiyr.pdf.

[4] Karlo Barrios Marcelo and Mark Hugo Lopez, "How Young People Expressed Their Political Views in 2006," Center for Information & Research on Civic Learning & Engagement, University of Maryland, November 2007; Scott Keeger, "Politics and the 'DotNet' Generation," Pew Research Center, May 30, 2006; Lori J.

Vogelgesang and Alexander W. Astin, "Post-College Civic Engagement Among Graduates," Higher Education Research Institute, University of California, Los Angeles, April 2005.

[5] U.S. Federal Trade Commission, "The FTC's Environmental Cases," www.ftc.gov/bcp/con-line/edcams/eande/contentframe_environment_cases.html.

[6] Intergovernmental Panel on Climate Change, *Climate Change 2007: The Physical Science Basis, Summary for Policymakers* (2007), pp. 3, 5. For background, see Marcia Clemmitt, "Climate Change," *CQ Researcher*, Jan. 27, 2006, pp. 73-96, and Colin Woodard, "Curbing Climate Change," *CQ Global Researcher*, February 2007, pp. 27-50.

[7] For background see Peter Katel, "Consumer Safety," *CQ Researcher*, Oct. 12, 2007, pp. 841-864, and Jennifer Weeks, "Factory Farms," *CQ Researcher*, Jan. 12, 2007, pp. 25-48.

[8] For background see Tom Price, "The New Environmentalism," *CQ Researcher*, Dec. 1, 2006, pp. 985-1008, and Tom Price, "Corporate Social Responsibility," *CQ Researcher*, Aug. 3, 2007, pp. 649-672.

[9] For details on these companies' pledges, see www.dupont.com/Sustainability/en_US/; www.monsanto.com/who_we_are/our_pledge.asp; and www.thinkgreen.com.

[10] For background see Marcia Clemmitt, "Climate Change," *CQ Researcher*, Jan. 27, 2006, pp. 73-96.

[11] U.S. Climate Action Partnership, *A Call for Action* (2007), p. 3, www.us-cap.org/US-CAPCallForAction.pdf.

[12] Joel Makower, *et al.*, *State of Green Business 2008* (2008), p. 3, www.stateofgreenbusiness.com/.

[13] CarbonCounter.org, www.carboncounter.org/offset-your-emissions/personal-calculator.aspx; Union of Concerned Scientists, "What's Your Carbon Footprint?" www.ucsusa.org/publications/greentips/whats-your-carb.html.

[14] *A Consumer's Guide to Retail Carbon Offset Providers* (2006), www.cleanair-coolplanet.org/ConsumersGuidetoCarbonOffsets.pdf.

[15] Ted Williams, "As Ugly As a Tree," *Audubon*, September/October 2007.

[16] Ben Elgin, "Another Inconvenient Truth," *Business Week*, March 26, 2007.

[17] Kevin Smith, *The Carbon-Neutral Myth: Offset Indulgences for Your Climate Sins* (2007), pp. 10-11, www.carbontradewatch.org.

[18] Federal Energy Regulatory Commission, "Electric Market Overview: Renewables," updated Jan. 15, 2008, www.ferc.gov/market-oversight/mkt-electric/overview/elec-ovr-rps.pdf.

[19] *Congressional Record*, Aug. 4, 2007, p. H9847.

[20] "Send Your Underwear to the Undersecretary," Competitive Enterprise Institute news release, May 16, 2007.

[21] "Washers and Dryers: Performance For Less," *Consumer Reports*, February 2008.

[22] For details see www.energystar.gov.

[23] U.S. Environmental Protection Agency, "Energy Star and Other Climate Protection Partnerships 2006 Annual Report," September 2007, p. 15, www.energystar.gov/ia/news/downloads/annual_report_2006.pdf. According to the Department of Energy, the average U.S. household uses about 11,000 kilowatt-hours of electricity annually; see www.eere.energy.gov/consumer/tips/appliances.htm.

[24] For details see www.usgbc.org.

[25] Jenny Shank, "An Interview With 'Lazy Environmentalist' Josh Dorfman," July 2, 2007, www.newwest.net/topic/article/an_interview_with_lazy_environmentalist_josh_dorfman/C39/L39/.

[26] Bill McKibben, "Reversal of Fortune," *Mother Jones*, March/April 2007.

[27] Michael Brower and Warren Leon, *The Consumer's Guide To Effective Environmental Choices* (1999).

[28] LOHAS Forum, "About LOHAS," www.lohas.com/about.htm.

[29] Patrrick O'Driscoll and Elizabeth Weise, "Green Living Takes Root But Habits Die Hard," *USA Today*, April 19, 2007; Anjali Athavaley, "A Serious Problem (But Not My Problem), *Wall Street Journal Classroom Edition*, February 2008.

[30] John Muir, "God's First Temples: How Shall We Preserve Our Forests?" reprinted in John Muir, *Nature Writings* (1997), p. 629.

[31] For background see Tom Arrandale, "Public Land Policy," *CQ Researcher*, June 17, 1994, pp. 529-552.

[32] Massachusetts Foundation for the Humanities, "Mass Moments," www.massmoments.org/moment.cfm?mid=262.

[33] Michael D. Johnson, "A Sociocultural Perspective on the Development of U.S. Natural Resource Partnerships in the 20th Century," USDA Forest Service Proceedings (2000), p. 206.

[34] *Scenic Hudson Preservation Conference v. Federal Power Commission*, 354 F. 2d 608 (1965).

[35] Philip Shabecoff, *Earth Rising: American Environmentalism in the 21st Century* (2000), p. 7.

[36] Republican Party Platform of 1980, adopted July 15, 1980, online at The American Presidency Project, www.presidency.ucsb.edu/show-platforms.php?platindex=R1980.

[37] Joanne Omang, "Reagan Criticizes Clean Air Laws and EPA as Obstacles to Growth," *The Washington Post*, Oct. 9, 1980.

[38] See "Acid Rain: New Approach to Old Problem," *CQ Researcher*, March 3, 1991.

[39] Environmental Defense, *From Obstacle to Opportunity: How Acid Rain Emissions Trading Is Delivering Cleaner Air* (September 2000); Robert N. Stavins, "Experience with Market-Based Environmental Policy Instruments," Discussion Paper 01-58, Resources for the Future, November 2001, pp. 27-29.

[40] See Charles T. Driscoll, *et al.*, *Acid Rain Revisited: Advances in Scientific Understanding Since the Passage of the 1970 and 1990 Clean Air Act Amendments* (2001); Mary H. Cooper, "Air Pollution Conflict," *CQ Researcher*, Nov. 14, 2003, pp. 965-988; and Jennifer Weeks, "Coal's Comeback," *CQ Researcher*, Oct. 5, 2007, pp. 817-840.

[41] Michael E. Porter and Claas van der Linde, "Toward a New Concept of the Environmental-Competitiveness Issue," *Journal of Economic Perspectives*, vol. 9, no. 4, fall 1995, p. 99.

[42] For background see Tom Price, "The New Environmentalism," *CQ Researcher*, Dec. 1, 2006, pp. 985-1008.

[43] Daniel C. Esty and Andrew S. Winston, *Green to Gold: How Smart Companies Use Environmental Strategy to Innovate, Create Value, and Build Competitive Advantage* (2006), pp. 70-71.

About the Author

Jennifer Weeks is a *CQ Researcher* contributing writer in Watertown, Mass., who specializes in energy and environmental issues. She has written for *The Washington Post*, *The Boston Globe Magazine* and other publications, and has 15 years' experience as a public-policy analyst, lobbyist and congressional staffer. She has an A.B. degree from Williams College and master's degrees from the University of North Carolina and Harvard.

[44] Mark Dowie, *Losing Ground: American Environmentalism at the Close of the Twentieth Century* (1995), pp. 114-124.

[45] For details see "Exxpose Exxon," www.exxposeexxon.com.

[46] See Mary H. Cooper, "Energy Policy," *CQ Researcher*, May 25, 2001, pp. 441-464, and Mary H. Cooper, "Bush and the Environment," *CQ Researcher*, Oct. 25, 2002, pp. 865-896.

[47] Steve Nadis, "Non-Government Organizations (NGOs) Mini-Reviews," New England BioLabs, www.neb.com.

[48] For more information, see www.epa.gov/stateply/index.html and www.pewclimate.org/companies_leading_the_way_belc.

[49] For more information see www.ftc.gov/bcp/workshops/carbonoffsets/index.shtml

[50] U.S. Federal Trade Commission, "Complying With the Environmental Marketing Guides," www.ftc.gov/bcp/conline/pubc/buspubs/green guides.pdf.

[51] Auden Schendler and Randy Udall, "LEED Is Broken; Let's Fix It," *Grist*, October 26, 2005; Stephen Del Percio, "What's Wrong With LEED?" *Green Building*, spring 2007.

[52] McGraw Hill, *Green Building Smart Market Report 2006*, cited in "Green Building by the Numbers," U.S. Green Building Council, February 2008.

[53] U.S. Environmental Protection Agency, Office of the Inspector General, "Energy Star Program Can Strengthen Controls Protecting the Integrity of the Label," Aug. 1, 2007, www.epa.gov/oig/reports/2007/20070801-2007-P-00028.pdf.

[54] U.S. Government Accountability Office, "Energy Efficiency: Opportunities Exist for Federal Agencies to Better Inform Household Consumers," GAO-07-1162 (September 2007).

[55] American Council for an Energy-Efficient Economy, "Assessment of the Renewable Electricity Standard and Expanded Clean Energy Scenarios," Dec. 5, 2007, http://aceee.org/pubs/e079.htm.

[56] The full letter is posted online at http://gristmill.grist.org/images/user/8/White_House_letter_on_CAFE.pdf.

[57] For background see Peter Katel, "Oil Jitters," *CQ Researcher*, Jan. 4, 2008, pp. 1-24, and Adriel Bettelheim, "Biofuels Boom," *CQ Researcher*, Sept. 29, 2006, pp. 793-816.

[58] Joseph Fargione, *et al.*, "Land Clearing and the Biofuel Carbon Debt," *Sciencexpress Report*, Feb. 7, 2008; Timothy Searchinger, *et al.*, "Use of U.S. Croplands for Biofuels Increases Greenhouse Gases Through Emissions from Land Use Change," *Sciencexpress Report*, Feb. 7, 2008.

[59] Randy Schnepf, "Agriculture-Based Renewable Energy Production," Congressional Research Service, Oct. 16, 2007, pp. 16-20; Colin A. Carter and Henry I. Miller, "Hidden Costs of Corn-Based Ethanol," *The Christian Science Monitor*, May 21, 2007; "Food Prices: Cheap No More," *The Economist*, Dec. 6, 2007.

[60] Bernard Simon, "Prius Overtakes Explorer in the U.S.," *Financial Times*, Jan. 11, 2008.

[61] "Chevy Volt FAQs," www.gm-volt.com/chevy-volt-faqs.

[62] Campaign for Safe Cosmetics, "A Poison Kiss: The Problem of Lead in Lipstick," October 2007, www.safecosmetics.org.

[63] Sheela Sathyanarayana, *et al.*, "Baby Care Products: Possible Sources of Infant Phthalate Exposure," *Pediatrics*, February 2008.

[64] Brangien Davis and Katharine Wroth, eds., *Wake Up and Smell the Planet: The Non-Pompous, Non-Preachy Grist Guide to Greening Your Day* (2007), p. 20.

[65] "Compare the Candidates," *Grist*, www.grist.org/candidate_chart_08.html.

[66] Jeffrey Ball, "Exxon Mobil Softens Its Climate-Change Stance," *The Wall Street Journal*, Jan. 11, 2007.

[67] Jenny Shank, *op. cit.*

[68] For example, see Mindy Pennybacker, "Local or Organic? I'll Take Both," *The Green Guide*, September/October 2006, www.thegreenguide.com/doc/116/local, and John Cloud, "Eating Better Than Organic," *Time*, March 2, 2007.

FOR MORE INFORMATION

American Council for an Energy-Efficient Economy, 1001 Connecticut Ave., N.W., Suite 801, Washington, DC 20036; (202) 429-8873; www.aceee.org. Supports energy-efficiency measures to promote economic prosperity and environmental protection.

Clean Air-Cool Planet, 100 Market St., Suite 204, Portsmouth, NH 03801; (603) 422-6464; www.cleanair-coolplanet.org. A nonprofit organization that partners with businesses, colleges and communities throughout the Northeast to reduce carbon emissions and educate the public and opinion leaders about global warming impacts and solutions.

Climate Trust, 65 SW Yamhill St., Suite 400, Portland, OR 97204; (503) 238-1915; www.climatetrust.org. Created to implement an Oregon law that requires new power plants to offset some of their carbon emissions, the Climate Trust produces greenhouse gas offset projects for energy companies, regulators, businesses, and individuals.

Competitive Enterprise Institute, 1001 Connecticut Ave., N.W., Suite 1250, Washington, DC 20036; (202) 331-1010; www.cei.org. A public policy research center dedicated to advancing the principles of free enterprise and limited government.

Consumers Union, 101 Truman Ave., Yonkers, NY 10703; (914) 378-2000; www.consumersunion.org. A nonprofit expert group that promotes a fair and safe market for all consumers; activities include testing and rating products and publishing *Consumer Reports* magazine, as well as GreenerChoices.org, a Web site focusing on green products.

Federal Trade Commission, 600 Pennsylvania Ave., N.W., Washington, DC 20580; (202) 326-2222; www.ftc.gov. Protects consumers' interests, promotes competition and advises businesses on eco-labeling; it is currently reviewing its green marketing guidelines.

LOHAS Forum, 360 Interlocken Blvd., Broomfield, CO 80021; (303) 822-2263; www.lohas.com. An annual business conference focused on the marketplace for goods and services related to health, the environment, social justice, personal development and sustainable living.

National Association of Manufacturers, 1331 Pennsylvania Ave., N.W., Washington, DC 20004; (202) 637-3000; www.nam.org. Promotes legislation and regulations conducive to economic growth and highlights manufacturers' contributions to innovation and productivity.

TerraChoice Environmental Marketing Inc., 1706 Friedensburg Road, Reading, PA 19606; (800) 478-0399; www.terrachoice.com. Conducts market research and advises on strategy, communication and policy issues.

Bibliography

Selected Sources

Books

Brower, Michael, and Warren Leon, *The Consumer's Guide to Effective Environmental Choices*, **Three Rivers Press, 1999.**

Although somewhat dated, this guide prioritizes consumer actions according to the scale of their environmental impacts based on extensive data and analysis. Brower and Leon, both senior environmental experts, draw on research by the Union of Concerned Scientists, a national environmental advocacy group.

Esty, Daniel C., and Andrew S. Winston, *Green To Gold: How Smart Companies Use Environmental Strategy to Innovate, Create Value, and Build Competitive Advantage*, **Yale University Press, 2006.**

Two Yale experts on business and the environment show how green strategies can help companies manage environmental challenges and gain an edge over competitors.

Szasz, Andrew, *Shopping Our Way to Safety: How We Changed from Protecting the Environment to Protecting Ourselves*, **University of Minnesota Press, 2007.**

Szasz, a sociologist, warns that the current green consumption boom could have negative impacts if it turns people away from broader political action.

Articles

"Climate Business/Business Climate," *Harvard Business Review*, **October 2007.**

A special report on the business challenges posed by climate change offers views from a dozen corporate and academic experts.

Davenport, Coral, "A Clean Break in Energy Policy," *CQ Weekly*, **Oct. 8, 2007.**

A national renewable electricity portfolio standard would trigger widespread changes in the ways that utilities produce power and state regulators oversee them.

Elgin, Ben, "Little Green Lies," *Business Week*, **Oct. 29, 2007.**

Auden Schendler, environmental director for Aspen Skiing Co., argues that many corporate greening actions are misleading and empty feel-good gestures.

Farenthold, David A., "Value of U.S. House's Carbon Offsets is Murky," *The Washington Post*, **Jan. 28, 2008.**

Critics say Congress wasted money by buying carbon offsets that funded activities already occurring.

Finz, Stacy, "Food Markets Getting Greener, More Sensual," *San Francisco Chronicle*, **Jan. 27, 2008.**

Consumers want healthier food raised using eco-friendly methods, and the grocery industry is responding.

Koerner, Brendan I., "Rise of the Green Machine," *Wired*, **April 2005.**

Koerner explains how Toyota made it cool to own a hybrid car.

Lynas, Mark, "Can Shopping Save the Planet?" *The Guardian* **(United Kingdom), Sept. 17, 2007.**

Numerous corporations are entering the green product market, but observers argue that at heart green marketing is all about sales, not sustainability.

Schultz, Abby, "How To 'Go Green' on a Budget," MSN **Money, June 29, 2007.**

Many green products are more expensive than conventional options, but consumers can make a difference if they choose their purchases carefully.

Underwood, Anne, "The Chemicals Within," *Newsweek*, **Feb. 4, 2008.**

Many common household products contain chemicals that could be harmful to humans. Concerns about health effects are driving many shoppers to seek alternatives.

Williams, Alex, "Don't Let the Green Grass Fool You," *The New York Times*, **Feb. 10, 2008.**

Many suburban Americans would like to shrink their carbon footprints, but skeptics argue that a lifestyle centered on big houses and multiple cars is inherently unsustainable.

Reports and Studies

A Consumer's Guide to Retail Carbon Offset Providers, **Clean Air-Cool Planet, 2006, www.cleanair-coolplanet. org/ConsumersGuidetoCarbonOffsets.pdf.**

An advocacy group that helps businesses, universities and cities and towns reduce greenhouse gas emissions describes key factors that contribute to the quality of carbon offsets and identifies some of the most credible offset providers.

Makower, Joel, *et al.***,** *State of Green Business 2008*, **January 2008, www.stateofgreenbusiness.com.**

A report on the spread of green business practices finds that companies are gradually becoming more eco-friendly, but economic growth is offsetting many of the gains, and that the trend is very hard to quantify.

TerraChoice Environmental Marketing, "The Six Sins of Greenwashing," November 2007, www.terrachoice.com/ Home/Six%20Sins%20of%20Greenwashing.

A study of environmental claims in North American consumer markets finds that virtually all purportedly eco-friendly products mislead consumers to some degree. More accurate green marketing, it asserts, will benefit consumers, businesses and the environment.

The Next Step:

Additional Articles from Current Periodicals

Activism

DeBare, Ilana, "Clorox Introduces Green Line of Cleaning Products," *San Francisco Chronicle*, Jan. 14, 2008.

Longtime bleach producer Clorox is launching a new series of natural, biodegradable household cleaning products called Green Works, co-branded with the Sierra Club.

McKibben, Bill, "Global Warming Knocking At Your Door," *The Boston Globe*, March 20, 2007.

Environmental activist McKibben sees growing support at the grass roots for global action to slow climate change.

Somaiya, Ravi, "Possum Fur, Eco-Limos, and Solar Sensations," *The Guardian* (U.K.), Sept. 20, 2007.

Creative examples of green marketing, from sustainable nightclubs to wind-up MP3 players.

Greenwashing

Mitchell, Dan, "Being Skeptical of Green," *The New York Times*, Nov. 24, 2007, p. 5C.

A recent report by a green marketing firm finds that although many of the claims associated with green products are not "demonstrably false," they are vague and misleading to consumers.

Yap, Chuin-Wei, "When It Comes to a "Green" House, Buyer Beware," *St. Petersburg Times*, Dec. 30, 2007, p. 1B.

With no legal definition of building "green," builders can market themselves as eco-friendly without third-party certification.

Green Consumption

Davidson, Paul, "Getting gold of green; Companies learn eco-friendliness helps bottom line," *USA Today*, April 19, 2007, p. 7A.

U.S. companies' investments in environmentally conscious programs prove lucrative, as the "green' lifestyle continues to attract Americans.

Higgins, Adrian, "A Bumper Crop of Organic Items for the Green Consumer," *The Washington Post*, Jan. 17, 2008, p. H1.

Larger corporations are increasingly staking a claim in the organic marketplace traditionally defined by smaller enterprises.

Williams, Alex, "Buying into the Green Movement," *The New York Times*, July 1, 2007, p. D1.

Critics question whether the damage of massive consumption will counteract the benefits of green consumerism.

Carbon Emissions

Clayton, Mark, "Senate energy bill: first skirmish over US greenhouse-gas regulation," *The Christian Science Monitor*, Dec. 14, 2007, p. 1.

The Environmental Protection Agency's authority to regulate greenhouse gas emissions is in jeopardy.

Davidson, Paul, "States take on global warming; More than half target emissions," *USA Today*, Jan. 21, 2008, p. 1A.

Although national legislation to cap greenhouse gas emissions has stalled, many states are acting individually by setting their own targets.

Carbon Offsets

Daley, Beth, "Carbon Confusion: Buying emission offsets is a challenge for consumers," *The Boston Globe*, March 13, 2007, p. A1.

In an unregulated market, consumers do not know what offsets projects their money is funding.

Deutsch, Claudia H., "Attention Shoppers: Carbon Offsets in Aisle 6," *The New York Times*, March 7, 2007, p. 1H.

Environmentalists fear offsets will allow consumers to make environmentally irresponsible decisions without guilt.

Story, Louise, "F.T.C. Asks If Carbon-Offset Money Is Winding Up True Green," *The New York Times*, Jan. 9, 2008, p. C1.

The Federal Trade Commission is growing increasingly concerned that many carbon-offset programs have the potential for deception.

Zimmerman, Eilene, "Undoing Your Daily Damage to the Earth, for a Price," *The New York Times*, Nov. 11, 2007, p. C5.

The market for carbon offsets continues to grow.

Citing CQ Researcher

Sample formats for citing these reports in a bibliography include the ones listed below. Preferred styles and formats vary, so please check with your instructor or professor.

<u>MLA STYLE</u>

Jost, Kenneth. "Rethinking the Death Penalty." <u>CQ Researcher</u> 16 Nov. 2001: 945-68.

<u>APA STYLE</u>

Jost, K. (2001, November 16). Rethinking the death penalty. *CQ Researcher, 11,* 945-968.

<u>CHICAGO STYLE</u>

Jost, Kenneth. "Rethinking the Death Penalty." *CQ Researcher,* November 16, 2001, 945-968.

11

REDUCING YOUR CARBON FOOTPRINT

BY THOMAS J. BILLITTERI

Excerpted from the CQ Researcher. Thomas J. Billitteri. (December 5, 2008). "Reducing Your Carbon Footprint." *CQ Researcher*, 985-1008.

Reducing Your Carbon Footprint

By Thomas J. Billitteri

The Issues

When Karen Larson, a mother of two in Madbury, N.H., took the "New Hampshire Carbon Challenge" she couldn't believe the results. [1]

The statewide effort to help residents reduce their environmental impact includes an online calculator to help consumers measure their "carbon footprint" — the amount of carbon dioxide (CO_2) created by their activities and consumption patterns. Carbon dioxide is the main greenhouse gas (GHG) that scientists believe leads to global warming. [2]

The calculator showed that Larson could save $700 a year and cut her carbon footprint by some 4,400 pounds by taking such actions as replacing her lightbulbs with compact fluorescents, cutting back on showers by a few minutes, putting electronics on power strips that she turned off when not in use, lowering her furnace a few degrees and getting an annual tune-up on her heating system.

"It blew my mind when I finished and it told me how much money per year I could save," Larson says. But the financial savings were only a "side benefit" to a larger objective, she adds. "The main goal was to find out how to be more earth-friendly and not use so many resources."

From voluntary actions like Larson's to emerging federal policies promoted by President-elect Barack Obama, more and more attention is shifting to the impact consumers have on the environment and how individuals can lower the amount of carbon dioxide emitted

Early-bird shoppers crowd into a Best Buy store in Los Angeles at 5 a.m. on Nov. 28 for post-Thanksgiving bargains. Concern about climate change, coupled with the nation's economic woes, is causing many Americans to ratchet back on their consumption of goods and services. But many environmentalists say government must also do its part to reduce carbon emissions by enacting tough, environmentally friendly policies.

by the production, transportation, use and disposal of the goods and services they consume.

Studies show the amount of CO_2 emitted by individuals can rival that of industry and commerce. For example:

- American consumers control — directly or indirectly — about two-thirds of the nation's greenhouse-gas emissions, compared to 43 percent for consumers elsewhere. [3] Passenger cars account for 17 percent of U.S. emissions, as do residential buildings and appliances. [4]
- Transportation accounts for a third of carbon output in the United States, 80 percent of it from highway travel. [5]

- The average home creates more pollution than the average car, according to the Environmental Protection Agency (EPA). [6] If every U.S. home replaced a single incandescent lightbulb with a compact fluorescent, it would prevent the equivalent of the GHG emissions from more than 800,000 cars. [7]

"Individual actions can make a significant difference," says Bill Burtis, communications manager for Clean Air-Cool Planet, an environmental group in Portsmouth, N.H. "We're not going to be able to solve the problem without a concerted and unified effort at reaching individuals and changing behavior in residential sectors, whether it's the compact fluorescent lightbulb or changing our use of two-cycle, highly polluting gas-powered lawn equipment."

However, environmental advocates argue that while individual carbon-fighting actions are important, putting too much attention on voluntary actions by individuals can undermine efforts to pass mandatory government policies to control emissions.

"Every time an activist or politician hectors the public to voluntarily reach for a new bulb or spend extra on a Prius, ExxonMobil heaves a big sigh of relief," Mike Tidwell, director of the Chesapeake Climate Action Network, a grassroots group in Takoma Park, Md., wrote last year. [8]

"While . . . we have a moral responsibility to do what we can as individuals, we just don't have enough time to win this battle one household at a time," he wrote. "We must change our laws. I'd rather have 100,000 Americans phoning their U.S. senators twice per week demanding a prompt phase-

Big Metro Areas Have Smaller Carbon Footprints

Big-city areas like Los Angeles emit less carbon per capita than smaller areas like Knoxville, Tenn., in large part because urban areas have high-density development patterns or depend on low-polluting mass transit.

Highest and Lowest Carbon Emissions in Metro Areas

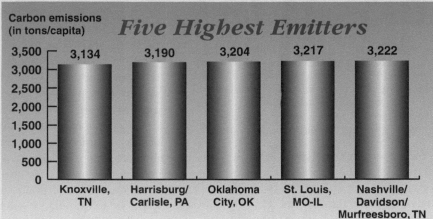

Source: "Shrinking the Carbon Footprint of Metropolitan America," Brookings Institution, May 2008

motivated by a combination of the two. (*See graph, p. 991.*) [9]

And according to a Harris Poll last spring, more than 60 percent of respondents said they had cut their home energy use to offset their carbon footprint or reduce their emissions, and 43 percent said they'd bought more energy-efficient appliances. [10]

But such findings should be viewed warily, say public opinion researchers. Soaring energy prices, reflected last summer in $4-per-gallon gasoline prices, are likely a stronger motivator for many Americans to cut their carbon appetites than concern about climate change, they say. And only about one in 10 respondents to the Harris Poll said they had ever calculated their personal or household carbon footprint, and more than one in four said they were doing nothing to reduce emissions.

"Experience suggests that we should be somewhat skeptical of claims people make about doing the 'right thing,'" Harris noted in an analysis of its poll results. While the polling company found it "encouraging" that so many Americans feel it is important to reduce their carbon emissions, U.S. energy consumption keeps rising, leading Harris to conclude that "whatever actions people are taking are probably modest ones." [11]

The trend toward larger homes, for instance, spurs greater energy use and higher personal carbon output. The average size of a new single-family home in the United States more than doubled from 1950 to 2005 — from 983 square feet to 2,434 — while the average number of occupants fell 22 percent between 1950 and 2000. [12] Between 1990 and 2006, residential carbon dioxide emissions jumped 26 percent, outpacing the 20 percent growth in population. [13]

For many consumers — even those committed to improving the environment — cutting back on carbon isn't always easy. For one thing, figuring out one's environmental footprint can

out of inefficient automobile engines and lightbulbs than 1 million Americans willing to 'eat their vegetables' and voluntarily fill up their driveway and houses with the right stuff."

Nonetheless, experts say concern about climate change, coupled with worries about the faltering economy, are prompting more and more consumers to make environmentally friend-

ly choices. An ABC News/Planet Green/Stanford University poll in July found that 71 percent of respondents said they were trying to reduce their carbon footprint, mainly by driving less, using less electricity and recycling. A fourth of them said saving money was their primary goal, a third said improving the environment was their chief aim and 41 percent said they were

be a challenge. Should a footprint consist only of carbon-based fossil fuels, such as those that run most cars and power plants, or should it also include non-carbon-based greenhouse gases like nitrous oxide emitted by fertilizers? Should it include methane gas released by landfills and from the digestive tracts of cattle? How should people measure their share of emissions stemming from the manufacture and transportation of consumer goods? And how does one best measure his share of emissions stemming from travel in airplanes or trains?

"Despite its ubiquitous appearance, there seems to be no clear definition of [carbon footprint] and there is still some confusion [over] what it actually means and measures," noted a research firm in Great Britain, where studies on climate change and consumers' environmental choices have been robust for years. [14]

Consumers also must figure out which actions actually help the environment and which only seem to. For example, in a list of 10 "green heresies," *Wired* pointed out that conventional agriculture can be more environmentally friendly than organic farming. "Organic produce *can* be good for the climate, but not if it's grown in energy-dependent hothouses and travels long distances to get to your fridge," the magazine noted. [15]

And of course, the geographic realities of American life — homes and jobs separated by miles — can thwart the good intentions of even the most environmentally conscious consumer.

"When we don't live close to where we work, there's not much choice in the matter," says Duane T. Wegener, a professor of psychological sciences at Purdue University who studies the social aspects of energy policy. The ways most cities are built "tie us into a particular level of energy use, and there's not a lot of control over it."

As consumers and policy makers continue to seek ways to reduce

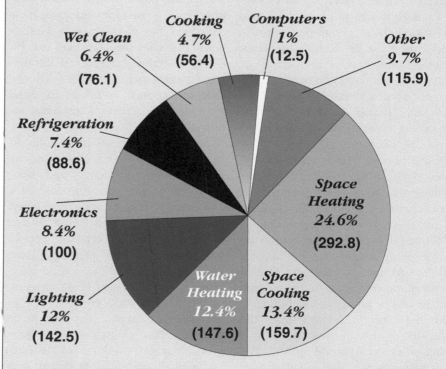

Home Heating and Cooling Emit Most Carbon

Space heating and cooling alone account for more than a third of all the energy used in residential buildings in the United States and more than 450 million metric tons of carbon dioxide.

Carbon Dioxide Emissions From Residential Energy Use
(by percentage and in million metric tons)

Cooking 4.7% (56.4)
Computers 1% (12.5)
Wet Clean 6.4% (76.1)
Other 9.7% (115.9)
Refrigeration 7.4% (88.6)
Electronics 8.4% (100)
Space Heating 24.6% (292.8)
Lighting 12% (142.5)
Water Heating 12.4% (147.6)
Space Cooling 13.4% (159.7)

Source: "Buildings Energy Data Book," U.S. Department of Energy, November 2008

harmful, globe-warming pollutants, here are some of the questions they are asking:

Are measures of individual carbon emissions valid?

Measuring one's carbon footprint is both art and science.

A growing number of agencies and organizations — from the federal Environmental Protection Agency to the nonprofit Nature Conservancy to the Berkeley Institute of the Environment at the University of California — offer online calculators intended to help people measure their greenhouse-gas emissions.

The calculators gauge the environmental impact of various activities, from heating and lighting a home to flying across the country, and guide consumers on how to reduce their carbon output. But the calculators vary widely in their detail and conclusions.

Air travel, a significant emission source, is a case in point. Calculating a per-passenger carbon footprint is a highly complex task that must take into account such factors as aircraft type, weather conditions, the number of landings and takeoffs along a given route and whether an aircraft flies with a full or partial load.

In June the Montreal-based United Nations International Civil Aviation Organization (ICAO) introduced a carbon calculator that "responds to the wish of many travelers for a reliable and authoritative method to estimate the carbon footprint of a flight," said Robert Kobeh González, president of the ICAO Council. [16]

But the calculator can produce misleading data, according to an official of a company that produces fuel data and supplied the aircraft-performance model used for the ICAO's estimates of airline emissions.

"Producing a single number is crude," Dimitri Simos, a director at Lissys Limited, said. "If you go from Heathrow [airport in London] to Athens, ICAO gives 217 kilograms [478 pounds] of CO_2 [per person]. That hides huge variations. Fly in a full [Boeing] B767 and it's nearer to 160 kg [352 pounds] per person, or in a half-empty [Airbus] A340 it's more like 360 kg [793 pounds]." [17]

While carbon calculators help make people more aware of their individual environmental impact, Daniel Kammen, a professor in the energy and resources group at UC Berkeley, told the *Chicago Tribune* that "the downside is that the methodology is being worked on as we speak." [18]

Still, many see the calculators as useful. Elise L. Amel, director of environmental studies and associate professor of industrial and organizational psychology at the University of St. Thomas in Minneapolis, says that while most carbon calculators are "not very fine-grained," they still serve a valuable purpose.

"Usually the options for somebody's response [to a calculator's questions] don't necessarily perfectly match any one individual situation," she says. Most calculators "use general categories" and give a "rough estimate . . . so it's no wonder they all give you a little something different. But I don't think they should be used to necessarily diagnose each minute activity that you should

adjust. They're just to get people aware that, 'Wow, we're using resources at a rate that's really hard to replace.' "

Making people think about their carbon footprint — and doing something about it — is the idea behind the New Hampshire Carbon Challenge and its calculator. [19]

The challenge, which has been online for a year, enables New England residents to measure their carbon footprints, set goals for reducing their emissions and pledge action. Households that take the challenge can be linked to other households through organizations such as churches, schools, civic group and businesses to show the collective action of residents' individual efforts.

So far, about 1,000 households in New Hampshire and Massachusetts have taken the challenge, and thousands more have used the calculator and other Web-based tools developed by the challenge, says Denise Blaha, the organization's codirector. People who have taken the challenge have reduced their home and vehicle energy use by an average of 17 percent, saving roughly $850 annually in fuel and electricity costs, she says. More than 5 million pounds of CO_2 have been pledged for reduction as a result of the challenge, with an energy cost saving of $700,000, she says.

"Households have really been the overlooked sector," Blaha says. "Most of our actions are very, very concrete, real-world and simple to make: changing lightbulbs, putting electronic devices on power strips and turning down your thermostat."

Still, while swapping energy-hogging lightbulbs for miserly compact fluorescents may be simple, some efforts at environmental responsibility can be maddeningly complex.

For example, some consumers try to factor in "food miles" — the distance food travels from farm to table — into their carbon-footprint calculations. Conventional wisdom says food

travels an average of 1,500 miles from farm to plate, Jane Black, a *Washington Post* food writer, noted in the online journal *Slate*. [20] But, she wrote, the ubiquitous figure is based on a university study of how far 33 fruits and vegetables grown in the United States traveled to a Chicago produce market. That figure, though it perhaps raises consumers' awareness of the environmental issues involved in food choices, oversimplifies the complex and global nature of the food industry, Black argued.

"If we all think in food miles, the answer is obvious: Buy local. But new studies show that in some cases it can actually be more environmentally responsible to produce food far from home."

Black cited a 2006 report from Lincoln University in New Zealand, which concluded that it was four times more energy efficient for people in London to purchase New Zealand lamb, which feeds on grass, than grain-fed lamb from England.

She also noted that Tesco, the British grocery giant, has begun adding carbon labels to its products. The labels — which reveal how much carbon was emitted by an item's production, transportation and consumption — are now on an initial 20 products, including orange juice and detergent. [21]

But, Black wrote, "Like food miles, these new numbers raise as many questions as they answer."

Should government do more to encourage individuals to reduce their carbon footprints?

While individuals can do much to reduce their carbon consumption and conserve precious resources, environmental advocates say policy makers should provide more incentives to further those efforts.

"Government needs to take a holistic approach to the whole carbon footprint issue," says Eric Carlson, founder and president of Carbonfund.org, a nonprofit organization in Silver Spring, Md., that sells "carbon offsets" — certificates that consumers and businesses can purchase voluntarily to help compensate for their own carbon footprints. The certificates are used to subsidize "green" efforts such as renewable-energy, reforestation and energy-efficiency projects.

Carlson says the federal government has a number of piecemeal consumer programs that provide incentives for greater energy efficiency, such as tax credits for home insulation and the purchase of hybrid cars. But he says Washington should adopt much broader policies aimed at encouraging business and consumers to transition from carbon-emitting energy to green power.

He suggests that the government adopt a "cap-and-trade" climate policy that would set limits on how much carbon dioxide energy producers, utilities or manufacturers could emit. (*See Current Situation, p. 1000.*) The European Union and a handful of other countries have adopted similar carbon-trading systems since 2005. [22] Companies that emit less than allowable amounts could sell their excess "rights to pollute" to others. In some versions of the scheme the rights are auctioned to companies, and the auction revenue is returned to taxpayers or used for renewable-energy development.

Legislation sponsored by Sens. Joseph I. Lieberman, I-Conn., and John W. Warner, R-Va., featured a cap-and-trade system, but the Senate rejected it earlier this year. President-elect Obama favors such a system.

Some have argued that cap-and-trade plans could raise energy costs, squeeze consumers and hurt the economy more than help the environment. Cap-and-trade bills are "nothing short of government re-engineering of the American economy," wrote Ben Lieberman, a policy analyst at the conservative Heritage Foundation. [23]

Yet others say the approach would be good for the environment, partly because it would tend to make

Americans Seeking to Reduce Footprint

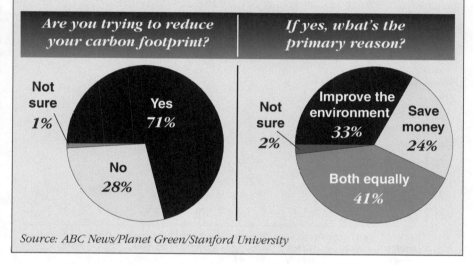

More than 70 percent of Americans say they are trying to reduce their carbon footprints. Of those, a third say they want to improve the environment, while a quarter are seeking to save money.

Are you trying to reduce your carbon footprint?

Not sure 1%
Yes 71%
No 28%

If yes, what's the primary reason?

Not sure 2%
Improve the environment 33%
Save money 24%
Both equally 41%

Source: ABC News/Planet Green/Stanford University

goods and services created and transported in carbon-intensive ways more expensive and lower the relative price of items produced in less carbon-intensive ways. Producers and consumers would have incentives to make more environmentally friendly choices.

Including the cost of carbon emissions in the price of goods and services would give consumers an idea of the carbon intensity of the items they buy and motivate them to make greener choices, says Eric Haxthausen, director of U.S. climate change policy for the Nature Conservancy.

He gives the example of buying fruit in a grocery store. "There's a lot of emphasis on buying local," he says. "It's tricky, but the beauty of the cap-and-trade approach is that it's all worked out through the pricing system. You don't have to calculate whether this fruit was shipped on an ocean-going vessel" and therefore had a bigger carbon footprint to reach the grocery shelf than fruit grown closer to home. The price "would distinguish between a high and low" footprint.

Consumers need more information about the environmental consequences of the goods and services they use, according to Rep. Brian Baird, D-Wash., chairman of the House Subcommittee on Research and Science Education. For instance, he told the panel at a hearing last year, households consume more than a third of the energy used in the United States each year, 60 percent of which is used at home. Yet consumers typically lack the information they need to factor energy use into purchases and behaviors, he continued, and both government and industry have "fallen far short in providing the needed information to the public in a way that will result in behavior changes." [24]

At the hearing — on how the social sciences can help solve the nation's energy problems — John A. "Skip" Laitner, senior economist for technology policy at the American Council for an Energy-Efficient Economy, said more funding was needed "to expand our understanding of the social dynamics of energy consumption, energy conservation and energy efficiency."

Efforts Focus on Home Energy Use

Nearly two-thirds of Americans are reducing home energy usage, including 43 percent who have bought more energy-efficient appliances, according to a recent survey. Nevertheless, more than a quarter are doing nothing to offset their carbon footprints.

Which of the following have you done in an attempt to offset your carbon footprint or reduce your emissions?

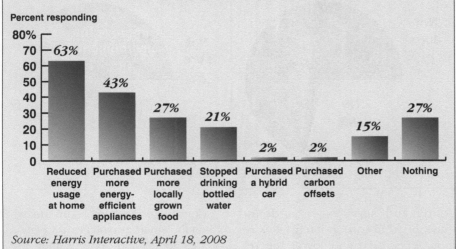

Source: Harris Interactive, April 18, 2008

(*See sidebar, p. 996.*) Funding for non-economic social-science research on energy consumption has fallen sharply since the 1980s, he said. [25]

Advocates also say part of the answer to shrinking individual environmental footprints is for government authorities, including those on the state level, to adopt stronger energy-efficiency building standards for such things as thermal insulation and heating and cooling systems. While such codes are "unsexy," Carlson says, they are highly effective in cutting energy consumption and reducing carbon emissions.

Many environmental advocates also call for a wider array of subsidies and tax incentives for energy-efficient appliances and alternative-energy equipment.

California is widely viewed as one the states most determined to promote energy efficiency and clean-energy technology. For example, under a 10-year "Go Solar California" rebate program, the state plans to inject more than $3 billion into financial incentives

for Californians to adopt solar energy. The money, which comes from utility bills, is expected to pay for about 3,000 megawatts of new solar energy. [26]

But the program has sparked controversy in some quarters, as the *San Francisco Chronicle* noted last summer. The rebates are aimed at cutting solar prices over time by creating a big and competitive solar industry in the state, the newspaper said, but "solar remains far more expensive than many other forms of power generation."

Moreover, a consumer watchdog group complained that wealthy companies or homeowners have received too many of the rebates, the *Chronicle* reported. Mark Toney, executive director of The Utility Reform Network, said "subsidies should go to the most needy, not the most wealthy." [27]

A spokesman for the California Public Utilities Commission said in mid-November that a program to direct some solar-rebate money to low-income housing was in "the talking stage."

Can individual action significantly reduce global climate change?

In September the nonprofit Pew Center on Global Climate Change and the Alcoa aluminum company foundation kicked off a partnership aimed at encouraging the company's U.S. employees and their local communities to reduce their energy consumption and shrink their carbon footprints.

Pew and Alcoa initially are rolling out the program at nine of the company's 120 U.S. locations, with plans to expand it. The Make an Impact program, modeled after a similar Alcoa effort in Australia, includes an interactive Web site that offers a carbon calculator and other tools to help employees cut their energy bills and emissions.

Concern about energy efficiency and climate change "is a huge focus inside our business, and this is about translating that into a tool that can be used to increase awareness and actions on climate change" among Alcoa's employees and their communities, says Libby Archell, director of communications for the Alcoa Foundation.

Katie Mandes, vice president of communications at the Pew center, says efforts focused on individuals can make a significant difference in curbing greenhouse-gas emissions. "Are individuals going to be able to solve this problem? No," she says. "But is there a role for individuals? Yes. . . . It's got to be an across-the-board change."

Individuals and households can play a larger role in cutting energy use and emissions than many consumers realize. "U.S. households account for about 38 percent of national carbon emissions through their direct actions, a level of emissions greater than that of any entire country except China and larger than the entire U.S. industrial sector," wrote Gerald T. Gardner, professor emeritus of psychology at the University of Michigan-Dearborn, and Paul C. Stern, director of the Committee on the Human Dimensions of Global

Climate Change at the National Research Council. [28] "By changing their selection and use of household and motor-vehicle technologies without waiting for new technologies to appear, making major economic sacrifices or losing a sense of well-being, households can reduce energy consumption by almost 30 percent — about 11 percent of total U.S. consumption."

A study by graduate students at the University of Michigan's School of Natural Resources and Environment a decade ago underscored how cost-effective changes in construction methods and other choices can reduce residential energy consumption.

A conventionally built 2,450-square-foot home in Ann Arbor, Mich., had a global-warming potential equal to 1,013 metric tons of CO_2 over its life span — from construction to use and, ultimately, to demolition — with 92 percent of that occurring during the home's use phase. An energy-efficient home, meanwhile, had a global-warming potential of only 374 metric tons of CO_2 — 67 percent of it occurring in the home's use phase. And although the energy-efficient home cost about 10 percent more than the conventional home, over its lifespan the more efficient home wound up costing about the same as the less efficient one. [29]

And at today's energy prices, the life-cycle costs are substantially lower for the efficient house, according to Gregory Keoleian, codirector of the Center for Sustainable Systems at the University of Michigan.

Refrigeration accounts for 8 percent of average U.S. residential energy consumption. [30] A separate study by the center found that households can save significant costs and energy by replacing refrigerators made before 1995.

Though many consumers still use old energy-hungry appliances, many others own units carrying the government's "Energy Star" label denoting energy efficiency. Last year alone, the joint program between the Envi-

ronmental Protection Agency and Energy Department helped Americans save enough energy to avoid greenhouse-gas emissions equivalent to the output of 27 million cars while saving $16 billion on utilities. [31]

Even so, the program has drawn some criticism recently. *Consumer Reports* magazine in October questioned the Energy Star label on several refrigerator models and said the program's qualifying standards are lax and its testing procedures dated. Moreover, the article pointed out, the program allows companies to test their own products, with the government relying on manufacturers to evaluate appliances made by competitors and to report suspicious claims about energy use. [32]

Many environmentalists say that while individual carbon-reducing actions, including purchasing energy-efficient appliances, are important in fighting GHG emissions, government action is crucial.

"We need fundamental change and strong policy to drive those changes," says Joe Loper, vice president of policy and research at the Alliance to Save Energy, an advocacy group in Washington. "If I take all those actions and reduce my footprint and no one else does it, it's not going to matter."

Tidwell, of the Chesapeake Climate Action Network, says living a low-carbon lifestyle is good, but "if that becomes the dominant core of our response to climate change, then we fail. We might as well do nothing."

"Instead of wagging our finger at Aunt Betty and telling her she needs to change all her lightbulbs, what we need to do is pass laws in this country so the next time Aunt Betty goes to buy lightbulbs, there are only energy-efficient lightbulbs. You do that with . . . statutory changes that send the right clean-energy market signals to the economy."

In fact, Congress and the Bush administration did take a step in that direction last year with passage of an energy bill that, among other things,

requires that traditional incandescent lightbulbs be phased out in favor of more-efficient ones by 2020. [33] But Tidwell asks, "Why wait 12 years?"

He compares the effort to fight GHG emissions to the 1960s-era battle for civil rights, noting that the government didn't rely on voluntary action to curb racial discrimination but instead banned it outright. "When it comes to climate change, we have to start thinking of this as the moral and economic immediate crisis that it is," he says. With that in mind, he adds, "you quickly begin to realize that we have to set strict energy standards that are fair but mandatory."

"We don't have time to green our neighborhoods one house at a time. Nature has chosen its own deadline." ∎

BACKGROUND

Environmental Awakening

Concern about society's impact on the environment has deep historical roots. In ancient Rome, garbage, sewage and the detritus of the metal-working and tanning industries polluted the water and air, turning the fetid atmosphere into what the Romans sometimes called *gravioris caeli* — "heavy heaven." [34]

In the 18th and 19th centuries, economists and political philosophers such as John Stuart Mill worried about the impact of population growth, industrialization and urbanization on food supplies, resources and natural beauty.

In 20th-century America, environmentalism and conservationism grew deep roots, nurtured by the preservation of lands for parks and monuments under Presidents Theodore Roosevelt (1901-1909) and William Howard Taft (1909-1913) and the formation of the Civilian Conservation Corps under President Franklin D. Roosevelt (1933-1945). [35] The Corps put hundreds of thousands of unemployed young adults, mostly men, to work planting trees, developing lakes and ponds, building logging roads, improving campgrounds and performing other tasks.

In the early 1960s, Rachel Carson, an environmental writer and marine biologist with the U.S. Fish and Wildlife Service, became what many regard as the chief catalyst of the modern environmental movement with publication of her 1962 book *Silent Spring*, which described the threats that the pesticide DDT posed to the environment and food chain. In an earlier work, *The Sea Around Us*, published in 1951, she presciently warned of a "startling alteration of climate" and "evidence that the top of the world is growing warmer." [36]

Other environmental milestones in the second half of the 20th century reinforced the idea that choices made not only by government and industry but also by individuals could have a profound influence on climate change, the sustainability of natural resources and, ultimately, the future of the human race. On the first Earth Day, April 22, 1970, millions of Americans participated in rallies and forums that helped thrust environmental concerns onto the political stage.

"It was on that day that Americans made it clear that they understood and were deeply concerned over the deterioration of our environment and the mindless dissipation of our resources," Earth Day's founder, the late Sen. Gaylord Nelson, D-Wis., wrote on the event's 10th anniversary. "That day left a permanent impact on the politics of America." [37]

Throughout the 1970s, a tide of environmental legislation flowed from Washington, including the Clean Air Act of 1970, Endangered Species Act of 1973 and Clean Water Act of 1977. Also, the Environmental Protection Agency was formed in the early 1970s under the Republican administration of President Richard M. Nixon.

But environmentalism was not always as durable as the millions who gathered on Earth Day 1970 had hoped. In the 1980s, for example, the Reagan administration drew heavy criticism for the actions of Interior Secretary James G. Watt and EPA Administrator Anne Gorsuch, both of whom resigned under fire.

Still, the 1980s saw a growing awareness of the consequences of both individual and industrial behavior on the global climate. For example, the Montreal Protocol, an international treaty that the Reagan administration and scores of other governments signed, called for the elimination of ozone-depleting chemicals called chlorofluorocarbons, found in such items as air conditioners and consumer aerosol products.

But getting a handle on carbon emissions has been a much tougher task. Richard A. Benedick, the Reagan administration's chief representative in the Montreal Protocol talks, noted in *The New York Times* in 2005 that tackling carbon dioxide emissions was different than phasing out chlorofluorocarbons, which the newspaper noted were produced by only a small number of companies in a few countries. [38]

"Carbon dioxide is generated by activities as varied as surfing the Web, driving a car, burning wood or flying to Montreal," *Times* environmental reporter Andrew C. Revkin pointed out. "Its production is woven into the fabric of an industrial society, and, for now, economic growth is inconceivable without it." [39]

'Planetary Emergency'

By the late 1990s and early 2000s the scientific community widely agreed that carbon dioxide and other greenhouse gases spelled a global climate catastrophe in the making. But many consumers and policy makers seemed to ignore warnings of catastrophic climate change. Domestic energy use continued to soar at the turn of the 21st century, gas-gulping sport-utility vehicles crowded the highways and conservative pundits belittled the notion that human activity was altering the climate in perilous ways.

In 2001 the Bush administration backed away from the Kyoto Protocol, an international treaty setting targets for industrialized nations to reduce their greenhouse-gas emissions, which the Clinton administration signed but never sent to the Senate for ratification. President George W. Bush complained that the treaty would damage the U.S. economy and that it didn't require developing countries to cut emissions. [40]

Still, other voices have persisted in raising alarms about climate change, none so urgently as former Vice President Al Gore. In a celebrated 2006 documentary film and book — *An Inconvenient Truth* — Gore warned of global peril stemming from rising greenhouse-gas emissions. "Not only does human-caused global warming exist," Gore wrote, "but it is also growing more and more dangerous, and at a pace that has now made it a planetary emergency." [41]

While industry and governments are key targets of Gore's message, individual Americans increasingly are hearing it too — sometimes from unlikely sources, including religious leaders. The Vatican announced this year that "polluting the environment" was among seven new sins requiring repentance. [42] And in March, 44 Southern Baptist leaders, including the current president of the conservative Southern Baptist Convention and two past presidents, signed a declaration supporting stronger action on climate change. [43] The year before, the convention had questioned the notion that humans are mainly responsible for global warming.

While individual action is widely viewed as important to protecting the environment, many still question how much impact consumers can have on

Continued on p. 996

Chronology

1890-1960
Conservation movement emerges, nurtured by government action, including passage of Air Quality Act in 1967.

1970s *Under pressure from environmentalists, Congress enacts landmark anti-pollution measures.*

1970
Millions gather for Earth Day on April 22. . . . Congress establishes Environmental Protection Agency and expands Clean Air Act.

1972
Government bans DDT.

1973
Endangered Species Act enacted.

1974
Safe Drinking Water Act enacted.

1978
New York State Department of Health declares public health emergency at Love Canal hazardous waste landfill site.

1979
Accident at Three Mile Island in Pennsylvania, the most serious in the history of U.S. commercial nuclear power plants, leads the Nuclear Regulatory Commission to tighten oversight.

1980s-1990s
Concern about greenhouse gas (GHG) emissions and climate change grows worldwide.

1983
Interior Secretary James Watt and EPA chief Anne Gorsuch resign amid environmentalists' criticism.

1987
Twenty-four nations initially sign Montreal Protocol, pledging to phase out ozone-depleting chemicals; 169 other countries eventually sign on.

1988
NASA scientist James Hansen warns Congress that global warming is occurring.

1989
Oil tanker *Exxon Valdez* runs aground, contaminating more than 1,000 miles of Alaskan coastline and killing hundreds of thousands of birds and other wildlife in the biggest U.S. oil spill in history.

1992
First international pledge to reduce greenhouse-gas emissions emerges from Earth Summit in Rio de Janeiro, Brazil.

1997
Kyoto Protocol limiting GHG emissions is approved, but wealthy nations are allowed to meet obligations by purchasing "offset" projects in developing countries. United States signs but does not ratify the agreement.

1998
U.S. Green Building Council starts Leadership in Energy and Environmental Design (LEED) program to rate energy-efficient buildings.

2000-Present
Bush administration relaxes strict environmental controls.

President-elect Barack Obama pledges to make environmental issues a central focus.

2001
President George W. Bush reneges on campaign pledge and rejects controls on greenhouse emissions.

2005
Kyoto Protocol takes effect. . . . Members of European Union begin trading carbon credits.

2006
"An Inconvenient Truth," a documentary film featuring former Vice President Al Gore, focuses national attention on global warming.

2007
Federal Trade Commission begins review of environmental marketing guidelines to include carbon offsets, renewable energy certificates and other so-called green products.

2008
British supermarket giant Tesco begins to print "green scores" on some items to show their environmental impact. . . . Senate rejects global warming bill that features a cap-and-trade system on greenhouse gas emissions. . . . Alcoa Aluminum Co. and Pew Center on Global Climate Change form partnership to help Alcoa employees and their communities reduce their carbon footprints. . . . President-elect Barack Obama says his presidency will "mark a new chapter in America's leadership on climate change." He pledges to develop a two-year economic-stimulus plan to save or create 2.5 million jobs, including "green" jobs in alternative energy and environmental technology.

Using Psychology to Influence Consumers' Behavior

Peer pressure proves potent.

Policy makers and environmentalists aren't the only ones trying to figure out how to coax consumers to shrink their carbon appetites. So too are psychologists.

Last summer the American Psychological Association formed a task force to address the role that psychology can play in helping individuals embrace environmentally sustainable practices and cope with the consequences of climate change.

"There hasn't been as much focus on the psychological impacts [of climate change], and we have reason to believe they'll be very serious," says task force member Susan Clayton, a professor of psychology and chair of environmental studies at the College of Wooster in Ohio.

If people have less access to "green, natural, healthy settings," she says, the result could be increased stress and aggression and diminished social interaction. Climate change also could lead to increased competition for dwindling resources such as food and water, sparking social conflict, she says.

When it comes to influencing individual behavior, Clayton says it is "more effective to change the structure of the situation" than trying to change people's minds through preaching.

That might mean providing recycling bins to households, not simply lecturing them on the merits of recycling. Letting consumers know how many of their neighbors are replacing their lightbulbs with compact fluorescents also can be effective, she says. "Social norms matter a lot. . . . Peer pressure never goes away" as an effective catalyst for influencing behavior, she says.

And, Clayton adds, providing the means for feedback on individual behavior — say, putting separate electric meters in apartment houses so tenants can monitor their individual power usage — can give consumers an incentive to conserve.

"Psychologists are increasingly becoming involved in helping alleviate environmental problems," according to Douglas Vakoch, an associate professor in the department of clinical psychology at the California Institute of Integral Studies in San Francisco.

"Most people recognize [that] we face a severe environmental crisis, but it's hard to deal with that head-on because many people feel helpless to do anything about it. . . . Psychologists are very experienced in dealing with denial and in helping to frame messages in ways that people can hear the bad news without being paralyzed by it."

Vakoch says policy experts and government leaders can learn from the psychology field. "The most important lesson . . . is that there is no 'one size fits all' solution to environmental problems," he says. "To create effective public policies, leaders need to recognize that different people are willing to adopt more environmentally sound behaviors for different reasons. What's compelling for one person will fail for another."

Psychological research has helped shed light on individuals' understanding of environmental issues and their willingness to make changes in their personal consumption habits.

Continued from p. 994
GHG emissions through their daily choices. Research offers competing views.

In a 2007 study provocatively titled "The Carbon Cost of Christmas," European researchers concluded that in the United Kingdom total consumption and spending on food, travel, lighting and gifts over three days of festivities — Christmas Eve, Christmas Day and the traditional U.K. holiday of Boxing Day — could result in as much as 650 kilograms [1,433 pounds] of CO_2 emissions per person — 5.5 percent of Britons' total annual carbon footprint and equivalent to the weight of 1,000 Christmas puddings. [44]

The researchers said consumers could cut their carbon emissions by more than 60 percent by taking such steps as cutting out unwanted gifts, buying "low-carbon" presents such as recycled wine glasses, reusing Christmas cards or using the phone or e-mail to send greetings, composting vegetable peelings when cooking Christmas dinner and reducing holiday lighting.

"In this time of seasonal goodwill, we should all spare a thought for the planet," they wrote.

But another study by researchers at the Massachusetts Institute of Technology pointed to "very significant limits" that voluntary lifestyle choices can have on energy use and carbon emissions. The researchers — MIT students under the direction of Timothy Gutowski, a professor of mechanical engineering — studied 18 different lifestyles ranging from that of a Buddhist monk, a retiree and a 5-year-old to a coma patient, pro golfer and investment banker. They found that even the most constrained lifestyle has an environmental impact far larger than the global average and that none of the lifestyles — including the most modest, that of a homeless person — ever resulted in CO_2 emissions below 8.5 metric tons annually, according to Gutowski. [45]

"The takeaway [from the study] is that we have a very energy-intensive system" that limits how much voluntary actions by individuals can affect climate change, Gutowski says. Still, he says, "I wouldn't want to say people shouldn't take voluntary action. It's a complicated path. Groups may take voluntary action, government agencies notice and then take actions that change the system."

Even so, individuals can sometimes innocently make matters worse by trying to be "green." An example is consumers' efforts to recycle old computers

For example, Stanford University psychologist Jon Krosnick found that as people's knowledge about climate change grew, the more concerned about it they became, though political affiliation and trust in science were also important factors.

The link between knowledge about climate change and concern about it "was especially true for respondents who described themselves as Democrats and those who said they trusted scientists," Krosnick told the annual convention of the American Psychological Association in August. "But for Republicans and those who had little trust in scientists, more knowledge did not mean there was more concern." [1]

In another study, Robert B. Cialdini, a professor of psychology and marketing at Arizona State University, found that a small change in message cards asking guests at an upscale Phoenix hotel to reuse their towels had huge potential environmental consequences.

One card exhorted residents to "Help Save the Environment" and was followed by information stressing respect for nature, Cialdini told a House subcommittee last year. [2] Another card stated "Help Save Resources for Future Generations," followed by information stressing the importance of saving energy for the future. A third card asked guests to "Partner With Us to Help Save the Environment," followed by information urging them to cooperate with the hotel in preservation efforts. A fourth card, Cialdini noted, said "Join Your Fellow Citizens in Helping to Save the Environment," followed by information that the majority of hotel guests reuse their towels when asked.

The outcome was striking, he said. Compared with the first three messages, the final one — based on a "social norm," or the perception of what most others were doing — increased the reuse of towels by an average of 34 percent, Cialdini said.

Not everyone in the psychological field agrees that trying to change consumer habits can ever do enough to make a significant difference in nationwide or worldwide carbon emissions or climate change. But many believe the effort is worthwhile.

"There's a huge debate going on among psychologists over whether it's just futile to even bother talking about individual behavior" or whether action by policy makers and corporations is the key to solving the nation's environmental problems, says Elise L. Amel, director of environmental studies and associate professor of industrial and organizational psychology at the University of St. Thomas in Minneapolis.

"I think it's got to be both. This is such a crucial problem coming so quickly that we can't leave any stone unturned."

---

[1] "Climate Change, Global Warming, Among Environmental Concerns Discussed at Psychology Meeting," press release, American Psychological Association, Aug. 15, 2008.

[2] Robert B. Cialdini, testimony to the Subcommittee on Research and Science Education, House Committee on Science and Technology, Sept. 25, 2007.

and other electronic items, which contain a toxic brew of chemicals, heavy metals and plastics such as mercury, lead and polyvinyl chloride.

In November the CBS News program "60 Minutes" followed recycled computer parts from Denver to Guiyu, China, which reporter Scott Pelley described as "one of the most toxic places on Earth . . . a town where the blood of the children is laced with lead." In Guiyu, computer parts are stripped, melted down and recycled by impoverished Chinese risking their health and lives for $8 a day. [46]

"This is really the dirty, little secret of the electronic age," said Jim Puckett, founder of the Seattle-based Basel Action Network, a watchdog group named for a treaty intended to keep wealthy nations from exporting toxic waste to poor countries. [47]

One way to keep harmful products from adding toxins to the environment is to produce them in an environmentally conscious way in the first place. Forrester Research Inc. found in a survey of 5,000 U.S. adults that 12 percent were willing to pay more for electronics that consume less energy or are made by an environmentally friendly manufacturer. [48]

In addition, computer manufacturers have been creating products that are more energy efficient than past models, companies are creating software to help older computers use less energy and some are making new models out of recycled materials, The Wall Street Journal reported. [49]

Katharine Kaplan, a product manager in the Energy Star program, said computer makers have sought to improve energy efficiency for years and that "the newer focus has been on toxins and recycling." [50]

But corporate efforts to make their products and services green — and to induce consumers to sign on to those efforts — can be challenging. For instance, in Britain, critics say carbon labels on retail goods often confuse consumers or give them information they don't want. [51]

Forum for the Future, a London think tank, expressed concern this year about giving consumers information without proper context. "Only a handful of our focus group participants associated carbon emissions [and climate change] with what they buy in the shops," the group stated in a report. "The majority knew that carbon emissions are linked with cars, airplanes and factories. They made that connection because they can 'see' the emissions, which makes them

How to Shrink Your Carbon Footprint

Driving habits and home energy use are key factors.

Here are several ways environmentalists, behavioralists and climate scientists say individual consumers can shrink their own carbon footprints:

Alter driving habits. Use of private motor vehicles accounted for more than 38 percent of total U.S. energy use in 2005, according to calculations by Gerald T. Gardner, professor emeritus of psychology at the University of Michigan-Dearborn, and Paul C. Stern, director of the Committee on the Human Dimensions of Global Climate Change at the National Research Council. [1]

Carpooling to work with another person can potentially save as much as 4.2 percent of individual and household energy consumption, they estimated. Other savings can come from avoiding sudden accelerations and stops (up to 3.2 percent); combining errand trips to a half of current mileage (up to 2.7 percent); cutting speeds from 70 to 60 miles per hour (up to 2.4 percent) and getting frequent tune-ups (3.9 percent).

Buying a car that gets an average of 30.7 miles per gallon rather than 20 can save an estimated 13.5 percent of household energy use, the authors estimated.

Reduce home energy consumption. Home space heating accounts for 34 percent of a typical homeowner's utility bill, according to the U.S. Department of Energy. Appliances and lighting account for the same proportion, followed by water heating (13 percent), electricity for air conditioning (11 percent) and refrigeration (8 percent). [2]

Residential buildings not only soak up money for utilities but also emit carbon dioxide. Heating accounts for 25 percent of CO_2 emissions, according to the Energy Department, followed by cooling (13.4 percent), water heating (12.4 percent), lighting (12 percent) and electronics, including color televisions (8.4 percent). Refrigerators and freezers account for more than 7 percent or residential carbon emissions. [3]

Gardner and Stern estimated that replacing 85 percent of all incandescent bulbs with compact fluorescents of equal brightness would reduce total individual and household energy consumption in the United States by 4 percent. Turning down the heat from 72 to 68 degrees during the day and to 65 degrees at night, and turning up the air conditioning from 73 to 78 degrees, would save 3.4 percent of energy consumption, they estimated. [4]

Many other steps can reduce a consumer's energy consumption — and carbon footprint — as well. For example, a home-energy checklist assembled by the American Council for an Energy-Efficient Economy suggests turning down the water-heater temperature to 120 degrees, cleaning or replacing furnace and air-conditioner filters, caulking leaky windows, improving attic and wall insulation and replacing inefficient appliances, among other tips. [5]

Downsize and scale back on consumption. Bigger homes typically use more energy, emit more carbon dioxide and produce more waste (which winds up in methane-emitting landfills) than smaller homes. Bigger cars tend to use more fuel. Bigger consumption patterns, from purchases of furniture, food and electronics to car and airplane travel, add to consumers' emissions.

The median size of a new single-family home rose more than 60 percent between 1970 to 2006, according to data from the National Association of Home Builders, and in 2006 nearly a fourth of new homes contained 3,000 square feet or more. Also that year, more than a fourth of the new homes had three or more bathrooms and about one in five had garage space for three or more cars. [6]

Know that geography can matter. Researchers at the Brookings Institution, a think tank in Washington, found that the carbon footprints of the nation's 100 largest metropolitan areas vary significantly and that development patterns and the availability of rail transit play a key role in the differences.

"[R]egions with high density, compact development and rail transit offer a more energy- and carbon-efficient lifestyle than sprawling, auto-centric areas," Brookings said. In addition, it

---

easy to interpret as being 'bad for the environment.' However, the link between products and climate change was less intuitive to them." [52]

In the United States, most companies are willing to embrace only "incremental change" on carbon labeling, Joel Makower, a sustainability consultant and cofounder and executive editor of Greener World Media Inc., in Oakland, Calif., told *The Christian Science Monitor*. [53]

Wal-Mart, the world's biggest retailer, kicked off a broad sustainability strategy in 2005 aimed at reducing the com-

pany's environmental impact. But Matt Kisler, the company's senior vice president of sustainability, told *The Monitor* that he has doubts about current carbon-labeling methodologies and customers' ability to link carbon with consumer goods. "I'm not sure the consumer will ever make a purchase based on the carbon footprint, especially the mass consumer," he told the newspaper. [54]

Green goods can indeed be a tough sell, but some analysts say a more concerted effort by business is needed to guide consumers on the

merits of environmentally friendly items.

A report by McKinsey & Co. consultants this fall cited a 2007 consumer survey by the trade publication *Chain Store Age*, which found that only 25 percent of respondents reported having bought a green product other than organic foods or energy-efficient lighting. The McKinsey researchers also noted that most green items have small market shares. For instance, green laundry detergent and household cleaners account for less than 2 percent of U.S. sales. [55]

said, while carbon output from urban centers continues to grow, the carbon footprint of a resident of a large metro area is 14 percent smaller than that of the average American and has grown in recent years by only half as much. [7]

Beware of "greenwashing." Along with an avalanche of "green" products and information on how to cut personal carbon emissions has come a steady tide of "greenwashing" — what TerraChoice Environmental Marketing Inc., based in Philadelphia, calls "the act of misleading consumers regarding the environmental practices of a company or the environmental benefits of a product or service." [8]

In a paper titled "The 'Six Sins of Greenwashing,' " the firm said it surveyed six "category-leading big-box stores" and identified 1,018 consumer products making 1,753 environmental claims. Of the total products examined, it said, "all but one made claims that are demonstrably false or that risk misleading intended audiences." [9]

Based on the survey, the firm identified what it said were six patterns of greenwashing:

- **The Sin of the Hidden Trade-Off** — Basing the suggestion that a product is "green" on only one environmental attribute without paying attention to other factors that may be more important, such as impacts on global warming, energy or water use or deforestation. An example is paper marketed as having recycled content without attention to the air, water and global-warming impact of its manufacture.

- **The Sin of No Proof** — Making an environmental claim that can't easily be backed up by supporting information or a reliable third-party certification.

- **The Sin of Vagueness** — Making poorly defined claims or ones that are so broad that consumers are likely to misunderstand the true meaning. An example is claiming a product is "chemical free." "[N]othing is free of chemicals," TerraChoice said. "Water is a chemical. All plants, animals and humans are made of chemicals as are all of our products."

- **The Sin of Irrelevance** — Making claims that might be true but are unimportant or not helpful. The most common example concerns chlorofluorocarbons, a key factor in depletion of the ozone layer, TerraChoice said. "Since CFCs have been legally banned for almost 30 years, there are no products that are manufactured with it."

- **The Sin of the Lesser of Two Evils** — Claims that, while they may be true within a product category, can distract consumers from the category's broader environmental impact. Organic cigarettes are an example, TerraChoice said.

- **The Sin of Fibbing** — Making false claims, such as saying a detergent is packaged in "100% recycled paper" but whose container is made of plastic.

---

[1] Gerald T. Gardner and Paul C. Stern, "The Short List: The Most Effective Actions U.S. Households Can Take to Curb Climate Change," *Environment*, September/October, 2008.

[2] "Your Home's Energy Use," U.S. Department of Energy, www1.eere.energy.gov/consumer/tips/home_energy.html.

[3] "2008 Buildings Energy Data Book," U.S. Department of Energy, http://buildingsdatabook.eren.doe.gov/.

[4] Gardner and Stern, *op. cit.*, http://buildingsdatabook.eere.energy.gov/TableView.aspx?table=2.4.3.

[5] "Home Energy Checklist for Action," American Council for an Energy-Efficient Economy, www.aceee.org/consumerguide/checklist.htm.

[6] "Selected Characteristics of New Housing," National Association of Home Builders, April 3, 2008, www.nahb.org/page.aspx/category/sectionID=130.

[7] "Brookings Institution Ranks Nation's 100 Largest Metro Areas for Carbon Footprint," press release, Brookings Institution, May 29, 2008, www.brookings.edu/~/media/Files/rc/papers/2008/05_carbon_footprint_sarzynski/press-release.pdf. The report is by Marilyn A. Brown, Frank Southworth and Andrea Sarzynski, "Shrinking the Carbon Footprint of Metropolitan America," Brookings Institution, May 8, 2008, www.brookings.edu/reports/2008/05_carbon_footprint_sarzynski.aspx.

[8] "The 'Six Sins of Greenwashing,' " TerraChoice Environmental Marketing Inc., November 2007.

[9] *Ibid.*

"Consumers in the United States and other developed countries have . . . done little to lighten their carbon footprint," the McKinsey consultants wrote. "Some of this lag between talking and walking could reflect insincerity, laziness or posturing. But much more of it stems from the failure of business to educate consumers about the benefits of green products and to create and market compelling ones." [56]

Patagonia, the clothing and outdoor-gear retailer, tracks the environmental footprint of more than a dozen of its products and shares both the good and bad on its Web site, though it warns that its environmental examinations are "partial and preliminary." [57]

For example, a down jacket's footprint, from origin of the fiber to the garment's distribution, spans more than 20,000 miles, touching California, Hungary, Japan, China and Nevada. The jacket's manufacture and transportation created nearly seven pounds of CO_2 emissions and enough energy to burn an 18-watt compact fluorescent bulb continuously for 22 days, according to Patagonia. [58]

The jacket uses "high-quality goose down . . . [that] comes from humanely raised geese," and the garment's light shell is made from recycled polyester, Patagonia says. But the zipper is treated with a substance "that contains perfluorooctanoic acid (PFOA), a synthetic chemical that is now persistent in the environment."

Says Patagonia, "We're investigating alternatives to the use of PFOA in water repellents — and looking for ways to recycle down garments." [59] ∎

CURRENT SITUATION

Creating Incentives

Beginning in 2009 buyers of the fuel-sipping Honda Civic hybrid will no longer receive one of the most popular incentives for going green: a tax credit. [60]

Tax incentives on hybrids phase out after an automaker sells 60,000 of them, a benchmark Honda reached in 2007 and that Toyota — maker of the hybrid Prius — hit in 2006. With gas prices having fallen sharply since peaking at $4-plus per gallon last summer, "it's getting a lot more expensive to be an environmentally conscious driver," *The Wall Street Journal* noted in an article about dwindling tax incentives on hybrid cars. [61]

Even so, the tax code remains one of the most powerful tools in the environmental-policy arsenal.

Consumers have long received tax breaks on everything from hybrid cars to energy-efficient appliances and home weatherization. The Emergency Economic Stabilization Act of 2008 — the so-called bailout bill passed this fall to deal with the rapidly deteriorating economy — included, extended or amended several such incentives as well as others aimed at businesses and public utilities. [62] For example, the measure included tax breaks for solar systems to generate electricity or heat hot water, energy-efficient home improvements and even bicycle commuting. [63]

As legislators and policy analysts contemplate how climate change and energy consumption are likely to unfold in coming decades, they are weighing other ideas to reduce carbon emissions.

"The most efficient approaches to reducing emissions involve giving businesses and individuals an incentive to curb activities that produce CO_2 emissions, rather than adopting a 'command-and-control' approach in which the government would mandate how much individual entities could emit or what technologies they should use," the Congressional Budget Office said in a report on policy options for reducing carbon-dioxide emissions. [64]

Cap-and-trade systems, which can be structured in a variety of ways, provide such incentives. They allow companies that emit CO_2 and other polluting gases to buy (or are allocated) emission credits that allow them to continue emitting a certain amount of the pollutant. Companies emitting less can save their allowances for the future or sell them at a profit to other companies.

So-called "cap-and-dividend" or "cap-and-cash-back" schemes also have been suggested. Entrepreneur and writer Peter Barnes, the author of *Who Owns the Sky?* and *Climate Solutions: A Citizens Guide*, described the idea this way: The caps would be placed "upstream — that is, on the small number of companies that bring carbon into the economy. An upstream cap could be administered without monitoring smokestacks, without a large bureaucracy and without favoring some companies over others. . . . If carbon doesn't come into the economy, it can't go out."

Caps would be auctioned rather than given away free, and the revenue would go to taxpayers to help offset the higher price of fuel and other carbon-intensive products. [65]

"This can be done through yearly tax credits, or better yet through monthly cash dividends wired . . . to people's bank accounts or debit cards."

Because that income would be taxed, the government would recoup about 25 percent of the money and could use it "as it sees fit," Barnes wrote. "More importantly, ordinary families would get the lion's share of the auction revenue, and get it in a way that rewards conservation. Since everyone would get the same amount back, those who use the most carbon would lose, and those who use the least would gain — their dividends would exceed what they pay in higher prices."

The approach would benefit low-income families because they use less energy than wealthier ones and would pay little if any tax on their dividends, Barnes argued.

Barnes wrote that while a carbon cap would raise fuel prices "for years to come," the cap-and-dividend approach would protect families' finances "by permanently linking dividends to carbon prices. As carbon prices rise, so — automatically — do dividends. If voters scream about rising fuel prices . . . politicians can truthfully say, 'How you fare is up to you. If you guzzle, you lose; if you conserve, you gain.' "

The Chesapeake Climate Action Network's Tidwell favors the cap-and-dividend idea and says that while it would lead to higher prices for fossil-based fuels, it would create an incentive for industry and consumers to conserve and switch to alternative forms of energy. The cap-and-dividend approach, he argues, would "lead to more energy-efficient cars, the discontinued use of energy-inefficient lightbulbs, and it will overnight [provide an incentive for] energy audits for homes, weatherization" and other carbon-reducing actions.

"Let's let the invisible hand of [economist] Adam Smith take over, but first we make carbon fuels more expensive as they come out of the ground and give the money back to all Americans in a way that protects the poor, so it's fair and effective and market-driven."

Roger W. Stephenson, executive vice president for programs at Clean Air-Cool Planet, says his group also favors an approach that generates dividends, but rather than sending a check to individuals, he says it would be better to recycle the revenue through the tax system for such purposes as reducing payroll taxes, corporate tax rates and other business and consumer taxes and

Continued on p. 1002

At Issue:

Will President-elect Obama's clean energy plan work?

BRACKEN HENDRICKS
SENIOR FELLOW, CENTER FOR AMERICAN PROGRESS

WRITTEN FOR *CQ RESEARCHER*, DECEMBER 2008

barack Obama is not yet sworn in as president, and it is far too early to know the details of his policies, let alone their effectiveness. Yet it is clear from the recent election campaign that the energy road map the president-elect and Democrats in Congress have laid out will move the nation forward. After years of inaction and obstruction from the White House, it is time for leadership. We must place clean energy center stage in America's economic renewal.

Clean energy means jobs and hundreds of billions in investment. The country faces a collapsing housing market, record unemployment and a fiscal crisis that hurts communities. New demand for goods and services from an energy transition can stimulate the economy. Retrofitting millions of homes for energy efficiency will put construction workers back on the job. Rebuilding our infrastructure for transit, alternative fuels and a renewable electricity grid will jump-start local economies. And retooling industry to serve the growing market for a new generation of cars and clean technology is our best hope for restoring manufacturing jobs.

A recent study by the Center for American Progress showed that investing $100 billion in smart incentives for energy efficiency and renewable energy would create 2 million jobs. These "green" jobs are in familiar professions in manufacturing and construction, driven by new technology and innovation. Clean and efficient energy investments have more local content and are harder to outsource, and they redirect spending from wasted energy into skilled labor. As a result, they create more jobs at better wages.

Climate solutions also mean fixing broken markets. Inaction in the face of global warming is not costless. Global warming is the biggest market failure the world has ever known, and if left unchecked will cost the economy trillions of dollars in lost productivity. We need smart policies that cap emissions and help businesses respond to the real costs of waste and pollution. Designed properly, smart climate policies can cut energy bills, increase consumer choice and create new markets and desperately needed demand for the ingenuity of American companies and workers.

We cannot drill our way out of our oil dependence, and we cannot deny our way to a stable climate. Barack Obama and congressional leaders instead have offered a vision that invests in innovation, that faces tough challenges squarely and that finds opportunity in crisis. This is real leadership. It is long overdue, and it will put America back to work.

KENNETH P. GREEN
RESIDENT SCHOLAR, AMERICAN ENTERPRISE INSTITUTE

WRITTEN FOR *CQ RESEARCHER*, DECEMBER 2008

barack Obama campaigned on an energy agenda of greenhouse-gas pricing, vehicle-efficiency standards and a fleet of plug-in hybrid cars. His plan is supposed to increase energy independence, lower greenhouse-gas emissions and create 5 million "green" energy jobs. Will it work? It's doubtful.

Obama's proposed cap-and-trade program, which would reduce greenhouse-gas emissions 80 percent by 2050, is 10 percent more stringent than the variation (S 2191) proposed by Sens. John Warner and Joseph Lieberman that was killed in 2007. The Congressional Budget Office estimated their proposal would cost $1.2 trillion from 2009-2018. The Environmental Protection Agency projected it would raise gas prices by $0.53 per gallon and hike electricity prices 44 percent in 2030. Economists at CRA International estimated the Warner-Lieberman proposal would result in a loss of 4 million jobs by 2015, growing to 7 million by 2050. In the face of a global financial crisis and a long, deep recession, passage of such a plan is both unlikely and undesirable.

The Obama plan also calls for reducing oil imports by tightening vehicle fuel-economy standards and subsidizing a fleet of 1 million plug-in hybrid vehicles. But there's a problem here: U.S. automakers are teetering on the brink of bankruptcy, Americans are strapped for cash, and plug-in hybrids are considerably more expensive than currently available cars. The National Renewable Energy Laboratory estimates the additional costs of plug-in hybrid vehicles that slightly outperform conventional hybrids at between $3,000 and $7,000. For the really fuel efficient ones, the laboratory estimates a premium of $12,000-$18,000.

Hybrids also cost more to insure. Are Americans going to shell out that kind of cash in a recession? I don't think so. Government fleets might buy some, but even they are strapped for cash and will have to cut costs elsewhere to afford plug-in hybrids.

As for creating green jobs, "job creation" is simply a myth. Governments do not create private-sector jobs, or wealth. They can only curtail jobs in one way (through taxation or regulation) and generate other jobs with subsidies and incentives. But since they impose costs in "managing" such programs — and because the market has already rejected the goods that the government is pushing (or there would be no need for intervention) — there are invariably fewer jobs and less wealth creation at the end of the day.

Markets create jobs, as markets create wealth: All the government can do is move it about to suit its priorities.

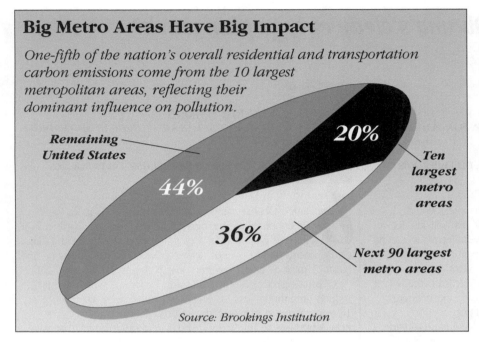

Big Metro Areas Have Big Impact

One-fifth of the nation's overall residential and transportation carbon emissions come from the 10 largest metropolitan areas, reflecting their dominant influence on pollution.

Remaining United States — 44%

Ten largest metro areas — 20%

Next 90 largest metro areas — 36%

Source: Brookings Institution

Continued from p. 1000

modifying the earned-income tax credit for low-income working people. Some money could also be used for social services, such as giving a "carbon boost to the food-stamp program" to help compensate poor people who are outside the tax system but nonetheless affected by higher energy costs.

"We have to be sure that non-taxpayers are taken into account — the bottom quintile of the population — people who don't pay taxes but who can be affected disproportionately" by a carbon-capping system, he says.

While the cost of the emission allowances would raise consumers' energy prices, Stephenson says, the government could limit the auction price of the credits to avoid extremes in consumers' utility bills.

Stephenson says in addition to a carbon-cap regime, new technology and greater energy efficiency — especially in transportation — also are needed to solve the climate-change problem. "Transport is 30 percent or more of emissions," he noted. "We cannot and are not going to stop driving. Therefore technology and efficiency solutions are essential."

Obama's Plan

President-elect Obama's energy plan calls for a wide range of efforts to reduce the carbon footprints of both industry and consumers, including an "economy-wide cap-and-trade program" aimed at cutting greenhouse-gas emissions by 80 percent by 2050. Under the plan, all pollution credits would be auctioned and the proceeds used for investments in "clean" energy, habitat protection and "rebates and other transition relief for families." [66]

The Obama plan also calls for other energy- and carbon-cutting efforts, including weatherizing at least a million low-income homes annually for the next decade, cutting electricity demand 15 percent from projected levels by 2020, putting a million plug-in hybrid cars on the road by 2015 and requiring a fourth of the nation's electricity to be generated by renewable sources by 2025. [67]

Obama even wants the federal government to practice what he is preaching. He has said he would transform the White House fleet to plug-in vehicles within a year of his inauguration "as security permits" and ensure that half of the new vehicles the government purchases are plug-ins and battery-powered cars by 2012, according to *The New York Times*. [68]

In mid-November Jason Grumet, the Obama campaign's chief energy and environment adviser, told a conference on carbon trading that Obama "will move quickly on climate change" but offered no specifics. [69]

Obama said during his campaign that spending $150 billion over the next decade to increase energy efficiency would help to create 5 million "green-collar" jobs in such industries as insulation installation, wind-turbine production and construction of energy-efficient buildings. [70] But some policy experts have questioned the job-creation number and underlying assumptions behind it, and the numbers have even been debated by the president-elect's own advisers. [71] "U.S. automakers are teetering on the brink of bankruptcy, Americans are strapped for cash and plug-in hybrids are considerably more expensive than currently available cars," according to Kenneth P. Green, resident scholar at the American Enterprise Institute. "As for creating green jobs," he continues, " 'job creation' is simply a myth." (*See "At Issue," page 1001.*)

While many details of Obama's energy plan remain to be fleshed out, he is wasting no time in making climate change and cutting CO_2 emissions a priority. In November, in a four-minute video message to the Governors' Global Climate Summit, he said his presidency would "mark a new chapter in America's leadership on climate change." [72] Separately, he also pledged to develop a two-year economic-stimulus plan to save or create 2.5 million jobs, including "green" jobs in alternative energy.

"I promise you this: When I am president, any governor who's willing to promote clean energy will have a partner in the White House," Obama told the climate summit. "Any company that's willing to invest in clean energy will have an ally in Washington. And any nation that's willing to join the cause of combating climate change will have an ally in the United States of America." [73]

Fred Krupp, president of the Environmental Defense Fund, said Obama was "clearly rejecting the timid, business-as-usual approach" to climate and energy issues. [74]

Eileen Claussen, president of the Pew Center on Global Climate Change, said in a statement that the president-elect's remarks were "exactly the kind of leadership the country and the world have been waiting for. . . . We urge the bipartisan leadership in Congress to work closely with the new president to quickly enact an economy-wide cap-and-trade system."

As policy makers in Washington turn their attention toward shaping a new energy strategy, concerns about a global climate catastrophe stemming from carbon-dioxide and other GHG emissions have never been stronger. Greenhouse-gas emissions from developed nations rose 2.3 percent between 2000 and 2006, according to the United Nations Framework Convention on Climate Change. [75] And individuals, as well as industry, are the source of those emissions.

The U.N. also reported in November that a poisonous, brownish haze of carbon dust from cars, coal-fired power plants, wood-burning stoves and other sources was blocking out sunlight in Asia and other continents, altering weather patterns and making people sick. [76] The thick, atmospheric blanket sprawls from the Arabian Peninsula to the Yellow Sea, goes past North and South Korea and Japan in the spring, and sometimes drifts to California. [77]

"We used to think of this brown cloud as a regional problem, but now we realize its impact is much greater," said Professor Veerabhadran Ramanathan, the leader of the U.N. scientific panel. "When we see the smog one day and

not the next, it just means it's blown somewhere else." [78]

Achim Steiner, executive director of the U.N. Environment Program in Beijing, said, "The imperative to act has never been clearer." [79] ∎

OUTLOOK

Economic Imperatives

Several developments could lead to significant changes — both positive and negative — in how consumers use energy and leave their carbon imprints on the environment.

One is the financial crisis. In the short term, many consumers are feeling pressured financially to lower their thermostats, forgo vacations and take other steps to conserve money — all of which help reduce carbon emissions.

In its annual Thanksgiving survey, AAA said the number of Americans planning to travel by car, plane, bus or train was down 1.4 percent from last year's record and marked the first decline since 2002. Air travel was expected to fall by 7.2 percent, the group said. [80]

The nation's economic woes also are causing consumers — even the wealthiest ones — to ratchet back on consumption of goods and services. "People are saying, 'We are going to save money, and we are going to save the environment,'" said Wendy Liebmann, chief executive of WSL Strategic Retail, a consulting firm in New York. [81]

But the economic slowdown could also drive up consumers' carbon emissions. With state and federal budgets under severe pressure, less money may be available for everything from tax credits for energy-saving home retrofitting to subsidies for mass-transit services.

In November, financial pressure stemming from the credit crisis propelled officials from 11 major transit

agencies — including those in New York, Boston, Washington, Los Angeles and Chicago — to seek help from Congress to avoid cuts in bus and subway service for millions of riders. [82]

In New York, the Metropolitan Transportation Authority called for raising fare and toll revenues by 23 percent in 2009 because of a huge budget gap, and it also drafted proposals for service reductions. [83]

Such moves would likely put more cars on the road and lead to greater emissions.

In the longer term, the incoming Obama administration could make dramatic changes to U.S. energy policy that would affect not only industry but consumers as well.

"Few challenges facing America, and the world, are more urgent than combating climate change," Obama told the governors' climate summit. [84]

Still unknown is whether Congress will pass a climate bill and how such legislation might be structured to provide incentives for consumers to lower their carbon emissions. "I think we have to wait and see" what Congress and the Obama administration do on energy policy, says Haxthausen of the Nature Conservancy.

"A reasonable conjecture is that [the administration] would want to get out in front of this issue, whether with legislation or a set of principles. In both houses of Congress, members are ready to work on this issue. They'll probably be looking to the administration for signals." ∎

Notes

[1] See New Hampshire Carbon Challenge, http://carbonchallenge.sr.unh.edu/.

[2] For background, see Marcia Clemmitt, "Climate Change," *CQ Researcher*, Jan. 27, 2006, pp. 73-96; and Colin Woodard, "Curbing Climate Change," *CQ Global Researcher*, February 2007, pp. 27-50.

[3] Jeffrey Ball, "A Big Sum of Small Differences," *The Wall Street Journal*, Oct. 2, 2008.

[4] *Ibid.*

5 Marilyn A. Brown, Frank Southworth and Andrea Sarzynski, "Shrinking the Carbon Footprint of Metropolitan America," Brookings Institution, May 2008, p. 8, www.brookings.edu/reports/2008/05_carbon_footprint_sarzynski.aspx.

6 "What's your EnviroQ? Answer Page," Environmental Protection Agency, www.epa.gov/epahome/enviroq/index.htm.

7 "Compact Fluorescent Light Bulbs," www.energystar.gov/index.cfm?c=cfls.pr_cfls.

8 Mike Tidwell, "Consider Using the N-Word Less," Grist, Sept. 4, 2007, http://grist.org/feature/2007/09/04/change_redux/index.html.

9 Gary Langer, "Fuel Costs Boost Conservation Efforts; 7 in 10 Reducing 'Carbon Footprint,' " ABC News, Aug. 9, 2008, http://abcnews.go.com/PollingUnit/story?id=5525064&page=1.

10 Harris Interactive, "For Earth Day: Two-thirds of Americans Believe Humans are Contributing to Increased Temperatures," The Harris Poll #44, April 18, 2008, www.harrisinteractive.com/harris_poll/index.asp?PID=898.

11 Ibid.

12 Center for Sustainable Systems, University of Michigan, "Residential Buildings," http://css.snre.umich.edu/css_doc/CSS01-08.pdf.

13 Ibid.

14 Thomas Wiedmann and Jan Minx, "A Definition of 'Carbon Footprint,' " ISA [UK] Research & Consulting, June 2007, www.isa-research.co.uk/docs/ISA-UK_Report_07-01_carbon_footprint.pdf.

15 Joanna Pearlstein, "Surprise! Conventional Agriculture Can Be Easier on the Planet," in "Inconvenient Truths: Get Ready to Rethink What It Means to Be Green," Wired, May 19, 2008, www.wired.com/science/planetearth/magazine/16-06/ff_heresies_intro.

16 "Universal, Neutral and Transparent Method for Estimating the Carbon Footprint of a Flight," press release, International Civil Aviation Organization, June 18, 2008, www.icao.int/icao/en/nr/2008/PIO200803_e.pdf.

17 Gerard Wynn, "Critics [say] air travel carbon offsetting too crude," Reuters, Aug. 21, 2008, www.reuters.com/article/environmentNews/idUSLK20281120080821.

18 Tim DeChant, "Calculating footprint often uses fuzzy math; Results vary, but give idea of environmental impact," Chicago Tribune, Aug. 10, 2008, p. 1.

19 The New England Carbon Estimator is at http://carbonchallenge.sr.unh.edu/calculator.jsp.

20 Jane Black, "What's in a Number?" Slate, Sept. 17, 2008, www.slate.com/id/2200202/.

21 See Eric Marx, "Are you ready to go on a carbon diet?" The Christian Science Monitor, Nov. 10, 2008, www.csmonitor.com/2008/1110/p13s01-wmgn.html.

22 For background, see Jennifer Weeks, "Carbon Trading," CQ Global Researcher, November 2008, pp. 295-320.

23 Ben Lieberman, "Beware of Cap and Trade Climate Bills," Web Memo No. 1723, Heritage Foundation, Dec. 6, 2007, www.heritage.org/Research/Economy/wm1723.cfm.

24 Opening Statement of Chairman Brian Baird before the Subcommittee on Research and Science Education, House Committee on Science and Technology, Sept. 25, 2007, http://science.house.gov/publications/OpeningStatement.aspx?OSID=1293.

25 Testimony of John A. "Skip" Laitner before the Subcommittee on Research and Science Education, House Committee on Science and Technology, Sept. 25, 2007, p. 11, http://science.house.gov/publications/Testimony.aspx?TID=7922.

26 David R. Baker, "State rebates lead more people to go solar," San Francisco Chronicle, July 15, 2008, www.sfgate.com/cgi-bin/article.cgi?f=/c/a/2008/07/14/BUNL11OVEF.DTL.

27 Quoted in ibid.

28 Gerald T. Gardner and Paul C. Stern, "The Short List: The Most Effective Actions U.S. Households Can Take to Curb Climate Change," Environment, September/October 2008, www.environmentmagazine.org/

Archives/Back%20Issues/September-October%202008/gardner-stern-full.html.

29 Steven Blanchard and Peter Reppe, "Life Cycle Analysis of a Residential Home in Michigan," University of Michigan Center for Sustainable Systems, September 1998, www.umich.edu/~nppcpub/research/lcahome/homelca.PDF. The project was submitted in partial fulfillment of requirements for the degree of master of science in natural resources.

30 "Residential Buildings," Center for Sustainable Systems, University of Michigan, http://css.snre.umich.edu/css_doc/CSS01-08.pdf.

31 "About Energy Star," www.energystar.gov/index.cfm?c=about.ab_index.

32 "Energy Star has lost some luster," Consumer Reports, October 2008, p. 24.

33 Paul Davidson, "A new twist for light bulbs that conserve energy," USA Today, April 22, 2008, p. 10B, www.usatoday.com/money/industries/energy/environment/2008-04-21-light-bulbs_N.htm.

34 Environmental timeline, Radford University, www.runet.edu/~wkovarik/envhist/about.html.

35 Jennifer Weeks, "Buying Green," CQ Researcher, Feb. 29, 2008, pp. 193-216.

36 Rachel Carson, The Sea Around Us, Illustrated Commemorative Edition, Oxford University Press, 2003, pp. 223, 225.

37 Gaylord Nelson, "Earth Day '70: What It Meant," EPA Journal, April 1980, www.epa.gov/history/topics/earthday/02.htm.

38 Andrew C. Revkin, "On Climate Change, a Change of Thinking," The New York Times, Dec. 4, 2005.

39 Ibid.

40 "Q&A: The Kyoto Protocol," BBC News, Feb. 16, 2005, http://news.bbc.co.uk/2/hi/science/nature/4269921.stm.

41 Al Gore, An Inconvenient Truth (2006), p. 8.

42 Daniel Stone, "The Green Pope," Newsweek, April 17, 2008, www.newsweek.com/id/132523.

43 Neela Banerjee, "Southern Baptists Back a Shift on Climate Change," The New York Times, March 10, 2008, www.nytimes.com/2008/03/10/us/10baptist.html?scp=1&sq=southern%20baptists%20back%20a%20shift&st=cse.

44 Gary Haq, Anne Owen, Elena Dawkins and John Barrett, "The Carbon Cost of Christmas," Stockholm Environment Institute, 2007, www.climatetalk.org.uk/downloads/CarbonCostofChristmas2007.pdf.

45 Timothy Gutowski, et al., "Environmental Life Style Analysis," IEEE International Symposium on Electronics and the Environment, May 19-20, 2008. All other authors were graduate or undergraduate students at the Massachusetts Institute of Technology during the 2007 spring term.

About the Author

Thomas J. Billitteri is a *CQ Researcher* staff writer based in Fairfield, Pa., who has more than 30 years' experience covering business, nonprofit institutions and public policy for newspapers and other publications. His recent *CQ Researcher* reports include "Campaign Finance," "Human Rights in China" and "Financial Bailout." He holds a BA in English and an MA in journalism from Indiana University.

[46] "Following The Trail Of Toxic E-Waste," "60 Minutes," CBS News, Nov. 9, 2008, www.cbsnews.com/stories/2008/11/06/60minutes/main4579229.shtml.

[47] Ibid.

[48] Joseph De Avila, "PC Movement: How Green Is Your Computer?" The Wall Street Journal, Sept. 4, 2008, http://online.wsj.com/article/SB122048465164497063.html.

[49] Ibid.

[50] Ibid.

[51] Marx, op. cit.

[52] Ibid.

[53] Ibid.

[54] Ibid.

[55] Sheila M. J. Bonini and Jeremy M. Oppenheim, "Helping 'green' products grow," The McKinsey Quarterly, October 2008.

[56] Ibid.

[57] "Environmentalism: Leading the Examined Life," Patagonia, www.patagonia.com/web/us/contribution/patagonia.go?assetid=23429&ln=267. See also www.patagonia.com/web/us/footprint/index.jsp.

[58] Ibid.

[59] www.patagonia.com/web/us/footprint/index.jsp.

[60] Mike Spector, "The Incentives to Buy Hybrids Are Dwindling," The Wall Street Journal, Nov. 6, 2008, http://online.wsj.com/article/SB122593537581103821.html.

[61] Ibid.

[62] U.S. Department of Energy, "Consumer Energy Tax Incentives," www.energy.gov/taxbreaks.htm. See also, "P.L. 110-343/The Emergency Economic Stabilization Act of 2008: Energy Tax Incentives," www.energy.gov/media/HR_1424.pdf.

[63] Ashlea Ebeling, "The Green Tax Gusher," Forbes, Nov. 24, 2008, p. 150.

[64] "Policy Options for Reducing CO$_2$ Emissions," Congressional Budget Office, February 2008, p. vii, www.cbo.gov/ftpdocs/89xx/doc8934/02-12-Carbon.pdf.

[65] Peter Barnes, "Obama's 'number 1 priority,'" Reuters, http://blogs.reuters.com/great-debate/2008/11/11/obamas-number-1-priority/.

[66] "New Energy for America," http://my.barackobama.com/page/content/newenergy_more#emissions.

[67] Ibid.

[68] Jim Motavalli, "The Candidates' Clean Car Plans," The New York Times, Oct. 23, 2008, http://wheels.blogs.nytimes.com/2008/10/23/the-candidates-clean-car-plans/.

[69] Deborah Zabarenko, "Obama will act quickly on climate change: adviser," Reuters, Nov. 12, 2008, www.reuters.com/article/environmentNews/idUSTRE4AB84K20081112?feedType=RSS&feedName=environmentNews.

[70] Jeffrey Ball, "Does Green Energy Add 5 Million Jobs? Pitch Is Potent; Numbers Are Squishy," The Wall Street Journal, Nov. 7, 2008, p. A13.

[71] Ibid.

[72] The Associated Press, "Obama Promises Leadership on Climate Change," The New York Times, Nov. 18, 2008, www.nytimes.com/aponline/washington/AP-Obama-Climate-Change.html?sq=climate%20summit&st=nyt&scp=2&pagewanted=print.

[73] Quoted in ibid.

[74] Ibid.

[75] Richard Black, "Obama vows climate 'engagement,'" BBC News, Nov. 18, 2008, http://news.bbc.co.uk/2/hi/science/nature/7736321.stm.

[76] Andrew Jacob, "Report Sees New Pollution Threat," The New York Times, Nov. 14, 2008, www.nytimes.com/2008/11/14/world/14cloud.html?hp.

[77] Ibid.

[78] Ibid.

[79] Ibid.

[80] Oren Dorell and Alan Levin, "Economy sets travel back a bit for holiday," USA Today, Nov. 18, 2008, www.usatoday.com/travel/news/2008-11-18-aaa-holiday-travel-forecast_N.htm.

[81] Jennifer Saranow, "Luxury Consumers Scrimp for Sake of Planet, and Because It's Cheaper," The Wall Street Journal, Nov. 4, 2008, http://online.wsj.com/article/SB122575617614495083.html.

[82] Lena H. Sun, "U.S. Transit Agencies Ask Congress for Help in Averting Service Cuts," The Washington Post, Nov. 19, 2008, p. 2D, www.washingtonpost.com/wp-dyn/content/article/2008/11/18/AR2008111803174_pf.html.

[83] William Neuman, "M.T.A. Said to Plan 23% Increase in Fare and Toll Revenue," The New York Times, Nov. 19, 2008, www.nytimes.com/2008/11/19/nyregion/19transit.html.

[84] Juliet Eilperin, "Obama Addresses Climate Summit," The Washington Post, www.washingtonpost.com/wp-dyn/content/article/2008/11/18/AR2008111803286.html.

FOR MORE INFORMATION

Alliance to Save Energy, 1850 M St., N.W., Suite 600, Washington, DC 20036; (202) 857-0666; www.ase.org. Group that promotes energy efficiency worldwide.

American Council for an Energy-Efficient Economy, 529 14th St., N.W., Suite 600, Washington, DC 20045-1000; (202) 507-4000; www.aceee.org. Fosters energy efficiency to promote economic prosperity and environmental protection.

Carbonfund.org, 1320 Fenwick Lane, Suite 206, Silver Spring, MD 20910; (240) 247-0630; www.carbonfund.org. Provides certified carbon offsets to help individuals, businesses and organizations reduce their carbon footprints.

Chesapeake Climate Action Network, P.O. Box 11138, Takoma Park, MD 20912; (240) 396-1981; www.chesapeakeclimate.org. Grassroots group that fights global warming in Maryland, Virginia and Washington, D.C.

Clean Air-Cool Planet, 100 Market St., Suite 204, Portsmouth, NH 03801; (603) 422-6464; www.cleanair-coolplanet.org. Nonprofit organization that partners with businesses, colleges and communities in the Northeast to reduce carbon emissions.

Energy Star, US EPA, Energy Star Hotline (62202J), 1200 Pennsylvania Ave., N.W., Washington, DC 20460; (888) 782-7937; www.energystar.gov. Environmental Protection Agency and Department of Energy program promoting energy-efficient products.

Nature Conservancy, 4245 North Fairfax Dr., Suite 100, Arlington, VA 22203-1606; (703) 841-5300; www.nature.org. Conservation organization that works worldwide to protect ecologically sensitive land and water.

New Hampshire Carbon Challenge, 39 College Road, CSRC, Morse Hall, Durham, NH 03824; (603) 862-3128; http://carbonchallenge.sr.unh.edu. Works to help households and communities reduce their energy consumption.

Pew Center on Global Climate Change, 2101 Wilson Blvd., Suite 550, Arlington, VA 22201; (703) 516-4146; www.pewclimate.org. Seeks new approaches to dealing with climate change.

Bibliography

Selected Sources

Books

Barnes, Peter, *Climate Solutions: A Citizens Guide*, Chelsea Green Publishing, 2008.
An entrepreneur and writer blames global warming on market failure and misplaced government priorities.

Brower, Michael, and Warren Leon, *The Consumer's Guide to Effective Environmental Choices*, Three Rivers Press, 1999.
Though nearly a decade old, this book by veteran environmental experts helps consumers determine what impact their decisions will have on the environment, backed by research from the Union of Concerned Scientists.

Gore, Al, *An Inconvenient Truth*, Rodale, 2006.
The former vice president and Nobel Peace Prize winner argues that exploding population growth and a technology revolution have transformed the relationship between humans and the Earth.

Articles

Ball, Jeffrey, "Six Products, Six Carbon Footprints," *The Wall Street Journal*, Oct. 6, 2008, http://online.wsj.com/article/SB122304950601802565.html.
Companies calculate the carbon footprints of their products in different ways, making it hard for consumers to compare goods.

Bonini, Sheila M., and Jeremy M. Oppenheim, "Helping 'green' products grow," *The McKinsey Quarterly*, October 2008.
Two consultants contend that the failure of business to educate consumers about green products has helped discourage them from doing more to reduce their carbon footprints.

Ebeling, Ashlea, "The Green Tax Gusher," *Forbes*, Nov. 24, 2008, www.forbes.com/forbes/2008/1124/150.html.
As part of this fall's $700 billion bailout bill, Congress enacted a new round of tax breaks that benefit consumers who embrace energy-saving home improvements and alternative energy.

El Nasser, Haya, " 'Green' efforts embrace poor," *USA Today*, Nov. 23, 2008, www.usatoday.com/news/nation/2008-11-23-green-poor_N.htm.
Cities and community groups are trying to help low-income households reduce their energy consumption.

Gardner, Gerald T., and Paul C. Stern, "The Short List: The Most Effective Actions U.S. Households Can Take to Curb Climate Change," *Environment*, www.environmentmagazine.org/Archives/Back%20Issues/September-October%202008/gardner-stern-full.html.

A professor emeritus of psychology and the director of the National Research Council's Committee on the Human Dimensions of Global Climate Change argue that households often lack accurate and accessible information on how to reduce carbon emissions and mitigate climate change.

Knight, Matthew, "Carbon dioxide levels already a danger," CNN, www.cnn.com/2008/TECH/science/11/21/climate.danger.zone/index.html.
International scientists — led by James Hansen, director of NASA's Goddard Institute for Space Studies — conclude that carbon dioxide concentrations in Earth's atmosphere are in the danger zone, threatening food shortages, more intense storms and other calamities.

Specter, Michael, "Big Foot," *The New Yorker*, Feb. 25, 2008.
An excessive carbon footprint is the modern equivalent of a scarlet letter, but calculating one's environmental impact of modern life "can be dazzlingly complex," the writer says.

Wald, Matthew L., "For Carbon Emissions, a Goal of Less Than Zero," *The New York Times*, March 26, 2008, www.nytimes.com/2008/03/26/business/businessspecial2/26negative.html?scp=14&sq=carbon&st=cse.
Researchers around the world are searching for so-called carbon-negative technologies that remove carbon dioxide from the atmosphere.

Reports and Studies

"Policy Options for Reducing CO_2 Emissions," Congressional Budget Office, February 2008, www.cbo.gov/ftpdocs/89xx/doc8934/02-12-Carbon.pdf.
The congressional agency analyzes incentive-based options for reducing greenhouse-gas emissions, especially carbon dioxide.

Brown, Marilyn A., Frank Southworth and Andrea Sarzynski, "Shrinking the Carbon Footprint of Metropolitan America," Brookings Institution, May 2008, www.brookings.edu/reports/2008/~/media/Files/rc/reports/2008/05_carbon_footprint_sarzynski/carbonfootprint_report.pdf.
The researchers quantify transportation and residential carbon emissions for the 100-largest U.S. metropolitan areas.

Haq, Gary, Anne Owen, Elena Dawkins and John Barrett, "The Carbon Cost of Christmas," Stockholm Environment Institute, 2007, www.climatetalk.org.uk/downloads/CarbonCostofChristmas2007.pdf.
Three days of Christmas festivities could result in 650 kilograms (1,433 pounds) of carbon dioxide emissions per person, according to the United Kingdom-based researchers.

The Next Step:

Additional Articles from Current Periodicals

Calculations

Ball, Jeffrey, "New Concept of Carbon Footprints Means Calculating," The Associated Press, Oct. 6, 2008.

Companies calculate their carbon footprints differently, making it difficult for consumers to effectively compare goods.

De Chant, Tim, "Calculating Footprint Often Uses Fuzzy Math," Chicago Tribune, Aug. 10, 2008, p. A1.

Carbon footprint calculators help raise awareness for global warming, but the methodology has yet to be perfected.

Lohr, Steve, "On the Web, an Advanced Carbon Calculator for Personal Use," The New York Times, May 15, 2007, p. C4.

A new Web service allows people to figure out their individual carbon dioxide production

Scherer, Ron, "Wall Street Adds Climate Change to Bottom Line," The Christian Science Monitor, Feb. 27, 2007, p. 1.

Calculating carbon footprints allows analysts to forecast the potential risks of climate-change lawsuits and future costs of any emission regulations.

Government Action

"Fargo Officials Want to Figure 'Carbon Credit,' " The Associated Press, Aug. 10, 2007.

City officials in Fargo, N.D., want to calculate the city's carbon footprint by inventorying carbon emissions to see if the city is eligible for carbon credits.

Geewax, Marilyn, "Data Sought on 'Carbon Footprint,' " Atlanta Journal-Constitution, Sept. 30, 2007, p. 1H.

Some members of Congress want corporations to release more information on their carbon footprints so investors can better judge whether they are exposing themselves to lawsuits.

Schwarzen, Christopher, "County Tackles Greenhouse-Gas Issue," Seattle Times, Aug. 4, 2007, p. B4.

A Washington county has appointed a committee to assess the environmental impact of county government operations.

Vega, Cecilia M., "S.F. Aims for Greenest Building Codes in U.S.," The San Francisco Chronicle, March 20, 2008, p. A1.

New regulations in San Francisco would require new, large commercial buildings to contain environmentally friendly features by 2012, reducing carbon emissions by 60,000 tons.

Impact of Individuals

Beal, Tom, "Measure Your Carbon Footprint and Reduce It," Arizona Daily Star, Sept. 18, 2007, p. B1.

The Environmental Protection Agency estimates that the average household of two emits 41,500 pounds of carbon per year.

Karp, Greg, "Spending Green Helps Reduce the Size of Your Carbon Footprint," Morning Call (Pennsylvania), Sept. 21, 2008, p. D1.

Many critics say individuals throw money at the global warming problem without modifying their behavior to actually reduce their carbon footprint.

Streater, Scott, "Calculate Your Carbon Footprint — And Figure Out Why It Matters," Fort Worth Star-Telegram (Texas), Sept. 23, 2007.

There is significant scientific evidence that global warming has been caused by man-made activities.

Reduction Methods

"Awnings Reduce Homeowners' Carbon Footprints," Business Wire, March 7, 2008.

Awnings are a proven method for reducing heat gain and air-conditioning loads, which typically results in a reduced carbon footprint.

"National Parks Ask Visitors to Reduce 'Carbon Footprint,' " The Associated Press, June 16, 2008.

Zion National Park in Utah is asking its visitors to reduce their carbon footprints by reducing driving, using energy-efficient appliances and purchasing clean-power sources.

Carroll, Cindy, "Reducing Carbon Footprint from the Comfort of Home," Union Leader (N.H.), Jan. 6, 2008, p. D1.

Carbon footprints can be reduced by using less energy, often without having to sacrifice comfort.

CITING CQ RESEARCHER

Sample formats for citing these reports in a bibliography include the ones listed below. Preferred styles and formats vary, so please check with your instructor or professor.

MLA STYLE

Jost, Kenneth. "Rethinking the Death Penalty." CQ Researcher 16 Nov. 2001: 945-68.

APA STYLE

Jost, K. (2001, November 16). Rethinking the death penalty. CQ Researcher, 11, 945-968.

CHICAGO STYLE

Jost, Kenneth. "Rethinking the Death Penalty." CQ Researcher, November 16, 2001, 945-968.

Excerpted from the CQ Researcher. Peter Katel. (January 4, 2008). "Oil Jitters." *CQ Researcher*, 1-24.

Oil Jitters

BY PETER KATEL

THE ISSUES

On a recent trip to Beijing, David Sandalow saw the world's energy future, and it wasn't pretty. "They tell me there are almost 1,000 new cars a day on the streets," says Sandalow, a senior fellow at the Brookings Institution think tank. "If those cars and trucks use oil in the same way the current fleet does, we're in trouble, for a lot of reasons."

Sandalow, author of the 2007 book *Freedom From Oil*, isn't alone. [1] The top economic researcher at the International Energy Agency (IEA) recently gave oil industry representatives in London a dire warning. "If we don't do something very quickly, and in a bold manner," said Fatih Birol, "our energy system's wheels may fall off." [2]

Demand for the key fuel of modern life is shooting up, especially in the developing world, but production isn't keeping pace, the IEA reports. Within the next seven years, Birol predicted, the gap will exceed 13 million barrels of oil a day — or 15 percent of the world's current output. [3]

"Rising global energy demand poses a real and growing threat to the world's energy security," said the IEA's 2007 annual report. "If governments around the world stick with current policies, the world's energy needs would be well over 50 percent higher in 2030 than today. China and India together

Heavy traffic in Beijing last August reflects the rising demand for energy in China and other developing nations. China had just 22 million cars and "light-duty vehicles" in 2005, with 10 times as many projected by 2030. By comparison, the United States, with a quarter of China's 1.3 billion population, had 250 million motor vehicles.

account for 45 percent of the increase in demand in this scenario." [4]

The IEA delivered its message when intense oil jitters had pushed crude oil prices as high as they've ever been: close to $100 a barrel in December 2007 and more than $3 per gallon at U.S. gas pumps.* The in-

creases reflected a variety of concerns, including worries that supplies would be interrupted by possible U.S. or Israeli military strikes against Iran or a potential Turkish incursion into northern Iraq. "The latest run-up . . . has to do with fear," said Lawrence J. Goldstein, an economist at the Energy Policy Research Foundation. [5]

Fears of Middle East war choking off oil flow have hit several times since 1973, when Arab nations launched an oil embargo against the United States and other countries in retribution for their support for Israel in a war with its neighbors. Iraqi dictator Saddam Hussein prompted another scare when he invaded Kuwait in 1990.

This time, though, the headline-induced jitters have emerged along with deeper worries about a variety of developments: rising oil demand from rapidly industrializing China and India; depletion of oil reserves in the United States, Europe and possibly the Middle East and the fact that since the 1960s, most of the world's oil has switched from corporate to government ownership, as in Iran, Venezuela and Russia. [6]

"Nationally owned companies are less efficient, and the traditional international majors [big oil firms] don't control as much of the resource as they used to," says Kenneth B. Medlock III, an energy studies fellow at the James A. Baker III Institute for Public Policy at Rice University in Houston.

Meanwhile, the world's total production of about 84 million barrels a day is spoken for. There is virtually no spare oil — "excess capacity," in

---

* The IEA was founded in Paris in 1974 during the first post-World War II oil crisis to help ensure a steady supply of reasonably priced fuel for the world's industrialized nations.

* On Jan. 2, 2008, crude oil prices hit the milestone $100-a-barrel mark for the first time. Violence in Nigeria's oil-producing region and speculative trading were blamed for the jump. In April 1980, during the turmoil that followed the 1979 Iranian revolution, prices were actually higher when adjusted for inflation: $102.81 a barrel.

industry jargon. [7] The United States alone consumes nearly a quarter of today's world production — about 20 million barrels a day.

Concern about rising demand for oil by industrializing nations is compounded by the fact that oil is a non-renewable resource and plays such a major role in other parts of the global economy.

"Oil (and natural gas) are the essential components in the fertilizer on which world agriculture depends; oil makes it possible to transport food to the totally non-self-sufficient megacities of the world," writes Daniel Yergin, an oil historian and chairman of Cambridge Energy Research Associates, a consulting firm. "Oil also provides the plastics and chemicals that are the bricks and mortar of contemporary civilization — a civilization that would collapse if the world's oil wells suddenly went dry." [8]

Oil's central role in the world marketplace means that an economic slowdown can push down demand for oil, while an economic boom raises demand. With the subprime mortgage crisis slowing down the U.S. economy, oil prices are likely to fall somewhat, says J. Robinson West, chairman of PFC Energy, a Houston-based consulting firm. "Then the economy rebounds, and oil demand picks up again. That's when you're going to see prices go through the roof. There's going to be a crunch, where demand outstrips supply."

West, who directed U.S. offshore oil policy during the Reagan administration, doesn't think the world is about to run out of oil altogether. He is a member of the chorus of oil-watchers who generally fault state-owned oil companies (except Saudi Arabia's and Brazil's) for not reinvesting at least some of their oil income in exploration and equipment maintenance — so they can keep the oil cash pouring in. "Politicians don't care about the oil industry, they care about the money."

Some other experts question how

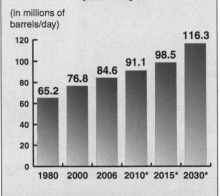

Global Oil Production to Continue Rising

Global daily oil production is expected to hit 116 million barrels by 2030, nearly 40 percent over 2006 levels.

World Oil Production, 1980-2030

(in millions of barrels/day)

Year	Production
1980	65.2
2000	76.8
2006	84.6
2010*	91.1
2015*	98.5
2030*	116.3

Average annual rate of growth, 2006-2030: *1.3%*

* *predicted*

Source: "World Energy Outlook 2007: China and India Insights," International Energy Agency, 2007

much of the run-up in oil prices is driven by doubts over supply capacity, and how much by financial speculators who benefit financially if prices move in the direction they have forecast. "The biggest thing that traders are now playing is the fear card," says Fadel Gheit, head energy analyst at Oppenheimer & Co., a Wall Street investment firm. "Commodity traders are spinning every piece of information that can embellish their position."

A 2006 report by the Senate's Permanent Subcommittee on Investigations of the Committee on Homeland Security and Governmental Affairs traced an energy futures trading boom to congressional action in 2000 that freed energy commodity trading on

electronic exchanges from regulatory oversight. [9]

But traders make a convenient target, one economics writer argues. "They are speculating against real risks — the risk that oil from the Persian Gulf could be cut off; that hurricanes in the Gulf of Mexico could damage U.S. oil rigs and refineries; that political events elsewhere (in Russia, Nigeria, Venezuela) could curtail supplies," columnist Robert Samuelson wrote in *The Washington Post*. "High prices reflect genuine uncertainties." [10]

Further complicating the oil-supply picture, Saudi Arabia and other big producers are devoting an increasing percentage of their petroleum to expanding their own economies — effectively withholding oil from the market (*See "Current Situation," p. 18.*) Advocates of the "peak oil" thesis argue that major global oil reserves — including those in Saudi Arabia — have hit the point at which about half of the oil they can yield has already been produced.

"Over the years, we've just always assumed that over time we always find more oil — because over time we always found more oil," says Houston energy consultant Matthew R. Simmons, a leading proponent of the peak oil theory. But the world seems to have run out of mega-fields, he says.

In fact, Simmons says, major discoveries since the late 1960s can be counted on the fingers of one hand. In 1967 came Prudhoe Bay in Alaska; about 10 billion of its 13 billion barrels of recoverable oil already have been pumped, according to BP, which operates the field. [11]

Since then, exploration has yielded a 13-billion-barrel Caspian Sea reservoir owned by Kazakhstan, a 3-to-5-billion-barrel U.S. field 3,500 feet under the Gulf of Mexico and a 5-to-8-billion-barrel field off the Brazilian coast. [12]

Some experts say such recent discoveries suggest that new exploration and production technology will supply the world with oil into the indefinite future. "What's really happening is the opening up of a whole new horizon in the ultra-deep waters of the Gulf of Mexico, and it looks like the upside is very significant," said Yergin, a critic of peak oil theory. But, he added, "It will take time and billions of dollars to get there." [13]

Indeed, recent discoveries come nowhere near the spectacular discoveries that launched the oil age. The Middle East set the standard for mega-discoveries. Even after decades of production, its reserves were estimated at 266.8 billion barrels in 2006. Oil behemoth Saudi Arabia, for example, was producing 8.6 million barrels a day in August 2007. [14] U.S. production for 2006 was 5.1 million barrels a day.

But because Saudi Arabia doesn't release detailed figures on oilfield-by-oilfield production, Simmons questions the country's reserve estimates. "I don't think there's a shred of evidence" to back up Saudi reserve numbers, he says. More cautiously, the Government Accountability Office (GAO) notes the "potential unreliability" of reserve data from members of the Organization of Petroleum Exporting Countries (OPEC), among whom Saudi Arabia leads in reported reserves. That issue is "particularly problematic," the GAO reports, because OPEC countries together hold more than three-quarters of the world's known oil reserves. [15]

Cambridge Energy Research Associates estimates Middle Eastern reserves at 662 billion barrels as of November 2006 — or about 15 percent of the world's total reserves of 4.82 trillion barrels. "Key producing countries such as Saudi Arabia have a vast reserve and resource base," the firm reported. "There is no credible technical analysis that we are aware of that demonstrates its productive capacity will suddenly fall in the near term." [16]

Other experts see production problems even if the peak oil theorists are wrong. Edward L. Morse, chief ener-

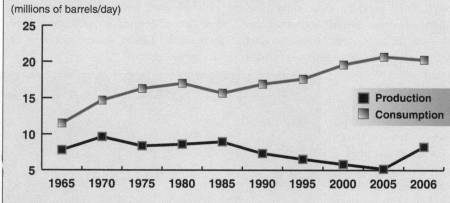

U.S. Oil Consumption Exceeds Production

The U.S. has continually consumed more oil than it has produced. The disparity between consumption and production exceeded 12 million barrels per day in 2006, forcing America to import more oil.

U.S. oil consumption and production, 1965-2005

(millions of barrels/day)

Source: U.S. Energy Information Administration

gy economist at Lehman Brothers, a New York investment bank, calculates that if Venezuela, Nigeria, Iraq and Iran were producing oil more efficiently, 6 million barrels a day more would be available to the world market. By contrast, the peak debate centers on technical issues, including the geology of oil reservoirs. Consequently, "Why should I believe in peak oil?" asks Morse, a deputy assistant secretary of State for energy policy during the Carter administration. The International Energy Agency guardedly shares Morse's skepticism about supply. "New capacity additions . . . are expected to increase over the next five years," the agency said. "But it is far from clear whether they will be sufficient to compensate for the decline in output at existing fields." [17]

To compensate, Birol said China, India and other big energy consumers need to step up energy efficiency efforts "right away and in a bold manner. We want more action, instead of more targets, more meetings and more talks." [18]

As oil-watchers monitor trends and conservation plans, here are some of the issues in debate:

Have global oil supplies peaked?

When a country's oil resources peak — or hit the point where half the oil is gone — it happens without warning, said a veteran energy company executive and researcher, Robert L. Hirsch, who conducted a peak-oil study in 2005 for the U.S. Department of Energy. That's what happened in North America, Britain, Norway, Argentina, Colombia and Egypt, Hirsch said. "In most cases, it was not obvious that production was about to peak a year ahead of the event. . . . In most cases the peaks were sharp, not gently varying or flat-topped, as some forecasters hope. In some cases, post-peak production declines were quite rapid." [19]

But Cambridge Energy Research Associates argues that a global peak — when it is reached many decades from now — will not mark the beginning of a precipitous drop-off. "Global production will eventually follow an "undulating plateau" for one or more decades before declining slowly," the firm said. [20]

A study by the nonpartisan GAO adds to the uncertainty over oil reserves. "The amount of oil remaining in the ground is highly uncertain," the agency concluded, "in part because the Organization of Petroleum Exporting Countries controls most of the estimated world oil reserves, but its estimates . . . are not verified by independent auditors." [21]

In part, debate turns on the extent to which oil producers can turn to so-called "unconventional" sources. Shale oil — a form of petroleum extracted by applying very high temperatures to certain types of rock formations abundant in parts of the American West — has been viewed for decades as an alternative to conventional crude oil. The GAO reported that one-half million to 1 million barrels a day could be extracted from U.S. shale within 10 years, though the process is expensive and energy-intensive. [22]

Oil can also be extracted from tar sands which have become one of Canada's major sources of petroleum. An oil-sands boom is under way in Canada, which is producing about 1.2 million barrels of oil from sands in Alberta Province, though the process requires burning so much natural gas that emissions have done considerable environmental damage. [23]

Cambridge Energy Research Associates said in its rebuttal of the "peak" thesis that oil sands and other unconventional sources may account for 6 percent of global production by 2030. [24]

Peak oil thesis advocates argue that unconventional sources won't suffice for the world's needs. They turn the argument back to the region still considered the globe's main petroleum reservoir, the Middle East. Houston

investment banker Simmons says that a lack of verifiable information about Middle Eastern reserves lies at the heart of peak oil theory.

"These optimists — I'm happy they're so happy about things — but they have no data to base their case on," Simmons says. "We have passed peak oil, and demand is not going to slow down." Simmons' 2005 book, *Twilight in the Desert*, is a major text of the hypothesis. [25]

A gas station burns in Tehran on June, 27, 2007, during protests against efforts by the Iranian government to reduce consumption by imposing gas rationing.

Simmons insists his projections and forecasts are more data-driven than those of peak oil critics. "That's one of the reasons I boldly predicted in 1995 that the North Sea was likely to peak by 1998-2000," Simmons says. "The major oil company people said I was nuts. All I did was look at the reports."

Experts agree the North Sea passed its high point and that the industry is doing its best to pump out the remaining crude. "Oil and gas production has peaked, [and] the industry is concentrating on managing the decline," said Trisha O'Reilly, communications director of Oil and Gas UK, the trade association of North Sea oil producers. "There's still a sizable prize out there." [26]

Some oil insiders accept parts of the peak oil argument, but others dismiss it as panic-mongering that only drives up prices for the benefit of price speculators. "Peak oil theory is a lot of baloney," says energy analyst Gheit, at Oppenheimer & Co. "We are consuming more, but we are finding more than we consume; reserves continue to bulge."

Vastly improved technology has facilitated the discovery of new reservoirs even in well-developed fields, Gheit and others argue. For example, he says, "In the old days, when they built the first platform in the North Sea, it was like a very big table made of concrete with hollow legs. Now there is something called sub-sea completion, where all the equipment is sitting on the ocean floor, and everything is robotically controlled."

West of PFC Energy agrees that while onshore U.S. fields and the North Sea have peaked and been squeezed "dry" thanks to technological advances, "there are parts of the Middle East and Russia that are virtually unexplored."

But peaking may be more widespread than some industry insiders say, another oil expert argues. "People are asking the right questions about peak oil, but they're asking about the wrong country," says David Pursell, managing director of Tudor Pickering, a Houston-based investment firm. "We know that Mexico has peaked. When does Russia peak?"

Will the rising energy needs of India, China and other developing countries keep oil prices high?

The newest twist in the volatile world of global oil economics is growing petroleum demand by Earth's two population giants — China (1.3 billion people) and India (1.1 billion people) — which together account for more than a third of the planet's 6.6 billion population. The two huge nations have been maintaining annual economic growth of about 10 percent a year, sparking intense demand among new members of their rising middle classes for cars and other energy-intensive consumer goods.

China had 22 million cars and "light-duty vehicles" on the road in 2005, with 10 times as many projected in 2030. In India, a tenfold increase is also expected — from 11 million to 115 million, according to the Paris-based International Energy Agency (IEA). [27] By comparison, there are about 250 million cars and other motor vehicles in the United States, or slightly more than one for each of the approximately 240 million adults in the population. [28]

If all countries maintain their present energy policies, the IEA says developing countries will account for 74 percent of the increase in worldwide energy use from all sources between 2005 and 2030, with China and Indian accounting for 45 percent of that boost. Developing countries now make up 41 percent of the global energy market. By 2030, if no policies change, those countries would account for 54 percent of world consumption. [29]

All in all, the IEA concludes, "The consequences for China, India, the OECD [Organization for Economic Co-operation and Development] and the rest of the world of unfettered growth in global energy demand are alarming . . . the world's energy needs would be well over 50 percent higher in 2030 than today."

The report goes on to recommend international efforts to reduce demand — for environmental reasons as well as to conserve oil and keep prices from skyrocketing. But some experts say the growing presence of China and India in the world energy market will keep prices high no matter what measures are taken.

"You're talking about economic development in two countries that comprise a little over one-third of the world population," says Medlock at the Baker Institute. "It's going to be difficult for the energy supply to expand production at a significant enough pace to drive down prices." Only an international economic slowdown could have that effect, he adds.

But Gheit of Oppenheimer & Co. argues that major price increases generated by continued growth in demand will force China and India to adapt, just as other nations do. "Energy conservation accelerates when prices go higher — even in China and India," he says. "That's the mitigating factor. Any developing economy becomes more energy-efficient with time."

Gheit adds that the Chinese and Indian governments have a highly efficient tool at their disposal if they want to curb demand: Both countries keep gasoline prices low through subsidies. "If gasoline subsidies were to cease, demand would crash," he says. "Roads will be half-empty."

But some oil experts say China's energy demands reflect far more than stepped-up car use. "The big thrust on Chinese demand is really on production of energy-intensive goods for their export industry," says Morse at Lehman Brothers. China's policy of keeping its currency undervalued to make exports cheaper is maintaining that effort — and causing the high energy demand that results.

Communist Party leaders in China oppose ending gasoline price subsidies, according to a Lehman Brothers analysis. And even if they were eliminated, "Chinese motorists might dip into their savings, and businesses might borrow more from banks to foot higher energy bills." In the long run, however higher prices likely would force down demand, the analysis says. [30]

In the United States, meanwhile, high gasoline prices, perhaps combined with wider economic troubles, have reduced demand. Normally, lower demand would send prices down. But some experts say the high oil demand from China and India has changed the outlook. Long term, says Pursell of Tudor Pickering in Houston, prices are going to be higher in the next 10 years than in the past 10 years.

Nonetheless, the market system continues to function, some economists point out. "At these prices, an enormous incentive exists to develop new [oil] sources," says Robert Crandall, a senior fellow at the Brookings Institution and former director of the Council on Wage and Price Stability in the Ford and Carter administrations. "My guess is that after three-four-five years, new pools will be found."

But, Crandall cautions, new oil fields may sit in regions that are difficult to reach, for geographical or political reasons.

Can the federal government do anything to significantly reduce energy demand?

American worries about oil dependence and its effects on the global environment reached critical mass in December, when Congress passed, and President Bush signed, an energy bill designed to force major reductions in U.S. petroleum consumption. Bush, a former oilman, had previously acknowledged that the political climate now favors energy conservation. In his 2006 State of the Union address, he said, "America is addicted to oil." [31]

The new energy law includes tougher corporate average fuel efficiency (CAFE) requirement for cars and light trucks (including SUVs). They will have to meet a fleetwide average standard of 35 miles per gallon by 2020, compared with the present 27.5 miles per gallon for cars and 22.2 miles per gallon for light trucks. The bill also requires the production of 36 billion gallons of

Most Oil Belongs to OPEC Nations

Members of the Organization of Petroleum Exporting Countries (OPEC) held more than three-quarters of the world's 1.2 trillion barrels of crude oil reserves in 2006 (left). Most OPEC oil reserves are in the Middle East, with Saudi Arabia, Iran and Iraq holding 56 percent of the OPEC total (right).

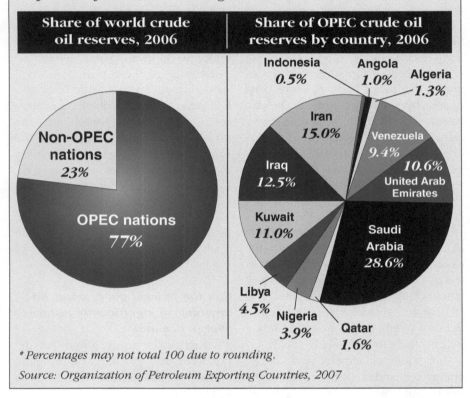

Share of world crude oil reserves, 2006

Non-OPEC nations 23%

OPEC nations 77%

Share of OPEC crude oil reserves by country, 2006

Indonesia 0.5%
Angola 1.0%
Algeria 1.3%
Iran 15.0%
Venezuela 9.4%
10.6% United Arab Emirates
Iraq 12.5%
Kuwait 11.0%
Saudi Arabia 28.6%
Libya 4.5%
Nigeria 3.9%
Qatar 1.6%

* Percentages may not total 100 due to rounding.

Source: Organization of Petroleum Exporting Countries, 2007

---

ethanol, the plant-based gasoline substitute, by 2022 — five times more than present production levels. [32]

Due to the combined effects of the fuel efficiency standard and the ethanol production boost, "We will save as much oil as we would import from the Persian Gulf — 2.59 million barrels a day," says Brendan Bell, Washington representative of the Union of Concerned Scientists, citing projected oil demand if the law hadn't been enacted.

Some disappointment could be heard amid the cheers, however, because lawmakers balked at dealing with renewable electricity. "It's really unfortunate that we didn't have the renewable electricity standard or the incentives for wind and solar," Sen. Barbara Boxer, D-Calif., chairwoman of the Senate Environment and Public Works Committee, said. "But we'll fight for those another day." [33]

Still, opposition hasn't entirely died away. Rep. F. James Sensenbrenner Jr., R-Wis., who voted against the bill, had argued at a Nov. 14 hearing of the House Select Committee on Energy Independence and Global Warming that the new energy-use standards were unlikely to have much global effect on auto efficiency or on tailpipe emissions. "These regulations may work if everybody all over the world agreed to them and then actually complied with them," Sensenbrenner said. [34]

In addition to skepticism about the likely impact of a U.S. law on world energy use, critics are also asking whether markets can be relied on, without government involvement, to resolve supply-demand imbalances. That is, will prices rise in response to scarcity? In the classic supply-demand scenario, higher prices encourage companies to find and produce more oil, because they'll make more money — though if prices rise too much, demand drops.

Some experts who hold that world oil supplies are diminishing argue that the resulting problems are too big for the market alone to handle. "Intervention by governments will be required because the economic and social implications of oil peaking would otherwise be chaotic," said the report to the Energy Department directed by former energy executive Hirsch in 2005. [35]

But two years later, Hirsch warns that government action to reduce demand won't produce immediate results. "We have to do it, but we can't be unrealistic in our expectations," Hirsch says. "If you pass a dramatic increase in CAFE now, a significant number of new cars will not show up for about three years. It takes that long to get prepared with parts suppliers, assembly lines and so forth. And people may not buy the cars unless they're feeling pained or are required to by the government."

Government intervention would do far more harm than good, argues Jerry Taylor, a senior fellow at the Cato Institute, a libertarian think tank. "One thing markets are good at doing is allocating scarce resources among competing users, based on ability to pay," he says. "A peak in global oil production would send a very strong signal to consumers that oil is going to become scarce. If government decides to help steer the economy through a peak scenario, its main mission will be to dull that price signal to make sure consumers don't get it in the teeth."

But even some Wall Street energy experts argue that tougher fuel efficiency standards are long overdue. "If we'd had [the 35-miles-per-gallon] standard in place in 1990," says Morse of Lehman Brothers, "we'd be consuming 2 million barrels a day less now, and we'd be consuming 3 million barrels a day less if we had imposed the fuel efficiency standards on trucks that we have on cars."

No new standards or sudden consumer preference for fuel-efficient gasoline-electricity hybrid cars in the near future will have a dramatic effect on oil demand, another industry expert says. "It's a feel-good measure in the near term," says Pursell of Tudor Pickering in Houston. "In the long term, it probably makes sense. But we have roughly 150 million cars on the road." With so many cars, he says, requiring better fuel efficiency for new cars would take years to show results.

"So what can you do in the near term?" Pursell asks. "Drive less." He adds that his fellow Texans, who favor big vehicles on the long roads they travel, don't cotton to the idea of cutting back on time behind the wheel.

Another veteran oil analyst argues that market reaction to changing conditions is already well under way. "We have reached a saturation point on cars," says Gheit of Oppenheimer & Co., citing anecdotal but plentiful evidence of jam-packed streets and highways throughout the country. "Everywhere you go you're stuck in traffic. You go out and you can't find parking. These things are beginning to take a toll on the number of cars sold in North America."

And the cars that are sold are more fuel-efficient than earlier models, Gheit says. Hence the market is coming up with its own solutions. "You're seeing more and more advancement. Economic advancement comes with much more energy efficiency." ∎

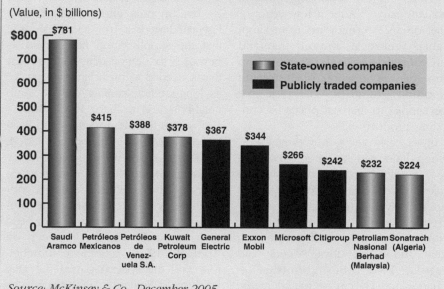

Top Oil Companies Are State-Owned

Six of the world's 10 biggest firms are state-owned oil companies from the Middle East or Asia. The largest is Saudi Arabia's Saudi Aramco, valued at $781 billion.

The world's 10 largest companies, 2005

(Value, in $ billions)

State-owned companies
Publicly traded companies

- Saudi Aramco — $781
- Petróleos Mexicanos — $415
- Petróleos de Venezuela S.A. — $388
- Kuwait Petroleum Corp — $378
- General Electric — $367
- Exxon Mobil — $344
- Microsoft — $266
- Citigroup — $242
- Petroliam Nasional Berhad (Malaysia) — $232
- Sonatrach (Algeria) — $224

Source: McKinsey & Co., December 2005

BACKGROUND

Energy Shock

In 1956, M. King Hubbert, a geologist for Shell Oil, told the American Petroleum Institute (API) that he had determined when U.S. oil production would hit its peak. After calculating the maximum reserves of U.S. oil fields (200 billion barrels) and the rates at which oil companies would keep pumping, he announced that the peak year would arrive in the 1970s. [36]

As it happened, U.S. production peaked in 1970, many experts say, when the United States was producing about 10 million barrels a day. Today, production has fallen by about half. [37] "We picked up again in the late '70s but still didn't go back to the previous high," says Ron Planting, an API economist.

But in 1956, Americans in general and the oil industry in particular believed American wells would be spouting oil and gas into the indefinite future. So when Hubbert announced his conclusion, "It was as if a physician had diagnosed virulent, metastasized cancer; denial was one of the responses," writes Kenneth S. Deffeyes, a retired professor of geology at Princeton University who was a protégé of Hubbert's at Shell. [38]

Some analysts take issue with the notion that Hubbert has been proved right. Technological advances have made it possible to probe oil and gas formations more accurately, leading to increased production in some cases, and recalculation — upwards — of reserves.

At the time of Hubbert's forecast, there was growing resentment among the oil-producing countries of the

Middle East and Latin America of the power wielded by the "big eight" foreign oil companies, half of them American (the so-called "Seven Sisters," plus France's state-owned oil company). While the foreign companies controlled the price of a resource that the world depended on, the supplying countries had little say. [39] After a few years of quiet discussion, ministers from Saudi Arabia, Venezuela, Kuwait, Iraq and Iran convened in Baghdad in 1960 to form the Organization of Petroleum Exporting Countries. The objective was simple: to manage prices by controlling production.

In its early years, however, OPEC swung little weight, largely because the big oil companies were making major discoveries in countries that weren't — yet — members of the new organization. Over the years, the membership expanded to include Qatar, Indonesia, Libya, Nigeria and Angola. Two other countries — Ecuador and Gabon — joined in the 1970s but dropped out in the '90s.

As soon as the United States began depending on foreign oil, events showed that dependency had made the country vulnerable.

Following the 1973 Yom Kippur War, which pitted Israel against Egypt, Syria and Iraq, Arab OPEC nations retaliated against the United States and other Israeli allies by launching an oil embargo against them.

The embargo began on Oct. 17, 1973, and almost overnight 4 million barrels of oil a day were removed from world supplies. Demand rose 7 percent above supply, and international prices quadrupled from $3 a barrel to $12. Some saw the boycott as vindication of predictions that the energy foundation on which Western civilization depended would dry up. "The party is over," declared E. F. Schumacher, a British economist who had long prophesied an end to cheap oil. [40]

To prevent the high oil price from rippling through the economy, President Richard M. Nixon imposed price controls on oil. And his successor, Gerald R. Ford, established the Strategic Petroleum Reserve, an emergency stockpile that today has about 695 million barrels. [41]

But the boycott that gave rise to those measures ended in March 1974 — five months after it had begun. Egyptian President Anwar Sadat declared that the supply cutoff had served its purpose: to demonstrate to the West that it needed to push Israel to resolve its longstanding conflict with its Arab neighbors and with the Palestinians.

Even as the memories — and frustrations — of the Arab boycott faded, Ford's successor warned the country about potential future emergencies. "This is the greatest challenge our country will face during our lifetimes," President Jimmy Carter said in a nationally televised speech on April 18, 1977. [42] Some commentators said the speech paved the way for Ronald Reagan's 1980 election as president. Reagan portrayed himself as the optimistic alternative to gloomy Democrats.

Meanwhile, however, the 1979 Iranian revolution renewed America's sense of energy vulnerability after Shah Mohammed Reza Pahlavi — a close U.S. ally who ruled an oil superpower — was toppled by Muslim radicals who made anti-Americanism a tenet of their doctrine.

On Nov. 4, 1979, Iranian revolutionaries seized the U.S. Embassy and took 52 employees hostage — holding them for 444 days. Panic took hold of energy markets again, and prices shot up to the $45-per-barrel range, as high as they'd ever been. [43] During both the 1973 and 1979 crises American motorists sat in long lines at service stations, and some station owners who were short on supplies began rationing gasoline. During the second "oil shock," Congress funded research into alternative fuels and encouraged Americans to conserve fuel.

Then, even before the hostages were released in early 1981, Iraqi dictator Saddam Hussein attacked Iran. Oil exports by the two countries virtually ceased for a time, as production facilities were bombed during the first months of fighting in 1980. About 4 million barrels a day vanished from the market, setting off a new round of panic buying. The eight-year war eventually had little lasting effect on oil markets.

Consumers Go Wild

After the disruptions caused by war and revolution, market forces restored stability to oil trade. The law of supply and demand got a big assist from Saudi Arabia, the world's biggest oil producer. Worried that a prolonged period of high prices would cut oil demand by newly conservation-minded Western countries — and force prices lower — Saudi Arabia had been steadily increasing its production. Non-members of OPEC followed suit.

Other developments were at work as well. After the oil shocks of the 1970s, energy companies stepped up exploration outside the turbulent Middle East. By the early 1980s, the results began pouring in.

At least 6 million barrels a day were added to world oil supplies by Britain and Norway's production in the North Sea, a new pipeline from Alaska's rich North Slope to the port of Valdez and a major discovery in Mexican waters in the Gulf of Mexico. The new output, coming from outside the OPEC circle, was within striking distance of Saudi Arabia, which in 1981 reached a top daily production level of about 9.8 million barrels. [44]

Meanwhile, the effects of conservation measures adopted during the 1970s kicked in. The most significant was a 1975 law setting tough corporate average

Continued on p. 12

Chronology

1950s-1970s
Oil imports ease concerns about decreasing U.S. oil supplies — until big oil-producing nations suspend their shipments.

1956
Shell Oil geologist forecasts that the U.S. oil supply will plateau in the early 1970s.

1960
Saudi Arabia, Venezuela and other oil giants form Organization of Petroleum Exporting Countries (OPEC).

1970-72
U.S. oil reserves peak, as predicted.

1973
Arab members of OPEC cut off oil exports to the United States and other allies of Israel, causing oil prices to skyrocket. The embargo ends in March 1974.

1975
President Gerald R. Ford signs the Energy Policy and Conservation Act, imposing fuel economy standards on carmakers.

1977
President Jimmy Carter calls energy conservation the country's biggest challenge.

1979
Revolution topples the U.S.-backed shah of Iran, prompting oil price spikes in the United States and other big oil-consuming nations.

- - - • - - -

1980s
New non-OPEC oil sources are discovered or come online, vastly expanding world oil supplies and causing prices to plummet.

1980
Non-OPEC oil supply expands by about 6 million barrels a day after Mexico's daily production rises, new North Sea sources come on-line and drilling is stepped up in Alaska. . . . Iraq attacks Iran.

1983
OPEC cuts prices from $34 a barrel to $29.

1985
Oil falls to $10 a barrel; Saudi Arabia steps up output and abandons efforts to prop up prices.

1988
Iran-Iraq War ends, removing source of potential oil-market disruption.

- - - • - - -

1990s
Steady supply of cheap oil sparks popularity of gas-guzzling sport-utility vehicles (SUVs), but trouble looms by decade's end.

1990
Automakers sell 750,000 SUVs; annual sales hit 3 million 10 years later. . . . Iraqi leader Saddam Hussein invades Kuwait, prompting fears about Saudi oil security.

1991
U.S.-led forces oust Iraq from Kuwait, maintain protective presence in Saudi Arabia.

1996
Russia begins developing oil production facilities in its Far East region. . . . Saudi Islamist Osama bin Laden releases manifesto attacking U.S. military presence in Saudi Arabia.

1998
Socialist Hugo Chávez is elected president of Venezuela.

2000s
Terrorism, war and fear of war disrupt oil prices in the Middle East.

Sept. 11, 2001
Arab terrorists crash hijacked U.S. jetliners into the World Trade Center and Pentagon, killing nearly 3,000 people.

2002-2003
Venezuelan oil workers strike against the Chávez government's efforts to reduce production, pushing prices up.

2003
U.S.-led coalition invades Iraq, topples Saddam.

2004
Insurgents attack Iraqi oil facilities, prompting price fluctuations.

2005
Gas hits $3 a gallon in the U.S. . . . Author of a report on "peak oil" tells lawmakers government should prepare for possible oil shortage.

2006
Saudi Arabia's oil consumption rises by 2 million barrels a day in one year. . . . Chávez pledges to sell China 1 million barrels of oil a day by 2012.

2007
International Energy Agency warns of looming oil shortfall. . . . Crude oil price nears $100 per barrel. . . . President George W. Bush signs new energy bill including a fuel-efficiency standard of 35 miles per gallon for cars and light trucks by 2020. . . . Environmental Protection Agency denies California and 16 other states the right to set auto emission standards.

Jan. 2, 2008
Crude oil price hits $100.

How Times Have Changed

Now petrostates are bailing out U.S. firms.

Only a few decades ago, American oil companies stood among the petroleum giants that controlled most of the world's oil, and their profits largely were recycled back into the United States.

But times have changed. "For some time now," writes former Treasury Secretary Lawrence H. Summers, "the large flow of capital from the developing to the industrialized world has been the principal irony of the international financial system." [1]

In today's world, a tiny Persian Gulf state can rescue a major American bank from financial catastrophe using money earned from selling millions of barrels of oil. And politicians in Europe and the United States are nervous about their nations' companies being bought up by cash-swollen petrostates.

"Their wealth is a reminder to our politicians that the West is no longer the force it once was in the world," wrote Michael Gordon, fixed-income director at Fidelity International, a giant investment-management firm. "And just maybe, business leaders are ahead of the politicians in welcoming this infusion of new money into the global financial system." [2]

Last year, U.S. lawmakers of both parties scuttled a deal that would have allowed a company owned by the government of Dubai* to run six major U.S. ports. "This proposal may require additional congressional action in order to ensure that we are fully protecting Americans at home," wrote House Speaker J. Dennis Hastert, R-Ill. [3]

Political jitters over the wide range of foreign government funds invested don't all center on the oil-rich countries. China, which

---

* Dubai is one of seven Arabian Peninsula city-states that constitute the United Arab Emirates.

---

has grown rich selling cheap goods to the rest of the world, has set alarm bells ringing on Wall Street over attempted investments in American and other Western companies. In 2005, a political firestorm forced China's state-owned oil company to abandon a bid to buy Unocal, a U.S. oil company. [4]

China's sheer size and strategic importance guarantee continuing interest in its investment projects. But high oil prices in 2007 have focused attention on efforts by oil-exporting countries to invest their profits — totaling a mind-boggling $3.4-$3.8 trillion — much of it in the West, according to the McKinsey Global Institute.

And the developing world's cash situation is expected to get even more dramatic in the future. "The most conservative assumptions you could think of, absent some catastrophic event, would have [these assets] double by 2012," Diana Farrell, the institute's director, said in December. [5]

In fact, even if the price of oil falls from current levels (now above $90 a barrel) to $50 a barrel, petrodollar assets would expand to $5.9 trillion by 2012, the institute says, fueling investment at a rate of about $1 billion a day. [6]

Political resistance to Middle Eastern oil profits buying up American companies surfaced even before oil prices skyrocketed in 2007. In 2006, Dubai PortsWorld bought a British firm that ran port operations in New York City, New Jersey, Philadelphia, Baltimore, Miami and New Orleans. Lawmakers of both parties lost no time in denouncing the deal with an Arab nation as a threat to national security, and the government of Dubai eventually sold its interest in the U.S. operations. [7]

But some international finance experts urge politicians and others to look at other implications, such as whether foreign-

---

Continued from p. 10

fuel efficiency (CAFE) standards for new cars of 27.5 miles per gallon by 1985. The measure would save about 2 million barrels of gasoline a day. [45]

Production from the new oil fields combined with new conservation efforts would have been enough by themselves to push oil prices down. But a third factor emerged as well: In a newly restabilized geopolitical environment, oil companies began selling off oil that they'd been holding in storage against the possibility of long-lasting shortages. Companies couldn't justify the considerable expense of warehousing the oil.

By March 1983, OPEC was feeling the pressure from an oil glut that it couldn't control by shutting down production, because much of the new supply came from outside the organization's control. So OPEC took the unprecedented step of cutting prices from about $34 per barrel to about $29. With the world oil supply still plentiful and with new CAFE standards reducing demand, prices kept falling even further. In 1985, with oil at $10 a barrel, Saudi Arabia gave up trying to limit OPEC output and stepped up its own shipments.

American automakers and consumers, meanwhile, reacted in their own ways. Unconcerned (for the moment) about oil prices and supplies, manufacturers began expanding their production of popular SUVs. Classified by the government as "light trucks," the gas-guzzling SUVs were subject to less rigorous fuel efficiency standards.

"Gasoline remained readily available, and its price stayed flat instead of soaring to the $20 a gallon level once predicted by energy forecasters," a journalist specializing in the auto industry wrote in 1996. [46]

In 1990, carmakers sold 750,000 SUVs nationwide. By 2000, annual sales were approaching 3 million.

---

owned companies could end up unduly influencing domestic policy. "What about the day when a country joins some 'coalition of the willing' and asks the U.S. president to support a tax break for a company in which it has invested?" Summers asked, using Bush administration terminology for U.S. allies in the Iraq War. "Or when a decision has to be made about whether to bail out a company, much of whose debt is held by an ally's central bank?" [8]

A firm owned by the government of Dubai backed out of a 2006 deal to run six U.S. ports after U.S. lawmakers protested.

AP Photo/Kamran Jebreili

the whole world goes in the tank," says Kenneth Medlock III, an energy research fellow at the James A. Baker III Institute for Public Policy at Rice University in Houston. "That would put a crimp in oil demand."

[1] See Lawrence Summers, "Sovereign funds shake the logic of capitalism," *Financial Times* (London), July 30, 2007, p. A11.

[2] See Michael Gordon, "Ignore the murk and myths on sovereign funds," *Financial Times* (London), Dec. 12, 2007, p. A13.

[3] Quoted in Jim VandeHei and Jonathan Weisman, "Bush Threatens Veto Against Bid to Stop Port Deal," *The Washington Post*, Feb. 22, 2006, p. A1.

[4] See Jad Mouwad, "Foiled Bid Stirs Worry for U.S. Oil," *The New York Times*, Aug. 11, 2005, www.nytimes.com/2005/08/11/business/worldbusiness/11unocal.html?_r=1&oref=slogin.

[5] See "Sovereign Wealth Fund Briefing," (transcript) Brookings Institution, Dec. 6, 2007, p. 16, www.brookings.edu/~/media/Files/events/2007/1206_sovereign_wealth_funds/1206_sovereign_wealth_funds.pdf.

[6] See "The New Power Brokers: How Oil, Asia, Hedge Funds, and Private Equity are Shaping Global Capital Markets," McKinsey Global Institute, October 2007, pp. 12-13, www.mckinsey.com/mgi/publications/The_New_Power_Brokers/index.asp.

[7] See Richard Simon and Peter Wallsten, "Bush to Fight for Port Deal," *Los Angeles Times*, Feb. 22, 2006, p. A1; "Dubai Firm Details Plans for U.S. Ports," *Los Angeles Times* (The Associated Press), March 16, 2006, p. C3.

[8] See Summers, *op. cit.*

[9] See Daniel Yergin, *The Prize: The Epic Quest for Oil, Money, and Power* (1992), pp. 511-512, 467-470.

In the 1950s, the oil-rich countries worried about foreign involvement in their economic and political affairs. For instance, the Iranians did not take kindly to the U.S.-organized 1953 coup in Iran that ousted Prime Minister Mohammed Mossadeq, who had nationalized a British-owned oil company. And oil-producing countries also resented foreign oil companies' control of petroleum pricing and marketing. Eventually, most countries nationalized their oil resources. [9]

Now the situation is almost reversed, with the industrialized countries coming to depend on the oil countries for oil as well as cash.

But that's not necessarily a bad thing, some experts note, because investments in the industrialized world give the oil countries a stake in maintaining stability and prosperity, not to mention a market for petroleum. "If the U.S. goes in the tank,

By the mid-1990s, however, there were warning signs that the latest cheap-oil era might be ending. The signs included a little-noticed 1996 anti-American manifesto by a Saudi Arabian millionaire and veteran of Afghanistan's U.S.-aided war against Soviet occupation in 1979-1989. By then, Osama bin Laden had developed a deep hatred for the United States, and he decried the presence in Saudi Arabia of American troops, which had been providing security for the oil giant ever since Saddam Hussein invaded Kuwait during the Persian Gulf War of 1990-1991. [47]

Tide Turns Again

The terrorist attacks of Sept. 11, 2001, for which bin Laden later claimed responsibility, might have been expected to cause a major disruption in the oil market. Indeed, only hours after the terrorists struck, prices on the International Oil Exchange in London rose by 13 percent, to $31.05 a barrel. And as rumors of major shortages swept through parts of the United States, some drivers in Oklahoma City saw prices at the pump surge to $5 a gallon. [48]

But the wholesale and pump price spikes proved momentary. No terrorists hit any oil facilities, and OPEC immediately issued a market-calming declaration that it would not use the oil weapon against the United States for whatever military action it took to answer the attacks. Overall, the average wholesale price paid by U.S. refineries in September was lower than they'd paid the previous month — a drop from $24.44 a barrel to $23.73. "By October 2001, the economy was having more effect on the price of oil — in terms of weakening oil demand and reducing oil prices — than the price of oil was having on the econ-

Plug-in Hybrids Offer Clean-Energy Future

New technology may enable motorists to burn less oil.

Its cities separated by hundreds of miles of windswept, open spaces, Texas may not be the place to start up a conversation about carpooling. "That's a very unpopular discussion to have here in Houston," says Kenneth B. Medlock III, speaking from his car.

"You look out on a freeway," says Medlock, an energy studies fellow at the James A. Baker III Institute for Public Policy at Houston's Rice University, "and all you see is car after car with a driver and no passengers."

Texans may be especially fond of their cars — but Lone Star State drivers aren't unique. Transportation (including airplanes and trucking) accounts for two-thirds of U.S. petroleum use, according to a July 2007 study by the National Petroleum Council. [1]

That's hardly a surprising statistic, given the size of the U.S. car and truck fleet: nearly 250 million vehicles in a nation of about 300 million. Relying on fuel efficiency standards alone to hold gasoline use in 2017 to what it was in 2005 would require improving average vehicle performance to 22 miles per gallon — a 25 percent improvement over today, researchers at the Baker Institute calculate. [2]

To reduce oil consumption to 2005 levels by conservation alone, every American would have to drive 45 miles a week less. "Basically, it's a lifestyle change," says Medlock, who worked on the study.

But some energy experts are arguing that new automotive technology will allow Americans to keep driving while burning less oil. They tout the plug-in hybrid electric vehicle (PHEV), a variant of the gasoline-electric hybrid car whose electric motor gets recharged from an ordinary wall socket. Limited-edition PHEVs — Toyota Priuses retrofitted by conversion companies or by enthusiasts — boast bigger batteries that allow drivers to cruise for about 20 miles on electric power alone, burning no gasoline. Unmodified Priuses can travel only about a half-mile on electricity alone, according to the Institute of Electrical and Electronics Engineers. PHEV advocates also say that recharging the cars at night uses surplus electricity that utilities hold in reserve for emergencies. [3]

The reliance on wall current, though, raises the question of whether the plug-ins wind up burning as much energy as the hybrid models now on sale. Alternative-energy advocates raise another objection. "If you start plugging in hundreds of cars all at once, you'll be finding out what the limits of the electricity grid are real quick," Paul Cass, a representative of Ballard Power Systems, a Canadian firm, told the *Los Angeles Times* at an alternative-vehicle convention. [4]

Ballard makes hydrogen fuel cells for use in cars. Hydrogen technology, attractive to many because it uses no fossil fuels at all, is getting a big push from the government — $195.8 million

---

omy," the Congressional Research Service concluded in a report a year after the attacks. Demand weakened in part because airplane travel dropped in the immediate aftermath of the attacks. [49]

As the decade wore on, however, a series of developments began to push prices higher. By late 2004, oil was commanding about $50 a barrel. Analysts cited the effects of the war in Iraq in reducing that country's production and export capacity, as well as the economic booms already under way in China and India.

By 2005, gas prices nationwide had passed $3 a gallon. Oil and marketing experts had long contended that the $3 price was a critical threshold. Rebecca Lindland, an automotive industry analyst for Global Insight, a research firm in Waltham, Mass., had told *The New York Times* in 2004 that

consumers would change their driving and car-buying behavior if prices at the pump exceeded $3 a gallon for at least six months. [50] The forecast proved accurate. As higher prices stayed steady, SUV lovers started shying away from sport-utility vehicles. "I really want my Explorer back, but I'm thinking about not getting it because of gas prices," said Angie Motylinski, a bank teller in Sylvania, Ohio, whose lease was expiring. "If they gave me an awesome, awesome deal, I might consider it. But who's going to want it when gas is $3.19 a gallon?" [51]

Other Sylvanians were thinking similar thoughts. "If I had a dollar for every time that somebody said I'm looking for something with better gas mileage, I'd be a wealthy man," said Bill Roemer, the manager of a local Chevrolet dealership. "With gas prices

the way they are, people just aren't looking at minivans, SUVs, trucks." [52]

Petro-Nationalism

As prices spiraled upwards, a trend that had begun decades earlier suddenly took on new importance for oil-watchers. In the 1970s, publicly owned firms — all in the West — owned roughly three-quarters of global petroleum; today, state-owned oil companies own three-quarters of the oil. [53] That poses a potential problem for the U.S because governments that control oil supplies may have economic and/or political reasons to limit their foreign sales. The 1973 OPEC oil embargo serves as a reminder of the potency of oil as a political weapon against the United States. And some

in research and development money from the Department of Energy. Electric and hybrid-electric car research is getting $50.8 million. [5]

Technical arguments aside, the PHEV is far closer to dealer showrooms. "There are no truly viable hydrogen fuel cells on the market today," acknowledged Bud DeFlaviis, government-affairs director of the U.S. Fuel Cell Council, a trade group. [6]

The argument that plug-in hybrids don't reduce energy consumption overall has been persuasive, because non-nuclear power plants burn fossil fuels. That issue is especially important on the environmental-protection side of the alternative-energy debate.

But a July 2007 report gives ammunition to the PHEV advocates. After an 18-month study, the Electric Power Research Institute and the Natural Resources Defense Council (NRDC), an environmental group, concluded that widespread use of plug-in hybrids would, in 2050, reduce oil consumption by 3-4 million barrels a day. It would also cut greenhouse gas emissions by 450 million metric tons a year — the equivalent of

Toyota and France's state-owned energy company plan to develop recharging stations for plug-in hybrid cars in major European cities.

taking 82.5 million cars off the road. "Our results show that PHEVs recharged from low- and non-emitting electricity sources can decrease the carbon footprint in the nation's transportation sector," said David Hawkins, director of the NRDC Climate Center. [7]

Those numbers might be persuasive, even in Texas.

[1] "Hard Truths: Facing the Hard Truths About Energy," National Petroleum Council, July 2007, p. 46, www.npchardtruth-sreport.org.

[2] See Kenneth B. Medlock III and Amy Myers Jaffe, "Gas FAQ: U.S. Gasoline Markets and U.S. Oil Import Dependence," James A. Baker III Institute for Public Policy, Rice University, July 27, 2007, pp. 3, 13, www.rice.edu/energy/publications/FAQs/WWT_FAQ_gas.pdf.

[3] See "Take This Car and Plug It," IEEE Spectrum, July 2005, http://ieeexplore.ieee.org/iel5/6/31432/01460339.pdf?arnumber=1460339.

[4] Quoted in Ken Bensinger, "The Garage: Focus on autos; 2 'green' technologies race for driver's seat," Los Angeles Times, Dec. 8, 2007, p. C1.

[5] Ibid.

[6] Ibid.

[7] See "EPRI-NRDC Report Finds Environmental Benefits of Deploying PHEVs," July 19, 2007, www.nrdc.org/media/2007/070719.asp. Report accessible at www.calcars.org/calcars-news/797.html.

producing countries may decide to increase the amount of oil they use at home. (See Current Situation, p. 18.) To be sure, the No. 1 international oil supplier, Saudi Arabia, still cooperates closely with the United States and other consuming countries. And Nigeria and Brazil invite foreign companies to help develop national petroleum resources.

But Venezuela is headed by an anti-American president who has threatened more than once to cut off sales to the United States by its state-owned oil firm. President Hugo Chávez also plans to sell less oil to the United States and more to China. In fact, in 2006 the pugnacious Chávez vowed to sell 1 million barrels a day to China by 2012. [54]

And then there is Russia. A major buildup of production capacities in the country's Far East region has turned Russia into an oil behemoth. As such,

it once again sees itself as a great power. And some of Russia's neighbors say it uses petroleum as a weapon. In winter 2006, vitally needed natural gas stopped flowing from Russia to the Republic of Georgia, headed by a president who tried to defy Russian supremacy in the region. Russia blamed a technical problem — an explanation Georgians rejected. [55]

But the major concern for private oil companies and oil-consuming countries such as the United States is not a cutoff in service by a state-owned oil firm. The big issues are access to oil fields and participation in production ventures. "Access really is a consideration," said oil historian and consultant Yergin. "Where can you go to invest money, apply technology and develop resources and bring them to market? Terms get very tough." [56]

In an ironic twist, some state-owned oil companies that have grown enormously wealthy recently have ridden to the rescue of some ailing U.S. companies. Notably, the Abu Dhabi Investment Authority spent $7.5 billion on a stake in Citigroup, bailing the big bank out of trouble. (A Saudi prince is also a major stockholder.) [57]

Abu Dhabi already owned shares in Advanced Micro Devices, a computer chip manufacturer, and bought a major American private-equity firm, the Carlyle Group. A "sovereign wealth fund" owned by Dubai, another Persian Gulf city-state, was forced to back out of a deal to manage some major U.S. ports. (See sidebar, p. 12.) The fund bought fashion retailer Barney's of New York in 2006, as well as a $1.2 billion stake in a U.S. hedge fund, the Och-Ziff Capital Management Group. These purchases are only the tip of the iceberg, and have prompted public worrying by Treasury Secretary Henry M. Paulson Jr. and finance ministers from the industrial countries about a lack of transparency in high-stakes global investing by petrostates. [58]

In fact, geopolitics experts including former Central Intelligence Director R. James Woolsey say profits from these investments could find their way into the coffers of terrorists, putting the United States in the ironic position of financing both sides in the war against terrorism. ■

CURRENT SITUATION

New Conservation Law

At year's end, environmentalists and the auto industry finally developed a fuel-efficiency standard they could agree on. The agreement opened the way for enactment on Dec. 18 of the first major petroleum-conservation law in decades. In addition to the new gasoline mileage requirements for cars and light trucks (including SUVs), the law demands a major increase in production of ethanol, the alcohol substitute made from corn or other plants.

Environmentalists and automakers alike say the new mileage standard is a breakthrough that ends a long standoff over different requirements

President George W. Bush celebrates with House Speaker Nancy Pelosi, D-Calif., Secretary of Energy Samuel Bodman, left, and other lawmakers after signing the 2007 Energy Act on Dec. 19. The legislation raises vehicle fuel economy standards for the first time in 32 years.

for cars and light trucks. The latter category includes SUVs — a favorite target of environmentalists who call them gas-guzzlers.

Under the new system, the 35-miles-per-gallon standard applies to the entire fleet of new cars and light trucks, by all makers, sold in the United States. Then, each manufacturer would have to meet an individual standard — one for each company — based on each of its models' "footprint," a size measurement based on a vehicle's wheel base and track width.

The legislation does demand that separate sets of standards be devised for cars and light trucks. "That wasn't our favorite provision," Hobson of the Union of Concerned Scientists says, "but since the overall target — the 35-miles-per-gallon standard — has to apply across both of those fleets, it was a compromise we accepted."

Industry leaders expressed support as well. "This tough, national fuel economy bill will be good for both consumers and energy security," Dave McCurdy, president of the Alliance of Automobile Manufacturers, said in a statement. "We support its passage." [59] McCurdy is a Democratic ex-House member from Oklahoma. The alliance is made up

of the Big 3 U.S. automakers and some of the biggest foreign-owned firms, including Toyota, Volkswagen and Mitsubishi.

Another endorsement came from the Association of International Automobile Manufacturers. "It's not perfect, but I think we're going to be pleased," said Mike Stanton, president and CEO of that trade group, which represents Honda, Nissan, Hyundai and others, including Toyota. [60]

EPA Blocks States

The era of good feelings between environmentalists and the Bush administration that opened with the Dec. 18 passage of the energy bill proved short-lived. The very next day, the Environmental Protection Agency (EPA) prohibited California and 16 other states from setting their own carbon dioxide emission standards for cars and trucks. Tougher state standards were designed to step up action against global warming.

But the new energy bill makes such moves by states unnecessary because cars will be polluting less because they'll burn less fuel, EPA Administrator Stephen L. Johnson told reporters. "The Bush administration is moving forward with a clear national solution, not a confusing patchwork of state rules," he said. "I believe this is a better approach than if individual states were to act alone." [61]

California Gov. Arnold Schwarzenegger, a Republican sometimes out of step with the Bush administration, immediately vowed to challenge the decision in court. "It is disappointing that the federal government is standing in our way and ignoring the will of tens of millions of people across the nation," Schwarzenegger said. "We will continue to fight this battle." [62]

Continued on p. 18

At Issue:

Are higher vehicle fuel-economy standards good energy policy?

MICHELLE ROBINSON
*DIRECTOR, CLEAN VEHICLES PROGRAM,
UNION OF CONCERNED SCIENTISTS*

WRITTEN FOR *CQ RESEARCHER*, DECEMBER 2007

*r*equiring automakers to build more fuel-efficient cars and trucks is the patriotic, common-sense thing to do. Strengthened corporate average fuel economy (CAFE) standards will reduce our dependence on oil, save consumers billions of dollars, create hundreds of thousands of domestic jobs and dramatically cut global warming pollution. And it can be done using existing technology. How could anyone argue with that?

The fuel-economy standards instituted in 1975, albeit outdated, worked. If our cars and light trucks still had the same fuel economy they did in the early 1970s, we would have burned through an additional 80 billion gallons of gasoline on top of the 140 billion gallons we will consume this year. That would have amounted to an extra 5.2 million barrels of oil per day. At an average price for regular gasoline of about $2.50 per gallon, we would have forked over an extra $200 billion to the oil companies.

After decades of inaction, Congress has strengthened the standard. Cars, trucks and sport utility vehicles (SUVs) will be required to average at least 35 miles per gallon (mpg) by 2020, a 10-mpg increase over today's levels. A Union of Concerned Scientists (UCS) analysis found this would save 1.1 million barrels of oil per day in 2020, about half of what the United States currently imports from the Persian Gulf. Consumers would save $22 billion in 2020 — even after paying the cost of the improved fuel-economy technology. It would prevent more than 190 million metric tons of global warming emissions in 2020, the equivalent of taking 28 million of today's average cars and trucks off the road. And the new fuel-economy standard would create jobs. According to a UCS study, the standard would generate some 149,300 new domestic jobs in 2020.

Clearly, requiring cars and trucks to average at least 35 mpg by 2020 is smart energy policy. However, a better standard by itself would not ensure that we would avoid the worst consequences of global warming or conquer our national addiction to oil. To tackle these problems, the federal government also must require utilities to generate more of their electricity from clean, renewable energy sources; enact a low-carbon fuel standard to ensure that alternatives to oil are produced in an environmentally friendly way and adopt an economy-wide cap-and-trade program. That said, improving fuel-economy standards is a big step in the right direction.

ROBERT W. CRANDALL
*SENIOR FELLOW, THE BROOKINGS
INSTITUTION*

WRITTEN FOR *CQ RESEARCHER*, DECEMBER 2007

*p*roponents of increases in mandated corporate average fuel economy (CAFE) standards often claim that they would be good for consumers, promote job formation and solve various environmental and energy-security problems. It is important, therefore, to disentangle these claims and to ask if there are not better options available.

First, any claim that raising fuel economy would be good for consumers and create additional jobs is surely incorrect. A highly competitive new-vehicles market delivers cars and trucks that are responsive to consumer demand. Any attempt to mandate greater fuel economy will lead to smaller, less powerful vehicles with more expensive fuel-saving technology than demanded by consumers. Inevitably, this will lead some consumers to hold their vehicles a little longer before trading them in. The result: lower consumer satisfaction, lower vehicle output and fewer auto industry jobs. Is it any wonder that auto producers oppose these proposals?

Second, any proposal to raise CAFE standards must be based on offsetting, non-market "externalities" associated with new-vehicle use. The current proposals are motivated in part by the desire to reduce carbon emissions, the precursors to potential global warming. But new U.S. vehicles generate a very small share of these greenhouse gases. To reduce carbon emissions efficiently, everyone on the globe should face a similar marginal cost of emitting a gram of carbon into the atmosphere.

Surely, it makes little sense to legislate mandatory reductions in carbon emissions (through CAFE) for new U.S. passenger cars while letting older cars, buses and trucks off the hook and — indeed — even encouraging the continued use of these older cars. More important, it is sheer folly to try to reduce global warming by setting high fuel-economy standards in California, Massachusetts or Hawaii while ignoring the much-lower-cost opportunities available in constraining emissions from coal-fired power plants or coke ovens in China, India, Europe or the U.S. Raising U.S. fuel-economy standards is a very high-cost approach, even by Washington standards, to reducing the threat of global warming.

Third, if the goal is to reduce oil imports for national-security purposes, increased fuel-economy standards are still an inefficient, blunt instrument. We burn oil in power plants, home furnaces, industrial boilers and about 250 million cars, trucks and buses already on the road. Any attempt to reduce oil consumption and, therefore, imports, should impose equal per-gallon costs on all of these alternatives. Higher CAFE standards will not do this and will even exacerbate the problem by encouraging Americans to use older gas-guzzlers more intensively.

Continued from p. 16

Twelve other states — New York, New Jersey, Connecticut, Maine, Maryland, Massachusetts, New Mexico, Oregon, Pennsylvania, Rhode Island, Vermont and Washington — had proposed the same standards as California. And the governors of Arizona, Colorado, Florida and Utah had pledged to follow suit. Had the EPA decision gone their way, an estimated one-half of the new vehicles sold in the United States would have had to meet the higher-than-federal air-pollution standards.

McCurdy saluted the EPA decision, tacitly referring to the potential widespread effect of the state-proposed standards. "We commend EPA for protecting a national, 50-state program," he said. "A patchwork quilt of inconsistent and competing . . . programs at the state level would only have created confusion, inefficiency and uncertainty for automakers and consumers." [63]

Environmentalists say they are confident the states will win out in the end. The EPA decision is a "short-term roadblock," says Eli Hobson, Washington representative of the Union of Concerned Scientists, which is active in energy-conservation issues. "The states will move forward."

Meanwhile, some lawmakers launched their own response to the EPA decision. Rep. Henry A. Waxman, D-Calif., chairman of the House Committee on Oversight and Government Reform, as well as Sen. Boxer, announced they had begun investigating the action. Waxman warned the EPA staff to "preserve all documents" relating to the decision. [64]

New Paradigm

China and India aren't the only countries that have some oil-watchers worrying about global oil sup-plies. Traditional oil-exporting countries are now using more of their petroleum for their own needs, shipping less to foreign buyers.

Saudi Arabia, for example, consumed 2 million barrels a day more in 2006 than in 2005, a one-year increase of 6.2 percent. Some projections have Saudi Arabia burning more than one-third of its oil by 2020. [65]

The Middle East isn't alone in putting its own oil to work in newly expanded fleets of cars, as well as homes and factories. Even countries such as Mexico, whose oil fields are said to be nearly played out, are consuming more and shipping less. "Production is declining in Mexico," says West of PFC Energy, in part because the national oil company has been lax in exploration and maintenance.

"One country that's making a huge investment is Saudi Arabia," West says. "They're going to raise production capacity to about 12.5 million barrels a day, with surge capacity to 15 million barrels a day. My people are skeptical they can do more."

Saudi Arabia also has an aggressive and ambitious industrial expansion program on the drawing boards or already under way, including aluminum smelters, petrochemical plants, copper refineries and new power plants. But Saudi industrialist Abdallah Dabbagh, director of the Saudi Arabian Mining Co., which is building a smelter, confessed some doubt to *The Wall Street Journal*. "I think the Saudi government will have to stop and think at some point if this is the best utilization of Saudi's crude." [66]

At street level, new cars are clogging the streets and highways of most of the world's oil giants, in large part because government subsidies keep gasoline prices low.

Saudi Arabians, whose home electricity costs are also subsidized, typically leave their air conditioners running when they go on vacation. Air conditioning accounts for nearly two-thirds of Saudi Arabia's electricity production. [67]

In Venezuela, motorists pay 7 cents a gallon. As a result, Hummers — perhaps the ultimate in gas-guzzling SUVs — are much in demand. The seeming disconnect between Venezuela's growing fleet of massive vehicles and President Chávez' plans for a socialist society prompted an outburst from the president. "What kind of a revolution is this — one of Hummers?" Chávez asked on his television show in October. [68]

And in Iran, where gasoline costs only slightly more, a government attempt in 2007 to cut back on consumption by rationing — instead of cutting or lessening the subsidy — caused violent street protests. Venezuelans predict the same thing would happen if their gasoline subsidy disappeared or shrank. [69]

Concern about consumption is an issue that also applies to China and India — can they be persuaded to moderate their taste for the same amenities that people in developed countries have been enjoying for decades? The International Energy Agency says the issue isn't one of fairness but of numbers. "A level of per-capita income in China and India comparable with that of the industrialized countries would, on today's model, require a level of energy use beyond the world's energy resource endowment." [70]

By comparison, the question of whether Saudi Arabia should be building more power plants to fuel more air conditioners seems like an easier question — at least for non-Saudis. ∎

OUTLOOK

Production Crunch?

Dire predictions invariably swirl around the question of Earth's energy resources. Oil historian and consultant Yergin counts present-day forecasts of imminent decline as the fifth set of such predictions since the petroleum industry began. "Cycles of shortage and surplus characterize the entire history of the oil industry," he wrote in 2005, dismissing the idea that this phase is inherently different. [71]

Still, even some oil-watchers who agree with Yergin on the fundamentals argue that the global panorama has changed enough to cause serious problems in the near term.

If world economic growth stays on track, says West of PFC Energy, "We believe in the likelihood of a production crunch coming between 2012 and 2014. The economic impact will be severe and the geopolitical impact will be severe."

The end result could be heightened competition for resources and "massive" transfer of wealth to oil-producing countries, West says.

To avoid such an outcome, research needs to focus — quickly — on finding technology that provides an alternative to petroleum as an energy source, West says.

"What energy research should do is prioritize limited numbers of areas, whether it's battery efficiency, or light materials with which to build automobiles." Up to now, he says, research has been unfocused.

The Brookings Institution's Sandalow argues that research has already developed a solution that's ready to go — "plug-in hybrids" — hybrid cars that are converted to recharge their electric motors on household current. Sandalow drives one himself. "In 10 years, all Americans will be aware of the option of buying a car that plugs into the power grid," he says. "We have a vast infrastructure for generating electricity in this country that does us almost no good for getting off oil. This is the breakthrough."

As he envisions it, the president could order all government vehicles to use plug-in technology. Overloads of the electricity system would be avoided by

Saudi Arabia, Canada Have Biggest Reserves

Saudi Arabia and Canada lead the world in oil reserves, with nearly 450 billion barrels — more than half as much as the next 10 nations combined.

Rank	Country	Barrels (in billions)
1.	Saudi Arabia	262.3
2.	Canada	179.2
3.	Iran	136.3
4.	Iraq	115.0
5.	Kuwait	101.5
6.	United Arab Emirates	97.8
7.	Venezuela	80.0
8.	Russia	60.0
9.	Libya	41.5
10.	Nigeria	36.2
11.	Kazakhstan	30.0
12.	United States	21.8

As of Jan. 1, 2007

Source: "World Proved Reserves of Oil and Natural Gas, Most Recent Estimates," Energy Information Administration, Jan. 9, 2007

drivers plugging in at night, using reserve capacity that utilities build into their systems.

But some energy experts sound a note of caution. Hirsch, who directed the peak-oil study for the Department of Energy, supports plug-ins but says they can create as many problems as they solve. "Imagine you have a lot of plug-in hybrids, enough to make a difference in U.S. oil consumption. Recharging them in off-peak hours — you can do that for a while. But if you're going to have a big impact, then you're going to have to build a lot of power plants."

In general, Hirsch sees an unhappy energy future not very far down the road. Oil supplies will shrink, he says. "I think there's not much question we will be in serious, long-term recession, deepening recession," he says. "With oil shortages, you'll have much higher prices — and shortages meaning you just simply won't be able to get it."

The world economy will adjust, Hirsch says, but until then, "It's not a pretty picture. Companies will be cutting back on employment; a lot of people will lose their homes because they can't afford to meet mortgages. International trade will go down."

Energy analyst Medlock at the Baker Institute is far less pessimistic. "I see conservation forces coming to bear over the next decade, which will tend to trim the growth of demand. I do see new supplies coming on line, and a major interest in developing unconventional oil."

Such developments would avoid the continued price spikes that some predict. "I think it's well within the range of possibility to see oil prices in the range of $60 to $70 a barrel," Medlock says.

Simmons, widely seen as the leading voice of the peak oil thesis, sees no grounds for such optimism. Oil producers can indeed use natural gas liquids and other unconventional sources of energy to make up a shortfall in crude oil supplies, but that will only hasten the day when the real crunch begins, he says. "We are basically living on borrowed time," he says. The gap between demand and supply "creates social chaos and war" by 2020.

Or, in the best of all possible worlds, Simmons says, a government-directed effort will come up with alternatives to petroleum. "But if we spend three more years arguing if it's time to get into a program like that," he says, the future is grim.

Gheit, the veteran oilman now on Wall Street, dismisses all such talk.

Oil-exploration and production technology isn't standing still and will enable oil companies to keep producing petroleum, he says. "I can assure people we are not going to run out of oil any time soon." ∎

Notes

[1] David Sandalow, *Freedom From Oil: How the next President can End the United States' Oil Addiction* (2007).

[2] Quoted in "Transcript: Interview with IEA chief economist," FT.com, Nov. 7, 2007, www.ft.com/cms/s/0/3c8940ca-8d46-11dc-a398-0000779fd2ac.html?nclick_check=1.

[3] *Ibid.*

[4] "World Energy Outlook 2007 — China and India Insights," International Energy Association, p. 41, www.worldenergyoutlook.org (only executive summary available to public).

[5] Quoted in Jad Mouawad, "Record Price of Oil Raises New Fears," *The New York Times*, Oct. 17, 2007, p. C1.

[6] For background, see Peter Behr, "Energy Nationalism," *CQ Global Researcher*, July 2007, pp. 151-180.

[7] For oil demand statistics, see "World Petroleum (Oil) Demand 2003-2007," Energy Information Administration, U.S. Department of Energy, updated Nov. 5, 2007, www.eia.doe.gov/ipm/demand.html.

[8] Daniel Yergin, *The Prize: The Epic Quest for Oil, Money and Power* (1992), p. 15.

[9] Gretchen Morgenson, "Dangers of a World Without Rules," *The New York Times*, Sept. 24, 2006, Sect. 3, p. 1.

[10] Robert J. Samuelson, "Is There an Oil 'Bubble,' " *The Washington Post*, July 26, 2006, p. A17.

[11] "Fact Sheet — Prudhoe Bay," BP, updated August 2006, www.bp.com/liveassets/bp_internet/us/bp_us_english/STAGING/local_assets/downloads/a/A03_prudhoe_bay_fact_sheet.pdf.

[12] Heather Timmons, "Oil Majors Agree to Develop a Big Kazakh Field," *The New York Times*, Feb. 26, 2004, p. W1; "Chevron Reports Oil Find in Gulf of Mexico," *The New York Times* [Bloomberg News], Dec. 21, 2004, p. C5; Alexei Barrionuevo, "Brazil Discovers an Oil Field Can Be a Political Tool," *The New York Times*, Nov. 19, 2007, p. A3.

[13] Quoted in Steven Mufson, "U.S. Oil Reserves Get a Big Boost," *The Washington Post*, Sept. 6, 2006, p. D1.

[14] "Crude Oil Production by Selected Country," Energy Information Administration, U.S. Department of Energy, November 2007, www.eia.doe.gov/emeu/mer/pdf/pages/sec11_5.pdf. For historical reserves figure, see Yergin, *op. cit.*, pp. 499-500. For Saudi Arabia reserve estimate, see "Crude Oil — Uncertainty about Future Oil Supply Makes It Important to Develop a Strategy for Addressing a Peak and Decline in Oil Production," Government Accountability Office, February 2007, p. 62, www.gao.gov/new.items/d07283.pdf.

[15] Government Accountability Office, *ibid.*, p. 20.

[16] Peter Jackson, "Why the 'Peak Oil' Theory Falls Down," Cambridge Energy Resource Associates, November 2006, pp. 2, 10.

[17] "World Energy Outlook 2007," *op. cit.*, p. 64.

[18] Quoted in "Transcript: Interview with IEA chief economist," *op. cit.*

[19] "Testimony on Peak Oil, Dr. Robert L. Hirsch, Senior Energy Program Advisor, SAIC," House Subcommittee on Energy and Air Quality, Dec. 7, 2005, http://energy-commerce.house.gov/reparchives/108/Hearings/12072005hearing1733/Hirsch.pdf.

[20] "Peak Oil Theory — 'World Running Out of Oil Soon' — Is Faulty; Could Distort Policy and Energy Debate," Cambridge Energy Research Associates, (press release), Nov. 14, 2006, www.cera.com/aspx/cda/public1/news/pressReleases/pressReleaseDetails.aspx?CID=8444.

[21] "Crude Oil — Uncertainty about Future Oil Supply . . .," *op. cit.*, p. 4.

[22] *Ibid.*

[23] See Tim Reiterman, "Canada's black gold glitters but tarnishes," *Los Angeles Times*, July 8, 2007, p. A1.

[24] Jackson, *op. cit.*, p. 6.

[25] Matthew R. Simmons, *Twilight in the Desert: The Coming Saudi Oil Shock and the World Economy* (2005).

[26] Quoted in Thomas Catan, "UK prepares for the day the oil runs out," *Financial Times* (London), May 27, 2005, p. A20.

[27] "World Energy Outlook," *op. cit.*, p. 122.

[28] "USA Statistics in Brief — Population by Age, Sex, and Region," U.S. Census Bureau, updated Nov. 6, 2007, www.census.gov/compendia/statab/files/pop.html.

[29] "World Energy Outlook," *op. cit.*, pp. 122, 77.

[30] "Olympic Trials: China's bout with $90 oil," Lehman Brothers, Fixed Income Research, Nov. 16, 2007, p. 3 (not publicly available).

[31] "President Bush Delivers State of the Union Address," The White House, Jan. 31, 2006, www.whitehouse.gov/news/releases/2006/01/20060131-10.html.

[32] John M. Broder, "House, 314-100, Passes Broad Energy Bill," *The New York Times*, Sept. 19, 2007, p. A16; Steven Mufson, "House Sends President an Energy Bill to Sign," *The Washington Post*, Sept. 19, 2007, p. A1.

[33] Quoted in Broder, *op. cit.*

[34] "House Select Committee on Energy Independence and Global Warming Holds Hearing on State Efforts Towards Low-Carbon Energy," *Congressional Transcripts*, Nov. 14, 2007.

[35] Robert L. Hirsch, *et al.*, "Peaking of World Oil Production: Impacts, Mitigation & Risk Management," Science Applications International Corp., February 2005, p. 5, www.projectcensored.org/newsflash/the_hirsch_report.pdf.

[36] Kenneth S. Deffeyes, *Hubbert's Peak* (2001), pp. 1-5.

[37] See "Crude Oil," *op. cit.*, Government Accountability Office.

[38] Deffeyes, *op. cit.*, p. 134.

[39] For background, see Behr, *op. cit.* Material in this sub-section is also drawn from Yergin, *op. cit.*

[40] Quoted in *ibid.*, p. 615.

About the Author

Peter Katel is a *CQ Researcher* staff writer who previously reported on Haiti and Latin America for *Time* and *Newsweek* and covered the Southwest for newspapers in New Mexico. He has received several journalism awards, including the Bartolomé Mitre Award for coverage of drug trafficking from the Inter-American Press Association. He holds an A.B. in university studies from the University of New Mexico. His recent reports include "Prison Reform," "Cuba's Future" and "Wounded Veterans."

[41] "Current SPR Inventory As Of Nov. 29, 2007," Strategic Petroleum Reserve, Department of Energy, www.spr.doe.gov/dir/dir.html.

[42] "Speeches by President J. Carter Outlining the Critical Nature of the Energy Crisis and Recommendations for Legislation to Deal with Issue," April 18, 1977, *CQ Public Affairs Collection*.

[43] Yergin, *op. cit.*, p. 702; "Imported Crude Oil Prices: Nominal and Real," Energy Information Administration, Department of Energy, undated, www.eia.doe.gov/emeu/steo/pub/fsheets/real_prices.html.

[44] Yergin, *op. cit.*, pp. 699-703.

[45] *Ibid.*, p. 718.

[46] Doron P. Levin, "How Ford Finally Found the Road to Wellville," *Los Angeles Times Magazine*, March 10, 1996, p. 16.

[47] See "The 9/11 Commission Report: Final Report of the National Commission on Terrorist Attacks Upon the United States," 2004, p. 48.

[48] Neela Banerjee, "After the Attacks: The Energy Market," *The New York Times*, Sept. 13, 2001, p. A7; Brad Foss, "Gas Prices Shoot Up," *The Washington Post* (The Associated Press), Sept. 12, 2001, p. E4.

[49] Gail Makinen, "The Economic Effects of 9/11: A Retrospective Assessment," Congressional Research Service, Sept. 27, 2002, p. 16, www.fas.org/irp/crs/RL31617.pdf.

[50] Simon Romero, "Laissez-Faire My Gas-Guzzler, Already," *The New York Times*, Sept. 7, 2004, p. C1.

[51] Quoted in Jeremy W. Peters, "On Auto-Dealer Lots, a Shift Away from Gas-Guzzling Vehicles," *The New York Times*, Sept. 1, 2006, C6.

[52] *Ibid.*

[53] Unless otherwise indicated, material in this sub-section is drawn from Behr, *op. cit.*

[54] "China seals oil deal with China," BBC News, Aug. 25, 2006, http://news.bbc.co.uk/1/hi/business/5286766.stm.

[55] "Millions in Georgia Without Heat," CNN, Jan. 24, 2006, www.cnn.com/2006/WORLD/europe/01/24/russia.gas/index.html. See also "Top World Oil Producers," Energy Information Administration, U.S. Department of Energy, www.eia.doe.gov/emeu/cabs/topworldtables1_2.htm.

[56] Quoted in Behr, *op. cit.*

[57] Steven R. Weisman, "Oil Producers See the World and Buy It Up," *The New York Times*, Nov. 28, 2007, www.nytimes.com/2007/11/28/business/worldbusiness/28petrodollars.html.

[58] "Sovereign Wealth Funds: A Shopping List," DealBook, *The New York Times*, Nov. 27, 2007, http://dealbook.blogs.nytimes.com/2007/11/27/sovereign-wealth-funds-a-shopping-list/.

[59] Quoted in John M. Broder and Felicity Barringer, "E.P.A. Says 17 States Can't Set Greenhouse Gas Rules for Cars," *The New York Times*, Dec. 20, 2007, p. A1.

[60] *Ibid.*

[61] *Ibid.*

[62] Waxman letter to Johnson, Dec. 20, 2007, http://oversight.house.gov/documents/20071220111155.pdf; and Janet Wilson, "EPA chief is said to have ignored staff," *Los Angeles Times*, Dec. 21, 2007, p. A30.

[63] See "Statement of President and CEO Dave McCurdy on National Fuel Economy Agreement," Alliance of American Automobile Manufacturers, Dec. 1, 2007, www.autoalliance.org/archives/archive.php?id=427&cat=Press%20Releases.

[64] Quoted in Dave Shepardson, "Auto industry backs CAFE deal," *Detroit News*, www.detnews.com/apps/pbcs.dll/article?AID=2007712010414.

[65] Neil King Jr., "Saudi Industrial Drive Strains Oil-Export Role," *The Wall Street Journal*, Dec. 12, 2007, p. A1.

[66] Quoted in *ibid*.

[67] *Ibid.*

[68] Quoted in Simon Romero, "Venezuela's Gas Prices Remain Low, But the Political Costs May Be Rising," *The New York Times*, Oct. 30, 2007, www.nytimes.com/2007/10/30/world/americas/30venezuela.html?n=Top/Reference/Times%20Topics/People/C/Chavez,%20Hugo.

[69] Ramin Mostaghim and Borzou Daragahi, "Gas rationing in Iran ignites anger, unrest," *Los Angeles Times*, June 28, 2007, p. A5; Najmeh Bozorgmehr, "Iran pushes on with fuel rationing in face of riots," *Financial Times* (London), June 28, 2007, p. A7. Also see Romero, *op. cit.*

[70] "World Energy Outlook," *op. cit.*, p. 215.

[71] Daniel Yergin, "It's Not the End of the Oil Age," *The Washington Post*, July 31, 2005, p. B7.

FOR MORE INFORMATION

American Petroleum Institute, 1220 L St., N.W., Washington, DC 20005; (202) 682-8000; www.api.org. Lobbies for the U.S. oil industry; supports loosening restrictions on oil exploration on public lands.

Cambridge Energy Research Associates, 55 Cambridge Parkway, Cambridge, MA 02142; (617) 866 5000; http://cera.com. A widely cited consulting firm that provides public summaries of studies performed for clients.

Energy Information Administration, U.S. Department of Energy, 1000 Independence Ave., S.W., Washington, DC 20585; (202) 586-8800; http://eia.doe.gov. The government's energy statistics division provides access to a wide range of data on all aspects of oil and gas production and use.

International Energy Agency, 9, rue de la Fédération, 75739 Paris Cedex 15, France, (011-33 1) 40.57.65.00/01; www.iea.org. An organization of industrialized countries, almost all in Europe, that studies energy trends and recommends policies on conservation and related topics.

James A. Baker III Institute for Public Policy, Energy Forum, Rice University, 6100 Main St., Baker Hall, Suite 120, Houston, TX 77005; (713) 348-4683; www.rice.edu/energy/index.html. Nonpartisan think tank that sponsors research and forums on oil-related topics.

The Oil Drum; www.theoildrum.com. A collective blog (with separate editions for the United States, Canada, Europe and Australia/New Zealand). Part of the "peak oil" community; provides a discussion forum on issues of conservation and alternative energy sources.

Organization of Petroleum Exporting Countries, Obere Donaustrasse 93, A-1020, Vienna, Austria; (011-43-1) 21112-279; www.opec.org. The cartel publishes statistics, forecasts and policy documents on global oil supplies.

Bibliography

Selected Sources

Books

Huber, Peter W., and Mark P. Mills, *The Bottomless Well: The Twilight of Fuel, the Virtue of Waste, and Why We Will Never Run Out of Energy*, Basic Books, 2005.
A lawyer and a physicist argue that energy in all its forms is plentiful.

Sandalow, David, *Freedom From Oil: How the Next President Can End the United States' Oil Addiction*, McGraw-Hill, 2007.
A former Clinton administration official lays out a plan for reducing U.S. oil usage.

Simmons, Matthew R., *Twilight in the Desert: The Coming Saudi Oil Shock and the World Economy*, John Wiley & Sons, 2005.
A leading "peak oil" proponent cites evidence that Saudi Arabia has vastly exaggerated the amount of its oil reserves.

Yergin, Daniel, *The Prize: The Epic Quest for Oil, Money and Power*, Simon & Schuster, 1992.
An American oil expert provides a classic history of the global oil industry and its role in contemporary geopolitics.

Articles

Bradsher, Keith, "Trucks Propel China's Economy, and Foul Its Air," *The New York Times*, Dec. 8, 2007, p. A1.
China's reliance on trucking is growing by leaps and bounds, far ahead of the government's ability to regulate the industry.

Hagenbaugh, Barbara, "Gas pump gulps more of family pay," *USA Today*, May 17, 2007, p. A1.
Average American consumers suddenly are shelling out appreciably more to fill their tanks.

Hoyos, Carola, and Demetri Sevastopulo, "Saudi Aramco dismisses claims over problems meeting rising global demand for oil," *Financial Times* (London), Feb. 27, 2004.
The Saudi oil company responds to the first major stirrings of the "peak oil" movement.

King, Neil Jr., "Saudi Industrial Drive Strains Oil-Export Role," *The Wall Street Journal*, Dec. 12, 2007, p. A1.
Saudi Arabia's rapidly expanding consumption of its major product is making even some Saudi industrialists nervous.

Morse, Edward L., and James Richard, "The Battle for Energy Dominance," *Foreign Affairs*, March-April 2002, p. 16.
Two Wall Street energy specialists presciently examine the geopolitical effects of Russia's sudden emergence as a major player in world energy markets.

Murphy, Kim, *et al.*, "Oil's Winners and Losers," *Los Angeles Times*, Nov. 24, 2007, p. A1.
High oil prices spell progress for some and disaster for others as new petroleum-market dynamics play out globally.

Rosenberg, Tina, "The Perils of Petrocracy," *The New York Times Magazine*, Nov. 4 2007, p. 42.
Focusing on Venezuela, a veteran writer reports that state-owned oil companies tend not to be models of efficient performance, nor reliable explorers for new energy deposits.

Weisman, Steven R., "Oil Producers See the World and Buy It Up," *The New York Times*, Nov. 28, 2007, p. A1.
Wall Street and Washington try to grasp the implications of oil-producing countries buying big chunks of major U.S. and European companies.

Reports and Studies

"Crude Oil: Uncertainty about Future Oil Supply Makes It Important to Develop a Strategy for Addressing a Peak and Decline in Oil Production," Government Accountability Office, February 2007, www.gao.gov/new.items/d07283.pdf.
The government should begin planning now for world oil supplies to peak, even if that moment is several decades away.

"Hard Truths: Facing the Hard Truths About Energy," National Petroleum Council, July 2007, http://downloadcenter.connectlive.com/events/npc071807/pdf-downloads/NPC_Facing_Hard_Truths.pdf.
The United States needs to rapidly prepare for a world in which oil is more difficult and more expensive to obtain, according to top energy experts and executives.

Medlock, Kenneth B. III, and Amy Myers Jaffe, "Gas FAQ: U.S. Gasoline Markets and U.S. Oil Import Dependence," James A. Baker III Institute for Public Policy, Rice University, http://bakerinstitute.org/Pubs/WWT_FAQ_Gas2.pdf.
Two experts from a think tank in the U.S. oil capital explain the basics of energy use in the United States.

Rosen, Daniel H., and Trevor Houser, "China Energy: A Guide for the Perplexed," Center for International and Strategic Studies, Peterson Institute for International Economics, May 2007, www.iie.com/publications/papers/rosen0507.pdf.
China's manufacturing expansion, not automobile fleet growth, accounts for most of the country's rising oil demand.

The Next Step:

Additional Articles from Current Periodicals

China and India

Klare, Michael T., "Kicking the Habit, All Over the World," *Los Angeles Times*, Feb. 11, 2006, p. B17.
China and India are pursuing oil and other energy deals with countries whose policies threaten global stability.

Mouawad, Jad, "Cuts Urged in China's and India's Energy Growth," *The New York Times*, Nov. 7, 2007, p. C3.
The International Energy Agency has urged advanced economies to work with India and China to reduce overall growth in energy consumption.

Timmons, Heather, "Citing Oil Prices, Asia Starts Reducing Fuel Subsidies," *The New York Times*, Nov. 2, 2007, p. C5.
Asian governments have begun to roll back subsidies that have kept costs for gasoline and other fuels artificially low.

Government Intervention

Adams, Rebecca, "Gas Prices Rise Along With Ethanol Use," *CQ Weekly*, April 24, 2006, p. 1070.
The Bush administration is touting fuel-grade alcohol as a way to wean the United States from its dependence on foreign oil and is handing out millions of dollars in loans and grants to producers.

Block, Sandra, "Tax Credits Can Reduce Premium You Pay for Hybrid Vehicle," *USA Today*, May 9, 2006, p. 3B.
The Energy Tax Incentives Act allows taxpayers to claim a credit up to $3,400 for purchasing hybrid vehicles.

Blum, Justin, "No Way Found to Cut Need for Foreign Oil," *The Washington Post*, Dec. 22, 2005, p. A10.
The Senate's refusal to allow oil drilling in Alaska's Arctic National Wildlife Refuge marks the latest failure in forming a strategy to reduce U.S. dependence on foreign oil.

Hebert, H. Josef, "Congress Approves Auto Fuel Economy Increase, More Ethanol Use," The Associated Press, Dec. 18, 2007.
President Bush has signed an energy bill requiring a 40 percent increase in fuel efficiency for cars, SUVs and small trucks within the next 12 years.

Hybrid Vehicles

Pender, Kathleen, "Sale and Manufacture of Hybrids Hit Some Potholes," *The San Francisco Chronicle*, May 25, 2006, p. C1.
Many second-generation hybrid vehicles — such as the Honda Accord Hybrid and Toyota Highlander Hybrid — provide minimal improvements in fuel economy.

Popely, Rick, "Hybrid Effort: Trio Takes on Toyota," *Chicago Tribune*, Sept. 1, 2006, p. C1.
General Motors, Chrysler and BMW have formed an alliance that will accelerate the development of hybrid systems that work on a variety of vehicles.

Wilson, Jon, "Hybrid Buses May Roll By Mid 2008," *St. Petersburg Times*, March 7, 2007, p. 1.
Florida's Pinellas County may introduce hybrid buses by summer 2008 if the county's bus agency approves the plan.

Peak Oil

"Steady As She Goes," *The Economist*, April 22, 2006, p. 65.
The U.S. Geological Survey concluded the world has 3 trillion barrels of recoverable oil in the ground and that global oil supplies won't peak until after 2025.

Carroll, Joe, "Firm: Peak-Oil Theory Is Bogus," *The Philadelphia Inquirer*, Nov. 15, 2006, p. C8.
Global oil production will increase for at least 25 more years as new drilling and refining techniques tap previously inaccessible reserves, according to Cambridge Energy Research Associates.

Fox, Justin, "Peak Possibilities," *Time*, Dec. 3, 2007, p. 52.
The world isn't running out of oil, but much of it is difficult to extract and even harder to refine.

Francis, David R., "Why 'Peak Oil' May Soon Pique Your Interest," *The Christian Science Monitor*, Aug. 6, 2007, p. 15.
The decreasing world output of oil may soon put peak-oil concerns ahead of those of global warming.